LIFE SCIENCE

— *Campion Integrated Studies Series* —

LIFE SCIENCES

ANATOMY AND PHYSIOLOGY
FOR HEALTH CARE PROFESSIONALS

L. Grégoire

Edited by
P. Gallagher

Illustrator
A. van Horssen

CAMPION
PRESS

—Campion Integrated Studies Series —

British Library Cataloguing-in-Publication Data

Grégoire, L.
Life sciences: anatomy and physiology for health
care professionals.
I. Title II. Gallagher, P.
611

ISBN 1-873732-01-5

Cover design:	Artisan Graphics, Edinburgh
Layout:	Studio Blanche, Noordwijk
Illustrations:	Ad van Horssen, Bussum
Photographs:	Academic Medical Centre, Amsterdam Bonnier Fakta, Stockholm Martinus Nijhoff, Leiden Spaarnestad Photo-archives, Haarlem

Printed by Groen bv, Leiden, The Netherlands
Bound by Jansenbinders, Leiden, The Netherlands

First published 1992   ISBN 1-873732-01-5

© 1990 SMD EDUCATIONAL PUBLISHERS I LEIDEN I THE NETHERLANDS

© 1992 CAMPION PRESS LIMITED
384 Lanark Road
Edinburgh   EH13 0LX

Ludo Grégoire　　MSc, is a graduate of the Tilburg Academy of Physical Education. He obtained his master's degree, majoring in Life Sciences at the Department of Kinesiology at the Free University of Amsterdam. Thereafter, he studied Labour and Oganizational Psychology at post-graduate level.
From 1976 to 1986 he taught Life Sciences in several nursing colleges. Since 1986 he has been working as Director of the Office for General Practitioners in the North Holland District of the Royal Dutch Society of Health Care.

Peter Gallagher　　MA RGN RMN DN DMS RNT, is currently Assistant Principal within the Sheffield & North Trent College of Nursing and Midwifery Education.
He has been directly involved in nurse education since 1980 and prior to his current post was responsible for the implementation of the Project 2000 Common Foundation Programme in Sheffield School of Nursing, which was one of the regional schools selected as a demonstration site for Project 2000.

Ad van Horssen　　was educated at art college and graduated from the Faculty of Illustration and Design at Utrecht Academy of Arts. This was followed by a course in anatomy and embryology directed towards morphological design at post-graduate level at the Anatomy Department of the University of Utrecht.
Since 1971 he has been Head of the Central Medical Illustration Department in the Medical Faculty of the University of Amsterdam. He has illustrated many text-books for medical and paramedical training colleges and has recently illustrated a reference book on liver surgery.

Preface

Nurse education in the United Kingdom is undergoing major changes with the introduction and implementation of Project 2000. These changes are consistent with improvements in the teaching of nursing throughout the European Community as health care provision prepares for the 21st century and nursing curricula adapt to the new demands which will be made on staff. The change in direction in nurse education aims to prepare the student for a new role as a practitioner who is: orientated towards health promotion and health education as well as care of the ill and disabled; capable of meeting the needs of patients or clients in ever changing circumstances, in hospital or community settings; educated rather than trained, with practice founded on a sound knowledge base and a capacity to approach nursing problems in a critical and reflective manner. The new nurse will be informed by a broad range of scientific, philosophical and humanities-based disciplines.

Life Sciences is one of the titles in the *Campion Integrated Studies Series,* a series which attempts to help students to develop an interdisciplinary view and break down barriers within the curriculum. The text and illustrations in this book will be suitable for students undertaking a detailed study of the subject for the first time. Learning outcomes for each chapter are explicitly stated to facilitate the learning process and to ensure that study can be properly targeted and evaluated.

One of the significant features of Project 2000 is the emphasis on the perception of the client as a whole person, stressing the interconnections and interdependencies which run through a client's physical and psychological condition and background. This holistic approach is reflected in the structure of this book, where the parts and functions of the body's various systems are examined within their context, as components within a larger organism.

Life Sciences deals with all of the material which anatomy and physiology texts traditionally cover and the subject matter is presented in a methodical and structured way. From the basic components at cellular level to overviews of the essential body systems, the structure and functions of the human body are investigated and explained in detail. In addition, the book enlarges on the scope of traditional texts by placing the subject matter within two new frames of reference. It examines anatomy and physiology as dynamic aspects of the person, changing throughout the life cycle from fetal development to the ageing processes. It further sets the human being within an evolutionary context, investigating and explaining the processes and selections which may have led to the human race as we know it. In common with the other books in the series, Life Sciences is presented in modular format and has clearly defined learning outcomes for each chapter.

Life Sciences is an appropriate text for students of nursing following the Common Foundation Programme, though it can also be used as a source of reference for students on pre-registration or post-registration courses. It will also be suitable for students of the caring professions allied to medicine such as physiotherapy, occupational therapy, speech therapy, radiography and chiropody.

Spring 1992   Ludo Grégoire
                              Peter Gallagher

Contents

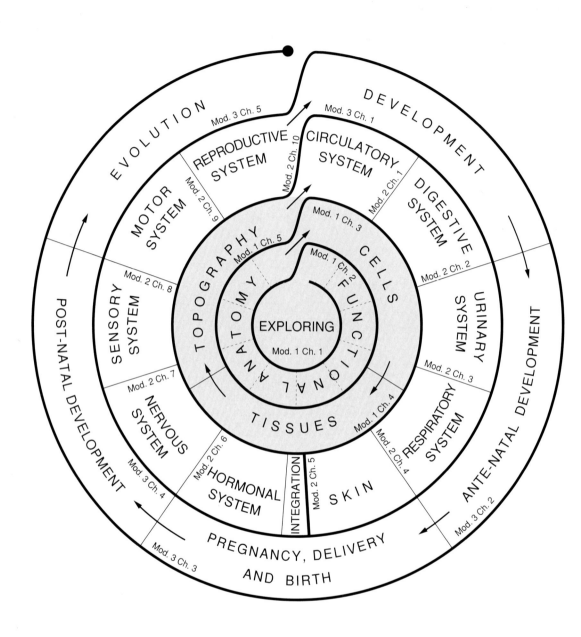

Module 1
FOUNDATION

Introduction to Module 1

Module 1 lays the foundation for a comprehensive and coherent study of life sciences.

Exploring the field
Chapter 1 considers your choice of a career as a health care practitioner. After an examination of various aspects of this, we discuss the role of life sciences as a subject for study within a health care course. This is followed by an explanation of the framework of this book, a more detailed description of life sciences and an analysis of some of the specific characteristics of this field of study.

Functional anatomy
Chapter 2 explores a holistic and integrated approach to the study of the human body. Functional anatomy is divided into functional systems or body systems and there are ten such systems in the human body.

Cells
Chapter 3 examines the cell as the basic constituent part and metabolic unit of the organism.

Tissues
Chapter 4 examines the structure, composition and function of tissues. The four main types of tissues are defined and described in detail.

Topography
Chapter 5 deals with the topography of the human body, describing the positions of the organs and other structures. The chapter includes a general description of man's internal and external appearance.

Exploring the field

Introduction

The choice of a career in health care is not, in most cases, a decision that is made lightly. You will have thought about the various aspects of the profession, about the satisfaction it will give you and the difficulties you will encounter. You will have found out detailed information about the content of the course and the amount of the financial support you can expect to receive. You may have applied for admission to one or more of the health care colleges.

Some people find the process of selecting a career both lengthy and difficult yet others find this an easy decision to make. Whether your choice of career was the correct one is something that only time will tell.

One thing is certain: the expectation of a satisfying professional career is greatly enhanced if you take full advantage of the time you spend on the course and use this time to the best of your of ability.

When you have successfully completed your course of study you will have the right to practise as a health care practitioner. You will be considered to be a person capable of undertaking the responsibilities and performing the skills with which the profession is concerned. 'Project 2000: a new preparation for practice' states: "The practitioner of the future should be both a 'doer' and a 'knowledgeable doer'.... competent to assess the need for care, provide care, monitor and evaluate and to do this in institutional and non-institutional settings"*. The purpose of a health care course of study is to properly prepare you to fulfil the career you have chosen in the caring professions.

People experience a far greater degree of job satisfaction if they know the reason why they perform certain tasks and activities. This observation certainly applies to the caring professions, where an understanding of the purpose of care will help you to be aware of both the possibilities and the limitations of a person's behaviour.

Pathology is the branch of medical science relating to the causes, nature and effects of disease. A pathologist will be able to assess how far an

---

* United Kingdom Central Council (1986) *Project 2000: a new preparation for practice*. London: UKCC.

infection (e.g. influenza), injury (e.g. a broken bone), or handicap (e.g. deafness) affects the normal functioning of a human being. Pathology will be able to establish what the limitations are and what possibilities for improvement can be offered by medical science supported by nursing and paramedical care.

It is, however, only possible to make a sound assessment about deviations when the 'normal' state is fully understood. Life sciences is the subject concerned with the 'normal' functioning of the human body. It is essential that the health care profession fully understands the normal state and functioning of the body before making judgements regarding deviations from the norm. The decision as to whether a person can be helped to eat or drink depends on the carer's knowledge of how the digestive system functions under normal circumstances. Similarly, an individual can be helped to maintain his body temperature within the accepted limits, only if the carer knows what is normal body temperature, what it depends on and how it is regulated.

These two simple practical situations are given as illustrations of the role of life sciences as an area of study applied in caring situations. A life sciences course gives insight into sciences such as pathology, physiology, anatomy and medicine. It serves these sciences by helping the carer to understand the structure and function of the human body.

Learning outcomes

After studying this chapter you should have sufficient knowledge and understanding of:
- the role of the study of life sciences;
- the structure of the book and the method of study to be followed;
- the content and the nature of life sciences;
- the limitations of this field of study.

1. Structure of the book

The composition of this book is represented in Figure 1.1.1. The starting point of the book and the beginning of the route to follow in your studies is in the middle of the circle. The circle is used in the diagram to represent the close cooperation and interdependence between the body systems. The basis of our study and discussions in Module 1 is the holistic approach to the understanding of the human being.

The chapters in this book are best studied in the given order. Module 2, however, has been written in such a way that, in principle, you could start with any chapter in relation to your current curriculum. We have tried to keep the number of cross references as concise as possible. An index of key words is given at the end of the book so that you can refer to the page dealing with a certain topic. When more than one page number is listed after the keyword, the bold text indicates the page with the most important reference to that topic.

2. Anatomy + physiology = functional anatomy

Functional anatomy, as indicated in the heading of this section, combines the sciences of anat-

Figure 1.1.1
Structure of the book and
the study route

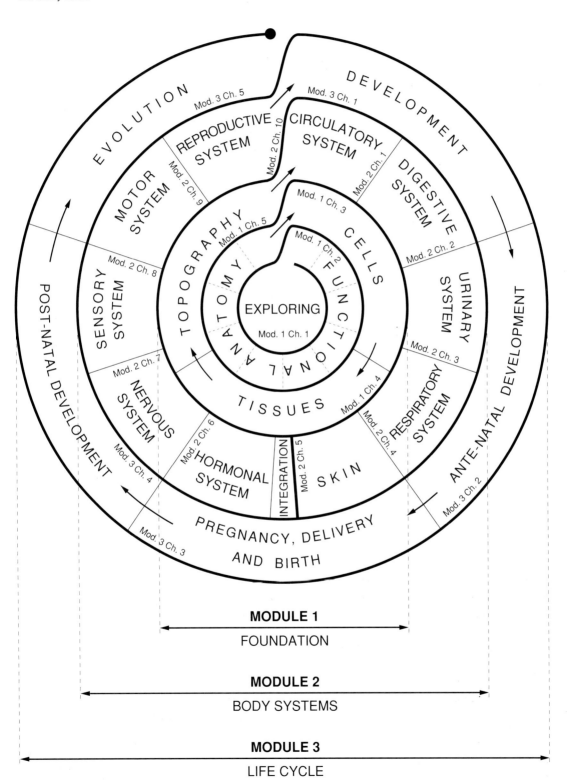

MODULE 1
FOUNDATION

MODULE 2
BODY SYSTEMS

MODULE 3
LIFE CYCLE

omy and physiology. The aim of functional anatomy as a subject area in health care education is to help the student to attain a level of knowledge and understanding of the normal structure and function of the human body. This knowledge and understanding can then be applied to care of the patient or client encountered in clinical practice.

One of the best ways to find out how something works is to take it apart. This principle has constantly applied in the field of scientific investigation, and in medical science the human body has always been the most obvious subject for research.

Anatomy is concerned with the structure and relationship of structures within the human body. The word 'anatomy' is derived from the Greek word meaning 'taking apart by cutting', but the dissection of dead bodies has obvious limitations with respect to finding out how the living body functions (Fig. 1.1.2).

Physiology deals with the functions of the organs and systems of the body. This word is also derived from Greek, meaning 'the study of the functions of life'. How the body functions can be determined by running tests and taking measurements (Fig. 1.1.3).

Shape and function

The shape of a part of the body can determine the function that it performs. If we take the hand as an example, we observe that the thumb can be placed opposite each finger and this gives the hand the ability to perform various skilful operations. The shape of the foot, however, does not allow such skill and flexibility. In these examples the shape dictates the functions that a body part can perform.

Conversely, the function of a body part can influence its shape. When we go running regularly our leg muscles get bigger, the capacity of the lungs increases, and the walls of the heart become thicker. The new changed shape of the legs, lungs and heart, in turn has a positive effect on their functional capacity.

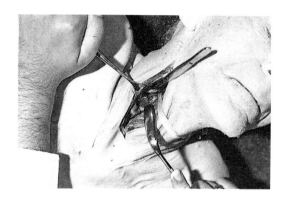

Figure 1.1.2
Dissecting a human
body in the dissecting-
room

We can see therefore, that shape and function are not static, but dynamic, factors. Though they can be distinguished, they cannot be separated as the shape and function of the human body are very closely interrelated. The field of functional anatomy combines both: *the study of the structure of the human body with an immediate link to its functioning.*

Methods of obtaining medical information

When studying the shape and function of human beings, functional anatomy makes full use of technology (Fig. 1.1.4):
- *X-rays* enable us to take photographs of the interior of the body for diagnostic purposes;
- *echography (ultrasound)* projects cross-sections of the interior of the body on a screen, using ultrasonic soundwaves;
- a *CT-scan (computerized tomography)* draws a computerized map of the interior of the body by analyzing X-rays or ultrasonic soundwaves;
- by using *optic fibres* we can look into the spaces or cavities within the body which contain internal organs;
- by interpretation of *electric impulses* received by electrodes placed on the surface of the body we can learn about the activity of the internal organs:
  • the electrocardiogram (ECG) gives information about the activities of the heart;

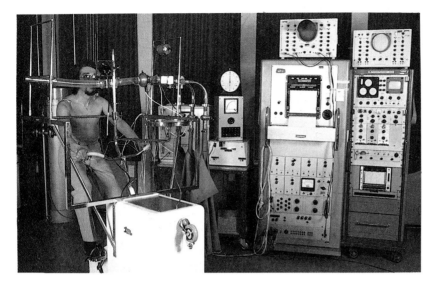

Figure 1.1.3
Measuring the
consumption of oxygen
during strenuous
exercise

- the electroencephalogram (EEG) tells us about the activities of the brain;
- the electromyogram (EMG) reports on the activities of the muscles.

In addition to these, there are a number of older but certainly no less valuable techniques for gathering functional-anatomical information (Fig. 1.1.5):
- by an *examination* of the body various questions can be answered.
  Is the skin discoloured? Is the person standing upright? Is the person able to move normally?
- in *palpation* the surface of the body is touched in such a way that information can be acquired about the internal working of the body parts;
- in *percussion* the practitioner taps on certain parts of the body.
  The quality of sound heard gives information about the nature and condition of deep-lying structures;
- in *auscultation*, the sounds produced by parts of the body can be heard, normally by use of a stethoscope.

We have outlined here some of the methods used to obtain information. These techniques and others, together with traditional dissection, have helped medical science to gather a large body of knowledge about the structure and function of the human body.

3. The Nomina Anatomica

Latin and Greek terms
You will certainly have noticed that doctors, pharmacists, physiotherapists and nurses among others, use specific names: a kind of trade jargon. In your study of life sciences, you will often meet with such jargon.

In order to avoid confusion when people of different professions or with different first languages communicate on a medical basis, there is an internationally recognized code for medical terminology. This is the *Nomina Anatomica*. Every five years it is checked to see if the list, in which all the anatomical names are registered, needs changing or supplementing. This Nomina Anatomica is largely based on classical Latin, but there are also many words derived from classical Greek.

We recommend that you use a good medical dictionary which will enable you to gain an understanding of the terminology, limiting the amount of words to be 'learned by heart'.

Figure 1.1.4
Some technical methods for
diagnosis

a. X-ray of the hand
b. Ultrasound image of the
   womb of a pregnant
   woman

   1. *placenta*
   2. *skull*
   3. *spinal column*
   4. *heart*

c. CT-scan of the brain

   1. *cerebrum*
   2. *skull*
   3. *ventricles*

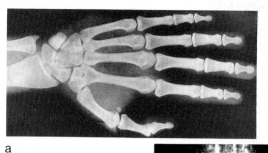

a

b

d. Arthroscopy: looking
   inside a joint
   I Place of entry of the
     arthroscope
   II Looking inside the
      knee-joint
      1. *knee cap
         (patella)*
      2. *thigh bone (femur)*
      3. *cruciate ligament*
      4. *end of the femur*
      5. *meniscus*
      6. *shinbone*
      7. *fibula*
      8. *arthroscope*
   III The image
      1. *end of the
         thigh bone*
      2. *front cruciate
         ligament*

e. Electrocardiogram
   (ECG)
   I ECG monitor
   II Position of the
      electrodes
      1. *electrode*
      2. *lead to monitor*
   III Recording

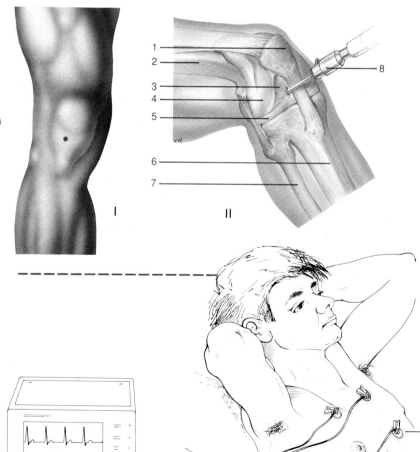

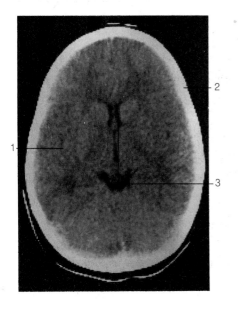

In Appendix 1 you will find a list of common Latin and Greek affixes which have been widely used in the formation of longer words. A knowledge of these affixes will help you to make an informed guess at the meanings of unfamiliar words.

Eponyms, that is, terms derived from the names of people (in this case normally from researchers in medical science) are used in anatomical and physiological terminology. Circle of Willis, Merkel's disc, Bainbridge reflex, islet of Langerhans, are all examples of these. This terminology has been used in the text where appropriate. Again, a good medical dictionary will help where there is alternative equivalent terminology.

Abbreviations

Abbreviations are used for names which are used frequently. The human body has for example hundreds of muscles. It would be unnecessarily time consuming to write the word muscles in full whenever a muscle is to be named. We use the abbreviation: m. and add the further particulars, e.g.: m. pectoralis major = musculus pectoralis major (the great chest muscle).

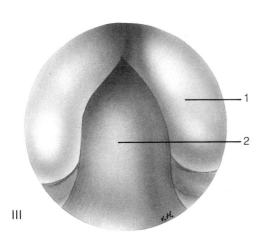

III

The abbreviations used in this book are:
- a. = arteria (artery);
- v. = vena (vein);
- n. = nervus (nerve), used for (branches of) spinal nerves;
- N.= nervus (nerve), for cranial nerves;
- m.= musculus (muscle).

Appendix 2 gives a more complete survey of the abbreviations used and their meanings.

4. The holistic approach

The entity: more than the sum of the parts

A square is more than the sum of four straight lines.

A chemical compound is more than just atoms. The human body is not the same as the parts of which it is composed. The entity is far more than the sum of its parts (Fig. 1.1.6).

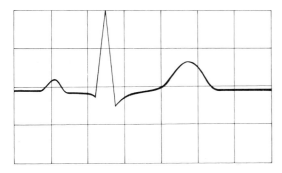

III

Figure 1.1.5

a. Examination: the posture of the spinal column and pelvis
b. Palpation: taking the pulse
c. Percussion: determining the boundary line between lung and liver
d. Auscultation: listening to the flow of air in the lungs

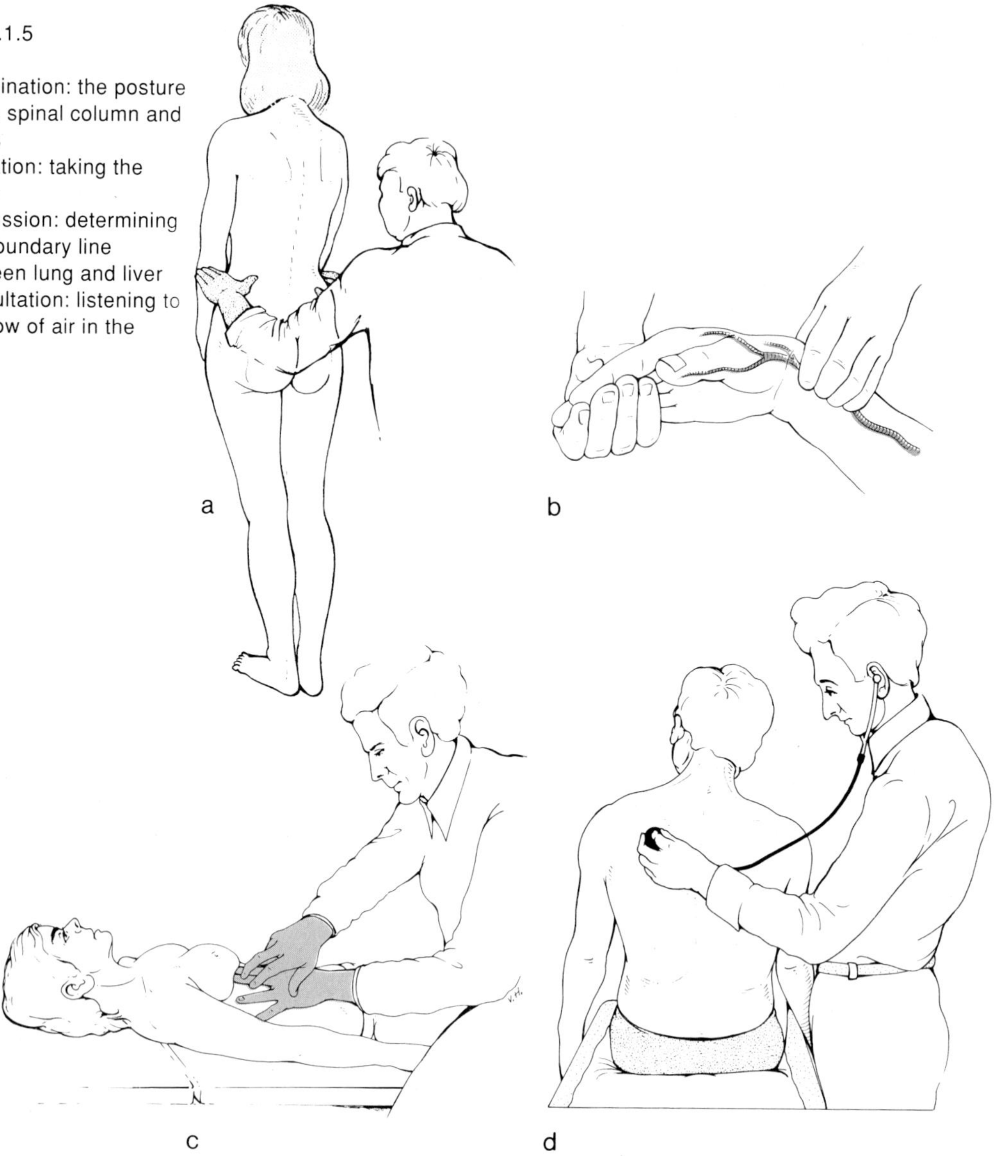

a

b

c

d

Yet in chemistry and other areas of medical science we often see an analysis being applied that singles out and separates the constituent parts. Such an analysis is often necessary but what is missing is the relationships and interdependencies between the discrete parts, so it is important to realize that it is impossible to deduce the whole simply from the sum of its parts.

From the fact that it is possible to live when a part of the body is missing you could come to the conclusion that this part therefore has no function. It is, however, quite possible that another part of the human body takes over that function partly or completely. In a similar way, it is incorrect to arrive at any conclusions about the functioning of a connection between organs from the phenomena that occur after that organ has been removed or that connection has been severed. Yet this is exactly what often happens, notably in research work on the brain.

Man is not a machine
A machine such as the engine of a car can be taken to pieces then reassembled to become a working machine once again. In a similar way,

the parts of the human body can be identified and separated, but the reassembled body could never work again like the car engine: death is death forever.

Given that, a definition of 'life' proves to be very difficult. The best we can do is to point out a number of characteristics of 'life' in comparison with death.

- We speak of 'life' when nutritious substances are absorbed to generate energy and to maintain the metabolic processes; 'life' is strongly inclined to *self-preservation*. A machine does not provide itself with energy and when something breaks down there is no 'self-repair'. A living organism indeed has the capacity to perform (within certain limits) these two functions.
- 'Life' has the tendency to *preserve the species*, making sure that there will be progeny by means of procreation. This characteristic is as important as the previous one. Self-preservation of individuals is futile if in the end there are no offspring. This feature is very important notably in the evolution of life.
- Though a living being has been constructed out of lifeless building materials it is very *malleable*. It has the capacity to adjust to many circumstances.
- Where there is life there is the capacity for *reaction, movement, development* and *learning*.

The human body has originated out of one cell, out of which have developed, after innumerable cell divisions, all the tissues, organs and body systems. In spite of the differences which arose during this development, the parts in the living organism have maintained their high level of interrelationship and interdependence.
Moreover the body is not a 'thing', a wrapper in which the 'I' lives. Your body is who you are, something you experience as 'yourself', and at the same time it is a part of yourself. The mysterious entanglement, of body (soma) and soul (psyche) has engaged philosophers throughout history.

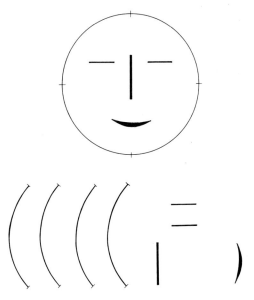

Figure 1.1.6
The entity: more than the sum of the constituent parts

Matters get even more complex when we consider that man does not live only for himself, but lives together with and for others in a community. The social aspect of life and living is an essential component of man's make-up. Furthermore, people live within the environment, and manipulate it to their own ends; we are constantly changing the appearance and the quality of the earth on which we live.
Man is a physical and psychological unity as well as pre-eminently a social being. Therefore it is actually quite improper and artificial to use dissection as the main tool of analysis. At best it gives us only a very incomplete impression of reality. Thus when specific data become available by means of this analysis, it must constantly be related to the functioning of the unimpaired living body.
Studying the 'parts' of the human body must always be complementary to understanding man as a psychic, somatic and social totality. We call this the *holistic approach*.

Functional anatomy

2

Introduction

In the previous chapter it was pointed out that analysis as a method
of gathering knowledge about functional anatomy is unavoidable. The
limitations of analysis have also been discussed.
In the scientific field of functional anatomy the object is to gain knowledge
about and insight into the normal somatic functioning of man as a whole.
This field is so large that for educational purposes it is necessary to divide
it up. In this chapter the method of subdividing the total field within the
human body is explained.
After that the various body systems are presented one by one on the basis
of a comparison between a unicellular organism and a multicellular
organism such as man.

Learning outcomes

After studying this chapter you should have sufficient knowledge and
understanding of:
– the possibilities and the limitations of the various reduction levels of
  functional anatomy;
– the basic functions of the ten body systems in the human body;
– the limitations of the learning method chosen for this subject.

1. Reduction

Levels of reduction
Reduction can be useful when designing a
curriculum in order to make it easier to handle.
Reduction can be described as the conversion
of data to simpler forms for clarity of understanding.
In functional anatomy the following levels can
be distinguished:

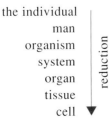

the individual
man
organism
system
organ
tissue
cell

reduction

– In the first step of reduction we act as if all
  human beings are alike.
– In the next step, man is considered as an
  organism in relation to other comparable
  organisms.
– Within the organism 'man' there are a
  number of systems, each one fulfilling a
  specific function which can be evaluated
  independently.
– Each of these systems is composed of a
  number of organs and the connections between these organs.
– In their turn these organs consist of several
  kinds of tissue.
– Finally the cell is the basic 'building-block'
  and the fundamental metabolic unit of any
  tissue.

Something is lost in every step of reduction but there is also an important gain: the subject matter becomes easier to handle, less complex in structure, and principles can be more easily explained and memorized.

Moderation
Every level of reduction deserves specific attention and will be covered to some degree in this book.

The level of the function systems (physiology) and related body systems (anatomy) has been adopted here as the basis for arranging the subject matter. In terms of reduction it is an intermediate level.
This has the advantage that the division has led to a limited number of subfields and at the same time the conceptual distance to man as a totality, our actual aim, does not become too great.

In addition it is possible to make excursions from the functional level into lower levels for necessary further detail.

In order to gain an understanding of the meaning and coherence of the various functional systems we shall compare the intricate human body, with its millions of cells, to a relatively simple unicellular organism. In this way it is possible to present the organ groups one at a time.

2. In the beginning there was water...

Water has played a vital role in the proliferation of 'life' on planet Earth and in the process of developing the countless organisms that lived and still live on earth. Indeed the greatest part of most organisms is water and this is also true of the human body.
The necessity of a 'watery' environment for the continued existence of unicellular organisms, and the importance of water for the cells of a multicellular organism, becomes apparent from the fact that all the materials which cells and their environment exchange must be dissolved in water. The pre-eminent exchange mechanism here is *diffusion*: the movement of dissolved particles from a space with a high

concentration of a certain substance to a space with a low concentration of that substance, (see Fig. 1.3.3a). Water is an excellent solvent. In addition water has a low viscosity, which assists the speed of diffusion. Another important quality of water for 'life' is that it forms an extraordinarily good 'heat-buffer'. Changes in temperature (e.g. as a result of the diurnal rhythms) generally have adverse effects on all kinds of metabolic processes in the cells. Water can absorb such changes of temperature to a very high degree. The reason for this is that water can absorb a lot of energy (joules[*]) without at the same time undergoing a substantial rise in temperature and vice versa. This characteristic of water is indicated with the term, *high specific heat*. Moreover a high degree of energy is needed for the evaporation of water, i.c. lots of energy can be removed without any great loss of water. This characteristic is called *high coefficient of heat*.

3. Unicellular organism

For a unicellular organism, (Fig. 1.2.1), the sea is an inexhaustible source of nutrition and at the same time an endless dumping-ground for waste materials. The sea is the outside environment (*milieu extérieur*) of the organism, basically an extremely tiny vesicle, filled with water, in which even smaller structures can be found, all with one or more functions of their own. The wall of the cell is called the *cell membrane*. This is a selective structure (usually called semipermeable), via which takes place the diffusion of dissolved substances into the intracellular space (the room within the cell membrane).

There is also a flow of materials passing through the membrane from within it, towards the outside. The unicellular system shows all the features of 'life'. Because of its metabolism the concentration of nutritious materials within the intracellular space decreases whereas the con-

---

[*]   The unit to measure energy is the joule (J). Formerly the calorie (cal) was used for this, or the kilocalorie (kcal); 1 kcal = 1000 cal = 4.186 kJ; 1kJ = 1000 J = 239 kcal (rounded off).

centration outside remains the same. Nutritious substances then diffuse from the milieu extérieur into the cell. At the same time the concentration of waste materials inside the cell increases (whereas the level outside remains stable).

Thus waste materials diffuse *out* of the cell into the milieu extérieur. Since there is in nature – under normal circumstances – a finely tuned cycle of substances, these nutritious materials will never stop being available and there will be no question of pollution of the environment in such a way as to threaten life. The unicellular organism can thrive in the sea and multiply freely by division.

4. Multicellular organisms and the importance of functional systems

Let us assume that we are dealing with a large number of cells gathered together into a clump (Fig. 1.2.2).

In this case only the cells situated on the very outside of the clump have direct contact (through one permeable layer) with the water (milieu extérieur). The cells 'in the middle' of the clump are more vulnerable. Their actual

surroundings (milieu extérieur) will quickly exhaust nutritious substances, and waste materials in the spaces between the cells will reach such a concentration that there would be a completely poisoned milieu extérieur.

The environment of such a cell (inside the clump) consists of extremely fine crevices between the cells filled with liquid. We call this inner environment the *milieu intérieur*.

Transport

For a multicellular organism (in our example the clump of cells), it is vital that the milieu intérieur is supplied with nutritious substances and that waste materials are transported out. Such an intensive refreshing is achieved with the help of a system of 'pipes' filled with 'refreshing-liquid' (or 'carrier liquid').

The pipe-systems branch off until they reach almost every crevice between the cells. This refreshing liquid and the milieu intérieur constantly exchange substances by means of diffusion, among other methods.

Transportation of nutritious substances to the cells and of waste material produced by the cells is established by pumping the 'liquid' between

Figure 1.2.1
Unicellular organism in the sea

1. cell wall, cell membrane
2. nucleus
3. milieu extérieur

Figure 1.2.2
Clump of cells

1. cell membrane
2. nuclei
3. cells
4. milieu extérieur

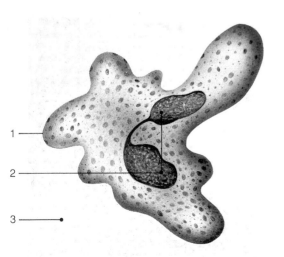

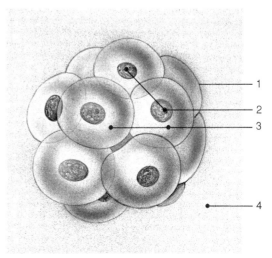

the cells through the 'network of tubes'. The system that executes this task is called **(1) the circulatory system**.

By the 'network of tubes' we mean the blood vessels, where 'the liquid' is the blood and the 'pumping' is the activity of the heart (Fig. 1.2.3).

Food supply

In order to provide the milieu intérieur (abbreviation: m.i.) with nutritious substances, the blood must be able to take in these nutritious' substances from the milieu extérieur (abbreviation: m.e.), for it is there that these substances are available. The circulatory system is in close contact with a 'channel' in which nutritious substances are being taken in, and then processed to be suitable for transfer into the blood. The system of which this 'channel' is a part is called **(2) the digestive system**.

As we can see from Figure 1.2.4, the contents of the alimentary canal belong to the m.e.

Excretion

The problem of the impending pollution of the liquid between the cells (sometimes called intercellular or interstitial fluid) has been shifted to the circulatory system. If the blood must take in waste material from the m.i. again and again, the blood in its turn must get rid of these waste materials.

In order to perform this task the circulatory system is connected with a 'purifying filter' (in the kidneys). This is an essential part of **(3) the urinary system**. By means of this system, the waste materials are transported to the m.e., together with a quantity of water (Fig. 1.2.5).

Gas exchange

Energy consumption is an important feature of life. Energy can be produced by burning energy-laden substances within the cell. Burning can take place with oxygen (aerobic combustion) and without oxygen (anaerobic combustion).

Aerobic combustion of e.g. sugar, yields as much as eighteen times the amount of energy

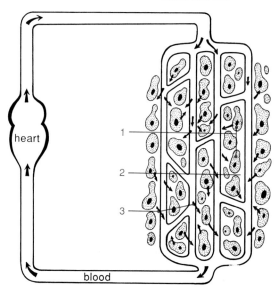

Figure 1.2.3
Heart and blood circulation

1. individual cells of a tissue
2. milieu intérieur
3. exchanging between blood and m.i.

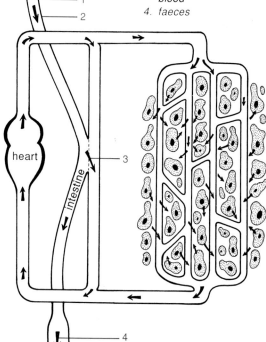

Figure 1.2.4
Digestive tract

1. ingestion and transport of nutriment
2. digestive tract
3. nutriment being passed over into the blood
4. faeces

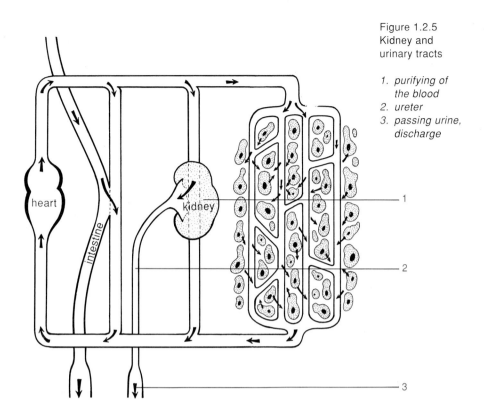

Figure 1.2.5
Kidney and
urinary tracts

1. purifying of
   the blood
2. ureter
3. passing urine,
   discharge

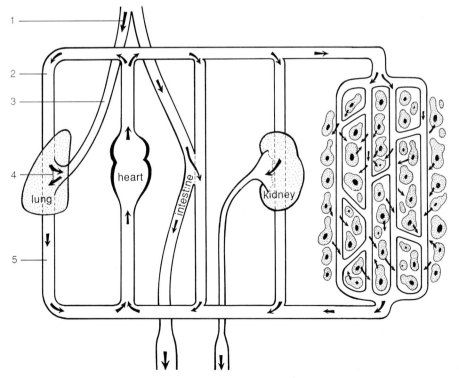

Figure 1.2.6
Respiratory tract
and lungs

1. breathing in,
   inspiration
2. blood:
   relatively
   poor
   in oxygen
3. air passage
4. exchange of
   gases in
   the lung
5. blood:
   relatively rich
   in oxygen

that is set free in anaerobic combustion of the same amount of sugar. Thus in aerobic combustion we use energy-laden substances far more efficiently than in anaerobic combustion.

Energy-laden substances come from certain types of nutriment, called energy sources. The digestive system processes them until they exist in such a form that they can be passed into the circulatory system. The circulatory system then conveys the energy-laden substances to the m.i. of the clumps of cells.

Oxygen must also travel a similar route, before it can take part in the process of combustion within the cell. Since oxygen is available in the m.e., it must be passed over into the blood. To do this the circulatory system is in contact with a gas-exchange membrane (in the lungs).
In aerobic combustion in the cell, a gas is produced – carbon dioxide, a waste product. Because of the difference in the concentrations of this gas, the carbon dioxide diffuses from the cell into the m.i. and from there into the blood. Later in the lungs, via the gas-exchange membrane, this waste product is then passed on to the m.e. (Fig. 1.2.6). The system responsible for the exchange of the gases oxygen and carbon dioxide is called **(4) the respiratory system**.

Marking the boundaries

In a unicellular organism the cell membrane is the boundary with the m.e. Multicellular organisms have an extra covering of the cells and the m.i., **(5) the skin** (Fig. 1.2.7).
The skin provides some protection against mechanical influences (e.g. pressure), chemical influences (e.g. acidity, pH) and influences of temperature changes. In the case of multicellular organisms living on land (such as man) the skin offers in addition a protection against dehydration.

Regulation

The functions of the five systems described above are collectively termed *vegetative functions*. They are functions which serve to keep the cells alive. In a unicellular organism control of such functions as taking in food and excretion is relatively simple. Here there are no specialized systems for the separate functions we have distinguished, and the distance which control-messages have to travel is relatively small.

Figure 1.2.7
Skin

1. skin
2. layers of the skin
3. body orifices

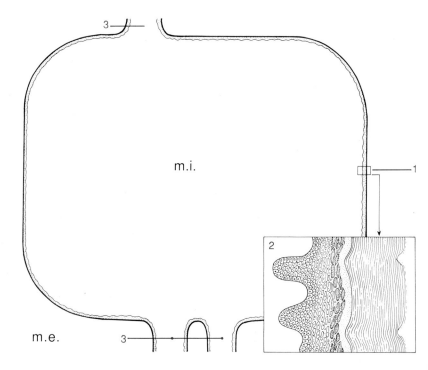

In the case of multicellular organisms, such as man, this is not as simple. The five vegetative functions have to be finely tuned. There needs to be an integration of the functions to ensure that they act as an entity. When a lot of energy is needed, the system must see to it that sufficient quantities of energy-laden substances and oxygen are transported to the active cells. The waste material then produced in larger quantities must be removed efficiently.

For this internal harmonization of the vegetative body systems we use the term *vegetative integration*. It is an integration *within* the individual.

Two systems are responsible for this internal integration (Fig. 1.2.8).

The first vegetative regulating system is called **(6) the hormonal system**.
Hormones are *chemical regulating control substances* which are produced by specialized cells

and they reach the cells via the circulatory system.
The second is the vegetative part of **(7) the nervous system**.
Electrical impulses are conducted to the organs which need regulating.

The two control systems must, of course, be well informed of the exact situation in the vegetative system, and the slightest changes in these situations. In the body systems themselves we find few organs (sensors) that are sensitive to specific changes. The sensors that supply the information needed to regulate or control activity are called *vegetative sensors*.
After the information has been received and processed, a stimulus can be sent out for regulatory activities (via the hormone system and/or the vegetative nervous system).
We call these stimuli *vegetative motors*.
This type of activity is effected by various operations such as the secretion of substances by

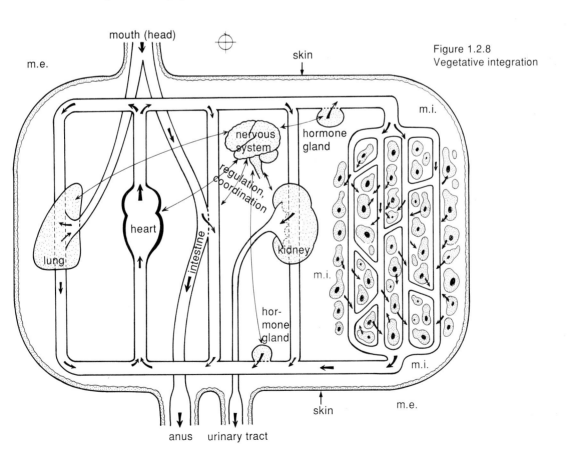

Figure 1.2.8
Vegetative integration

glands, and the increasing or decreasing of the bloodflow to and from an organ by the widening or narrowing of the blood vessels.

Interaction

In addition to vegetative 'life' we can identify animal 'life'. The literal meaning of animal is: with a soul, infused with a spirit, provided with a mind.

In functional anatomy we distinguish the vegetative system (and vegetative integration) from the animal system (and animal integration).

Animal functional systems take care of the active (and often conscious) interaction between the organism and its surroundings: *animal integration* (Fig. 1.2.9).

This interaction requires that the organism receives information about the surrounding environment and that it is alerted to possible danger. This task is performed by the animal part of **(8) the sensory system**.

The best known parts of this system are the five senses: smell, taste, sight, hearing and touch. The information produced by the senses is conveyed via the nerve tissue connections of the nervous system. In addition to these functions all kinds of intellectual activities take place in this animal part of the nervous system, e.g. thinking, wanting, remembering.

From the animal part of the nervous system there are activities initiated in the animal part of **(9) the motor system**.

The word 'motor' may immediately make you think of muscle activity, but the skeleton and the joints are also part of the motor system. We are not now involved with the 'clumps of cells' we talked about earlier, but with a human being, a real person.

In comparison with the unicellular organisms living in the sea, the multicellular organisms living on land are subject to gravity. Without a firm, solid framework, a multicellular organism such as man would collapse under its own weight and be pressed down into a flat disc on the surface of the earth.

The bones of the skeleton develop sufficient resistance against the force of gravity. Since the bones are linked together by means of joints, there is also the potential for making movements. These movements are brought about by the muscles.

Preservation of the species

Having distinguished the vegetative and the animal systems, we now look at the way in which self-*preservation* is achieved in multicellular organisms in relation to that in unicellular organisms.

A unicellular organism takes care of the preservation of the species by simply dividing itself into two 'daughters'. This type of creating offspring is called asexual procreation.

Figure 1.2.9
Animal integration

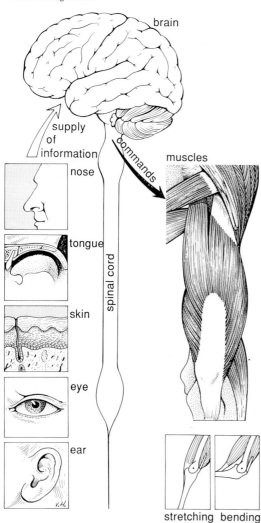

brain

supply of information

commands

muscles

nose

tongue

spinal cord

skin

eye

ear

stretching bending

Because of their specialized organ systems multicellular organisms cannot simply 'split up'.

With these organisms the preservation of the species is achieved by means of sexual pro-creation. Here a female reproductive cell (egg) and a male reproductive cell (sperm) must unite (the moment of fertilization), after which a new individual can then start to develop.

The complex of organs responsible for the pres-ervation of the species is: **(10) the reproductive system**.

5. A model is not the real thing

At the end of this brief survey of the ten body systems, it is of major importance to stress again that the step-by-step method used here has been chosen for educational purposes. The method has provided us with a model, but nothing more than that. Therefore all kinds of unintended simplifications are bound to occur. The division into body systems is a method to distinguish and discuss them, not to disconnect them.

The liver, for example, is an organ which can be considered to belong to the digestive system, since the nutritional substances from the intes-tines – having been absorbed into the blood – pass first through the liver and are processed or stored, before they are further distributed in the body. The liver, however, also plays an im-portant role in circulation; all blood proteins, for example, are produced by the liver.

The skin not only protects us and marks the boundaries of our body, it also offers oppor tunities for sensory activities and for dis-charging waste material (through sweating).

We could easily mention more examples of organs that fulfil two or more functions (multi-functionalism). In Module 2 – in the more detailed examination of the organ systems – they will be discussed in far greater depth.

6. From variation to standardization

When you ask someone 'How much blood do you think you have got?', most people will answer 'Five litres'. In most cases this answer is incorrect. It is an answer that is the result of the reduction of *a* human being to *the* human being. When an average value is being taken, every variation in human beings is lost.

A better answer to our question would have been: 'About 7.5% of my body weight'. Even then there would still be a certain degree of inaccuracy, because of individual variations.

Yet, in pinpointing such an average figure, such a standardisation of man is very understandable, We would hardly be able to register, let alone remember, any figures in view of the enormous variety which exists in reality.

In this book we mention a number of 'human measurements'. They have all originated from a non-existent 'ideal type of person'. We shall call this the *average human being* with the following characteristics:

– male
– 25 years of age
– height 1.75 metres
– weight 70 kilograms
– of average build
– in good health.

In the daily practice of our professional life, it is especially important to adapt the figures of the average to the actual patient.

7. Water and homeostasis

The average human body consists of approxi-mately 60% water, i.e. for the average body 42 litres, divided approximately as follows:

– intracellular water	25 litres
– tissue fluid (liquid in the m.i.)	12 litres
– blood	5 litres
	42 litres

Refreshing or changing the intracellular liquid takes place from the m.i. The composition of the component parts of this m.i. is in its turn com-pletely dependent on its being refreshed by the blood.

The circulatory system must offer the oppor-tunity of thoroughly refreshing the blood by the digestive, urinary and respiratory systems.

All this clearly shows the vital importance of the circulatory system.

Therefore:
5 litres of blood → refresh 12 litres of tissue fluid → refresh 25 litres of intracellular water.

The purpose of the refreshing of the m.i. is firstly to keep the composition of the m.i. constant so that there are the right amounts of nutritional material, oxygen, regulating substances etc. present, and secondly to prevent the accumulation of waste materials and carbon dioxide.

This process, this activity of keeping the m.i. constant, is called *homeostasis*. The outcome of homeostasis is a finely balanced composition of the contents of the liquid which is constantly in danger of being disrupted by the activity of the cells as they take out nutritive substances and discharge waste materials into the m.i. The balance is constantly being maintained by the activities of the circulatory system in co-operation with the other systems.

Cells

3

Introduction

The cell is the smallest living unit in the human body which still shows all the characteristics of 'life'. Therefore the cell is considered to be the basic building material (anatomical term) and the fundamental unit of metabolism (physiological term) of the organism. When we study the cell we find it consists of a large number of structures each with its own form and function. In this chapter we will discuss the functional anatomy of the cell. The study of cells is called cytology.

Learning outcomes

After studying this chapter you should have sufficient knowledge and understanding of:
- the cell as the basic unit for construction and metabolism in the human body;
- the structure and functions of cells;
- cell growth and normal cell division;
- the importance of the differences in construction and function of the various types of cells.

1. Differences and similarities

Everybody is aware of the fact that the nerve cells in the brain are not suitable for lifting and that muscle cells are not suitable for thinking. In the body we can distinguish different types of cells, each with its own form and function(s). Yet the differing cells show many similarities. This is not surprising when we consider that the complex human body, consisting of millions of cells, has after all developed from that one fertilized egg cell (the zygote).

The differences will be discussed later but for now we will primarily concentrate on the similarities. To start with we envisage some kind of standard cell, (Fig. 1.3.1), in which many common cell structures can be explained.

The cell consists of a jelly-like mass, called cell liquid or *cytoplasm*, enclosed by a very thin cell wall, the cell membrane (plasma membrane).

The cytoplasm consists of water in which are dissolved, among other substances, proteins, carbohydrates, fats and salts. In addition to this – with the help of an electron microscope – we can see a number of highly specialized structures which are called organelles (a collective term). In the *nucleus* (kernel) the codes have been stored for the manufacturing of all the protein molecules. The *endoplasmic reticulum* (ER) manufactures the proteins in accordance with the codes from the nucleus.

The protein molecules that were formed in the endoplasmic reticulum are processed in the *Golgi apparatus* in such a way that they become suitable for transportation by the circulatory system to the m.i.

Thus the nucleus, endoplasmic reticulum and Golgi apparatus are all organelles which are involved in the building of substances. Other types of organelles are for example the

Figure 1.3.1
Cell

1. *lysosome*
2. *cell membrane*
3. *mitochondrion*
4. *endoplasmic*
   reticulum
5. *cytoplasm*
6. *nuclear membrane*
7. *nuclear pore*
8. *nucleus*
9. *nucleolus*
10. *ribosomes*
11. *Golgi-body*
12. *centrioles*
    (developed out of
    centrosomes)

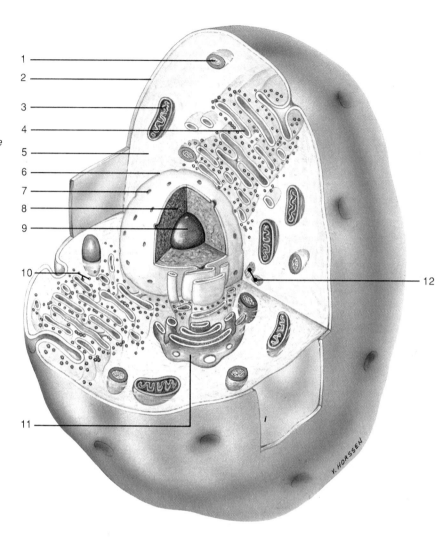

mitochondria and *lysosomes*. These are in- volved in the demolition of cells.

A mitochondrion can be seen as a power station, where substances are broken down to form energy-laden packages for the benefit of all the processes in the cell which require energy.

A lysosome splits up compounds to prepare them for processing elsewhere (e.g. in the mitochondria). At the same time they serve as rubbish dumps for damaged and redundant substances which have come into the cell or have developed there.

The *centrosome* is an organelle that plays an important part in the division of the chromo- somes during cell division. The first step to-

wards this is its splitting up into two *centrioles*. In the following sections we will take a closer look at cell structure.

2. Membrane transport

Within the cell it is important that activities are organized and that each organelle is able to perform its function without interference from other structures.

In addition, it is important that the organelles are able to communicate. The different structures are separated by semi-permeable walls (usually known as *membranes*) which perform this communicating function.

2.1. Membranes
Plasma membranes

The cell membrane, (Figs. 1.3.1 and 1.3.2), also called the plasma membrane or the *lipoprotein membrane* (lipid = fat; protein = protein), consists of a double layer of fatty molecules in which protein molecules 'float'.

The lipid molecules have a hydrophilic part (with a strong affinity to water) and a hydrophobic part (with an aversion to water). Both cytoplasm (the bulk of the contents of the cell) and m.i. consist mostly of water. The hydrophilic parts of one of the lipid layers are turned in the direction of the m.i.; those of the opposite layer are turned in the direction of the cytoplasm. So the hydrophobic parts of the lipid molecules point toward each other.

This construction of the cell wall gives a very high flexibility of shape. In the lipid layer there are protein molecules drifting about. Some protrude on the outside into the m.i., others are beneath the surface and protrude on the inside into the cytoplasm of the cell and there are some that stick out on both sides.

The protein molecules constantly move about. Not only can they protrude out of the double lipid layer, but they can also move horizontally.

The function of the cell membrane on the one hand is to keep apart the cytoplasm of the cell and the fluid of the m.i., with their different compositions, and on the other hand to enable an intensive transport of varied substances from the m.i. to the cytoplasm and vice versa.

The protein molecules drifting about can serve as a kind of antenna for receiving messages. When the protein is beneath the surface the message cannot be received. Proteins with such an antenna function are called *cell-receptors*.

Other types of membranes

The nucleus of the cell, the Golgi body, the mitochondria, the lysosomes and the parts of the centrosome all have a lipoprotein membrane as a separating wall. Even the endoplasmic reticulum consists to a large extent of membrane. Although the architecture of these membranes does show differences, compared with the plasma membranes discussed above, the simi-

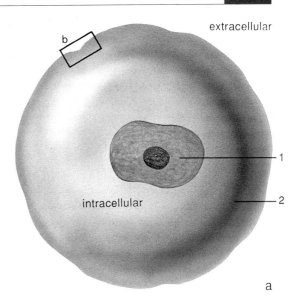
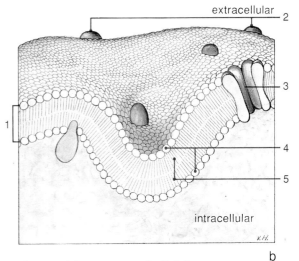

Figure 1.3.2
Cell membrane

a. Cell

   1. nucleus
   2. plasma membrane

b. Detail

   1. double layer of lipids
   2. proteins
   3. membrane pore
   4. hydrophilic parts
   5. hydrophobic parts

larities are far greater. In this section we will not discuss the differences further.

2.2. Forms of transportation

The lipoprotein membrane is suited for both *passive* and *active* transportation. Passive transportation requires no energy; active transportation on the other hand *does* consume energy.

Diffusion

There is only one type of mechanism of *passive*

transportation: diffusion (Fig. 1.3.3a). This is the movement of particles from a place where there is a high concentration of this type of particle to a place where there is a lower concentration.

If you place a little perfume on a plate in a corner of a room you will soon be able to notice the effects of diffusion throughout the room. If we divide the room into two parts by means of a screen with holes in it through which perfume particles can pass, then, after some time, there will be equal amounts of particles present in each part of the room. At first there is a greater chance of particles going from the side with the higher concentration of particles to the side with the lower concentration than vice versa. The end effect is the same as in the situation without such a permeable screen: the number of particles increases on that side of the screen where the concentration is smallest. When there is an equally strong concentration of the particles on both sides of the screen, the diffusion stops. That does not mean that the particles stop moving – on the contrary, they stay as mobile as ever. Now, per unit of time, there will be just as many particles going through the openings from A to B as there are going from B to A. The bigger the openings in the screen, the bigger the particles that can pass through. Particles that are bigger than the 'holes' in the screen will 'try' to get through, but will not be successful.

The plasma membrane can be compared with such a semi-permeable screen. On one side there is the m.i. with its concentrations of *dissolved* particles; on the other side there is the intracellular milieu with its own specific composition. The 'gaps' between the adjacent lipid molecules allow the passage of particles smaller than 0.7 nm (1 nm = 1 nanometre = 10^{-9} metre). A membrane of this type (with gaps of 0.7 nm) is generally called *semi-permeable*.

Oxygen molecules and carbon dioxide molecules are examples of particles smaller than 0.7 nm that can diffuse through this type of membrane. Although monosaccharides, fatty acids and amino acids are smaller than 0.7 nm they do not diffuse but are dependent on active transportation.

The speed of diffusion depends on a number of factors:
- *temperature:* the higher the temperature, the faster the diffusion;
- the difference between the *levels of concentration* of the substance: the greater this difference, the faster the diffusion;
- the *diffusion distance:* the smaller the distance, the faster the diffusion;
- the *diffusion surface:* the larger this surface, the faster the diffusion;
- the *size of the particles:* the smaller the size, the faster the diffusion;
- the *viscosity* (stickiness) of the solvent; the lower the viscosity, the faster the diffusion.

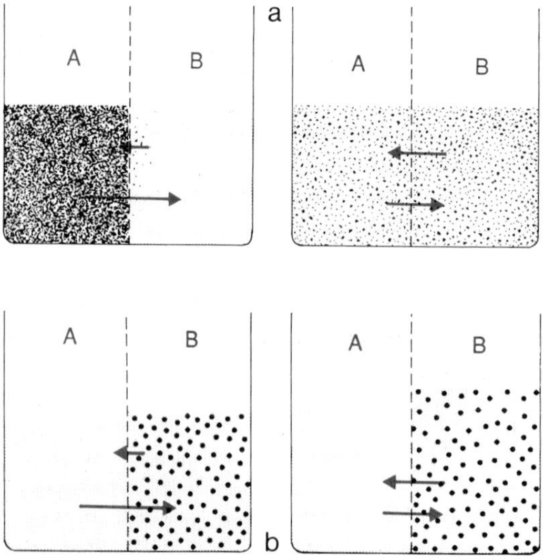

Osmosis

Osmosis is the diffusion of the *solvent* itself, but not its *solute* (Fig. 1.3.3b). Osmosis can occur when certain particles cannot pass the semi-permeable wall because they are too large to get through. In osmosis we see a shifting of solvent (usually water) from a place with a high density of this solvent to a place with a lower density of this solvent.

The particles which are too big to pass the semi-permeable membrane would go from side B to side A, but cannot because of their size. The presence of these larger particles on side B, e.g. proteins (colloids) or certain types of salts

Figure 1.3.3
Passive transportation

a. Diffusion
b. Osmosis

(crystalloids), and virtually no such particles on side A, has the effect that on side B the relative amount of solvent is lower than on side A, where there is a high density of water.

Water therefore diffuses from side A (where there is a higher concentration of solvent) to side B (where there is a lower concentration of solvent). It almost looks as if the colloids and the crystalloids *suck up* the solvent towards themselves; but that is only how it appears.
The quantity of colloids or crystalloids – in other words the concentration of the solution – determines the degree to which the solvent can diffuse, so we speak of the colloid-osmotic value (COV) or the crystalloid-osmotic value of a given solution. Often there are different names used: colloid- and crystalloid-osmotic pressure. In view of the direction in which the solvent moves the word 'pressure' is a misleading one. So in this book we will always speak of colloid- (and crystalloid-) osmotic *value*.

Pores
Particles that are bigger than 0.7 nm (e.g. certain types of enzymes) and electrically-charged particles (ions) can in many cases pass the lipo-protein membrane via the pores.
Pores are 'openings' in the double layer of lipids which are formed by certain specific proteins (Fig. 1.3.2). It goes without saying that the basis for 'transportation' via pores is similar to the basis for the mechanism of diffusion.

Active transportation
Active transportation requires energy; the size of the particles is basically unimportant and the transport can take place in a direction opposite to the one dictated by the difference of the concentrations. We can distinguish two types of active transport. Type 1 is called *vesicle transport* (Fig. 1.3.4).

There are two directions in vesicle transport:
– when particles are being brought from the m.i. into the cell via enclosure by the membrane: *endocytosis*. When solid particles are being ingested: *phagocytosis*. When liquid particles are being engulfed: *pinocytosis*;

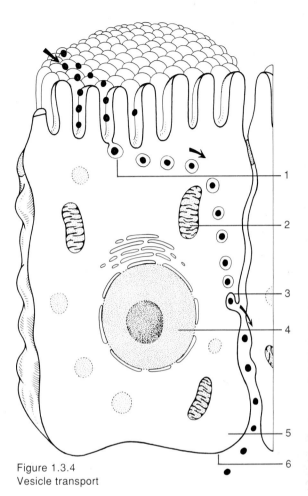

Figure 1.3.4
Vesicle transport

1. endocytosis
2. mitochondrion
3. exocytosis
4. nucleus of the cell
5. cytoplasm
6. cell membrane

– when particles are being removed from the cytoplasm into the m.i. by the reverse process we speak of *exocytosis*.

Both with endocytosis and exocytosis the particle that must be transported is completely engulfed in a double layer of lipoproteins. The engulfed substance itself does not come into direct contact with the intracellular milieu and the 'engulfing membrane' can easily unite with other membrane-structures within the cell.
The second form of active transport is the *enzymatic pump* (Fig. 1.3.5). We can visualize this pump as a kind of 'wheelbarrow molecule' (the enzyme). Within the cell (on the intracellular side of the cytoplasm membrane) this molecule

Figure 1.3.5
Enzymatic pump

1. cell wall
2. pump (schematic)

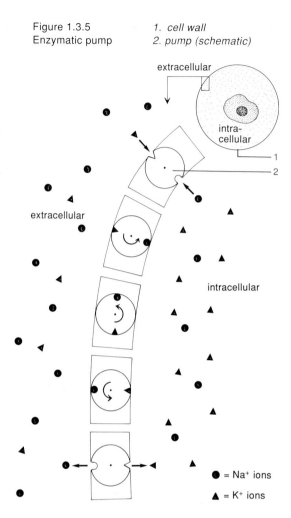

extracellular

intra-cellular

— 1
— 2

extracellular

intracellular

● = Na+ ions
▲ = K+ ions

chemical reactions in which bigger molecules are built up out of smaller ones (*assimilation*) or in which bigger molecules are split up into smaller ones (*dissimilation*).

As a rule, the biochemical reactions that take place within the cell involve energy conversion or metabolism: either the reactions demand energy (anabolism) or the reactions release energy (catabolism).

Material exchange is aimed at the growth and repair of the human body, and is accompanied by anabolic assimilation. It is often called *anabolism*.

Energy conversion aims at the release of energy by the cell components for the benefit of anabolic processes.

Energy metabolism is accompanied by catabolic dissimilation and is therefore often called *catabolism*.

To ensure that all the metabolic processes taking place in and outside the cell run smoothly, the body has at its disposal a large number of *enzymes*. Such 'biochemical catalysts' have the following features:

– they speed up or slow down a chemical reaction;
– they are not themselves consumed in a chemical reaction, nor permanently altered;
– they are proteins;
– they are 'reaction specific', which means that to a certain, specific chemical reaction there belongs a certain, specific enzyme;
– they are temperature specific: at a higher or lower temperature than their specific temperature they work less efficiently or even not at all. For most of the enzymes in the human body the optimal function temperature is exactly the body temperature (37°C);
– they are acid specific. For each enzyme there exists a pH-optimum; at that pH-level the enzyme works most efficiently;
– there are often other substances necessary in order to allow enzymes to work. These substances are called *co-enzymes*.

unloads substance A and then loads substance B, finds its way through the membrane (taking B with it) and so transports B to the extracellular side of the membrane. There the substance B is unloaded. After that a new particle of A is loaded. This loading and unloading takes place by means of a chemical binding and releasing reaction.

The best-known enzymatic pump is the sodium-potassiumpump. With the help of this 'mechanism' there is maintained a relatively high concentration of Na+ions on the outside of certain cells and within these cells a relatively high concentration of K+ions.

3. Digestion/Metabolism

The cell has been defined as the smallest unit of metabolism. *Metabolism* contains all the bio-

Enzymes are often named after the materials they split up or after the reaction over which they exercise an influence. Their names often end in the suffix -ase. For example: lipase splits up lipid, amylase splits up amylum (starch, a carbohydrate), the proteinases are the group of enzymes involved in splitting up proteins. For a number of enzymes this nomenclature does not hold good. Examples: ptyalin (salivary amylase), pepsin, trypsin.

3.1. The building up of substances

In the cells we find the building up of organic substances, the larger part of which are proteins; the greater percentage of the proteins being synthesized are enzymes.

The receptors in the cell wall consist of protein-molecules. In the m.i. we can find various types of fibres that have also been composed out of proteins.

Lipids and carbohydrates are also produced. In addition to this there are substances formed with energy-laden *compounds* in the cell.

Nucleus

The nucleus, (see Fig. 1.3.1), consists of a quantity of nuclear plasm (nucleoplasm) enwrapped by a membrane: the nucleus-envelope. In this nucleus membrane we can distinguish two layers. It is very permeable because of the unusual amount of rather large pores (nm 50 approx in diameter).

In the nucleoplasm – consisting largely of water – we notice one or more nucleoli and a number of networks of fine filaments: the chromatin network. Both these nucleus bodies consist mainly of nucleic acids. The nucleoli contain *ribonucleic acid* (RNA). The chromatin net-work, its name derived from the fact that it is so readily stained with dyes, contains *deoxyribo-nucleic acid* (DNA).

All inherited codes, which means all the 'recipes' for the production of all the proteins, are recorded within the strings of DNA in the chromatin network.

The process of protein synthesis is discussed later in this chapter.

Endoplasmic reticulum

The *endoplasmic reticulum* (ER), (see Fig. 1.3.1), is an extensive, closed network (reticu-lum) of flat cavities and linking tubes. The membranes forming the endoplasmic reticulum are basically folds of the two-layered nucleus membrane. On the outside of a part of the endoplasmic reticulum we find ribosomes, which are granules with a high grade of *ribonucleic acid* (RNA). This area is known as the *rough endoplasmic reticulum.* The part of the endoplasmic reticulum on which there are no ribosomes is called the *smooth endoplasmic reticulum.* This smooth endoplasmic reticulum plays a part in the manufacturing of lipids for membranes. The rough endoplasmic reticulum works together with the nucleus on the synthesis of proteins.

The code of life

The chromatin filaments consist of chains of DNA. This DNA is sometimes called 'the code of life' because it is the storage place for all the information necessary to conduct the countless processes required for the realization of each unique individual and for its functioning. The actual 'archives' are found in the order in which the nucleic acids are arranged in the DNA. It is the 'database' of the human body, copied in each cell nucleus.

The form of the *DNA molecule* is a kind of spiral-shaped ladder (Fig. 1.3.6). Both posts of the ladder consist of a succession of one deoxyribose (i.e. sugar) molecule (D) and one phosphoric acid molecule, a phosphate (P).

The rungs of the ladder consist of nitrogenous bases attached to the sugar molecules of the 'posts' of the ladder. Each rung is formed of either a combination of the nitrogenous bases *adenine* (A) and *thymine* (T) or a combination of *cytosine* (C) and *guanine* (G). The rungs of the ladder appear as if they have been 'sawn in half', i.e. there is a loose chemical connection be-tween the pairs of nitrogenous bases, so that they can readily zip together and unzip easily.

This fact, in combination with the fact that there are fixed couples of nitrogenous bases, is essen-tial in the function of DNA. When the ladder divides in the middle of the rungs we get a kind of mould. It is then possible to form a copy of the opposite ladderpost and its half rungs on this mould. This is of importance for both the protein

Figure 1.3.6
DNA molecule

1. *a pair of bases (1)*
2. *a pair of bases (2)*

 sugar

 phosphate

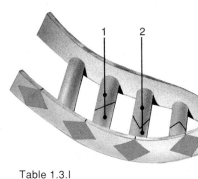

Table 1.3.I
Amino acids with
accompanying codons;
starting- and stopping codons

Amino acid	*Codon in messenger-RNA*
Alanine	GCU GCC GCA GGG
Arginine	CGU CGC CGA CGG AGA AGG
Asparagine	AAU AAC
Aspartic acid	GAU GAC
Cysteine	UGU UGC
Glutamic acid	GAA GAG
Glutamine	CAA CAG
Glycine	GGU GGC GGA GGG
Histidine	CAU CAC
Isoleucine	AUU AUC AUA
Leucine	CUU CUC CUA CUG UUA UUG
Lysine	AAA AAG
Methionine	AUG
Phenylalanine	UUU UUC
Proline	CCU CCC CCA CCG
Serine	UCU UCC UCA UCG AGU AGC
Threonine	ACU ACC ACA ACG
Tryptophan	UGG
Tyrosine	UAU UAC
Valine	GUU GUC GUA GUG
Start	AUG
Stop	UAA UAG UGA

Figure 1.3.7
Triplet

T = thymine
C = cytosine
G = guanine

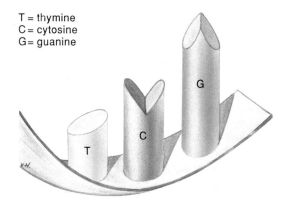

synthesis and the necessary copying of the 'DNA-archive' when the cell divides.

A succession of three half rungs with the parts of the ladderpost belonging to them (Fig. 1.3.7), is called a *triplet*. Each triplet presents the mould on to which an amino acid can be attached. The specific succession of triplets determines the specific order of amino acids resulting in the formation of a specific protein.

On the basis of four nitrogenous bases 64 different triplets can be formed. As there are only 20 amino acids in existence, for many of these acids there are several possible triplets available (Table 1.3.I). Moreover there are triplets which mark the separation to the mould for a following amino acid order; these are known as the *start and stop triplets*.

The *RNA molecule* (found in the nucleoli and the ribosomes) shows three differences compared with the DNA molecule:
- the RNA consists of *one* ladderpost and half rungs;
- we always find a ribose molecule (R) in the place of the deoxyribose molecule on the post;
- we have the chemically related uracil (U), that can also combine with adenine (A) in the place of the nitrogenous base thymine (T).

We will now look at the formation of a protein, step by step (Fig. 1.3.8).

The DNA molecule splits in two over a certain distance along the line of the half-cut rungs of the spiral ladder. The triplets of one side serve as a mould. A string of RNA is formed against this mould, beginning at a starting triplet and ending at a stopping triplet (transcription). This resulting RNA string disconnects from the DNA-mould and – now called *messenger-RNA* – leaves the nucleus via a nuclear pore. The messenger-RNA shows an exact negative picture of the DNA-mould involved. Such a negative of a DNA-triplet is called a codon.

The messenger-RNA is then attached to the ribosomes. The cytoplasm contains *transport-RNA*, consisting of separate triplets. Specific transport-RNA-triplets accord with specific triplets of the DNA-mould, which is why they can easily combine with the negative 'picture' of the DNA-mould as this has been formed by the codons of the messenger-DNA. This is why a transport-RNA-triplet is called an *anti-codon*. On the other hand each transport-RNA-triplet can take a specific amino acid 'in tow'. Such an amino acid must then be present in the cytoplasm. The messenger-RNA determines the order of the amino acids; the amino acids are coupled together with a peptide-compound and thus form the desired type of protein.

The transport-RNA in its turn breaks away from the amino acids and from the messenger-RNA. The stringing together of amino acids starts at a starting-codon; the desired type of protein has been formed when a stopping-codon is found in the messenger-RNA. The proteins which have been formed at the ribosomes and are destined for usage within the cell itself remain in the cytoplasm; the other proteins end up in the canal system of the endoplasmic reticulum. In this structure they are conveyed to the Golgi body.

The Golgi body
The Golgi body, (see Fig. 1.3.1), looks very much like a piece of the smooth endoplasmic reticulum taken apart. It consists of a curved pile of flat cisternae (little vesicles), which are in contact with one another via linking tubes. In the Golgi body the manufactured proteins are stored temporarily. Then they can be enveloped by means of membrane-enclosure and transported to a destination in accordance with the function of the protein or enzyme produced, e.g. to the cytoplasm membrane for exocytosis. It is probable that (a) carbohydrate synthesis and (b) production of lipids for the construction of membranes also take place in the Golgi body.

Compounds rich in energy
A substance called ATP (*adenosine triphosphate*) is formed in the cells. Every molecule of this substance contains a compound rich in energy. The amount of energy set free when ATP is split up is used for all the processes in the cell which demand energy (protein

Figure 1.3.8
a. Transcription

1. *nuclear pore*
2. *membrane of the cell nucleus, nuclear envelope*

b. Protein synthesis

transcription

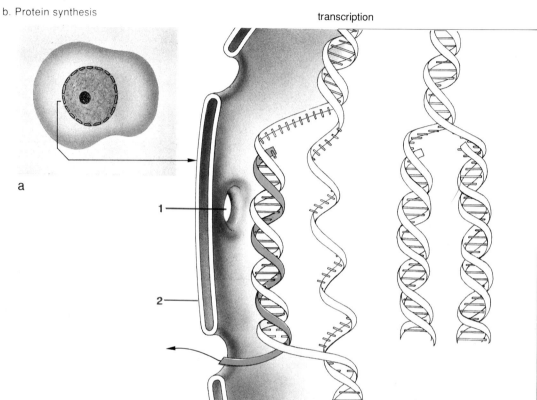

a

A = adenine
C = cytosine
G = guanine
U = uracil

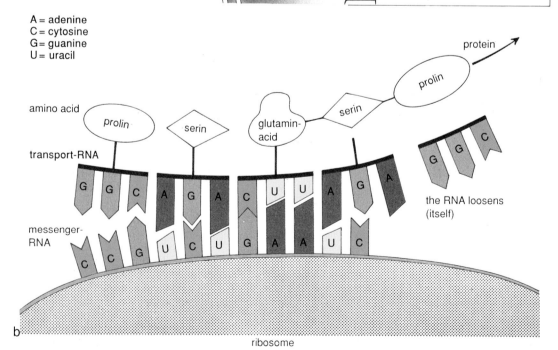

b

ribosome

production, movements, active membrane transport, etc.).
ATP is built up out of ADP (adenosine diphosphate) and phosphoric acid.

This process is represented in the simple formula below:

ADP + P + energy $\rightarrow$ ATP

The energy required is obtained by burning certain substances, e.g. glucose and fats.

3.2. Decomposition of substances

In a cell we not only find the building up of organic substances, we also find them being decomposed. This decomposition can be directed towards the management of energy in the human body or the removal of damaging substances.
The decomposition of glucose and fatty acid molecules is accompanied by the formation of substances with compounds that are rich in energy, the most important being the molecule adenosine triphosphate (ATP).
Decomposition of glucose takes place in the cytoplasm under the influence of enzymes that are present. In this process there are only two ATP molecules formed per glucose-molecule, and lactic acid is formed. There is no oxygen required for this manufacturing of 'energy packages'. It is a form of *anaerobic* oxidation. This anaerobic oxidation is far less efficient than the aerobic oxidation (burning with 'the help of' oxygen) in the mitochondria.

Mitochondria

Mitochondria, (see Fig. 1.3.1), are mostly oval in shape and can occur in large numbers in certain cells. Their walls consist of a double layer of membrane; the inner membrane layer is folded towards the inside. The contents of the mitochondrion consist mainly of a large number of enzymes. These enzymes have an influence upon the reactions within the *citric acid cycle*, in which glucose is burned with oxygen; carbon dioxide and oxidized water are produced as well as 36 ATP molecules for every glucose molecule.
Aerobic combustion in the mitochondria

delivers 18 times as many energy-packages as anaerobic combustion in the cytoplasm.
In addition to the glucose, muscle cells can also bring fatty acids into the citric acid cycle for the production of ATP. This possibility provides the cells concerned with an enormous stock of reserve fuel. The combustion of fatty acids is less 'clean' than that of glucose; besides carbon dioxide and water there are also other waste products released.
Remarkably enough the mitochondria possess nucleic acids (DNA and RNA) of their own which do not have the same composition as the nucleic acids within the nucleus. There is some speculation that originally mitochondria were independent organisms which at a certain point in evolution entered into a symbiotic relationship with a unicellular organism.

Lysosomes

The wall of the bulb-shaped lysosomes, (see Fig. 1.3.1), is three layers thick. The lysosome contains a number of fiercely active enzymes that can selectively enter into the intracellular milieu through openings in the membrane. There these enzymes set off the splitting up of proteins, of carbohydrates and of fats for further decomposition or processing elsewhere (e.g. in the mitochondria).
Lysosomes can also serve as 'dustbins' for superfluous and/or damaging substances that might be present in the cytoplasm. The older the cells are (e.g. nerve cells), or the more they have the specific function of tidying up (such as white blood cells), the more lysosomes we will find.

Energy releasing

Breaking down ATP sets energy free for all those processes in the body for which energy is required. The energy is released by the following (simplified) reaction:

ATP $\rightarrow$ ADP + P + energy

4. Growth and division of the cell

In the body of an adult there are millions of cells being destroyed every minute. Most of these cells are immediately replaced by the division of

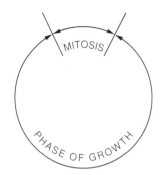

Figure 1.3.9
Mitotic cycle

living cells. During the development from fertilized egg cell to a mature human being the increase in the number of cells by division is no less than spectacular: out of *one* cell (the zygote) originates the multi-million cellular organism, man.

Prior to the phase of division there is always a phase of growth.

Growth

The growth of the cell is a quantitative increase of the materials from which the cell has been built. An increase of those structures which were halved in size as a result of cell division must occur.

Some of these halved structures are produced on command from the nucleus. These include the

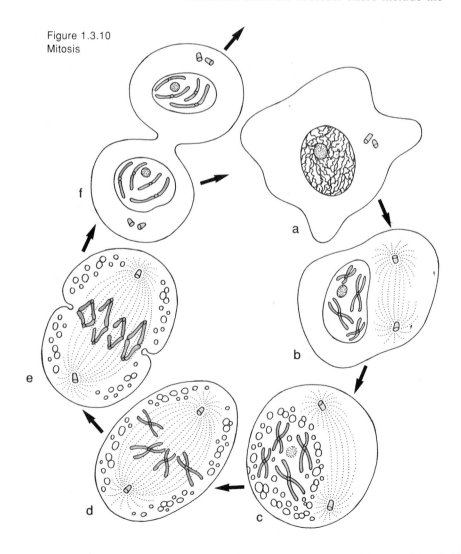

Figure 1.3.10
Mitosis

endoplasmic reticulum, the ribosomes, the Golgi body and the lysosomes. Others, such as the mitochondria and the centrosomes, possess a reproductive capacity of their own. The membranes are enlarged by building in lipoprotein and taking in liquids. The growth of the cell takes place between two divisions (Fig. 1.3.9).

The division itself is called *mitosis*. Both the period of mitosis and the period of growth are dependent on:
- the *type of cell involved* (blood-forming cells are quick dividers, skin cells divide much more slowly);
- the *circumstances in the environment* (the presence or absence of division-stimuli, presence of building substances etc.);
- the *phase of development* (rapid increase when a child, a very rapid increase during the growth spurt at approximately 15, hardly any increase or decrease between 20–60 years of age, in old age a marked decrease). The phase of growth is always much longer than the phase of mitosis. In a skin cell, for example, under normal conditions the phase of growth takes some 25 hours, whereas mitosis takes about 1 hour.

The record is copied

During the phase of growth, (Fig. 1.3.10a), the DNA strings of the chromatin network are copied (Fig. 1.3.12). To achieve this the DNA molecules split up at the same places as in the RNA-synthesis (i.e. the half-cut ladder rungs), but now along their entire lengths. On the basis of the nitrogenous bases of the resulting 'half-ladder' the complementary 'halves of the DNA-ladder' are being built up so that two completely identical DNA molecules come to exist. All the hereditary codes present are then doubled in the DNA-record. The beginning of mitosis is characterized by the beginning of the various doubled very long chromatin filaments forming a three-fold spiral (Fig. 1.3.11). As a result of this thickening they become visible under a microscope as *chromosomes* (Fig. 1.3.10b). A chromosome consists of two identical chromatids (because of

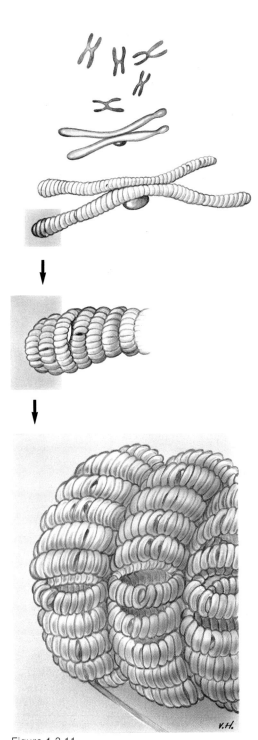

Figure 1.3.11
Chromosome structure

the copying of every DNA-string) which are linked by means of a centromere.
In the meantime the centrosome, (Figs. 1.3.1,

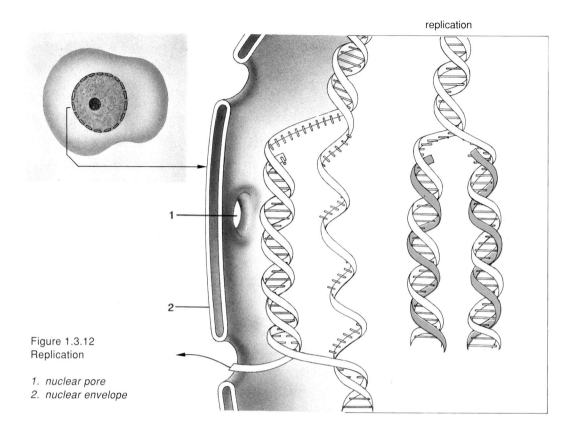

replication

Figure 1.3.12
Replication

1. nuclear pore
2. nuclear envelope

1.3.10a and b), has divided itself into two *centrioles*. Between the two centrioles there are *pulling cords* and the centrioles move to the poles (the opposite ends) of the nucleus. After that (Fig. 1.3.10c), the envelope of the nucleus and the nucleoli disappear and the centrioles move further on towards the poles of the cell.

The next phase of mitosis, (Fig. 1.3.10d), is characterized by all the chromosomes lining up in the plane of the greatest cross-section of the nucleus (equator plane); the centrioles span their pulling cords out from cell pole to cell pole. Then the pulling cords attach themselves to the centromeres of every chromosome.
The division itself can now start (Fig. 1.3.10e). The pulling cords shorten, they break in half in the middle and at that moment the centromere of every chromosome also splits up. Every chromosome, therefore, is split up into two completely identical *chromatides*. At almost exactly the same time the mother cell subdivides

Figure 1.3.13
Homologous chromosomes

a. Chromosome pair number 1
b. Chromosome pair number 17

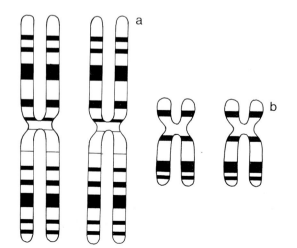

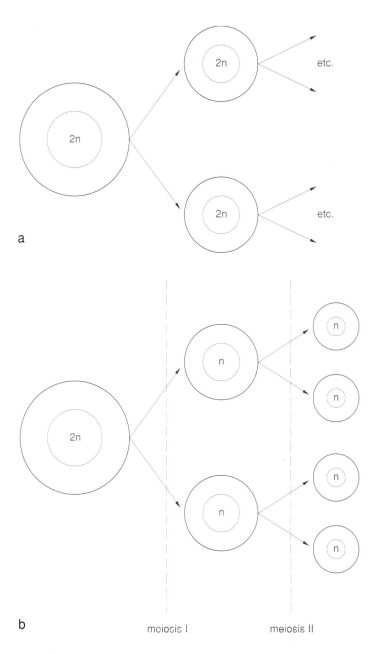

Figure 1.3.14

a. Mitosis
b. Meiosis

a

b   meiosis I   meiosis II

at the equator plane. Each of the two cells to be formed receives approximately half of the cytoplasm.

In the last stage, (Fig 1.3.10f), of cell division, there now exist two identical daughter cells; around the two now-separated groups of chromatids a new nucleus envelope has formed. We also see nucleoli appearing again, and the centrosome in each of the offspring cells dividing itself into two centrioles.

The chromatids unfold and uncoil from their hyper-spiralized state until they are again the very fine filaments of the chromatin network. Both daughter cells are now at the beginning of the phase of growth.

Chromosomes

In the majority of human cells we find 46 chromosomes. The chromosomes come in pairs; so in man there are 23 pairs. The 23rd pair consists of *gender chromosomes* (*heterosomes*). The other 22 pairs are called *autosomes*.

Figure 1.3.15
Differentiation
The stem cells, in the red
boxes, disappear

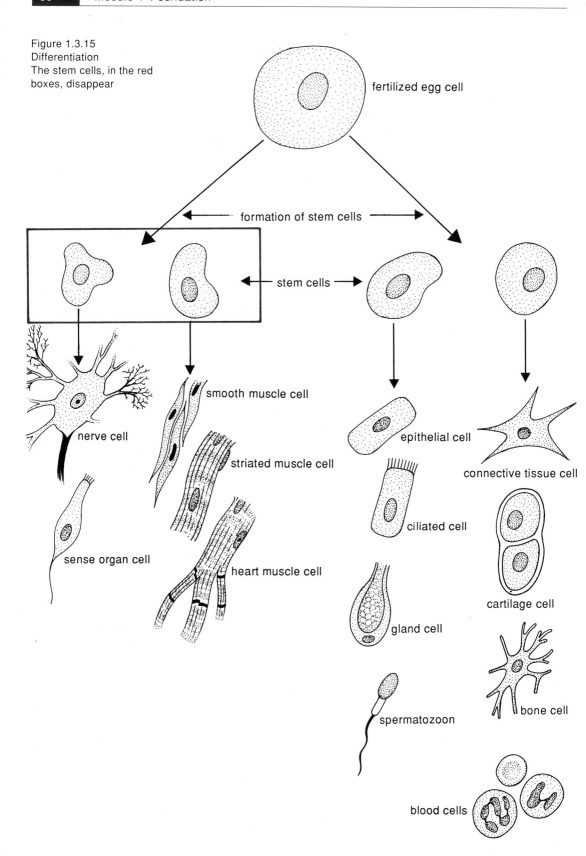

fertilized egg cell

formation of stem cells

stem cells

nerve cell

smooth muscle cell

striated muscle cell

heart muscle cell

sense organ cell

epithelial cell

ciliated cell

gland cell

spermatozoon

connective tissue cell

cartilage cell

bone cell

blood cells

Each pair of chromosomes can be regarded as a 'pair of twins'. 'Twin chromosomes' can be recognized by the identical spot of the centromere, by the identical lengths of the chromatids and by identical colouring if any (Fig. 1.3.13). These are known as *homologous chromosomes*.

The gender chromosomes of the female are called XX-chromosomes.
In the male there are 22 pairs of identical autosomes; the 23rd pair, the heterosome, the pair of the gender chromosomes, consists of two unequal partners. One is comparable with the gender chromosomes of the female (X). The other one is smaller; it almost looks as if the 'legs were cut off'. To distinguish them the gender chromosomes of the man are called XY-chromosomes.

When chromosomes occur in pairs we refer to this as *diploidy*. This is indicated by the symbol 2n. The number 2 refers to each chromosome occurring coupled; n refers to the number of pairs. Diploid human cells have 2n chromosomes; n = 23, so these cells contain 46 chromosomes. In mitosis the 'mother cell' with a number of 2n chromosomes divides itself into two 'daughter cells', each containing 2n chromosomes (Fig. 1.3.14a). In the reproductive cells, however, we do not find twin chromosomes.
As a result of a special kind of cell division (reduction division, meiosis (see Fig. 2.10.2)) the homologous chromosomes are separated from each other. When chromosomes occur singly in pairs it is termed *haploidy*. This is represented by the symbol n.
Reproductive cells are haploid. In human reproductive cells we find 23 different, single chromosomes. Reproductive cells arise by division of the mother cell with 2n chromosomes into four granddaughter cells, each having n chromosomes (Fig. 1.3.14b).
In a few types of cells (e.g. the red blood cells and the blood platelets (thrombocytes)) we do not find chromosomes. During the formation of these cells the nucleus was lost, and so the chromosomes are lost.

5. Differentiation and specialization

If it was really the case that at every mitosis the 'mother cell' gave way to two exact copies of itself, then multicellular organisms would only exist as a mass of copied zygotes.
In Module 3 we we will see that certain groups of cells already start to show small differences in shape and structure at an early stage of development. In later stages these differences become increasingly greater, in comparison with the successive mother cells. This *differentiation* is accompanied by an ever increasing *specialization*, i.e. the ability to perform a specific function.
Examples of cells with a high degree of differentiation are muscle cells, nerve cells and sense organ cells. The fertilized egg cell, of course, has a very low degree of differentiation and specialization whereas the opposite is true in the case of the intestine cell.
It is striking that there is a high correlation between the level of differentiation and the ability of division.
Highly differentiated cells (muscle cell, nerve cell, sense organ cell) cannot divide any more and there is also no possibility of replenishment from some kind of a (less differentiated) pre-stage of these types of cells.
The number of these cells manufactured out of pre-stages during the pre-natal period has to last us throughout our lifespan. When, in the course of life, cells of these types are destroyed, because of illness, damage or old age, they cannot be replenished. That this need not be a catastrophe can be deduced from the fact that there are many nerve cells destroyed every day without immediate serious consequences.
Other cells, such as skin cells, blood cells or bone cells, in their mature stage cannot divide any more, but, also, for each of these types of cells there are *stem cells* available, less differentiated and specialized cells of a pre-stage, which via mitosis take care of replenishing the cells (Fig. 1.3.15).
The replenishment of cells from stem cells will be discussed as appropriate in the chapters covering the body systems.

4

Tissues

Introduction

In defining the cell as the basic building block and the fundamental metabolic unit of the human body, we run the risk of diverting our attention away from the wider context in which the cells must normally be seen. A collection of cells which have the same shape and structure and perform a common joint function is called a tissue. Belonging to them are those substances which have been brought into the m.i. by the cells of certain tissues and from which the tissue very often derives its specific function. The m.i. with the substances it contains (e.g. fibres) is called *intermediary substance*.

The definition of what a tissue is does not always hold good. In a tissue there can be various types of cells present (e.g. in the blood) and a tissue can perform more than one function (e.g. fatty tissue).

Histology studies the structure and function of the tissues. In this chapter we give a sketch of the four main types of tissue: epithelial tissues, supporting tissues, muscle tissues and nerve tissue. More detailed discussion of the separate tissues of the main groups will take place – where necessary – when we deal with the body systems in Module 2.

Learning outcomes

After studying this chapter you should have sufficient knowledge and understanding of:
- the characteristic features of a tissue;
- the division of the tissues into four main groups;
- the structure and the functions of the different types of tissues.

1. Epithelial/Covering tissues

Epithelial tissues are tissues in which the cells form a closed layer. There is hardly any intermediary substance in between the cells of this layer. The covering tissues have no blood vessels, so that feeding the cells must be taken care of by diffusion from the underlying layer of connecting tissue, which *does* have blood vessels. Epithelial tissue forms bordering covers and glands. We find it on the outside of the body (epidermis) and it lines the inner side of hollow organs and body cavities (such as the heart, blood vessels, digestive tract, urinary tract, bronchial tubes, vagina and uterus). Covering tissues rest on a unicellular *basal membrane*, which belongs to the connective tissues. Covering tissues consist of one or more layers of cells. In the *unilayered covering tissues* we can identify four types according to the structure of the cells:
- *one-layered pavement epithelium* (Fig. 1.4.1a). The internal lining of the heart, the blood and lymph vessels consists of one layer of flattened cells that neatly fit together. Here we speak of *endothelium*. In the unilayered

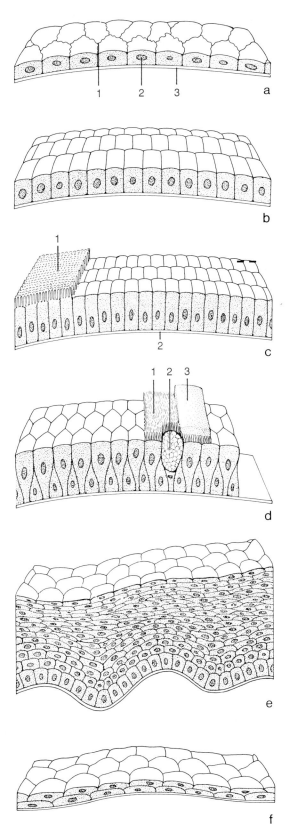

Figure 1.4.1
Covering tissues

a. One-layered
   pavement epithelium

   1. adjacent, closely
      connected cell
      membranes
   2. nucleus of the cell
   3. basal membrane

b. Cuboidal epithelium
c. Cylindrical epithelium

   1. folds directed
      outwardly
   2. basal membrane

d. Ciliated epithelium

   1. cilia
   2. cell, producing
      mucus
   3. coating consisting
      of a layer of
      mucus

e. Multilayered
   pavement epithelium
f. Transitional
   epithelium

alveoli in the lung it is called *plate epithelium*. The inner edges of the serous membranes (pleura, i.e. lung membrane, pericardium and peritoneum) also consist of one-layered pavement epithelium (here called *mesothelium*) resting on the basal membrane;

– *cuboidal epithelium* (Fig. 1.4.1b). The cells are cube-shaped. This type is found in the kidney tubes and as gland- or glandular cells (Fig. 1.4.2);

– *cylindrical epithelium* (Fig. 1.4.1c). The cells are long and contain many organelles. This type of epithelium is found in the coating of the intestines, the uterus and the gall bladder. In the small intestines each single epithelium cell has a large number of folds (microvilli, Fig. 1.3.4), which bring about a spectacular enlargement of the surface area;

– *ciliated epithelium* (Fig. 1.4.1d). The tall cells have several cilia, which can all be moved in the same direction in one forceful sweep and can then slowly return to their original position. Ciliated epithelium is found especially as the inside coating of the passageways for air (from the nose cavity to the alveoli) and it is found as the inner lining of the oviducts.

In the *multilayered epithelium* we can identify:
– *multilayered pavement epithelium* (Fig. 1.4.1e). The deeper-lying layers of this tissue are still cubic in form. The deep-lying cells divide themselves and a part of the resulting daughter cells moves towards the surface.

During this movement, these cells become more and more flattened.

In the type of epithelium which is *subject to keratosis* (formation of horn), the cells on the surface have lost their nucleus; they die and dry out. We find this type of cell in the outer skin. The type of epithelium that does *not* show the *process of keratosis* is found on the glans of the penis and as the mucous membrane in the mouth and the vagina;

– *transitional epithelium* (Fig. 1.4.1f), consists of two layers of cells which are very elastic and which still have the capacity of dividing themselves. This type is found as the coating in the urinary tract.

The names *epi*thelium, *meso*thelium and *endo*thelium tell us something about the positions of the covering tissues involved. The epithelium forms a boundary with the m.e.; the mesothelium of the serous membranes is never in contact with the outside world; the endothelium lies even deeper.

Covering tissue has four functions, each layer performing a separate function in addition to *bordering*. The ciliated epithelium also helps in *transportation*: for example many of the dust particles which are inhaled are removed from the air passageways by the cilia; in the oviducts the egg cell is conducted to the uterus by the ciliated epithelium.

The cylindrical epithelium of the intestines has not only the function of covering but also of *absorption* (i.e. taking in of nutritional substances). The epithelium of the intestines, for example, takes in nutritious substances from the m.e.; from here they will go through active and passive transport into the m.i. The capillary network then absorbs them for further transportation. The last function of covering tissue to be mentioned is *secretion*. In most non-keratosing epithelium we find large amounts of *mucosa (mucus cells)*. The mucus produced by them functions as a glue (for dust particles in the air passageways), as a lubricant (in the intestines), as a protection against drying out (in the mouth cavity), or as a protection against the effects of enzymes and/or acids (in the stomach, intestines and urinary tract).

Secretion is also performed by *glands*.

Glands with a discharging tube (duct) bring their secretions into the m.e.; they are therefore called *exocrine* glands (external secretion). Examples of this type of gland with external secretion are the sweat, mammary, prostate, lachrymal and salivary glands. Some glands with external secretion are *tube-shaped* (Fig. 1.4.2a), e.g. the peptic gland, others have the form of a bunch of grapes (*racemose*) (Fig. 1.4.2b), e.g. sebaceous glands. The exocrine glands are discussed with the organ systems involved.

Glands without an outlet tube (ductless glands) deposit their secretions in the m.i. via diffusion and/or active transport, and are therefore called *endocrine* glands. The secretory substances of such glands with internal secretion are called *hormones*. From the m.i. they are diffused into

Figure 1.4.2
Exocrine glands

a. Tube-shaped gland
b. Gland in the shape of a
   bunch of grapes
   (racemiform)

1. *connective tissue*
2. *feeder nerve fibre*
3. *feeder blood vessel*
4. *glandular cell with
   secretory product*
5. *muscle tissue*

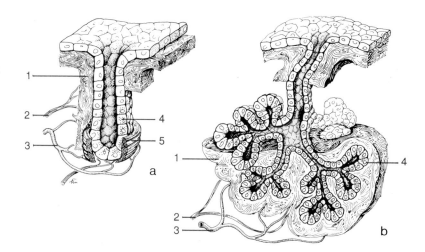

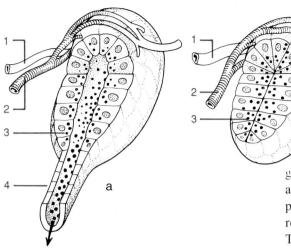

Figure 1.4.3
Differences between
exo- and endocrine
glands

a. Exocrine gland
b. Endocrine gland

1. *vein*
2. *artery*
3. *secretory product*
4. *discharging tube*

the bloodstream for further transportation. The hormone discharging glands (e.g. thyroid and adrenal glands) will be described according to their function with the separate body systems and also summarized in Module 2, Chapter 6.
Both the exocrine and endocrine glands obtain the raw material for their products from the feeder blood vessels (Fig. 1.4.3). Covering tissue itself is not provided with blood; the glands are positioned in deep-lying connective tissue. These deep-lying tissues contain blood vessels (i.e. they are vascularized).

2. Support tissues

The tissues belonging to this group are less homogeneous in structure and function than in the case of the covering tissues. In support tissues the intermediary substance is extremely important.
The nature of the intermediary substance largely determines the functions of the types of support tissue that we can distinguish: connective tissue, cartilage tissue, bony tissue and blood.

2.1. Connective tissue

Connective tissue is found in the human body in many places and performs widely varying tasks. In the m.i. we find type-specific fibres that have been produced by certain cells for a specific purpose (Fig. 1.4.4). Amongst the normal cells of connective tissue we also find 'specialized cells' that are capable of phagocytosis (i.e.

guarding the system against infection by absorbing microbes and other foreign particles), playing a part in combating inflammation or repairing damage.
The types of fibre are distinguished according to the proteins of which they are made:
– *collagenous,* consisting of the protein collagen: this substance forms unforked, inflexible fibres (with a high tensile force). The name of this substance refers to the fact that collagenous fibres (which often occur in bony tissue) release glue when heated;
– *elastic,* consisting of the protein elastin: this forms forked out (elastic) fibres;
– *reticular,* much thinner than the other two types of fibres; the substance reticulin forms very finely forked out fibres, which cannot be stretched.

The nature of the intermediary substance and the cell types that are present determine the type of connective tissue with which we are concerned. In *connective tissue* collagen is prevalent. The collagen fibres can be woven criss-cross, intertwined within themselves, as in the skin or in the hard cerebral membrane (hard meninx, dura mater). At other places these fibres have been placed in one specific direction as in tendons or in joint ligaments. Collagenous connective tissue has above all a mechanical function, the resisting of tensile forces.
Adipose or *loose connective tissue* (loosemeshed) is distinguished by its small quantity of fibres (collagen and elastin). It is a filling tissue (e.g. between and around the intestines) because of its many hollow spaces.
In *elastic connective tissue* the protein elastin prevails, e.g. in the wall of the arteries and in the lining of nerve cells.

Figure 1.4.4
Connective tissue

1. nerve
2. fibre-producing cell
3. elastin fibres
4. intermediary substance
5. capillary
6. collagen fibres
7. reticulin fibres
8. pericyte
9. white blood cells

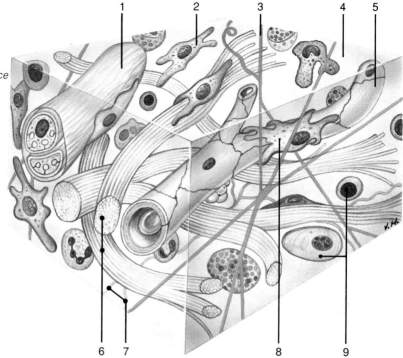

Fatty tissue occurs in certain cells of connective tissue where large amounts of fat can be stored. The drops of fat 'push out' the cytoplasm and the nucleus. As the fat-cells themselves swell, there is hardly any intermediary substance. The fatty tissue can fulfil several functions. Most of it serves as a reserve store of fuel. When the amount of energy sources taken in is smaller than the amount of energy required (e.g. during lengthy athletic exertion, or when dieting) then it is possible to mobilize fatty tissue from these deposits; this fat is then transported via the bloodstream to the active cells (especially muscles) to be burned in the citric acid cycle in the mitochondria. We find fat deposits in the ligaments on which the intestines hang and especially in the subcutaneous connective tissue, where it is also important as a *heat isolator*, as fat is a poor conductor of heat. Fat can also have a *supporting* function. The kidneys, for example, are wrapped in support fat which functions as a shock absorber. Something similar is found around the eyeball in the eye-sockets (orbits). The subcutaneous fat in the palms of the hands and the soles of the feet is not only very thick, but also marbled and enveloped with collagen fibres, in order to produce a firm and strong buffer zone. This type of construction allows a hard blow to be made with the palm of the hand whereas a similar blow made with the back of the hand would result in considerable pain. The fat cushions in the soles of the feet resist the enormous forces of pressure which occur during walking and jumping. Only a few metres of crawling can cause problems in the skin of your knees, since they lack these shock-absorbing cushions.

Reticular connective tissue consists mainly of reticulin fibres and reticulum cells. The latter are less differentiated. They can phagocytize and serve as stem cells for fibre producing connective tissue cells. In the lymphatic organs (the lymph nodes, spleen and thymus) and in the blood-producing cells of bone marrow, they function as stem cells for the blood cells.

2.2. Cartilage tissue

In cartilage tissue (Fig. 1.4.5), the matrix (the substance between the cells) is much firmer and more elastic than in connective tissue. The matrix is translucent and consists of *chondrin*, produced by cartilage cells. Chondrin is also called cartilage glue. The cartilage cells lie together in groups, surrounded by a thick coat

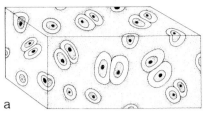

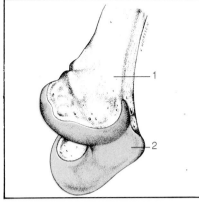

Figure 1.4.5
Types of cartilage

a. Hyaline cartilage

    1. *thigh bone*
    2. *cartilage of the*
       joint surface

b. Elastic cartilage

    1. *surface of the skin*
    2. *skull*
    3. *nasal bone*
    4. *nasal cartilage*

c. Fibrous cartilage

    1. *symphysis pubis*
    2. *pelvis*

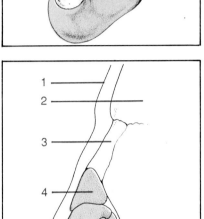

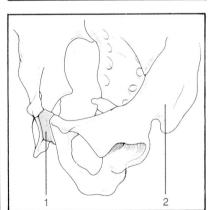

of collagen fibres. The following three types of cartilage are recognized:

– *hyaline cartilage*: contains a large number of evenly-spread collagen fibres in the chondrin. On the bone surfaces of joints, it forms an extremely smooth sliding plane, which can at the same time absorb shocks. Between the ribs and the breast bone (i.e. sternum) it forms a supple connection, making the chest a firm, but flexible structure.

– *elastic cartilage*: contains large amounts of elastin fibres in the chondrin which gives it a flexible shape. We find it in the external ear, the epiglottis, the auditory (Eustachian) tube, the nostrils and the internasal septum.

– *fibrous cartilage:* one large network of collagen fibres closely packed together in the chondrin. It thus develops a high tensile force and has a high resistance to pressure. It is found in articular discs, such as those be-

Figure 1.4.6
Bone tissue

a. Long bone
   (thighbone)
b. Detail

1. *periosteum
   (bone
   membrane)*
2. *bone cell*
3. *osteon*
4. *compact bone*
5. *the Volkmann or
   perforating canal*
6. *lammellae*
7. *central or
   Haversian
   canals with
   central blood-
   vessel*
8. *spongy bone*

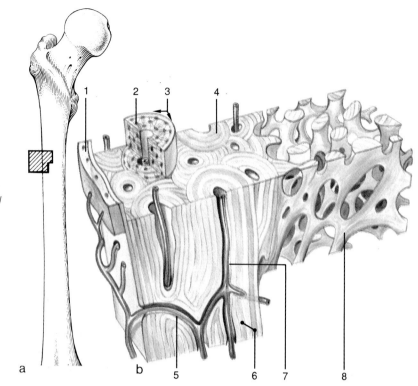

tween the vertebrae (see Fig. 2.9.18), and in the connection between the two pubic bones (symphysis pubica) (see Fig. 2.9.20a). The fibrous cartilages in these places also function as a flexible connection of bone to bone.

Cartilage tissue is not provided with blood vessels (avascular). This has the advantage that it can absorb direct force without causing bleeding at every instance, but the disadvantage is that the process of maintenance is very slow and, in case of damage repair, healing takes a very long time.

2.3. Bony tissue

Bony tissue (Fig. 1.4.6), is characterized by a solid matrix, due to the presence of calcium phosphates. At the same time it is also flexible, since it contains a great number of collagen fibres. The bone cells are evenly spread out in the bony tissue and are in contact with one another by means of fine threads.

Bony tissue has been built up into bones and then with the help of joints into a skeleton. This structure and the functions of the skeleton are discussed in Module 2, Chapter 9. It will be

stated that bony tissue has quite an intensive metabolism. It is continually being broken down, and at the same time continually being built up. For these purposes it is provided with an extensive network of blood vessels.

2.4. Blood

Blood is considered to belong to the support vessels, not so much because of any 'support' being rendered, but because in the blood the importance of the matrix (i.e. the substance between the cells) prevails. The matrix here is a fluid; there are no fibres. It can thus circulate in the blood vessels, and be pumped round by the heart, which enables it to fulfil a transport function related to the 'refreshing' of the m.i. In the liquid matrix (the plasma) there are large amounts of *blood cells* present, with widely varying functions. The constituent parts and the functions of the blood will be described in Module 2, Chapter 1.

3. Muscular tissue

In muscular tissue (Fig. 1.4.7), the cell component is of vital importance for the function of the tissue. The muscle cells are characterized by

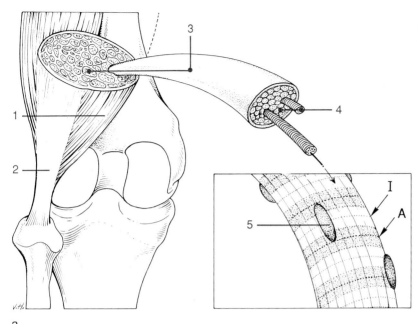

Figure 1.4.7
Muscle fibre with
contractile elements

a. Striated muscle
   (skeletal muscle)

   1. *the body, the*
      venter of the
      muscle
   2. *tendon*
   3. *muscle bundle*
   4. *muscle fibre*
   5. *nucleus*

   I = isotrope
      (light)
   A = anisotrope
      (dark)

b. Actin and myosin
   shifting in relation to
   each other

a

chains of protein   unit of shifting

actin
myosin

b

the presence of protein chains that have been arranged in series, the *myofibrils*. These myofibrils consist of the proteins *actin* and *myosin*, which can shift in relation to each other. A shifting which results in the shortening of the muscle cell takes energy (ATPs: adenosine triphosphate). This contraction takes place when the combination of actin and myosin, actomyosin, is activated by ATPs. The capacity of myofibrils to contract and of being contracted is called *contractility*. The reverse shifting away from each other (resulting in increased length) takes no energy in this muscle cell, but it can take energy in its 'counterpart'. In every living tissue, metabolic reactions take place. In muscle tissue this happens in order to bring about an outward mechanical effect: *movement*.

Muscle tissue is very active and so has an intensive blood circulatory system. We can identify three types of muscle tissue according to the differences in their structure and function:

– *smooth muscle tissue* (Fig. 1.4.8a) consists of spindle-shaped (i.e. fusiform) cells with one central nucleus, lying closely together. The cytoplasm is filled with myofibrils.
Smooth muscle tissue contracts rather slowly, reacts rather slowly on stimuli and is almost indefatigable. Smooth muscle tissue is found in all vegetative systems and it is also the vegetative nerve system which provides these muscles with impulses (vegetative innervations). Because of this it is sometimes known as *involuntary muscle tissue*;

– *striated muscle tissue* (Fig. 1.4.8b), derives its name from the fact that in the myofibrils the chains of actin and myosin are arranged in parallel in such a way that the actin of one series lies alongside the actin of the next series (and consequently the myosin lies alongside the myosin of the adjacent series). The separate myofibrils themselves are also ordered in an analogous way. As actin is more translucent than myosin, under the micro-

Figure 1.4.8
Types of muscle tissue

a. Smooth muscle tissue,
   e.g. in the wall of the
   stomach
b. Striated muscle tissue,
   e.g. in the leg
c. Cardiac muscle tissue,
   e.g. in the wall of the
   ventricle

a

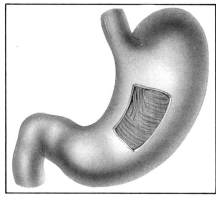

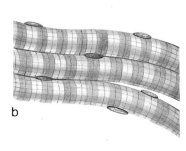
b

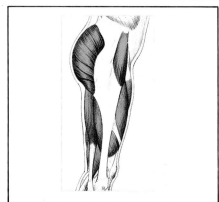

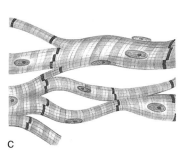
c

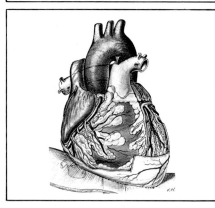

scope a crosswise striping pattern appears. The crosswise striped muscle cells are much longer than the smooth muscle tissue cells. These muscle fibres have several nuclei, which are pressed, as it were, against the outer cell membrane by the large number of myofibrils. This phenomenon of having several nuclei, and its relatively large size, arises during the embryonic stage of development by the fusing together of many cells (membrane fusion). This is called a *syncytium*. Striated muscle tissue can contract very quickly; it reacts very swiftly and is soon

exhausted because of the high level of metabolism. This type of muscle tissue has also been structured into striated muscles. These muscles are attached to parts of the skeleton by means of tendons. This explains its name: *skeletal muscle tissue*. Innervation (provision of nerve impulses) of the skeletal muscle tissue takes place via the animal nerve system, and that is why it is sometimes called *voluntary muscle tissue*. The structure and function of the skeletal muscles are discussed in Module 2, Chapter 9;

– *cardiac muscle tissue* (Fig. 1.4.8c) displays

the same pattern as the striated muscle tissue under the microscope, but the cells of the heart muscle do not form a syncytium. There is one nucleus that is positioned in the centre and the individual cells form – by forking – links with other cells. The heart is a hollow muscle which has been formed by networks of cardiac muscle tissue. Cardiac muscle tissue can contract very quickly, and responds very swiftly to a stimulus, but because of its regular alternation of activity and rest it is capable of performing the lifelong activity of pumping blood around. Cardiac muscle tissue works autonomously because of the presence of a stimulus automaton. This stimulus automaton which gives the heart an impulse is itself under the influence of regulating impulses from the vegetative nerve system.

4. Nerve tissue

In nerve tissue, two types of cells occur: the conducting cells or *nervous cells* (neurones) and the support cells or *glial cells* (neuroglia). Neurones are excitable. This means that they react to stimuli with a change in the Na^+/K^+ ratio that exists intra- and extracellularly. Excitability is not the monopoly of nerve cells; muscle cells and glandular cells are also excitable. Nerve cells specialize in the reception of stimuli and in conducting impulses. Through the change in the Na^+/K^+ ratio a small electric charge (action potential) can arise which via the often very long and widely branched processes (nerve fibres) can be conducted from the neurones to elsewhere. We will distinguish several types of neurones in Module 2, Chapter 7. The general structure of a neurone is illustrated in Figure 1.4.9 (a nerve cell in the brain).

The *cell body* contains not only nucleus and nucleoli but also a vastly extended endoplasmic reticulum and a number of Golgi apparatuses. Generally the *cell membrane* has several short offshoots: *dendrites*, which conduct the stimulus, the signal towards the cell body, and the neurite or axon, one long process, sometimes forked out, which conducts the signal away from the cell body. The processes are hollow and thus contain cytoplasm. Most long processes are surrounded by a thin insulating layer of *myelin*. This fatty substance forms the so-called myelin sheath, or marrow sheath, which prevents short-circuiting of adjoining neurones. In addition to these myelin sheaths most processes also have glial cells around them. Their main function is to feed the processes, but they also enhance the speed of signal conduction.

Nerve tissue is very active and in particular requires a lot of glucose and oxygen. There is a vast network of blood vessels in nerve tissue to deliver these substances (and to drain waste products away).

In Module 2, Chapter 7, we will enter more deeply into the structure and function of the nervous system.

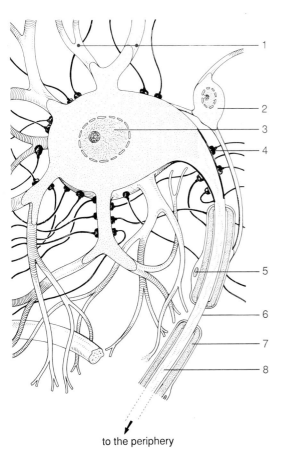

to the periphery

Figure 1.4.9
Neurone and glial cells

1. dendrites
2. glia cell
3. nucleus of the cell
4. synapse
5. nucleus of the Schwann-cell or neurolemmocyte
6. node of Ranvier or neurofibral node
7. myelin sheath
8. neurite; axon

5

Topography

Introduction

Topography literally means *description of the place* and it is a term we are familiar with in geography. It is important that there is an established method for finding a certain place on earth and in a similar way it is necessary to have an accepted frame of reference for describing and locating sites on and within the human body. This is obvious for a surgeon, but for health care professionals also, a knowledge of topography is essential.

If someone complains of pain halfway down his back to the left and the right of the spine, he is probably aware that that is where the kidneys lie. When the pulse is taken, the nurse must know that the artery concerned can be felt on the side of the thumb in the wrist area. We could mention many more clear examples.

Topography in functional anatomy describes the position of body structures (mostly organs) in relation to each other. All the major structures will be defined in their spatial relationships. When discussing the body systems in Module 2 we will also present topographical data.

Learning outcomes

After studying this chapter you should have sufficient knowledge and understanding of:
- the importance of topography in functional anatomy;
- the several planes, directions and sides which have been defined topographically;
- the broad topographic divisions of the human body;
- the position of the larger organs and structures in the human body.

1. Frame of reference

The anatomy

When two people talk about travelling to various places in the world, it is necessary that both parties have the same frame of reference, which means that they make use of the same terms for the same notions. People have agreed universally on many terms and their meanings, and many definitions of names, e.g. the names for the points of the compass, the notion of degrees of latitude, etc. Otherwise an enormous confusion would arise.

In functional anatomy, a frame of reference has also been defined; it is based on the *anatomical posture* (see Fig. 1.5.1). The person is standing upright, head erect, the palms of the hands turned to the fore, the feet somewhat apart.

In geography we use latitude and longitude to mark the position of places, and in a similar way in life science we use body planes for the same purpose. These planes make cross sections

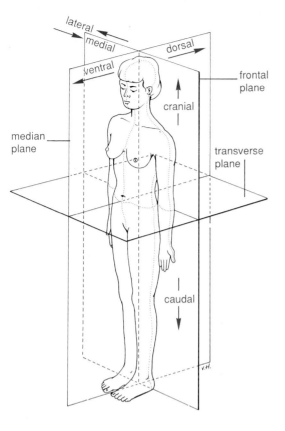

Figure 1.5.1
Anatomical posture and body planes

through the entire body, or through parts of the body, as if they were huge slices.

Illustrating the position, the composition and the function of body structures by means of such cross sections is very important and the many figures in this book, using sections, demonstrate this importance. Being able to 'read' these types of figures is thus essential.

- *Frontal planes* run parallel to the front, to the forehead. They describe the body or parts of it in relation to such a plane.
  A frontal plane makes a *frontal section* (see Figs. 2.9.13 and 2.7.30).
- *Transverse planes* run parallel to the base and divide the body or parts of it into upper and lower.
  A transverse plane makes a transverse section or cross section (see Figs. 2.9.12 and 1.5.4).
- *Sagittal planes* are at a 90° angle to a frontal plane and divide the body or parts of it into

left and right. A sagittal plane makes a sagittal section (Fig. 2.5.5b).

Because of the left/right symmetry in the human body (there are no other symmetries in the human body) the sagittal plane through nose and navel has received a special name: the medio-sagittal or medial plane (see Figs. 2.2.18 and 2.10.4).

When describing the structure of the many lengthy tubular organs, such as blood vessels and lymphatic vessels, intestines, ureters, sperm ducts and oviducts, we encounter a problem. Their position in relation to the three main planes is constantly different. Here we apply a slightly adapted set of names for sections (Fig. 1.5.2).

We speak of a *cross* section or a *transverse* section when the result is a circular end, or two circular ends (see Figs. 2.2.21 and 2.10.17b). The space or cavity which is enclosed by tubular organs, is called the lumen. After a transverse section therefore we can look at and see into the lumen.

When a section is made, which results in a kind of gutter or furrow arising, then we refer to a *lengthwise* or *longitudinal* cross section (see Fig. 2.10.9d).

The same set of names is applied to more massive structures such as nerves, muscles, bones and the like (see Figs. 2.9.6c, 2.7.17 and 2.8.12). All these structures are taken out of the body so to speak and are then put in the anatomical position; after this the cross section is made.

Directions and sides
Anatomists describing the positions of structures use specific terms for directions and for the sides of the body.

These terms are listed side by side in the Nomina Anatomica and consist of pairs of names which are the opposite of each other. Sometimes the pairs of names look very much like others as far as meaning is concerned. Yet they are not always used. In the list overleaf the pairs of names most used are mentioned, with their meaning, their application and an example.

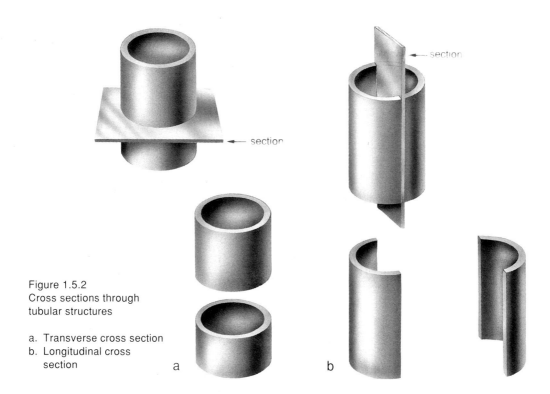

Figure 1.5.2
Cross sections through
tubular structures

a. Transverse cross section
b. Longitudinal cross
   section

a   b

It will become apparent that the name given is *always in relation to another structure.* In this way the position of every organ and every structure can be described accurately.

- • ventral = on the same side as the abdomen
  • dorsal = on the same side as the back
    These indications of place refer to the larger structures or greater distances.
    Example:
    'The gullet (= oesophagus) is dorsal to the windpipe (= trachea)' (see Fig. 2.1.33).
    'The gullet lies ventral to the spinal column' (see Fig. 2.2.18).
- • anterior = at the front(side), before
  • posterior = at the back(side), behind
    This pair of names is very similar in meaning to the previous one; it is however used for smaller structures or indicates a smaller distance.
    Example:
    'The anterior cerebral artery provides blood mainly for the foremost part of the brain; the posterior cerebral artery, mainly

the back-part of the brain' (see Fig. 2.7.45).
- • central = in the middle
  • peripheral = at the outer ends
    These names are used especially with extensive systems such as the nervous system and the circulatory system.
    Example:
    'The central nervous system comprises the brain and spinal cord. All the other nerves belong to the peripheral nervous system' (see Fig. 2.7.1).
- • cranial = above, towards the skull (cranium)
  • caudal = below, towards the tailbone (coccyx)
    This pair of names is used mainly in relation to the spinal column and the central nervous system.
    Example:
    'The thoracic vertebrae lie cranial to the lumbar vertebrae' (see Fig. 2.9.14).
- • superior = higher, above
  • inferior = lower, under, below.

These indications of place look very much like the former pair, but are used with smaller structures or to indicate a smaller distance.
Example:
'The inferior vena cava brings the blood from the legs and the abdominal organs to the heart; the superior vena cava does the same with the blood of arms and head' (see Fig. 2.1.32).
– • lateral = to the side
• medial = towards the middle
Example:
'The lungs lie lateral to the heart' (see Fig. 2.1.16).
'The stomach lies medial to the liver and the spleen' (see Fig. 1.5.3a).
– • proximal = close to the trunk
• distal = far away from the trunk
These terms are used to indicate places in the limbs.
Example:
'The proximal end of the last joint of the finger is the end nearer the hand; the nail is at the distal end' (see Fig. 2.2.9).
– • sinistra = to the left
• dextra = to the right
These names are necessary to name the symmetrically positioned structures. Left and right are always named from the point of view of the person or the structure depicted. So the reader has to reverse things: on a passport-photograph someone's right-hand eye is found on the left-hand side of the picture.
Example:
'The chest tube flows out into the vena subclavia sinistra' (see Fig. 2.1.47).
– • internus = internal
• externus = external
These names indicate the position in depth (from the skin) especially used with blood-vessels and nerves.
Example:
'The internal carotid artery enters into the skull; the external carotid artery divides repeatedly and spreads out on the outside of the skull' (see Fig. 2.1.34).
– • profundus = deep

• superficialis = superficial
These are practically identical with the names in the previous pair.

The directions and sides discussed above are *static* terms. They are *static indications of direction*, with the person standing still.
In topography, however, it must also be possible to describe movements, changes of place of parts of the body, such as limbs, trunk or head. The terms in which movements are described belong to *dynamics*; the *dynamic indications of direction* will be dealt with when the motor system is discussed in Module 2, Chapter 9.

Most words in the pairs of names are abbreviated. These abbreviations can be found in Appendix 1.

2. Division

Head, trunk, limbs
When we look at the human body, it is not surprising that we refer to a topographical division of *head, trunk and limbs* (Fig. 1.5.3). In a very broad sense we can assign different functions, which supplement one another, to these parts of the body.

The *head* contains the brain, which is the most important *command-post* of the body. In order to ensure that the body is regulated and coordinated the brain must be provided with information from the organism itself and from the outside world.

Information is supplied via the sensory system, important parts of which are situated in the head (e.g. hearing, sight, smell, taste, equilibrium). The high position of the head is especially advantageous for the range of the senses of sight and hearing.

In addition, the head contains entrances into the respiratory tract and the digestive tract (Fig. 1.5.3a and b).

The *trunk* contains those body systems which are directly concerned with the vegetative

Figure 1.5.3
Topography of the great organs

a. Front of the body

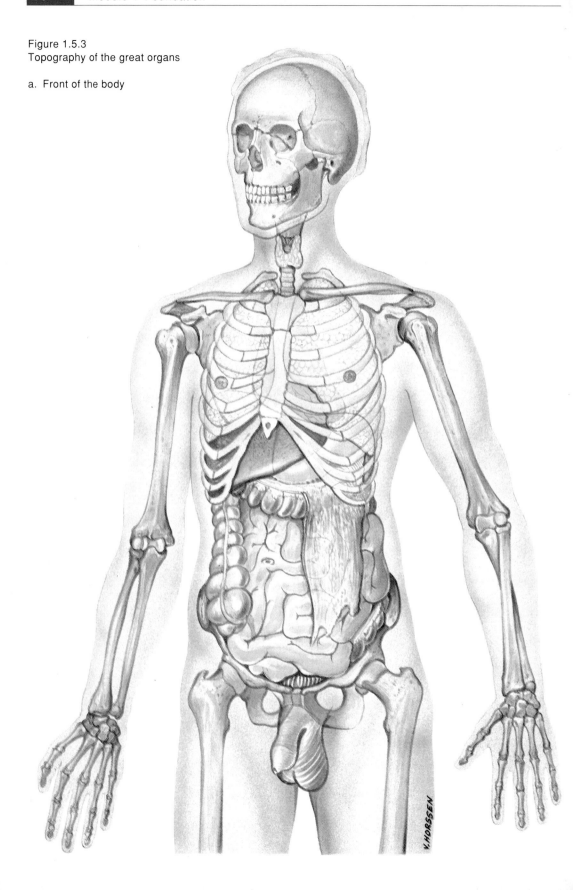

b. Back of the body

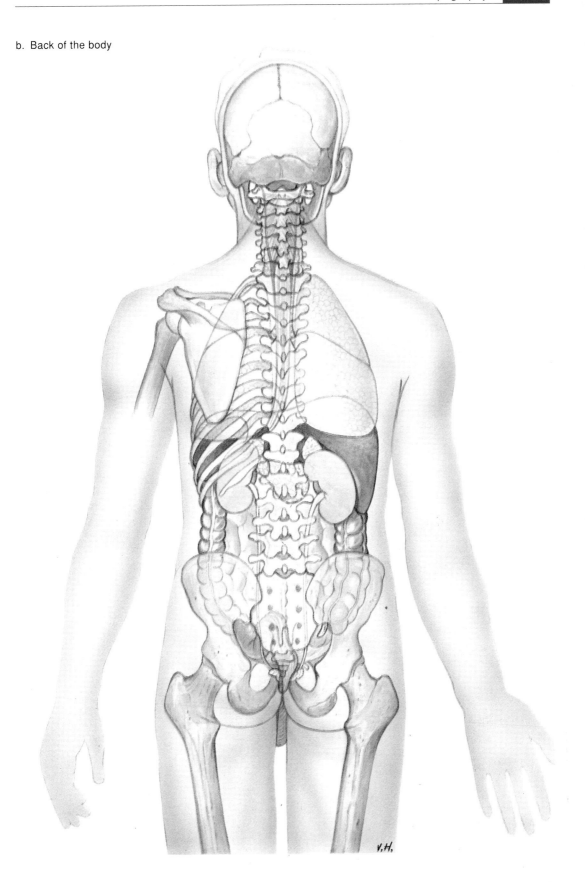

functions (taking in food, respiration, discharging waste material, transportation, metabolism and heat regulation).
Here we find the digestive tract, the respiratory system, the urinary system and the most important parts of the circulatory system (Figs. 1.5.3a and b). In a woman we also find here the important parts of the reproductive system.

The *limbs* have principally a movement function. Legs and feet are for standing, walking and jumping. They increase man's spatial range of action. The arms are there to serve the hands which specifically have a seizing function. Arms and hands have an even greater flexibility and mobility than legs and feet.

Structural components, cavities and intestines

This division of head, trunk and limbs is very basic and gives little more than an overall indication of the function. The topographic division into structural components, cavities and intestines is much more accurate.
The structural components consist of the

skeleton and the soft tissue connected to the skeleton. In broad outline we can divide the skeleton into skull, vertebral column, ribs with the sternum, shoulder girdle with the arms, and pelvic girdle with the legs.
In describing the topography of certain intestines and especially in describing the spinal cord and the nerves, we make use of a more detailed division of the vertebral column. The 8 cervical vertebrae are numbered as follows: C_1 up to and including C_8; the 12 thoracic vertebrae: T_1 up to T_{12}; and the 5 lumbar vertebrae: L_1 up to L_5. These 'names' of the vertebrae are often used in the topographical description of the place of a certain organ or structure. Example: 'The left kidney is lying at the same height as L_1/L_2' (see Fig. 2.3.1b).
The soft tissue round the skeleton (Fig. 1.5.4), and round the trunk consists of skeletal muscles, the muscle sheaths, (the fascia(e)), subcutaneous connective (or fibrous) tissue and the skin.

The skeletal muscles are attached to the bones by means of tendons.
The muscle sheath (the fascia) is a firm coating

Figure 1.5.4
Transverse cross-section through the upper leg

1. knee-cap
2. skin and subcutaneous connective tissue
3. thighbone
4. artery
5. vein
6. nerve
7. skeletal muscle
8. muscle sheath (fascia)
9. general sheath (fascia)

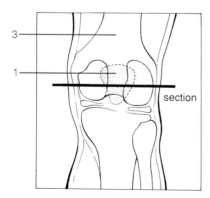

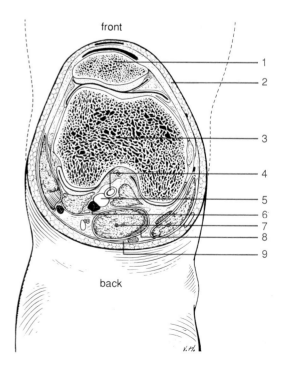

of connective tissue that wraps the muscle and becomes the tendon at the end of the muscle. The general fascia is a similar type of connective tissue sheath which, like a cloak, is wrapped round all the structures lying beneath it.

The subcutaneous connective tissue is a structure of collagen and elastin fibres in a rather large open network, filled up mostly with fatty tissue. The fat serves as a stock of fuel, as a heat-isolator and in some places as a buffer.

The skin forms the border with the milieu extérieur.

At the body-openings (mouth cavity, nose cavity, anus, urethra, vagina) the skin becomes a mucous membrane; it is also the transition to that part of the m.e. that is lying inside the body. The functions of the structural components can be summarized as follows:

– the skeleton has a mainly *supportive* function. It offers a counterforce against gravity. The individual bones are connected by means of joints (articulations). The muscles bring about the *movements* of parts of the skeleton in the joints.

With regard to the skin and the subcutaneous connective tissue, it is mainly the *protective* function that is most obvious (protection against pressure, drying up (desiccation), bacteriae, radiation, loss of heat, etc).

Body cavities are spaces which are enclosed by the head and trunk. These consist of the cranial cavity, the vertebral canal, the chest cavity and the abdominal cavity (Figs. 1.5.3a and b).

The *cranial cavity* and the *vertebral canal* contain the central nervous system. The hard bony envelope gives ample protection for these vulnerable structures.

The *chest cavity* is surrounded by the semi-hard envelope (ribs/muscles) of the chest cage (thorax). It is a more or less cone-shaped structure that is bordered at the bottom by the *diaphragm*. The diaphragm is a dome-shaped, tendinous partition. It consists of a central plate of tendinous fibres, and is attached to the wall of the trunk by a thin layer of striated muscle tissue (see Fig. 2.9.38). The relatively vulnerable organs such as the heart and the lungs are well protected in the thorax, yet movements are still possible. We also find the large blood-vessels, the windpipe and the gullet (oesophagus) here.

The space between the lungs – occupied mainly by the heart – is called the *mediastinum* (Fig. 1.5.3a). The mediastinum is filled with loosely meshed connective tissue.

The *abdominal cavity* is separated from the thoracic cavity by the diaphragm (see Fig. 2.5.5a). On the front and on the sides it is enclosed mainly by the soft tissue mentioned before: muscles, fasciae and skin. These structures provide sufficient protection for the organs lying in the abdomen, (Figs. 1.5.3a and b): stomach, intestines, liver, gall bladder, pancreas, spleen, kidneys and urinary tract, reproductive organs (in the female).

The lowermost part of the abdomen is often separately named the *pelvic cavity*. There is, however, no separation between this pelvic cavity and the rest of the abdominal cavity, as there is between the thoracic and abdominal cavity.

The lower, caudal border of the abdominal cavity – as of the pelvic cavity – is formed by the *pelvic floor* which consists mainly of muscles. In the pelvic floor we find the openings for the rectum, for the urethra, and in the female for the vagina.

In addition to the body cavities we can identify *serous cavities*, though there is not really much of a 'cavity' here as serous cavities are enveloped by *serous membranes*. These enwrap certain organs as a double-layered membrane. The best way to envisage the development of this 'double layeredness' is to imagine a growing organ which pushes itself more and more into a single walled 'balloon' (Fig. 1.5.5). The developing organ (heart, lungs, intestine) forces the membrane into enveloping it. The inner (visceral) layer of the serous membrane grows attached to the enfolded organ; the outer (parietal) layer is attached to the surrounding structures. The surfaces of the inner and the outer layer that are turned towards each other can glide along one another with hardly any friction at all. This is due to their structure and because of the presence in the serous cavity of a lubricating (serous) liquid. The place where the

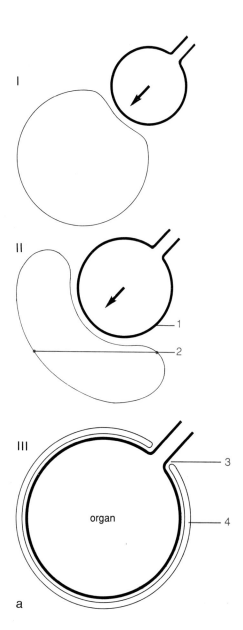

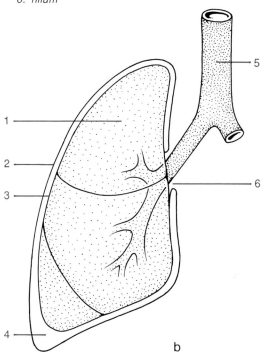

Figure 1.5.5
Serous membranes

a. Development (schematically)

    1. *organ-bud*
    2. *serous membrane*
    3. *hilum*
    4. *cavity*

b. Pleura

    1. *right hand lung*
    2. *outer (parietal) layer*
    3. *inner (visceral) layer*
    4. *pleural cavity*
    5. *windpipe (trachea)*
    6. *hilum*

visceral layer turns into the parietal layer is called the *hilum*. This forms the gateway to the organ. Through this opening, blood vessels, lymph vessels and nerves can pass to and from the enclosed organ, without having to pierce the serous membrane.

It is obvious that such a piercing (through the serous membrane) would annul the function of the serous membrane. The name 'hilum' can also be applied to organs which are not covered by a serous membrane, such as the kidneys.

If serous membranes are extraordinarily suitable to neutralize friction and its effects, we may expect such serous membranes round those organs which generate friction by their mere functioning: the heart, the lungs and the intestines.

Around the heart we find the *pericardium*: it consists of an inner and an outer wall enclosing the *pericardial cavity*.

Around the lungs we find the *pleura*. Each lung is enveloped by an inner and an outer covering which enclose the *pleural cavity*.

Around the intestines – that is to say more or less

around the abdominal cavity – we find the *peritoneum*. It is a double layered membrane with a very complicated outline. It encloses the *peritoneal cavity*.

The positions of the abdominal organs in relation to the peritoneum are indicated as follows:

Intraperitoneal organs are organs which are enclosed by the double-layered peritoneum, e.g. the stomach, pancreas and the liver.

Extraperitoneal are those organs which are not completely enclosed by the peritoneum. In this group we find the following: *retroperitoneal* organs are those that lie behind the peritoneum, e.g. kidneys; *subperitoneal* organs are those that lie under the peritoneum, e.g. the rectum; *preperitoneal* organs are those that lie in front of the peritoneum, e.g. the filled bladder.

The serous membrane will be dealt with again when discussing the related body systems.

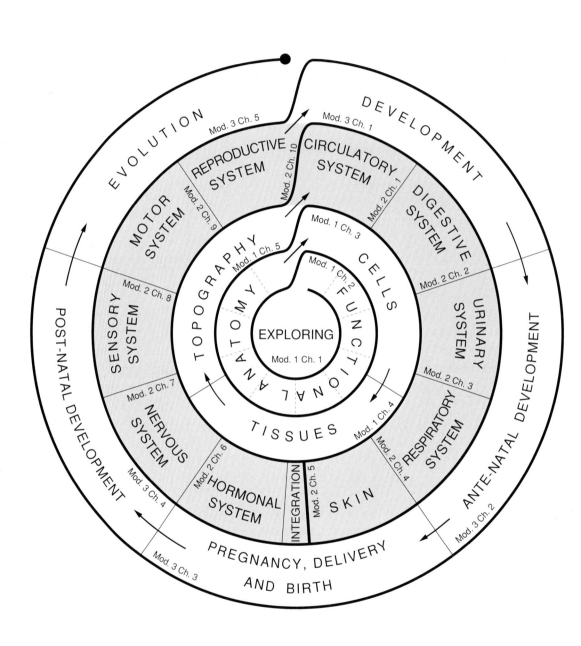

Module 2

BODY SYSTEMS

Introduction to Module 2

In Module 2 the ten body systems are discussed systematically and in greater detail. The order chosen corresponds with the first presentation of the body systems in Chapter 2 of Module 1.

In Chapter 1 the circulatory system is described. Transportation is the main task of this system. After examining the blood, we look at the structures and functions of the heart and blood vessels, followed by the lymphatic system.

In Chapter 2 the digestive system is discussed. Food is treated in such a way in the digestive tract that it can be absorbed into the blood. Firstly the composition of our food is examined, then the structures and functions of the digestive tract, the pancreas, the liver and the biliary tract are discussed.

In Chapter 3 the functional anatomy of the urinary system is described. The urinary system has a supporting task in regulating the composition and volume of the blood.

In Chapter 4 the respiratory system is discussed. During respiration oxygen is carried to the blood and waste gas (carbon dioxide) is transported to the outside world. The availability of oxygen in the cells enables energy to be supplied efficiently.

In Chapter 5 the structure and functions of the skin are discussed. The skin is the outside covering of the body and has a number of important functions.

Chapters 1 to 5 examine the five body systems with somatic (vegetative) functions. These are functions which are engaged in maintaining the life of cells. There is, however, much more to the functioning of human beings than these vegetative functions.

An introduction to the following five body systems is given in the inter-mezzo, the connecting section between Chapters 5 and 6. The key-word of Chapters 6 to 10 is integration. Integration means functioning as a whole, and is concerned with regulation of the somatic functions, inter-action with the outside world and preservation of the species.

In Chapter 6 the hormonal system is examined. This system regulates all kinds of body functions by means of hormones that are transported via the blood stream. Firstly the general working of hormones is discussed, then the hormonal glands and the effects of the hormones produced.

In Chapter 7 the nervous system is described and discussed. The nervous system is the most important body system for regulation, control, coordination and interaction. In this chapter the various parts of the nervous system are discussed, both from an anatomical and a physiological standpoint.

In Chapter 8 the sensory system is described and discussed. This body system gives voluntary and involuntary information from the external environment and from the internal functions of the body. Firstly the general mechanisms of the sensory system are discussed, then the senses and other sensors of the body.

In Chapter 9 the motor system is described and discussed. The motor system consists of all kinds of movements taking place inside the body or made by the body. Particular reference is made to the kinetic apparatus – the skeleton, the joints and the muscular system.

In Chapter 10 the reproductive system is described and discussed. The purpose of reproduction is the conservation of the species. The reproductive organs of both sexes, the reproductive act (coitus), and fertilization are all described.

Circulatory system

Introduction

There is a constant danger that, due to the metabolism in the cells and the accompanying intensive exchange of substances between cells and the internal environment (milieu intérieur), the composition of the m.i. will be disturbed.

If the cells are to function optimally, however, then *homeostasis* (or homeostatic equilibrium) of the internal environment (m.i.) is an absolute prerequisite.

To achieve homeostatic equilibrium, the circulatory system plays a key role of *transportation* for every body system.

- Nutrients are absorbed from the external environment or m.e. (milieu extérieur); after being digested, they pass through the membrane of the intestines to the circulatory system. The nutrients are then transported to the tissues which need them for metabolism.
- Waste matter, released by metabolism, is absorbed from the m.i. into the metabolism and transported to, for example, the kidneys, which are preeminently purifying organs.
- An efficient energy metabolism is dependent on the adequate supply of oxygen to the mitochondria. Oxygen is absorbed into the circulatory system in the lungs and transported to the tissues. The carbon dioxide, released in the energy metabolism, diffuses into the m.i. and is subsequently absorbed into the circulatory system that conveys it to the lungs and the m.e.
- The circulatory system distributes heat and conveys heat to the skin. Here the heat can be given off to the m.e.
- Endocrine glands produce a variety of regulating substances (hormones). The circulatory system absorbs the hormones from the endocrine glands and takes them to their destination.
- The circulatory system transports substances that protect against diseases, etc. all through the body.

In this chapter the circulatory system is discussed. Firstly, the 'replacement fluid', the *blood*, will be discussed. Then we shall talk about the heart.

The heart is a pump which transports the blood through the blood vessels by means of rhythmic contractions. *Arteries* conduct the blood away from the heart; they split into smaller arteries (arterioles) and then into capillaries. The capillaries nearly always form a network; therefore the term *capillary* will be used from now on whenever we talk about these networks.

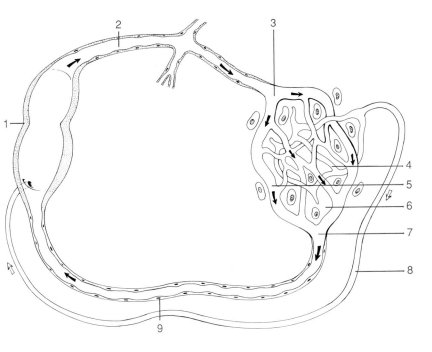

Figure 2.1.1
Circulation
(schematic)

1. *heart*
2. *artery*
3. *arterioles*
4. *lymph capillary*
5. *capillary*
6. *m.i.*
7. *venule*
8. *lymph vessel*
9. *vein*

The exchange of substances between the blood and the *tissue fluid* (m.i.) takes place in the capillary networks. The capillary networks converge into smaller *blood vessels* (venules); they, in turn, converge into larger blood vessels (veins) that conduct the blood to the heart (Fig. 2.1.1).
The *lymphatic system* also belongs to the circulatory system; among other things it includes lymph vessels that conduct an important part of the tissue fluid to certain blood vessels (Fig. 2.1.1). The lymphatic system also plays a part in defending the body against disease.

Learning outcomes

After studying this chapter you should have sufficient knowledge and understanding of:
– the general function of the circulatory system;
– the central position of the circulatory system in relation to homeostasis;
– the composition and function of the blood;
– the structure and working of the heart;
– the topography, structure and functions of the different blood vessels;
– blood pressure and the regulation of blood pressure;
– the exchange between the circulatory system and tissue fluid;
– the structure and function of the parts of the lymphatic system.

1. Blood

Blood is a fluid tissue. It is classified as a connective tissue because of the importance of the intermediate substance. Blood is fluid as there are no fibres in the intermediate substance. Blood is the body's medium of *transportation*, carrying oxygen and substances rich in energy – building materials, waste products, carbon dioxide, regulating materials, protecting substances and buffer substances.

The milieu in the blood vessels (intravascular milieu) does not belong to the m.e. and is not regarded as part of the m.i. either. Blood is the link between m.e. and m.i.

Blood consists mainly of water. It can easily absorb or give off heat and can play an important part in distributing and regulating heat as it is pumped through the body.

Volume and composition

The normal volume of blood in the 'average man' of about 70 kg is around 7.5% of his body weight, i.e. about 5 litres. The volume of blood in a person of 100 kg is considerably larger, as are the capillary networks, and therefore the heart must work harder.

The volume of blood in women is about 6.5% of body weight. This difference between men and women is due to a larger fat weight in women. When we consider the blood volume per kilogramme of fat-free mass there is almost no difference between the sexes.

Blood is a red-coloured, somewhat viscous fluid. Treated in such a way that it cannot clot, left in a test tube for some time and centrifuged, blood exemplifies the relation between plasma and the cells (Fig. 2.1.2). The cells are heavier and sink to the bottom, where they form a dark substance. The percentage of blood cells to the total volume is called the *haematocrit*.

Figure 2.1.2
Haematocrit
(Packed Cell
Volume)

1. *blood plasma*
2. *blood cells*

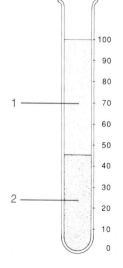

In women this volume is normally about 40%, in men 45%. The layer of clear, light yellow fluid over it is plasma. Strictly speaking, plasma is the m.i. of the blood cells as blood is after all a tissue. For the sake of clarity we will use the previously introduced term *intravascular milieu* for the whole blood.

1.1. Blood plasma

The plasma component of blood consists of water with electrolytes, plasma proteins and blood gases – oxygen (O_2), carbon dioxide (CO_2) and also nitrogen (N_2). Moreover, there are so-called transient substances in the plasma.

Water

90% of plasma consists of water. Water from our food (drinking and eating) is absorbed into the circulatory system in the intestines. Removal of water from the circulatory system takes place via the kidneys, the skin and to a lesser extent via respiration and defecation. Water plays the part of a *heat buffer*, i.e. it can easily absorb excess heat and give it off somewhere else. Moreover, water is an excellent medium for *electrolytes* in the blood.

Electrolytes

Electrolytes are salts separated into positive and negative ions (charged particles). The most important ions are sodium ions (Na^+), potassium ions (K^+), calcium ions (Ca^{2+}), magnesium ions (Mg^{2+}), phosphate ions (PO_4^{3-}), bicarbonate ions (HCO_3^-) and chloride ions (Cl^-). They bind water molecules in a solution, forming the water coat. The electrolytes form a *pH-buffer*; the mixture can bind both acid and alkaline substances. This keeps the acidity of the blood (pH) at a constant level (pH blood = 7.4) The bicarbonate ion is the main buffer. The amounts of electrolytes are kept at a constant level, principally by the kidneys, under the influence of regulating hormones.

The major part of positive ions is formed by Na^+. This ion is smaller than K^+, but its water coat is bigger (Fig. 2.1.3).

When the relation between sodium and potassium ions (the Na/K balance) changes, for instance, under the influence of a certain

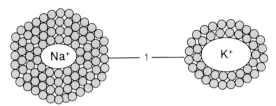

Figure 2.1.3
Sodium and potassium ion

1. water coat

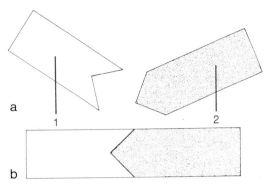

Figure 2.1.4
a. Antigen and antibody
b. Harmless complex

1. antigen
2. antibody

hormone, the amount of water in the blood is affected and consequently the blood pressure. The crystalloid-osmotic value of blood is determined principally by the amount of Na$^+$. Calcium ions also play a part in the coagulation of the blood and the ability of the muscles to contract. Supply and release of salts from food takes place in the digestive tract; removal of salts takes place via the kidneys and the skin (sweat).

Plasma proteins

There are also proteins dissolved in the plasma. These *plasma proteins* are mainly produced and broken down by the liver. There are three kinds with different characteristics:

– *albumin* is an elongated protein. The amount of albumin is almost twice that of the other two plasma proteins. The colloid-osmotic value of the blood is especially determined by albumin;
– α-, β- and γ-*globulins* each consist of protein chains tangled into a ball. The α- and β-globulins are primarily responsible for transportation; they can bind glucoses and lipids. Iron, vitamins, hormones and agglutinins are linked with the β-globulin and are thus transported.
The particular responsibility of the γ-globulin is the defence of the body. It is a mixture of a number of *antibodies* that are, in general, specifically aimed at foreign proteins (e.g. bacteria), which can harm the organism. The antibodies are produced in lymphatic organs. The antibody links with the harmful protein (antigen) and thus forms a basically harmless complex (Fig. 2.1.4).

A number of proteins (among others, anti-haemophilia globulin and prothrombin) are also classed among the group of globulins; they play a part in coagulation.
– *fibrinogen* is a relatively large molecule, the pre-phase of the protein fibrin, to be discussed later under 'coagulation'. There are relatively fewer fibrinogens than other plasma proteins. Blood plasma from which fibrinogen has been removed is called serum. Consequently serum cannot clot; it is used for blood transfusions.

Apart from the above-mentioned specific functions of the plasma proteins and the general function in the *determination of the colloid-osmotic value*, the plasma proteins form a small *food reserve* that can supply energy in emergency situations.
Separation of the proteins in the liver gives amino acids that can be built into glucose molecules when there is a temporary shortage. Finally, the plasma proteins have a *buffer function*; in an acid solution they bind H$^+$ ions, in an alkaline solution they release H$^+$ ions. They therefore help the electrolyte HCO$_3^-$ to maintain the pH of blood at 7.4.

Blood gases

The following gases are in solution in blood plasma: oxygen, carbon dioxide and nitrogen.

Oxygen is much less easily soluble in blood plasma than carbon dioxide. Although nitrogen does not play a part in the metabolic reactions, it is still present in the plasma as a result of diffusion.

Transient substances

There are always substances in the blood plasma that are being carried temporarily: *transient substances*. They are nutritive substances (glucose, lipids and amino acids) from the digestive tract or depots, waste material from metabolism on its way to be removed to, for instance, the kidneys, hormones (from endocrine glands) and vitamins (from food or produced in the body), necessary for the regulation of a vast number of processes.

1.2. Blood cells

There are three groups of blood cells: *erythrocytes* (red blood cells), *leukocytes* (white blood cells) which are divided into sub-groups) and *thrombocytes* (platelets – literally meaning clotting cells).

The *reticuloendothelial system* (RES) includes the red bone marrow and a number of organs with lymphatic tissue. All cells of this system may become phagocytes, but they may also differentiate into undifferentiated (root) cells of blood cells (Fig. 2.1.5). As a result of differentiation, specialization, growth and maturation the groups of blood cells are mutually divergent as to structure and function.

The main producer of blood cells is the red bone marrow (Fig. 2.1.6), which produces millions of blood cells every second.

Degradation of 'old' blood cells is done by the phagocytes of the RES. Production of the different kinds of cells by the undifferentiated cells is fairly equal to the degradation of those kinds of cells. Consequently, the number of each cell type in the blood remains constant.

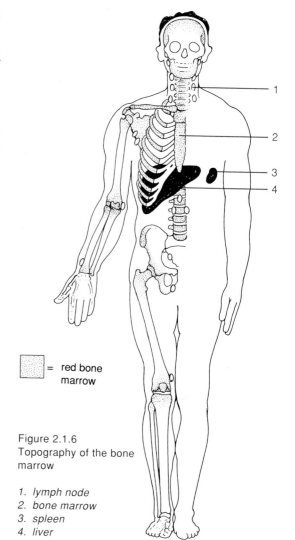

= red bone marrow

Figure 2.1.6
Topography of the bone marrow

1. *lymph node*
2. *bone marrow*
3. *spleen*
4. *liver*

Figure 2.1.5
Cell types from the reticuloendothelial system

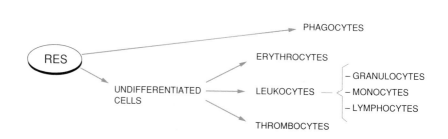

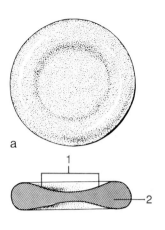
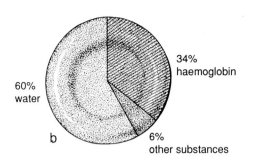

Figure 2.1.7
Erythrocyte

60%
water

34%
haemoglobin

6%
other substances

a. Shape

1. concavity
2. section

b. Composition

Erythrocytes

When blood is examined under a microscope it is noticeable that one kind of blood cell is in the majority. Nearly 95% of the cell component consists of *red blood cells*, or erythrocytes. A man has about 5 million erythrocytes/mm^3 of blood, a woman about 4.5 million/mm^3 of blood. The lifespan of erythrocytes is about 120 days, which means that there must be a continuous production of vast amounts of erythrocytes.

An erythrocyte is a rather small cell in the shape of a *bi-concave disc* that is indented on both sides (Fig. 2.1.7).

Erythrocytes have no nucleus and no mitochondria, so they have only a minor metabolic function. Their function is derived from their enormous amount of *haemoglobin* (Hb) (Fig. 2.1.7). The haemoglobin content of blood in men is 8-11 mmol/litre; in women it is 7-10 mmol/litre. The name haemo*globin* indicates that it is a protein. This red-coloured protein with its built-in iron atom has a large capacity for binding oxygen in an environment rich in oxygen (the lungs), but it easily releases the fused oxygen in an environment poor in oxygen (active tissues) (Fig. 2.1.8). Due to the flat, bi-concave shape of the erythrocyte, the diffusion distance (haemoglobin-oxygen) is short. Moreover, erythrocytes are transformable, which helps them to pass through the narrow capillary networks. The basic function of the erythrocytes is to supply the m.i. of the active tissues with oxygen via diffusion from the capillary networks. Haemoglobin, however, is also capable of binding a part of the carbon dioxide produced and of transporting it to the lungs. In Module 2, Chapter 4, Respiratory System, we shall return to this subject.

When erythrocytes are broken down in the reticuloendothelial system, the haemoglobin breaks down into *bilirubin* and *iron*. The bilirubin undergoes some more processes before it is secreted, mainly via defecation. The iron is recycled as much as possible and re-used to form new erythrocytes.

Leukocytes

White blood cells, or leukocytes, are considerably less numerous than red ones: 6000–8000 per mm^3 blood. There are several kinds of leukocytes (Fig. 2.1.9), but they all have something to do with the defence of the body. They clear away foreign substances (such as bacteria) and dead body cells; they also produce antibodies against foreign proteins (antigens). The lifespan of leukocytes is between several days and several weeks. They are larger than erythrocytes.

$$HHb \quad + \quad O_2 \quad \underset{\text{IN ACTIVE TISSUES}}{\overset{\text{IN LUNGS}}{\rightleftharpoons}} \quad HbO_2 \quad + \quad H^+$$

(REDUCED HAEMOGLOBIN)

(OXYHAEMOGLOBIN)

Figure 2.1.8
Reactions of haemoglobin (simplified)

Three kinds of leukocytes are:
- *Granulocytes* (Fig. 2.1.9), which characteristically contain many grains (granules) in their cytoplasm. Moreover, the segmented nucleus is evident. They are produced in the red bone marrow. Granulocytes are specialists in the phagocytosis of bacteria which reveal their presence by the waste matter they produce. Granulocytes wriggle themselves through the cells of the capillary membrane, get out of the bloodstream (leukodiapedesis) and move in the direction of the invaders (Fig. 2.1.10). Thus the phagocytizing reticulum cells get massive help from granulocytes which digest the surrounded bacteria.
- *Monocytes* (Fig. 2.1.9) are the largest leukocytes; they have a fairly large C-shaped nucleus. Like the granulocytes, they are produced in the red bone marrow. They are less active than the granulocytes and their main role is to phagocytize the remains of dead body cells. After about 6 hours they come to the assistance of the granulocytes, in cases of more serious infection.
- *Lymphocytes* (Fig. 2.1.9) are the smallest type of cell within the group of leukocytes. The nucleus is so large that only a little cytoplasm is left. They are produced throughout the reticuloendothelial system. Their number is highly dependent on the appeal for help made to them. After about 10 hours lymphocytes assist granulocytes and monocytes in their phagocytosis activity in cases of more serious infections.

Certain lymphocytes are not only able to phagocytize, but they can also produce *antibodies* in response to an invading antigen. Such activity of lymphocytes takes place in the lymphatic tissue and results in *immunity*.

Figure 2.1.9.
Types of blood cells

1. *lymphocyte*
2. *monocyte*
3. *granulocyte*
4. *erythrocyte*
5. *thrombocyte*

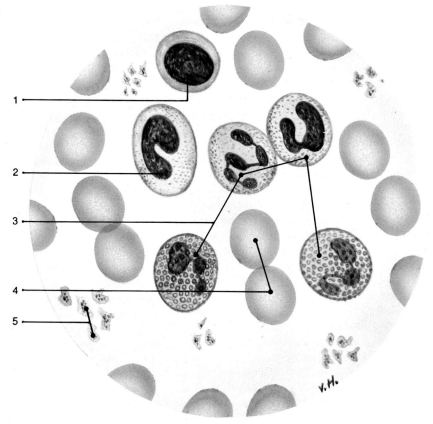

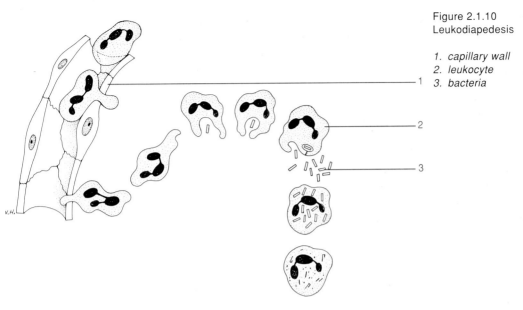

Figure 2.1.10
Leukodiapedesis

1. *capillary wall*
2. *leukocyte*
3. *bacteria*

Thrombocytes

The 'clotting cells' used to be called *platelets* as they only slightly answer to the description of the 'standard' cell of Module 1, Chapter 3. Thrombocytes are very small fragments of large stem cells in the red bone marrow (cytoplasm surrounded by a membrane). On average there are about 250,000/mm³ of blood. The life span of thrombocytes is slightly more than one week. The membrane of the platelets is not very stable and therefore the contents of the thrombocytes can easily be released when tissue is damaged. Thrombocytes contain *thromboplastinogen*, a substance which plays an important part in blood clotting.

2. Resistance

The body continuously offers *resistance* against potential pathogens (micro-organisms capable of causing disease). It is only a matter of disease or disorder (the field of pathology) when the normal protection (the field of functional anatomy) evidently fails. The resistance of the body includes a general *defence* component (aimed at all potential pathogens) and a specific *immunity* component (aimed at specific pathogens).

2.1. Defence

General resistance is offered by the covering tissues of the skin, the digestive tract, the bron-

chial tubes and the urinary passages. The majority of germs cannot penetrate these covering tissues. In certain places (for instance in the stomach and in the vagina), acids which are present offer extra protection against micro-organisms. The acid kills the micro-organisms. Apart from these mechanical and chemical defence systems, the general resistance also includes phagocytosis (engulfment) and the production of histamine.

Phagocytosis

Phagocytosis is the surrounding of germs (for instance a bacterium) by a part of the membrane, after which it is made harmless by digestion.

When the function of the leukocytes was described, phagocytosis was discussed. To summarize, the reticulum cells at the location of the infection take the lead; shortly afterwards they are supported by vast amounts of granulocytes. After about 6 hours the monocytes appear, if necessary, and after about 10 hours the lymphocytes.

The invaded bacteria have enzymes with which they can kill the cells that phagocytize them. Consequently, there may be a heap of dead ruptured phagocytes at the location of the infection. The yellowish fluid thus formed is called *pus*. Pus consists of tissue fluid with dead and still living bacteria, dead and still living phagocytes and the debris of the dead tissue cells. Small infections with a slight formation of pus regularly occur in the skin.

Histamine production

Damaged tissue cells produce the hormone *histamine*. This substance has a local vasodilating effect and increases the permeability of the capillary network. This can easily be established as damaged tissue is red and somewhat swollen. Histamine enables the affected tissue to stimulate the metabolism, thus helping both defence and repair.

2.2. Immunity

Immunity is a specific component of defence and is based on a totally different defence mechanism: *the antigen-antibody reaction*. We shall first discuss these antigen-antibody reactions in order to understand the concept of immunity.

When substances invade an organism, certain lymphocytes can react by producing and releasing specifically-aimed substances. These *antibodies* and the foreign body react, making the foreign body ineffective or harmless (see Fig. 2.1.4).

When 'foreign bodies' are mentioned, germs should not be the only ones to be considered. Each substance, identified by the body as an 'odd one out' and causing the production of antibodies, is called an *antigen*.

Generally speaking, antigens are proteins, but they may also be carbohydrates, lipids or other organic or inorganic substances. The body itself consists of proteins, carbohydrates, lipids and other organic substances, but it should be remembered that one organic molecule is not the same as another. Let us restrict ourselves to proteins. The twenty different amino acids may be arranged in almost innumerable combinations (as to number and sequence). That gives just as many kinds of proteins. In order to get an impression of the number of possibilities, one should imagine chains of two amino acids being built (dipeptides). When there are 20 amino acids available, there are 400 possible combinations. With chains of three amino acids 8000 combinations are possible.

The smallest body proteins are built of about 100 amino acids.

Figure 1.3.2 shows how proteins are built into the cell membrane. These membrane proteins expose themselves opposite the proteins in the cytoplasm. There is a large number of these body proteins in the tissues of an organism. Such proteins are also present in any other organism, but normally their structures are more or less aberrant, apart from those in identical twins.

All tissues therefore have their *protein specificity*. The thymus gland prevents the production of antibodies against the body proteins. This happens during embryonic development. If this did not happen the organism would probably destroy itself during the embryonic stage.

If the body proteins are known, all other proteins are foreign and contact between body and foreign proteins will lead to the production and release of antibodies.

Cellular and humoral immunity

An organism is *immune* when it is not susceptible or less susceptible to one or more antigens. Immunity is effected by the activity of two types of lymphocytes: the T-lymphocytes and the B-lymphocytes. Both types of cells are made from stem cells of the reticuloendothelial system (Fig. 2.1.11). These stem cells develop into T-lymphocytes or T-cells (T = thymus) in the thymus. Other stem cells are probably built into B-lymphocytes or B-cells (B = bursa) in the lymphatic tissue of the tonsils or in the aggregated lymphatic follicles (Peyer's plaques). The immunity brought about via the B-cells is based on the production of antibodies.

These antibodies circulate through the blood plasma and make the antigen-antibody reaction possible. This is called *humoral immunity*. The T-cells are responsible for *cellular immunity*. The different kinds of T-cells themselves are active in different ways in the specific defence. The activities of B and T-cells (Fig. 2.1.12), will be further discussed.

A B-cell, stimulated by an antigen which starts producing antibodies and releasing them, is called a *plasma* cell. All antibodies are proteins,

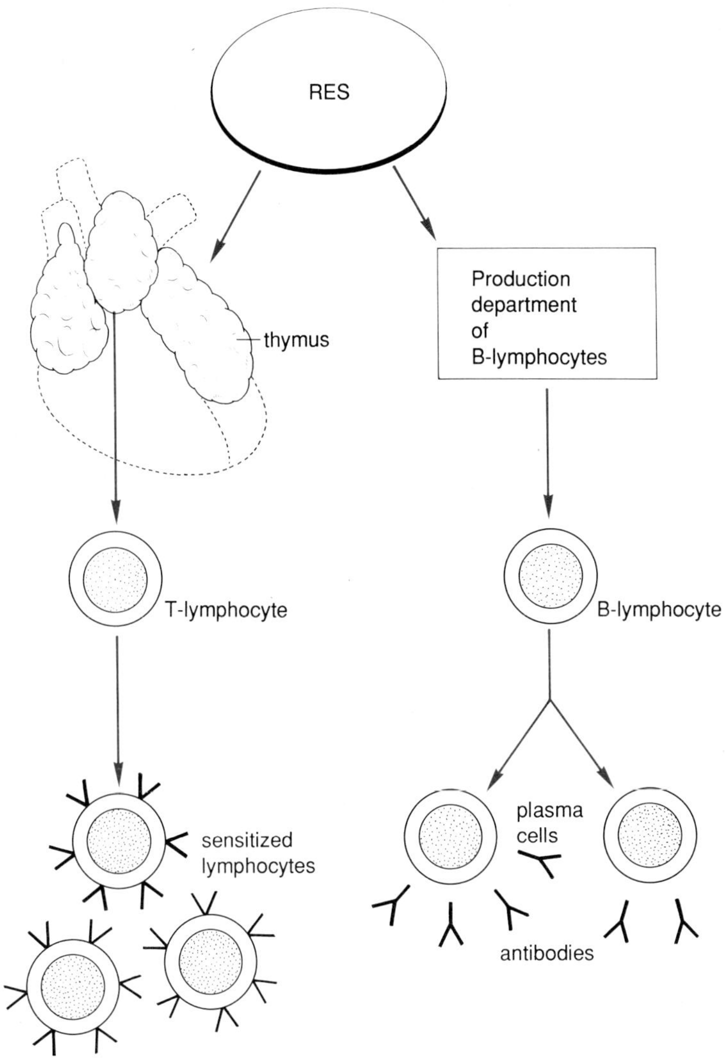

Figure 2.1.11
Cellular and humoral immunity

i.e. globulins. A specific antibody is produced for each intruding antigen. The γ-globulins, also called *immunoglobulins* (Ig), are divided into five classes: IgA, IgD, IgE, IgG and IgM.

Each class has its characteristics. IgA is found mainly in nasal secretions, mucus, saliva, tears and intestinal juice. IgE plays an important part in certain allergic reactions. IgG and IgM are the main antibodies against bacteria and viruses. Until recently, less was known about IgD. The antibodies do not overcome the antigens until there are sufficient antibodies in the blood plasma. It takes time to produce antibodies. Sufficient antibodies are not produced until about ten days following the first contact with the antigen. The antibodies, once they have been produced, usually stay in the plasma, either for a long time or forever, even after they have conquered the antigens. This means that if there is another infection by the antigen concerned, antibodies are readily available to fight the pathogenic organism.

The renewed production of a certain antibody is much faster than the first production. The B-lymphocytes start up a temporarily inactive production line. This is done by the *B-memory cells*; they are produced simultaneously with the plasma cells. The T-cells react differently when they make contact with an antigen. Moreover,

Figure 2.1.12
Reactions and linkage of
B-lymphocytes and
T-lymphocytes

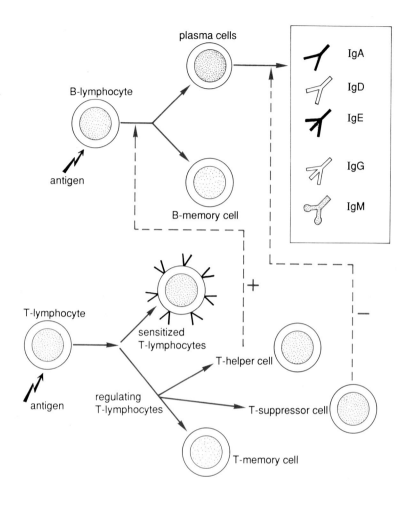

there are several types of T-cells. On the basis of their functions, T-lymphocytes can be divided into two classes: the *effector* T-cells and the *regulator* T-cells. The effector cells have a direct defence reaction; they react immediately with the antigen, for instance a body cell, in which a virus has lodged. The regulator T-cells play an indirect role in the defence; they influence the activity of the B-cell. There are two types of regulator T-cells which are mirror images: the *T-helper cell* and the *T-suppressor cell*. The T-helper cells make sure B-cells are activated; the T-suppressor cells make sure the B-cells stop making antibodies. We see, therefore, that humoral immunity is closely linked with cellular immunity. Whenever there is contact with an antigen, the T-cells produce *T-memory cells*, in much the same way as the

B-cells produce B-memory cells. The functions of T-memory cells are the same as those of B-memory cells.

Immunization
Immunity may be acquired in different ways. When the organism itself produces immuno-globulin and memory cells, it is termed *active immunization*. Active immunization is acquired in response to a *natural* infection by a pathogenic organism in the m.e.
The almost annual occurrence of influenza (a virus) is an example of active immunity. As the influenza viruses have repeatedly developed different protein specificities, as a result of mutations, a person may contract influenza again and again.
When the antibodies are not produced by the

organism itself, but given to it, we talk about *passive immunization*.

Passive immunization takes place in a *natural way* by transfer of immunoglobulin by a pregnant woman to her fetus. Only IgG is able to pass through the placenta. This gives an immediate protection against a number of germs.

Artificial types of active and passive immunization belong to the field of medicine.

3. Tissue typology

As mentioned before, each tissue has its own *protein specificity*. This means that the proteins concerned each have their own unique combination of amino acids.

Protein specificity is one of the main factors in that field of medicine which deals with *tissue transplantation*. The reaction of the body to the grafted organ causes rejection symptoms. The implanted antigens lead to antibodies being produced. The more the type of protein of the donor tissue is similar to that of the recipient, the less will be the rejection symptoms that can be expected.

Tissue typology is as accurately as possible a determination of the type of protein in a certain tissue. Tissue typology is important to all organ transplantations (heart, kidney, liver, pancreas, etc), but it has come into being thanks to the first and, still, the most common transplantation: the *transfusion of blood*.

Blood groups

A large number of proteins are built into the cell membrane of erythrocytes and leukocytes. Of course, the blood cells of a certain individual all have identical membrane proteins, as they came into being via mitosis (copying). Some of these proteins are notorious due to the fact that they cause violent reactions when the blood of a certain type of tissue is mixed with the serum of an individual with another type of tissue. The blood cells begin to clot. This process is called *agglutination*. A number of membrane proteins, however, do not lead to such a violent reaction.

Erythrocyte groups

The first proteins described are found in the erythrocyte membrane.

They are classified in the *A-B-0-system* (Fig. 2.1.13). In the description of blood typologies, the first discovered protein was given the first letter of the alphabet, the second protein was given the second letter – it was not unthinkable that the entire alphabet would be needed. The number of blood groups described at the moment almost justifies this premise. The proteins are called *agglutinogeñs*, as they cause agglutination when combined with certain antibodies. For the sake of using a homogenous terminology the term *agglutinins* is used for the antibodies. So: agglutinogen A + agglutinin anti-A gives agglutination.

It is remarkable that agglutinins apparently need not be produced. Each individual has in his plasma the antibodies (of the A-B-0-system)

Figure 2.1.13
A-B-0-system

AGGLUTINOGEN protein ● = A protein ○ = B			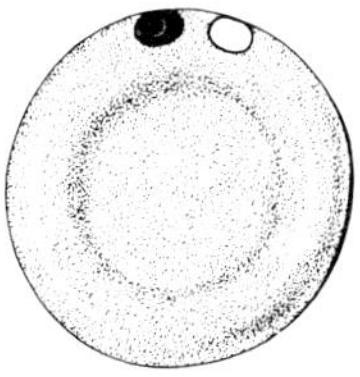	
AGGLUTININ	β	α	–	αβ
BLOOD GROUP	A	B	AB	0

that are directed at all agglutinogens, apart from those that would make his own blood clot (Fig. 2.1.13).

No agglutinins will be found in the blood group AB; in blood group 0, however, both anti-A (also called alpha) and anti-B (also called beta) will be found. The cell membrane has neither protein A nor protein B (no protein, consequently of blood group nought (0) – and not blood group 'oh' (O) as it is usually wrongly pronounced) and cannot provoke agglutination. A general *transfusion scheme* can be drawn with the help of these facts (Fig. 2.1.14).

It transpired that there was still another agglutinogen, after the A-B-0-system had been described, which can cause agglutination. It has been discovered by examination of the erythrocytes of the *Macaca rhesus*, a monkey species in South-East Asia. This agglutinogen is called the *rhesus factor* (Rh-factor) and was given the letter D.

positive (Rh+), when it is absent, one speaks of *rhesus-negative* (Rh−). The agglutinogen D exists in combination with all blood groups of the A-B-0-system.

It is expressed as blood group A−, AB+, 0+, etc. (Fig. 2.1.15).

It could have been possible to use A, ABD, 0D, etc., instead, but there is something special in the agglutinin that 'belongs' to factor D. Each individual evidently has agglutinins against all agglutiniogens in his plasma, except against his own, in the A-B-0-system. In all rhesus-negative blood groups the agglutinin anti-D (delta) could be expected. This, however, is not the case. Anti-D presence is only possible after it has been produced following contact with rhesus-positive blood (agglutinogen D). In this case one could say that there is something like active immunization against factor D.

In addition to its significance in blood transfusions, the rhesus-factor is extremely important in obstetrics.

Figure 2.1.14
Transfusion scheme

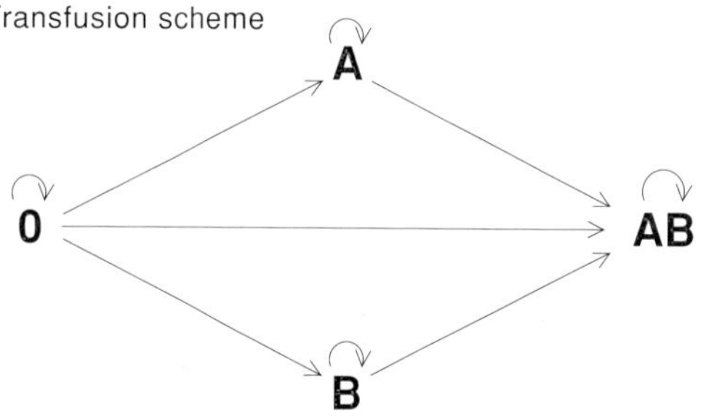

```
            A
      ↗         ↘
  0 ─────────────→ AB
      ↘         ↗
            B
```

When the agglutinogen D is present in the erythrocyte membrane, one speaks of *rhesus-*

Apart from the A-B-0-system and the rhesus-factor, there is the M-N-S-system. It deals with three possible antigens in the erythrocyte membrane, against which antibodies cannot be readily formed. As these proteins can be made visible via examination of the blood, they can be important. Each erythrocyte forms a personal, rather unique 'business card' due to the large amount of possible combinations. This can be used in forensic medicine, for instance to identify people, or to establish blood relationship. The latter can be important when biological parenthood of a particular child is being established.

Figure 2.1.15
Rhesus-factor

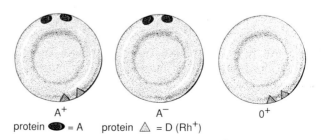

protein ● = A   protein △ = D (Rh+)

Finally, there are erythrocyte-agglutinogens which occur only in certain families or in certain, often isolated, regions. They are the so-called *rare blood groups*. They are particularly important when an individual from such a blood group needs a blood transfusion.

Leukocyte groups

Antigens are built in the membranes of leukocytes as well. This is the HLA-system (Human Leukocyte Antigen-system). In the case of a blood transfusion it is less relevant, as antibodies cannot be produced any more easily than in the M-N-S-system. Moreover, the number of leukocytes per unit of blood is comparatively small. The HLA system is, however, more important in a person who needs frequent blood transfusions, because in this case too many antibodies might have been produced. The HLA system is also very important in transplantation organs, as the rejection reactions are 'staged' by the leukocytes.

4. Blood clotting

When a blood vessel is damaged, a sequence of mechanisms is initiated in order to curb the loss of blood as far as possible and to repair the damaged tissue. This happens not only when it is a question of fairly large bleeding, such as an injury with a knife or a fracture, but also when there is smaller bleeding. Small bleeds are common, but they usually remain unnoticed. When a person bumps against something, some capillaries in deeper tissues will break. A small amount of damage is done to blood vessels and accompanying bleeding often occurs during chewing, teeth brushing or defecating. In each injury, accompanied by bleeding, a succession of processes occurs:
- local vasoconstriction (narrowing of the lumen of a blood vessel)
- clot formation
- coagulation (clotting)
- recovery of the tissue.

The first three processes are together called *haemostasis*.

Local vasoconstriction and clot formation

The smooth (involuntary) muscles in the wall of the small arteries (arteriolae), which supply the damaged capillary network with blood, contract soon after the injury. This local vasoconstriction curbs the loss of blood. At the site of the damage thrombocytes attach themselves to the edges of the 'hole', under the influence of anti-haemophilic globulin in the plasma, and then clot together. In this way a somewhat loose clot is formed, thus almost 'mending the hole'.

Coagulation

Coagulation itself includes a cascade of reactions resulting in a *clot*, a dense network of protein fibres, in which blood cells are captured, and with which the hole in the blood vessel is hermetically sealed. Initially the tissue damage over the blood vessel concerned is also mended by means of a protein network. In the whole process of coagulation 13 coagulating factors are involved, ultimately resulting in the conversion of the plasma protein fibrinogen into the insoluble fibrin.

A simplified description of the complicated coagulating process can be described as follows:
- the wall of the thrombocyte breaks; *thromboplastinogen* (= antithrombin, factor III) is released among the contents;
- this thromboplastinogen is activated by certain plasma factors to become *thromboplastin* (= thrombokinases, factor IV);
- the thromboplastin and the Ca^{2+} (calcium) in the plasma are the catalysts for the conversion of *prothrombin* (factor II, present in the plasma) into thrombin;
- thrombin in its turn activates the conversion of *fibrinogen* (factor I, present in the plasma) into fibrin;
- the protein fibrin makes a dense network of fibres in the hole of the wound. Blood cells are caught in this fibrin network as a result of which a clot is formed;
- by dehydrating (in a skin injury) the fibrin fibres shrink, as a consequence of which the edges of the wound are drawn together and a

smaller scar can be formed. The clot is wrung out, as it were, and a small crust is formed.

Recovery of the tissue

After the hole has been sealed, the local vaso-contriction disappears and the converse occurs. As a result of the tissue damage, the hormone histamine is produced. This hormone stimulates vasodilation (widening of the lumen of the vessels) and permeability (the skin becomes red). Thus the repair of the damaged tissue soon begins. Mitosis and metabolism are stimulated. After some days the crust falls off (or the scab is scratched off). The strong vascularization in and underneath the recovered tissue is discernible.

5. The heart

The beating heart is one of the most significant symbols of life. This is illustrated by the large number of proverbs and sayings in which 'heart' occurs. 'Out of the abundance of the heart the mouth speaketh', 'To devote oneself with heart and soul', 'To be heart broken' etc. In these sayings, other functions are attributed to the heart apart from its role in functional anatomy. Day after day, all through life, the heart pumps with every beat a quantity of blood into the closed circuit of arteries, capillary networks and veins. When at rest, about 5 litres of blood are pumped into the aorta every minute (Fig. 2.1.16); that is 7200 litres per 24 hours. When oxygen is required, the quantity of blood per minute increases greatly.

Functionally, the heart consists of two parts: the right half and the left half. The right half is responsible for blood circulation to the lungs, while the left half is responsible for the circulation of blood through the rest of the body. Each half is divided into an upper part and a lower part, respectively called *atria* (upper chambers) and *ventricles* (lower chambers).

Figure 2.1.16
Location of the heart
in the mediastinum
(interpleural space)

1. right subclavian artery
2. right subclavian vein
3. superior vena cava
4. diaphragm
5. carotid artery
6. jugular vein
7. first rib
8. aortic arch
9. edge of pleura
10. heart in pericardium

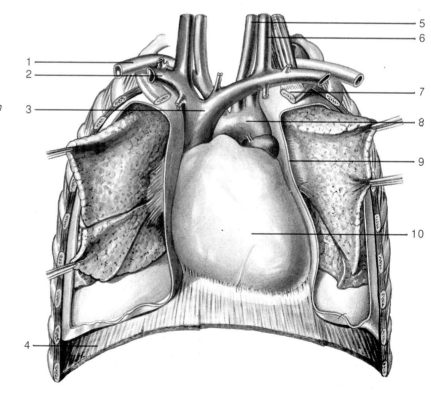

5.1. Location

The heart is located in the thorax, directly behind the breast bone (sternum) in the *mediastinum*. This is the cavity between the lungs in which, apart from the heart, a section of the oesophagus and part of the larger vessels run (Fig. 2.1.16). The heart is as large as a fist and rests on the diaphragm in a somewhat tilted position; the apex points left. The ventricles are lower left, the atria upper right. The right ventricle forms the ventral side of the heart and the left ventricle forms the dorsal side.

The heart cavity is right behind the centre of the sternum (Fig. 2.1.17), but as the left ventricle has a much thicker wall than the right ventricle, the heart protrudes more to the left and can be felt beating at the left side of the sternum.

The asymmetrical structure of the heart is accompanied by a difference in lung size: the left lung (two lobes) is smaller than the right lung (three lobes).

5.2. Structure

The muscle tissue of the heart consists of three hollow muscles, the superior vena cava and the inferior vena cava (atrial and ventricular muscle respectively), entirely separated by a fibrous ring, the *annulus fibrosus* (Fig. 2.1.18). Both the v. superior and the v. inferior are separated into a left and a right half by the dividing wall of the heart, the *septum cordis*. The atrial septum consists mainly of connective tissue, whereas the ventricular septum consists mainly of muscular tissue. So there are four compartments in the heart, separated by the annulus fibrosus and the septum cordis: the left and right atrium, and the left and right ventricle.

Wall

The structure of the heart wall is a variation of the general structure of blood vessels (see Fig. 2.1.30b). This is because the heart originates from a pulsating blood vessel. The lining of the heart consists of a single layer of squamous epithelial cells (endothelium) and a thin layer of elastic connective tissue. This is called *endocardium*.

Functionally the most conspicuous layer is called *myocardium*. The myocardium consists of heart muscle tissue. The muscular fibres are

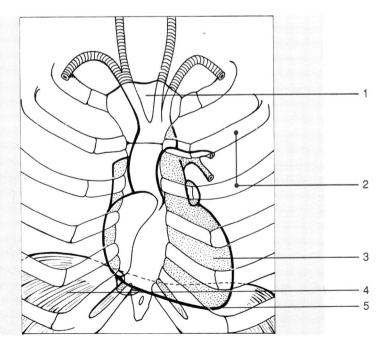

Figure 2.1.17
Projection of the heart
behind the breastbone

1. *breastbone*
2. *ribs*
3. *heart*
4. *diaphragm*
5. *apex of the heart*

Figure 2.1.18
The heart, frontal section

1. left atrium
2. superior vena cava
3. aorta with semi-lunar valves
4. back cusp of mitral valve
5. front cusp of mitral valve
6. right atrium
7. tricuspid valve
8. septum cordis
9. right ventricle
10. muscle beam
11. coronary artery
12. right ventricular wall
13. left ventricular wall
14. papillary muscles with tendinous cords
15. apex of the heart, consisting of the cone-shaped apex of the left ventricle

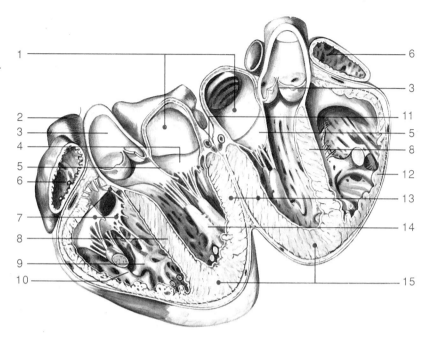

arranged in three-layered bundles which surround the heart cavities in bundles set in different directions, separated by the annulus fibrosus. The above-mentioned order of bundles enables an efficient reduction in the ventricular volume.

The variation in thickness of the myocardium in the different parts of the heart is apparent. The atrial myocardium is rather thin, the outside wall of the right ventricle is somewhat thicker. The myocardium of the ventricular septum, and the outer wall of the left ventricle are about three times as thick as the outside wall of the right ventricle. We shall learn that these differences in structure are connected with differences in function.

The inside of the ventricular myocardium is not smooth. Elevations (papillary muscles) and many small muscle beams can be seen at a number of places (Fig. 2.1.18). Sometimes such a muscle beam goes straight through the ventricle.

On the outside of the myocardium a great number of branched blood vessels can be seen (Fig. 2.1.19), the arteries and veins of the heart circulation. The heart circulation maintains the homeostasis of the heart muscle tissue itself.

The myocardium is lined with connective tissue: the *epicardium*. This is the inside lining of the heart sac (pericardium). The heart is after all a serous membrane.

The inferior vena cava and the superior vena cava, and the sinus coronarius terminate in the right atrium; the four pulmonary veins, two from each lung, terminate in the left atrium (Fig. 2.1.19).

Valves

The annulus fibrosus has two openings for the connection between atria and ventricles and another two openings for the connection between the ventricles and the major arteries (Figs.2.1.18 and 2.1.20). The four openings can be closed off by membranes of connective tissue, the *heart valves*. The valves between atria and ventricles are called the *atrioventricular valves*, those between ventricles and major arteries (aorta and pulmonary artery) are called *arterial valves*.

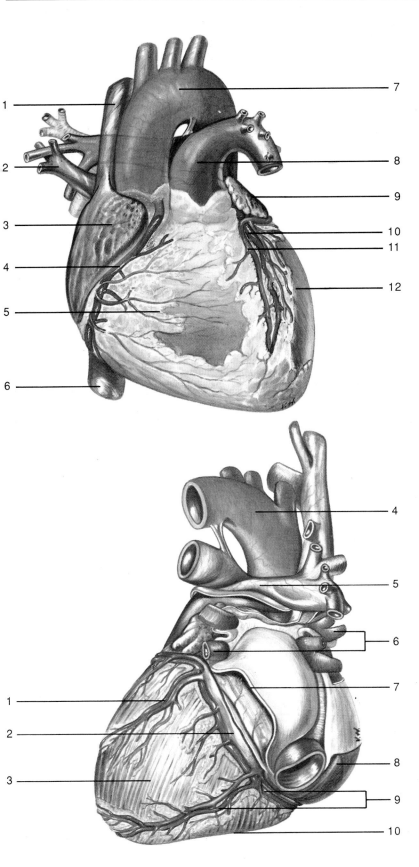

Figure 2.1.19
The outside of the heart

a. Front

1. superior vena cava
2. branch of right pulmonary vein
3. right atrium
4. right coronary artery
5. right ventricle
6. inferior vena cava
7. aortic arch
8. pulmonary artery
9. left atrium
10. terminal branch of left coronary artery
11. terminal branch of left coronary vein
12. wall of left ventricle

b. Back

1. left coronary artery
2. coronary sinus
3. left ventricle
4. aorta
5. right pulmonary artery
6. pulmonary veins, terminating in the left atrium
7. cutting edge of the pericardium (heart sac)
8. right atrium
9. right coronary artery
10. right ventricle

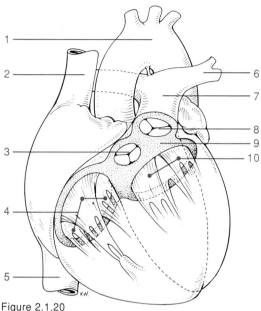

Figure 2.1.20
Annulus fibrosus

1. aortic arch
2. superior vena cava
3. aortic valves
4. tricuspid valve
5. inferior vena cava
6. left pulmonary artery
7. pulmonary trunk
8. pulmonary valves
9. fibrous tissue of annulus fibrosus
10. bicuspid valve

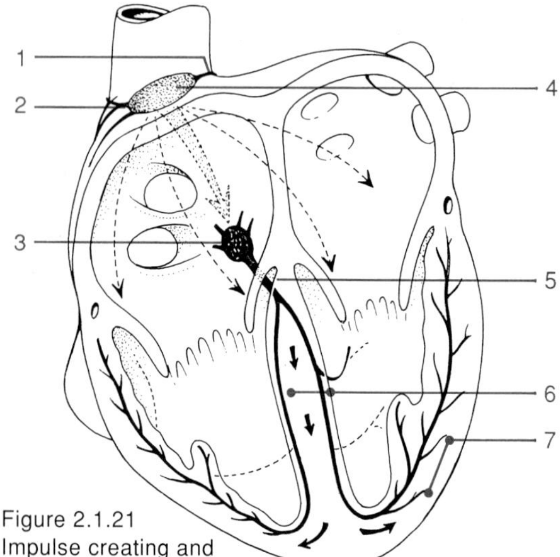

Figure 2.1.21
Impulse creating and
conducting tissue of
the heart

1. vagus nerve (–)
2. accelerator nerves (+)
3. atrioventricular node
4. sinoatrial node
5. atrioventricular bundle (Bundle of His)
6. bundle branches
7. impulse conducting fibres (Purkinje fibres)

The direction of the blood flow can be ascertained from the direction of the valves.

The atrioventricular valve between the right atrium and the right ventricle has three cusps of fibrous tissue. It is called the *tricuspid valve*. The atrioventricular valve between the left atrium and the left ventricle has two cusps of fibrous tissue. This is the *bicuspid valve*. This valve is also called the *mitral valve*, as it looks like the inside of a mitre.

The cusps of the atrioventricular valves are connected to the papillary muscles previously mentioned, by thin tendinous cords (Fig. 2.1.18).

The membranes of the arterial valves are in the shape of a half moon. For this reason they are called *semilunar valves*. The semilunar valves between the right ventricle and the *pulmonary artery* are together called pulmonary valve. Those between the left ventricle and the aorta are called *aortic valves*. The pulmonary trunk and the aorta (Fig. 2.1.20) arise from the annulus fibrosus and pass in front of the atrial septum, so they do not cut through the muscular tissue of the heart.

Impulse creating and conducting tissue

At two sites in the heart there is a small network of specialized fibres in between the 'ordinary' heart muscle fibres. They are the *impulse-creating fibres* of the *sinoatrial node* and the *atrioventricular node* (Fig. 2.1.21).

The sinoatrial node is located between the terminals of the inferior vena cava and the superior vena cava. The node generates a number of stimuli per minute – the *sinus rhythm*. When a person is at rest the sinus rhythm is about 70. A branch of the vagus nerve, part of the *parasympathetic* system and a number of nerves of the *sympathetic* system, the accelerator nerves terminate in the sinoatrial node. The vagus nerve has a decelerating influence on the sinus rhythm. The accelerator nerves, on the other hand, have an accelerating influence.

The atrioventricular node is also located in the wall of the right atrium, near the intersection of the annulus fibrosus and the atrial septum. The atrioventricular node generates, without any other influence, 50 impulses per minute – the *atrioventricular rhythm*.

Apart from this impulse-generating tissue there is also *impulse-conducting tissue* (Fig. 2.1.21), consisting of *conduction (Purkinje) fibres*. These fibres stand midway between heart muscle fibres and nerve fibres. A bundle of fibres goes from the atrioventricular node through the atrial septum and then downward into the ventricular septum: the *atrioventricular bundle (bundle of His)*. This separates into a left and a right bundle. Both limbs pass down

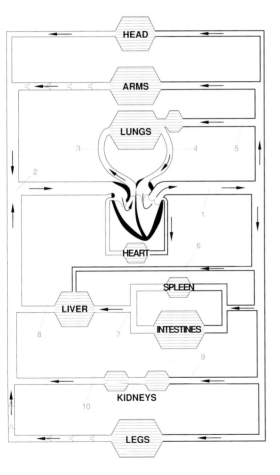

Figure 2.1.22
Diagram of circulation

1. aorta	6. hepatic artery
2. venae cavae	7. portal vein
3. pulmonary arteries	8. hepatic vein
4. pulmonary veins	9. renal artery
5. bronchial artery	10. renal vein

to the apex of the heart, turn around and ascend in the outside walls of the ventricular myocardium. The bundles branch out more and more, so that eventually the whole myocardium wall of the ventricles, papillary muscles included, is reached. Finally, the separate impulse conducting fibres assimilate with 'ordinary' myocardium fibres. The myocardium fibres have, just like the sinuatrial node and the atrioventricular node, a capability of generating impulses of their own. This *ventricular myocardium rhythm* is 40 per minute.

Normally, neither the atrioventricular node nor the ventricular myocardium ever reach their own rhythms. The sinoatrial node spreads its impulse over the entire atrial myocardium. The atrioventricular node also receives this impulse 70 times a minute, slows it down about 0.1 second and then passes it on to the ventricular myocardium via the impulse conducting fibres. Thus, the sinus rhythm accelerates the atrioventricular rhythm (from 50 to 70) and then the ventricular myocardium rhythm is accelerated (from 40 to 70). Therefore, the sinoatrial node is also called the *pacemaker*.

5.3. Function

When heart muscle tissue contracts, the heart chambers concerned become smaller; as a result, the pressure in the cavities grows ($P \times V = C$) and the blood is pushed into the direction where the pressure is lower. The heart valves ensure absolute *one-way traffic*.

With its left ventricle (Fig. 2.1.22), the heart pushes the blood, under high pressure, through the arteries of the circulatory system of the body (greater circulation) to the capillary networks in the periphery of the body. The blood returns to the heart via the veins and enters the right atrium. The right atrium pumps the blood into the right ventricle. From the right ventricle the blood is pumped into the capillary networks of the lungs via the arteries of the pulmonary circuit (lesser circulation). The blood returns to the heart via the pulmonary veins and enters the left atrium. The left atrium pushes the blood into the left ventricle. The circle is complete.

It should be realized that both atria are active *at the same time* and that later both ventricles are

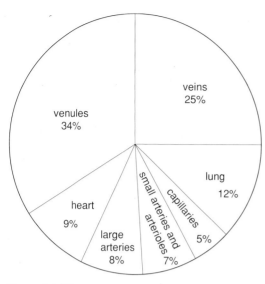

Figure 2.1.23
Average blood volume
distribution during rest

greater circulation (Fig. 2.1.23): 10%–20% of the total volume is in the pulmonary circuit. This is due to the fact that both the *length* and the *diameter* of the vessels are smaller.

Action – relaxation cycle

When at rest, the heart beats about 70 times per minute under the influence of sinus rhythm. Thereby an *action or contraction phase* (systole) is always followed by a *rest or relaxation phase* (diastole) (Fig. 2.1.24). In this alternation of systole and diastole a subdivision should be made (Fig. 2.1.25). The relaxation phase begins at the end of the ventricular contraction (t = 0 second); the atria are relaxed, the ventricles are relaxed. The blood flows into the atria from the periphery and then into the ventricles. This is the *passive filling phase* of the ventricles. The atrioventricular valves are open, the semi-lunar valves are closed (Fig. 2.1.26). At the end of the heart relaxation phase (t = 0.4 second), both atria as well as both ventricles are filled with blood.

The action phase begins with atrial contraction (t = 0.4 second). Both atria wholly contract under the influence of the impulse from the sinoatrial node, which has earlier spread over the entire atrial myocardium. The sinus impulse cannot reach the ventricular myocardium, as the atrium muscle and the ventricular muscle are totally separated 'electrically' by the annulus

active *at the same time*. The two ventricular chambers have the same capacities (volumes). Consequently, both ventricles pump the *same amounts* of blood into the corresponding arteries. Should the right ventricle, for instance, pump out more blood per minute than the left ventricle, the blood would pile up in the pulmonary circuit. The total average *blood flow rate* is also equal in both circulations.

The *total amount* of blood in the lesser circulation, however, is smaller than that in the

Figure 2.1.24
Heart cycle

a. Systole
b. Diastole

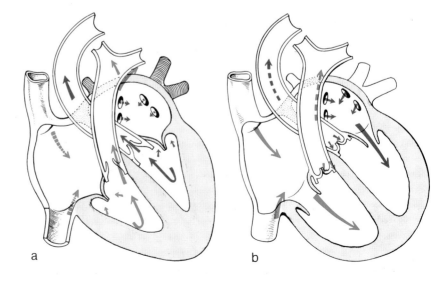

a   b

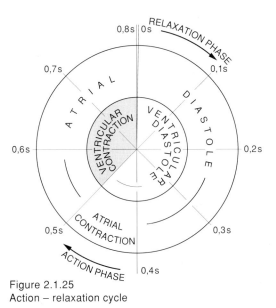

Figure 2.1.25
Action – relaxation cycle

the influence of the blood pressure in the large arteries. It is the end of one cycle and the beginning of the next one.

It can be deduced from the action–relaxation cycle that, although the heart seems to be working continuously, the rest phase is always longer than the action phase for each part of the heart (Fig. 2.1.25). In this way the heart gets sufficient rest. Moreover, during the relaxation phase, blood can flow through the heart, as during the action phase the capillaries are pressed shut.

Heart sounds

The closing of the heart valves can be heard by means of auscultation (for instance with a stethoscope or just by keeping an ear on someone's chest, left of the breastbone). The first heart sound is caused by the closing of the atrioventricular valves, the second one by the closing of the semilunar valves. The sequence of 'lub-dub'-rest, 'lub-dub'-rest, etc. can easily be heard.

Blood pressure in the heart and major arteries

Blood pressure is the pressure to which the blood in the vessels (of the heart) is subject. In science and technology the pressure must always be given in *pascal* units (Pa) or multiples, for instance *kiloPascal* (kPa), or *bar* (1 bar = 100 kPa). The use of other units of pressure, such as atmosphere (atm), water column meter (mH2O), millimeter of mercury (mmHg), has not been in common usage internationally for some years. An exception, however, has been made for the medical world. The *mmHg* unit may be used to give the pressure of the blood and other body fluids. In this book all pressures are given in mmHg, followed by the Pa value in brackets, using the conversion factor 1 mmHg = 133 Pa = 0.133 kPa. In most cases this will be rounded to the nearest decimal.

Blood pressure is not the same in all places. The value measured depends on the site being measured. There is atrial, ventricular, arterial, capillary and venous pressure.

fibrosus. (Connective tissue does not conduct impulses.)

The atrial contraction adds 10% extra to the ventricle volume. This is the *active filling phase*. In this phase the ventricular wall is stretched. The end of the atrial contraction coincides with the beginning of the ventricular contraction (t = 0.5 second). The atrioventricular node has slowed down the sinus impulse by about 0.1 second and thereafter distributed it over the whole ventricular myocardium via the conduction myofibres (Purkinje fibres).

The ventricular systole is characterized by strong contraction of the whole ventricular muscle. The blood in the ventricles becomes subject to high pressure. The atrioventricular valves slam shut. As soon as the pressure in the ventricles is higher than in the aorta, and higher than in the pulmonary trunk, the semilunar valves are pushed open. The ventricular volume decreases, whereby the annulus fibrosus is drawn downwards. Due to the high pressure, the atrioventricular valves could open in the direction of the atria, should the attached tendinous cords not prevent them from doing so. The tendinous cords are stretched by the papillary muscles contracting together with the ventricular myocardium. The ventricular contraction ends at t = 0.8 second. The ventricles are empty, the semilunar valves close under

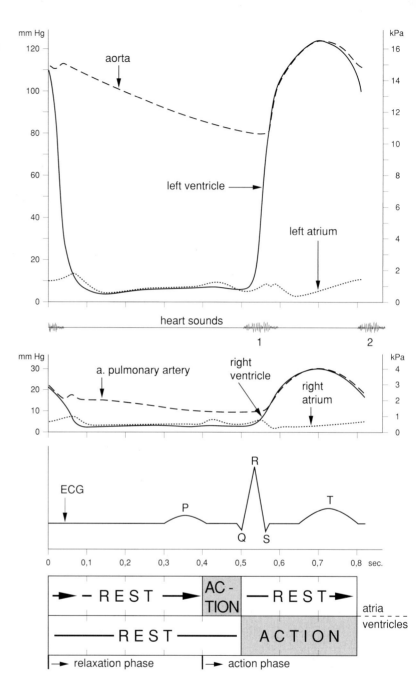

Figure 2.1.26
Diagram of the pressure in
the heart cavities and the
major arteries; the heart
sounds and the ECG

At the beginning of the relaxation phase (t = 0 second), the ventricles are virtually empty and the pressure in the atria drops considerably because of the opening of atrioventricular valves. Following this, the pressure in the atria gradually increases until the beginning of the atrial contraction (t = 0.4 second). The atrial contrac-

tion (systole) is characterized by a greater increase in pressure. Together with the beginning of atrial diastole (t = 0.5 second) the pressure in the atria goes down, to rise again after the atrioventricular valves have been closed. The increased pressure is caused by the blood carried from the periphery and also by the decrease in volume of the atria effected by the billowing of the annulus fibrosus with the atrio-ventricular valves due to ventricular pressure.

After this increase in pressure a decrease in pressure can be seen. The ventricles make their volumes smaller and push the blood into the major arteries. At the same time they draw the annulus fibrosus downwards, as a result of which the atrial volume grows and the pressure goes down (t = 0.6 second). Until the end of the ventricular contraction the atrial pressure gradually rises because of the supply of blood from the periphery.

During the relaxation phase (t = 0 second to t = 0.4 second) the atrioventricular valves are open: atrium and ventricle together form one space. In this phase the pressure course in a ventricle is similar to that of the corresponding atrium (Fig. 2.1.26).

At the beginning of the ventricular contraction the ventricular pressure rises fast. The atrioventricular valves have been closed, the semilunar valves are not yet open, as the pressure in the arteries is still higher than the pressure built up in the corresponding ventricles. This phase of rapid increase in pressure, with closed valves, is called the *isovolumetric phase*.

As soon as the ventricular pressures become higher than the corresponding arterial pressures (Fig. 2.1.26), the semilunar valves open and the blood is pushed out of the ventricles. This phase is called the *ejection phase*. This lasts until about t = 0.7 seconds. The ejection phase is followed by the *relaxation phase* (until t = 0.8 second). In this phase the blood pressure in the ventricles becomes lower than in the corresponding arteries and as a result the semilunar valves close. During the ejection phase the ventricular pressures initially rise (until 0.7 second), but in the relaxation phase thereafter they rapidly go down to their lowest levels (t = 0.1 second).

During the relaxation phase and atrial contraction, the blood pressure values in the aorta and pulmonary artery are always considerably

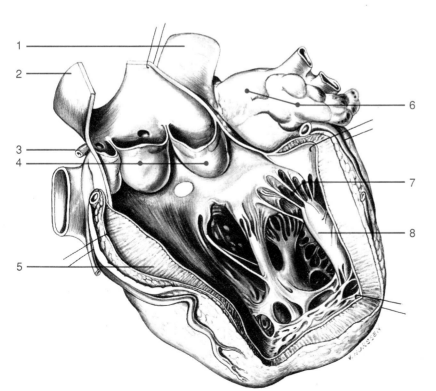

Figure 2.1.27
Aortic valve

1. pulmonary trunk
2. aorta
3. coronary artery
4. valves
5. ventricular wall
6. left atrium
7. tendinous cords
8. papillary muscle

higher than in the corresponding ventricles, as these large arteries do not become 'empty' like the ventricles.

During the ejection phase the blood pressure rates of the aorta and the pulmonary artery rise similarly to those in the ventricles concerned. At the same time the arteries are stretched. At the end of the ejection phase the outflow of blood stops. As the ventricles relax, the pressure falls. Immediately the blood wants to flow back, out of the arteries, but it 'bumps' into the semilunar valves, as a result of which these are filled with blood and the cusp edges are pressed together (Fig. 2.1.27); the semilunar valves are closed. As the stretched arteries

spring back to 'rest', a short-lasting increase in pressure can be seen after the semilunar valves have closed.

When the blood pressure curve of the left ventricle is compared to that of the right ventricle, differences in rates are evident. In the left ventricle we record (systolic/diastolic pressure): 120/0 mmHg (16/0 kPa), in the right ventricle 30/0 mmHg 4/0 kPa). This corresponds with the differences, discussed earlier, in the thickness of the myocardium between the left ventricle and the right ventricle. Functionally, the difference in pressure is logical; the resistance met by the blood is much weaker in the lesser circulation than in the greater circulation.

Figure 2.1.28
Electrocardiogram

a. Standard electrocardiogram
b. The relation between the ECG-curve and the course of the impulse through the heart

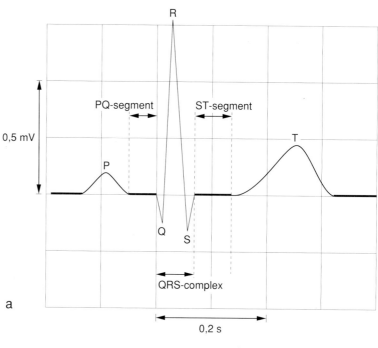

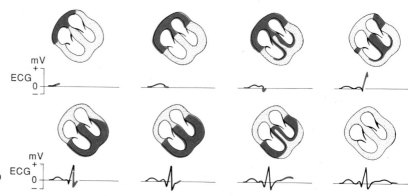

Electrocardiogram

There is a cycle of electrical potential in the heart which can be attributed to the activities of the impulse generating and the impulse conducting tissues. These electrical phenomena can be made visible on paper or on an oscilloscope by means of electrodes and amplifiers (Fig. 1.1.4e). The graphic recording is called an *electro-cardiogram (ECG)*. It gives important information about the electrical activity of the heart and thus, indirectly, about the functioning of the heart. The graphic recording of the ECG depends on the sites of the electrodes on the body surface.

The standard ECG is recorded in Figure 2.1.28:
- The *P-wave* corresponds with the conduction of the impulse in the myocardium;
- the *PQ-segment* records the slowed-down impulse conduction in the atrioventricular node.
- the *QRS-complex* represents the impulse intruding into the ventricular myocardium. Firstly, the bundle branches are reached, via the atrioventricular bundle (Q), then the impulse conducting fibres (R), and after that the impulse goes upward into the whole ventricular myocardium (S) to the annulus fibrosus. During the QRS-complex the atrial myocardium is in the (electrical) rest situation;
- the *ST-segment* records the disappearing of the ventricular impulse situation;
- the *T-wave* corresponds with the return to the (electrical) rest situation of the ventricular myocardium.

The ECG of Figure 2.1.26 has the same time scale as the records of the pressure course. It is evident that the electrical phenomena precede the mechanical phenomena.

Faster/slower

The heart rate is, of course, not always 70 beats per minute. 'Pumping according to need' could be the description applicable to the heart in exertion. The heart rhythm also changes when a person is excited, scared or angry.

Adaptation of the sinus rhythm is effected mainly by the sinoatrial node being influenced

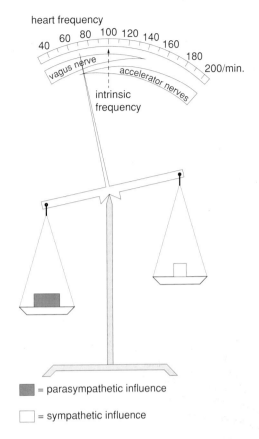

= parasympathetic influence

= sympathetic influence

Figure 2.1.29
Influence on sinus rhythm

via nerves which terminate there and belong to the autonomic nervous system. The vagus nerve (parasympathetic system) has a restraining influence; the accelerator nerves (sympathetic system) have a stimulating influence (Fig. 2.1.29). The sinoatrial node would generate 100 impulses per minute (intrinsic frequency) if it were not influenced by vegetative impulses. The normal heart rhythm (about 70 beats per minute) is effected by the predominating (restraining) influence of the vagus nerve. Acceleration of the heart rhythm appears to be a well-balanced harmony of a decrease in activity of the vagus and an increase in activity of the accelerator nerves. When the heart rhythm slows down, the opposite is true.

Apart from this *neural influence* on the heart rhythm, there is the *hormonal influence* of the hormone *adrenaline*. This hormone of the adrenal medulla is produced mainly during physical and/or mental strain. It influences the heart, via the blood, in much the same way as do the accelerator nerves.

Cardiac output

The capacity of a pump is evaluated mainly by the amount of fluid ejected per unit of time. Something similar is true for the heart.

The heart capacity is represented by:

$$HMV = f_H \times V_S,$$

in which HMV = *Heart/minute-volume* (cardiac output)
   f_H = *heart frequency*
   V_S = ejection rate

The cardiac output is the volume of blood ejected by the heart per ventricle in a unit of time. This volume (in litres) is determined by multiplying the number of ejection movements by the volume of blood which is pushed out of a ventricle per beat.
In an average human during rest the heart rate is 70–75 per minute and the ventricular output is 70 millilitres, so the cardiac output is 5 litres.

During gentle exercise the cardiac output can be a few times larger: the heart rate increases up to, for instance, 130 beats per minute. The ventricular output is also variable. When the supply of blood to the heart is large, the ventricles can be filled up to 120 ml because of the elasticity of the ventricular myocardium. The cardiac output is then about 15.5 litres. It is therefore obvious that a stretched myocardium is able to contract more strongly.

The cardiac output of the left ventricle is always distributed over the organs according to two principles: the '*vital importance*' *of the organs* and the *demand*.
The brain, with its vital functions, always receives preferential treatment in the blood supply, as, after a short time, lack of oxygen is fatal

to neurones. The heart itself is also given preferential treatment when supplied with blood (about 5% of cardiac output). A metabolic disorder in the heart, caused by a disrupted blood supply, immediately leads to serious circulatory disorders throughout the body.
The kidneys receive an average of 20% of the cardiac output. This relatively large share is not connected with their need for oxygen, but with the secreting function of the kidneys and their supervision of homeostasis.

During physical exercise not only does cardiac output increase, but it is also distributed over the organs in a different way. The skeletal muscles take more than 60% of the cardiac output.
During digestion the digestive system gets an increased percentage of the cardiac output.
The share of the cardiac output supplied to the skin, depends mostly on the temperature regulating function of the skin.

Once again it is stressed that the cardiac output applies to the cardiac output per ventricle. The cardiac output of the right ventricle is equal to that of the left ventricle, as the ventricular muscle as a whole contracts (in other words the same frequency rate holds good for the right as for the left) and the ventricular volumes are equal.
The total cardiac output of the right ventricle (de-oxygenated) always goes to the lungs and the same volume (oxygenated) arrives at the left atrium.

5.4. Heart sac

The heart sac (Fig. 2.1.30) belongs to the serous membranes. It consists of a visceral or inner sac, the *epicardium*, and a parietal or outer sac, the *pericardium*. The term pericardium may also be used for the heart as a whole. The potential space between the epicardium and the pericardium is called the *pericardial cavity*.
The epicardium is strongly connected with the myocardium and consists of mesothelium on a thin layer of elastic fibrous tissue. A relatively large amount of fat tissue can develop in the epicardium.

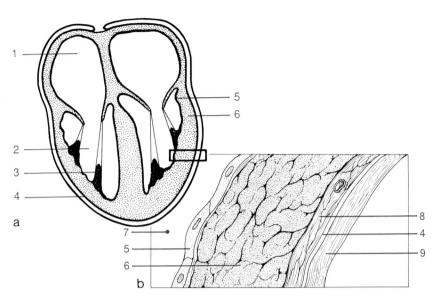

Figure 2.1.30

a. Heart wall and heart
   sac
b. Detail

1. *right atrium*
2. *right ventricle*
3. *papillary muscle
   with tendinous
   cords*
4. *pericardial cavity*
5. *endocardium*
6. *myocardium*
7. *ventricular lumen
   (cavity)*
8. *epicardium*
9. *pericardium*

The hole where the blood vessels enter the heart, with its edges from epicardium to pericardium, is the 'hilum' of the heart.

The walls of epicardium and pericardium which face each other are very smooth. The pericardial cavity contains serous fluid. Because of this construction, friction caused by the contraction of the heart is totally absorbed in the serous cavity. The pericardium adheres to the tendon of the diaphragm and to the outer sac of the serous membrane of each lung.

5.5. Vascularization

The endocardium belongs to the epithelial tissues and has no blood vessels. The endo-thelium is supplied mainly via diffusion from passing blood.

The myocardium has a very intensive metab-olism. The oxygen needed, nutrients, the waste matters produced, including carbon dioxide, are conveyed by the blood.

So far a double circulation (greater and lesser circulation) has been discussed. The circulation of blood to the heart is part of the greater cir-culation, just like that of other organs. There is, however, an important factor; the heart cir-culation begins with branches of the aorta: *the arteriae coronariae* (coronary arteries) (Figs. 2.1.31 and 2.1.27), but terminates direct

in the right atrium: *the coronary sinus* (see Fig. 2.1.19b).

The two coronary arteries originate in the space behind two of the three cusps of the aortic valve (see Fig. 2.1.27). Functionally, this construction is very effective. When the ven-tricular myocardium is active, the muscular tissue of the heart shuts all of the capillary network, so no blood can reach the heart muscle. Fortunately, the coronary arteries do not receive any blood either, as during the same ventricular contraction the cusps of the aortic valve are pushed against the inside of the aorta and thus the entrance to the coronary arteries is blocked.

When the myocardium relaxes, all capillary networks open; the blood in the aorta rushes in the direction of the aortic valve as a result of which the edges of the semilunar valves are pressed together. Under high pressure the blood is pushed into the two coronary arteries.

The left coronary artery has to supply the largest area: the outer wall of the left ventricle, the largest part of the ventricular septum and the left atrium. The right coronary artery is smaller and supplies the outer wall of the right ventricle, a small part of the ventricular septum and the right atrium. In the myocardium, the arteries branch into arteriolae and then into

Figure 2.1.31
Blood supply of the heart

1. *right coronary artery*
2. *coronary vein*
3. *terminal branch of the left coronary artery*
4. *branches in the ventricular myocardium*

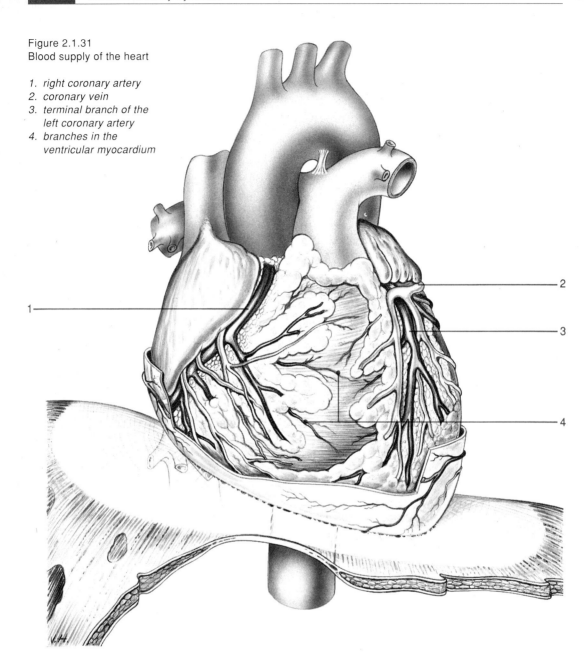

6. Blood vessels

extensive capillary networks. The m.i. of the myocardial fibres is liberally supplied with blood. The capillary networks converge into venules (minute vessels) and from these the blood is collected in a number of coronary veins. The veins join to form a large vein, situated in the posterior part of the atrioventricular groove of the heart, the *coronary sinus*, which terminates directly in the right atrium (see Fig. 2.1.19b).

In the greater circulation and the heart circulation, oxygenated blood is found in the arteries and de-oxygenated blood in the veins. In the lesser circulation the opposite is true. *Arterial blood* usually means *oxygenated blood* and *venous blood* means *de-oxygenated blood*: these concepts, therefore, are not correct for the lesser circulation.

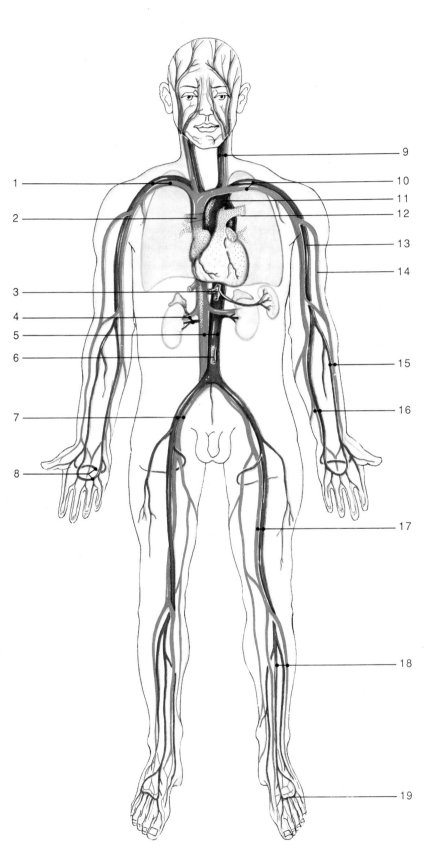

Figure 2.1.32
Topography of the major
blood vessels

1. right subclavian
   artery and vein
2. superior vena cava
3. celiac trunk (artery)
4. renal artery and
   vein
5. abdominal aorta
   and inferior vena
   cava
6. inferior mesenteric
   artery
7. iliac artery and vein
8. palmar arcades
9. common carotid
   artery and internal
   jugular vein with
   branches to head
   and face
10. subclavian artery
    and vein
11. aortic arch
12. pulmonary artery
13. axillary artery
14. brachial artery
15. radial artery and
    vein
16. ulnar artery and
    vein
17. femoral artery and
    vein
18. tibial artery and vein
19. arterial arcades in
    the foot

6.1. Topography

Hundreds of different arteries and veins can be named. The smallest vessels however, cannot be given names because of their quantities and their varieties.

The topography (Figure 2.1.32), and the relevant vascularization of the major vessels will be discussed. Quite often we see an artery, a vein and a nerve next to each other, surrounded by a protecting coat of elastic fibrous tissue. This is called a *vaso-neural cord* (Fig. 2.1.43).

Arteries

Two major arteries leave from the heart: the *pulmonary artery* and the *aorta*. The pulmonary artery arches across the aorta and divides directly below the aortic arch into two short but wide pulmonary veins, each going to a separate lung. For descriptive purposes the aorta is divided into different parts.
The first part of the aorta goes upwards (ascending aorta), the next part forms an arch and runs in a dorsal direction. Then the aorta runs ventrally to the vertebral column behind the heart in the direction of the diaphragm (descending aorta). The aortic arch arches across the lower part of the windpipe (trachea), the descending aorta runs dorsally from the oesophagus (Fig. 2.1.33). There are sensors in the aortic arch which are sensitive to changes in the pH and pO_2 of the passing blood.
All sections of the aorta in the thoracic cavity are called the thoracic aorta; after the aorta has passed through the muscular section of the diaphragm it is called the *abdominal aorta*.

From the aortic arch a number of arteries arise which supply the head and arms with blood (Figs. 2.1.33 and 2.1.32). The first artery soon divides into two branches: the *right subclavian artery* and the *right common carotid artery*. The *left subclavian artery* and the *left common carotid artery* each rise separately from the aortic arch. In the common carotid arteries are the same kind of chemoreceptors, sensitive to changes in pH and pO_2, as those found in the aortic arch.
The subclavian arteries each arch across the first

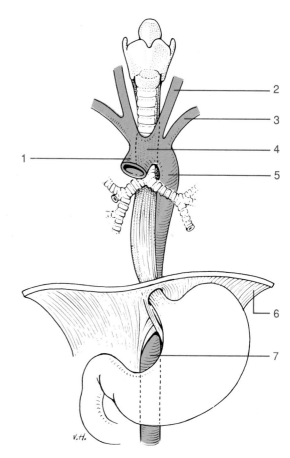

Figure 2.1.33
Location and
course of aorta

1. *ascending aorta*
2. *common carotid artery (left)*
3. *subclavian artery (left)*
4. *aortic arch*
5. *descending aorta*
6. *diaphragm*
7. *abdominal aorta*

ribs, extend via the armpits (axillae) and are then called *axillary arteries* (Fig. 2.1.32).
They then run along the medial sides of the upper arms and are called *brachial arteries*. Just distally to the elbows these arteries bifurcate into the *radial* (side of the thumbs) and *ulnar* (side of the little fingers) *arteries*.
These arteries continue into the palms of the hand, where they are connected by arterial links, the *palmar arcades* (Fig. 2.1.32). Branches of these arcades are the finger arteries.

The carotid arteries rise next to the windpipe and each divide into the *external carotid artery* and

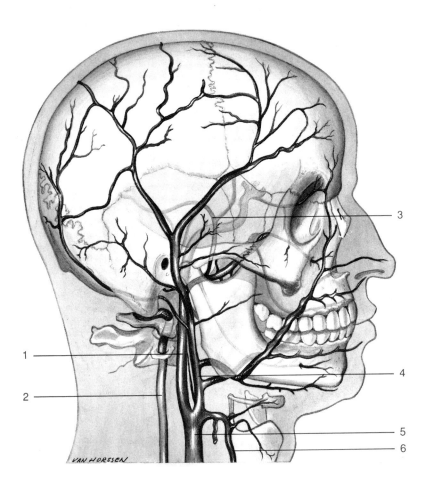

Figure 2.1.34
Arteries of the neck and
head

1. *internal carotid artery
   (to the brain)*
2. *vertebral artery (to the
   brain)*
3. *middle meningal artery
   with branch for dura
   mater*
4. *external carotid artery
   with branches to skull
   and face*

VAN HORSSEN

internal carotid artery. The two external carotid arteries provide the blood supply to the thyroid gland, the face and the lateral side of the head through branches that run between the skull and the dura mater (see Fig. 2.7.48).

The two internal carotid arteries (Figs. 2.1.34 and 2.7.45) each enter the skull via an opening in its base and divide into branches for the blood supply to the eyes and, mainly, the brain.

From each subclavian artery rises a branch which also supplies the brain with blood. These are the vertebral arteries that go upwards next to the vertebral column and enter the skull through the foramen magnum (Fig. 2.1.34).

The other arteries which arise from the aorta are divided into two groups: the *parietal arteries* and the *visceral arteries*.

The main arteries in the first group are the *intercostal arteries* in the thoracic cavity and the *lumbar arteries* in the lumbar region.

The above-mentioned *carotid arteries*, which rise from the base of the ascending aorta and branch towards the oesophagus, belong to the second group, as does the *bronchial artery*, which goes to the windpipe and each lung.

Directly below the diaphragm the left and right renal arteries rise from the aorta (Fig. 2.1.35). These renal arteries have a relatively large lumen, which corresponds with the function of the kidneys.

At the front of the aorta are, from top to bottom: the *celiac trunk*, with branches to the stomach, spleen, liver, pancreas and duodenum; the *superior mesenteric artery*, with branches to the ilium and the first half of the colon; and the

Figure 2.1.35
Blood supply of the
abdominal organs

a. Branches of the aorta

    1. celiac trunk
    2. superior mesenteric
       artery
    3. diaphragm
    4. splenic artery
    5. renal artery
    6. inferior mesenteric
       artery
    7. common iliac artery

b. Blood supply of stomach,
   liver, spleen, pancreas
   and duodenum

    1. oesophagus
    2. liver
    3. hepatic artery
    4. duodenum
    5. superior mesenteric
       artery
    6. abdominal aorta
    7. aorta
    8. projected line of
       diaphragm
    9. spleen
   10. pancreas
   11. stomach
   12. omentum

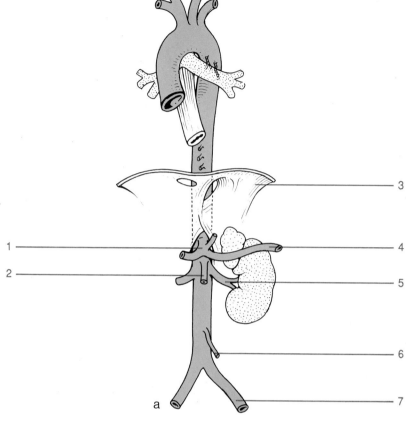

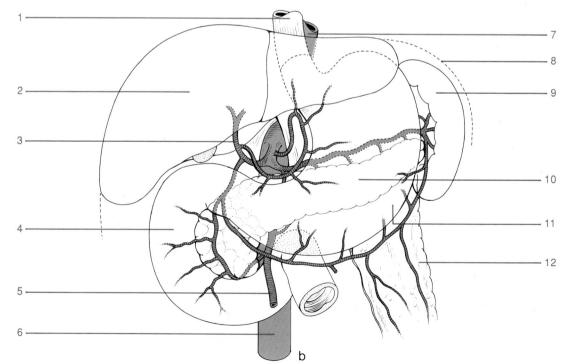

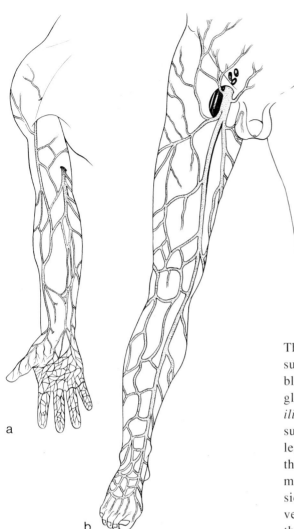

a

b

Figure 2.1.36
Subcutaneous venous network

a. Of the arm
b. Of the leg

inferior mesenteric artery with branches to the rest of the colon.

Somewhat higher than the inferior mesenteric artery, two smaller arteries rise which go to the ovaries (in women) or to the testes (in men). That there is such a large distance between the place of origin of these arteries and the organs they supply corresponds with their embryological development.
Distally, at the middle lumbar vertebra, the aorta bifurcates into the iliac arteries. Each iliac artery divides into an internal and an external branch.

The posterior branch (*internal iliac artery*) supplies the organs in the pelvis (especially the bladder) and also the muscles and skin in the gluteal region. The anterior branch (*external iliac artery*) leaves the pelvic cavity via the superior aperture and continues into the upper leg as the *femoral artery*. At the back of the knee this artery bifurcates into the *tibial artery* on the medial side and the *fibular artery* on the lateral side. This bifurcation is analogous to that of the vessels in the arm. There are arterial arcades in the foot with branches to the toes.

Arterioles, capillary networks, venules
When the arteries reach the organs to be supplied, they branch into arterioles that penetrate into the organ tissue and then branch into capillary networks. The homeostasis of the m.i. of the organ tissue is taken care of by the capillary networks. The capillary networks give their blood to the venules leaving the organ tissue.

Veins
The venules (or veinlets) converge into veins which, generally, leave the organ at the same place where the arteries have entered, especially when there is a *hilum*. Usually there are more venules then arterioles; moreover, they run

Figure 2.1.37
Portal vein

1. inferior vena cava
2. gall-bladder
3. liver
4. portal vein
5. superior mesenteric vein
6. colon
7. hepatic veins
8. stomach
9. spleen
10. pancreas
11. inferior mesenteric vein
12. ileum
13. rectum

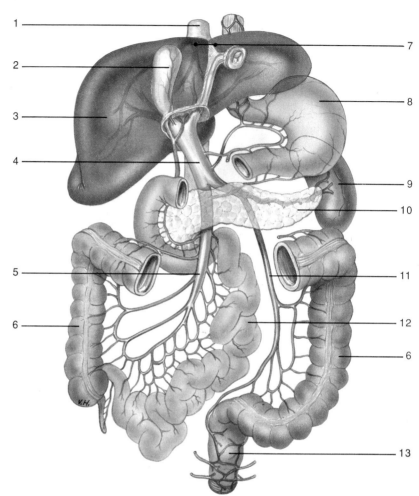

more at the surface in the limbs and wall of the torso (Fig. 2.1.36).

The topography of the major veins and their terminology are comparable to those of the arteries (see Fig. 2.1.31). The veins carry blood to the heart, apart from the veins from the largest part of the intestines and a number of other abdominal organs. These converge into the *portal vein*, which terminates in the *liver* (Fig. 2.1.37). Its importance will be discussed later, in Module 2, Chapter 2.

The *inferior vena cava* carries blood from the legs, pelvis and abdominal organs to the heart. The *superior vena cava* drains the arms, the head, neck and the organs in the chest. The superior vena cava, the inferior vena cava and

also the *coronary sinus* terminate in the right atrium.

The blood from the lungs returns to the heart via *two pulmonary veins per lung*, so four pulmonary veins terminate in the left atrium (see Fig. 2.1.19b).

6.2. Structure and function

The walls of blood vessels, except those of the capillaries, consist, principally, of three layers (Fig. 2.1.38a):

- the *tunica interna* (or intima), the innermost coat, comprising a smooth layer of endothelial cells and a thin outer layer of elastic fibres;
- the *tunica media*, consisting of connective

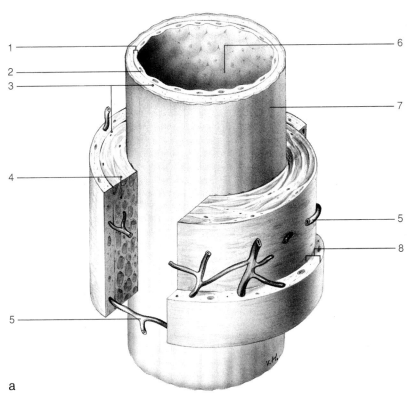

Figure 2.1.38
Structure of vascular wall

a. General structure:
   interna, media,
   adventitia

   1. tunica interna
   2. endothelium
   3. tunica media
   4. smooth muscle fibre
   5. lesser blood
      vessels
   6. lumen
   7. elastic connective
      tissue
   8. tunica adventitia

b. I   Elastic artery
   II Muscular artery
   III Arteriole
   IV Capillary
   V   Venule
   VI Vein

   1. connective tissue
   2. muscle layer
   3. endothelium
   4. lumen
   5. vein
   6. pericyte
      (perivascular cell)

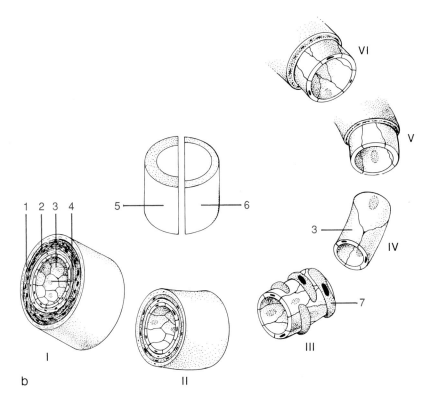

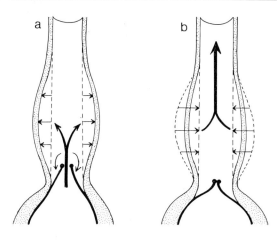

Figure 2.1.39
Air chamber function

a. The elastic arteries stretch
b. The elastic arteries recoil

tissue and smooth muscle tissue; different types of vessels can be distinguished, dependent on the structural circumstances in the tunica media;
– the *tunica adventitia* (or externa), relatively dense connective tissue.

Arteries and arterioles

The tunica media of the major arteries comprises mainly connective tissue and elastic fibres. All through the elastic connective tissue run smooth muscle fibres which are able to regulate the tension in the tunica media. They are called *elastic arteries* (Fig. 2.2.38b (I)). This elasticity is very important, as the heart intermittently pumps a large volume of blood into the arteries.

The blood, however, is conducted through the capillary networks to the venous system in a steady stream. Thereby the flow of blood experiences a resistance, not only because of the viscosity of the blood, but also because of the friction against the vascular walls. This is called *peripheral resistance.*

The blood, ejected by the left ventricle at each beat, must be received in the arterial system. This is possible due to the elasticity of the arterial vascular walls. The stretched arteries spring back because of their elasticity and thus

promote the steady flow of blood in the capillary networks. This intermittent stretching and springing back of the elastic arteries is called the *air chamber function* (Fig. 2.1.39). The widening of the arteries is a result of the sudden rise in pressure at each beat of the heart which goes as a wave through the arterial system to the periphery: the *pulse wave*. This concept is derived from the 'pulsating' of the arteries. These pulsations can be distinguished wherever arteries can be felt at the surface of the body. The heart rate can be determined by counting the pulsations. This is usually done at the radial artery near the pulse, from which it derives its name. The radial artery is softly pressed on the underlying tissue with the three middle fingers: the 'pulse' is 'felt' or 'counted'. Similarly, the heart rate can also be determined at the arteries in the neck (common carotid artery), at the temple (temporal artery), at the ankle, etc. After exercise the pulse wave which runs through the abdominal aorta causes the conspicuous 'pulsating' of the abdominal wall.

A change in structure of the tunica media can be observed during the gradual transition from major arteries to lesser arteries to arterioles (Fig. 2.1.38b (I and II)). The elastic connective tissue increasingly gives way to smooth muscle tissue. The *muscular arteries* can narrow the lumen of the blood vessel (vasoconstriction) or widen it (vasodilation), whatever the demand of the tissue concerned. They therefore have a *distributing function*. During physical exercise, for instance, the supply of blood to the skeletal muscles will increase, whereas the supply of blood to the intestines will decrease.

The smallest divisions of the arteries, the arterioles, are even more suitable for vasoconstriction or vasodilation because of their structure. There is a single layer of smooth muscle fibres around the tunica interna (Fig. 2.1.38b (III)). With this layer they can almost entirely shut their lumina. The relatively narrow passage formed by the arterioles together, adds to the peripheral resistance mentioned earlier. The arterioles actually determine the size of the peripheral resistance because of the regulating capability of their lumina.

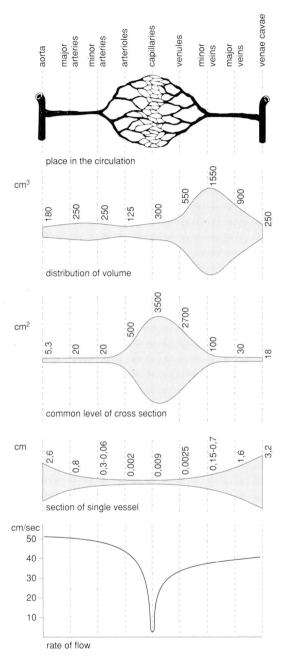

Figure 2.1.40
Combined curves

Capillaries

Capillary networks consist of veinlets without tunica adventitia and tunica media (Fig. 2.1.38b (IV)). The capillary wall is very thin and consists only of tunica interna, bordering endothelial cells coated with a thin fibrous membrane. The capillary wall has the characteristics of a semipermeable membrane and is therefore extremely suitable for the exchange of all kinds of matter with the m.i.

Furthermore, divisions between adjacent endothelial cells add to the exchange possibilities. The capillaries of the brain, however, lack these divisions and consequently they are impermeable to many substances. This is called the blood/brain barrier.

Venules and veins

The capillary networks turn into venules. The venules, in their turn, become veins. During the gradual transition to these kinds of vessels the tunica media appears first, and then, in the major veins, the tunica adventitia appears (Fig. 2.1.38 (V and VI)). In the tunica media the structural emphasis is on connective tissue. There are many collagen fibrils in this tissue. The muscle fibres are arranged more loosely than in the corresponding arteries.

The veins have much thinner walls and larger lumina than the corresponding arteries. This is comparable with the differences in blood pressure between the arteries and the veins. Moreover, the veins in the arms and legs have valves (Fig. 2.1.42). These are, in their structure, similar to the semilunar valves mentioned earlier. The major veins in arms and legs, and also those in the trunk, such as the inferior vena cava, have no valves. There is, however, at the place where the inferior vena cava enters the right atrium, a connective tissue cusp, similar to a valve, which prevents the back flow of blood from the atrium to the inferior vena cava. The valves in the venules and veins aid the transport of blood to the heart.

Figures

The distribution of the blood volume in the circulation is represented in Figure 2.1.23. The proportion in the veins, large compared to that of the arteries and capillaries, is conspicuous. When we limit the picture to the greater circulation (Fig. 2.1.40) and link this to a number of other data, a series of conclusions can be made. In the distribution of volume venules and veins have easily the largest share. These vessels serve as *blood reservoirs*. The curve

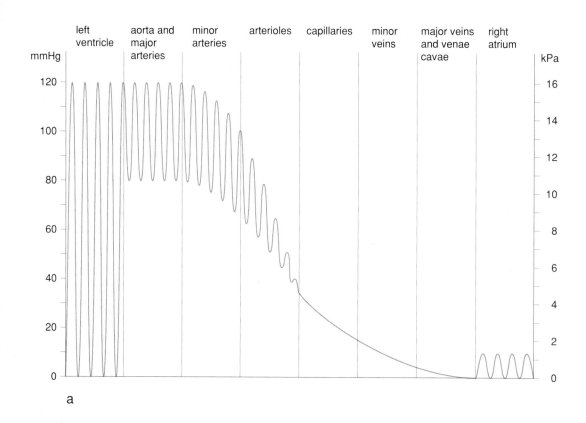

a

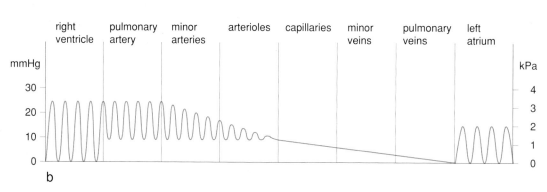

b

Figure 2.1.41
Course of blood pressure

a. Greater circulation
b. Lesser circulation

of the combined openings shows that all capillary networks together have a much larger diameter than the arteries and/or veins. This, combined with the distribution of the blood volume, means that only a few capillary networks 'are open'. The muscular arteries and the arterioles control the relevant distribution, depending on demand.

The flow rate of blood in the different types of vessels corresponds with the joint diameters of the corresponding open vessels. The smaller the joint opening, the greater the flow rate. Veins convey the same volume of blood to the heart as the arteries convey from the heart, but the flow rate in the veins is considerably lower. This is due to the fact that the joint diameter of the veins, compared to that of similar arteries, is larger. It is larger, not only because of the larger amount of veins, but also because of the larger diameters of the separate veins compared to similarly-named arteries.

As to the capillaries, the diameters of separate capillaries are minute, but the joint diameter is very large. This means that there is a large number of capillaries. The opposite holds good for arteries and veins.

The flow rate curve shows that the blood in the capillaries flows very slowly. The joint diameter of the open capillaries must therefore be larger than that of the distributing arteries. The curve of the total blood volume in arteries and capillaries shows smaller differences than the curve of the flow rate in these vessels. The low flow rate in the capillaries gives greater opportunities for exchange in the m.i. There the blood can be compared to a van being slowly driven through a factory, unloading raw materials and loading up waste matter and produce. An increase in the speed of the van would adversely affect the opportunities for loading.

7. Blood pressure

Blood pressure (tension) is the force exerted by the blood against the walls of the blood vessels. The blood pressure ratio is strongly dependent on the position of the blood vessel in the circulatory system (Fig. 2.1.41). Firstly, the pressures in the greater circulation will be discussed, then those of the lesser circulation and finally pressure regulation.

7.1. Greater circulation

Arterial pressure
There are rhythmic changes in the blood pressure curve in the major arteries as a result of the activity of the ventricular myocardium. The peak is reached during the systole, the lowest point during the diastole, so we talk about *systolic* and *diastolic pressure.*

The systolic pressure (upper pressure) has an average of 120 mmHg (16 kPa), the diastolic pressure (lower pressure) has an average of 80 mmHg (10.6 kPa). The difference between peak systolic pressure and the lowest diastolic pressure is called *pulse pressure.* So in a tension of 120/80 mmHg (16/10.6 kPa) the pulse pressure is 40 mmHg (5.4 kPa).

The level of arterial blood pressure is dependent on the filling of the arterial system. It can be compared with the pumping of a tyre: the pressure increases with the amount of air inflated. Arterial blood pressure is influenced mainly by the following three factors:
– *ventricular output*: the more blood that is pushed into the arteries per beat, the more the pressure increases. During exercise, ventricular output increases and arterial pressure rises;
– the *elasticity of the vascular wall*: when the arterial wall is supple it will easily stretch when pressure increases (air chamber function). The arterial vessel volume becomes larger and as a result, arterial pressure rises less than if the vessel were less elastic;
– the *peripheral resistance*: blood flows from a place with high pressure to a place with low pressure. By activities of the distributing arteries and arterioles the arteries are prevented from 'emptying' during diastole. They keep the vessels filled at the same level because of their selective vasoconstriction and vasodilation. Peripheral resistance, and with it blood pressure, decreases when a relatively large number of arterioles are open, and vice versa.

Systolic pressure is mainly dependent on the first two factors, and diastolic pressure on the third factor.

The intermittently changing blood pressure, resulting from the sequence of systole and diastole, is continuously changed into a steady blood pressure, which is reached – thanks to the air chamber function – at the beginning of the capillary networks. Not only the flow rate but also the blood pressure goes down as a result of the increased joint diameter (Fig. 2.1.40).

Capillary pressure
Blood pressure is about 35 mmHg (4.7 kPa) at the arterial side of the capillary networks and about 15 mmHg (2 kPa) at the venous side. It will be explained in Section 9 that these

capillary pressures as they relate to the colloid-osmotic value of the plasma largely determine the rate of exchange of substances between blood and m.i.

Venous pressure

The venous pressure at the end of the capillary networks goes down to less than 5 mmHg (0.7 kPa) in the hollow veins in the rest of the venous part of the greater circulation and is dependent on the extent to which the veins are filled. This *rest pressure*, supported by *gravity*, is still greater than that in the heart, thus allowing venous return. When a person stands, however, this pressure is too little to be able to carry the blood back to the heart, against gravity, from the parts of the body at levels lower than the heart. The *venous return* is stimulated by a number of cooperating factors:

– *valves* in the venules and veins of the arms and legs (Fig. 2.1.42). Due to these valves the blood can flow in only one direction, i.e. towards the heart. The function of the valves can easily be seen, for instance, in the superficial brachial veins. At first the veins are made more visible by creating stasis in the lower arm. This can be done by grasping the upper arm just above the elbow and squeezing. Then the under arm vein is pressed shut by the top of an index finger. At the same time the vein is stroked with the thumb nail of the same hand, in a proximal site. In this way the vein concerned is emptied past the place where there is a valve. When the thumb nail is removed, the blood does not flow back to the index, but to the last valve only. When the index is lifted, the blood flows distally into the empty vessel.

– *heart pump*. During ventricular systole the annulus fibrosus is pushed downwards (see Fig. 2.1.24), whereby the atria are stretched, then the pressure drops, and this stimulates the venous return;

– *respiration pump*. During inhalation the thoracic volume becomes larger; this causes the pressure in the mediastinum to drop. The pressure becomes considerably lower than that outside the thorax and because of the

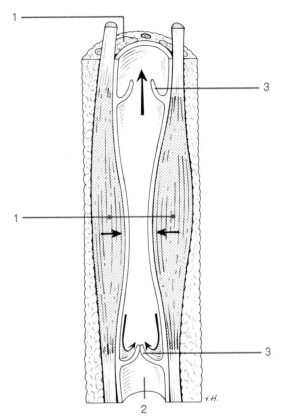

Figure 2.1.42
Muscle pump

1. *muscle*
2. *vein*
3. *valve*

pressure gradient, atrium and hollow veins are opened. The abdominal pressure also rises due to the flattening of the diaphragm, which increases the venous pressure difference between the mediastinal vessels and the vessels in the abdominal cavity even more;

– *muscle pump*. Muscular activity stimulates the venous return in the arms and the legs (Fig. 2.1.42). The veins in or between muscles are pressed shut as the muscles grow thicker when contracting. As there are valves in the veins, the blood can only be pushed in the direction of the heart;

– *arterial pump*. At many places, arteries and veins are next to each other in a vascular (nervous) cord (Fig. 2.1.43). The fibrous tissue tube does not yield to the pressure of

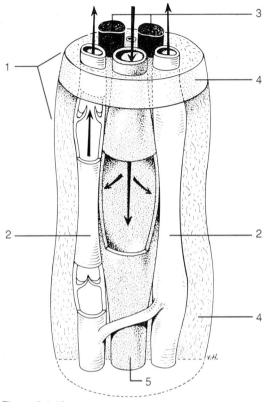

Figure 2.1.43
Arterial pump

1. vasculo-nervous cord
3. vein
3. nerves
4. fibrous tissue
5. artery

the pulse wave. The widening of the artery with each heartbeat narrows the lumen of the vein next to it and as a result the blood flow to the heart is stimulated.

7.2. Lesser circulation

Systolic and diastolic pressures are considerably lower in the lesser circulation than in the greater circulation, viz. 25/10 mmHg (3.3/3.1kPa) (see Fig. 2.1.39). This is due to the peripheral resistance in the lesser circulation being almost negligible. The arterioles and the capillary networks in the lungs are always open. Compared to the left ventricle the right ventricle has a considerably thinner myocardium.

The capillary pressure is more or less equal to the diastolic pressure, 10 mmHg (1.3 kPa). This is an important fact, in relation to the

colloid-osmotic value: see Module 2, Chapter 4, Section 5.

In the pulmonary veins the pressure further reduces to less than 5 mmHg (0.7 kPa).

7.3. Blood pressure regulation

Blood pressure regulation is mainly directed at the preservation of arterial blood pressure. An increase in blood pressure in the arterial system may be possible to allow increased blood flow to active organs.

Arterial pressure is – as discussed earlier – determined by the activity of the heart and the characteristics of the arterial system. In blood pressure regulation, therefore, both an adaptation of the cardiac function and a change in vascular characteristics can play a role. It is a question of both *neuronal* and *hormonal* regulation.

– *Neuronal regulation* is controlled by the *heart regulation centre* and the *vasomotor centre*, which lie in the spinal bulb (medulla oblongata), and are closely connected functionally (see Fig. 2.7.25). From here nerves of the autonomic nervous system stimulate or slow down the heart activity, and peripheral resistance can be regulated by vasoconstriction and vasodilation of the arterioles (vegetative motor system).

In order to be able to regulate blood pressure, the centres mentioned must be informed about changes in blood pressure. This is done by vegetative sensor system cells, sensitive to pressure (baro- or pressure centres), in the wall of the aortic arch and at the site where the left and right common carotid arteries bifurcate into internal and external carotid arteries.

– *Hormonal regulation* takes place by means of the hormones ADH, aldosterone, adrenaline and noradrenaline, renin and the locally effective histamine.

ADH (*vasopressin, antidiuretic hormone*) is produced in the hypophysis and stimulates the return reabsorption of water in the kidneys. This hormone prevents 'loss of water' and has the effect of increasing blood pressure. That is why this antidiuretic hormone is also called *vasopressin*.

Aldosterone is one of the adrenal cortex hormones. It regulates the Na/K-balance. In the kidneys Na^+ is selectively reabsorbed back and K^+ is secreted. The Na^+ ion combines with a larger water coat than the K^+ ion, so more water is held. The volume of blood and thereby the pressure is increased.

Adrenaline and *noradrenaline* are hormones produced by the adrenal medulla. Adrenaline stimulates cardiac activity and produces vasoconstriction in most of the arterioles. In spite of adrenaline causing vasodilation in the skeletal muscles and the muscular tissue of the heart, the final result is an increase in pressure.

Noradrenaline is similar to adrenaline, but it also causes vasoconstriction in the muscular tissue, as a result of which blood pressure rises even more.

Renin is a hormone which is produced in the kidneys. This hormone catalyzes the production of angiotensin in the bloodstream. This substance causes vasoconstriction in the arterioles, but also stimulates the production of aldosterone. Both effects lead to a rise in pressure.

Histamine is produced by damaged tissue cells. Histamine particularly causes vasodilation of the arterioles in the damaged area, resulting in a decrease in blood pressure.

8. Vascular systems

So far, the starting point for all discussion has been the situation in which arteries are successively followed by arterioles, capillaries, venules and veins (Fig. 2.1.44a).

There are, however, a number of different vascular systems:
- *portal circulation*. In this circulation two capillary networks are connected in series; there are two variations:
  - the first capillary network is arterial, the second is 'normal' (Fig. 2.1.44b). This situation is found in the filtration units in the kidneys (see Fig. 2.3.4);
  - the first capillary network is 'normal', the second is venous (Fig. 2.1.44c). This

situation is found in the portal venous system between the intestines and the liver (see Figs. 2.2.34 and 2.2.35), and also in the portal venous system between the

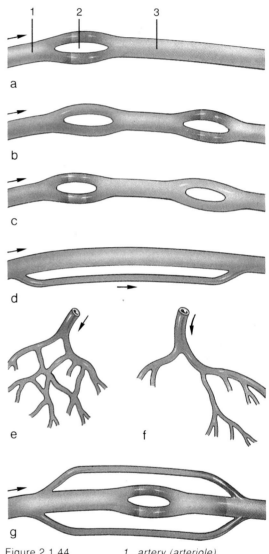

Figure 2.1.44
Vascular systems

1. artery (arteriole)
2. capillary
3. vein (venule)

a. Terminal artery
b. Arterial portal circulation
c. Venous portal circulation
d. Collateral circulation
e. Arterial anastomosis
f. Terminal artery
g. Arterio-venous anastomosis

hypothalamus and the hypophysis (see Fig. 2.6.4c);
- *collateral circulations*, in which blood vessels run parallel to and are connected with certain vessels (Fig. 2.1.44d). Basically, a collateral circulation provides an extra insurance for the transport of blood from or to the tissue concerned. Many collaterals are found, particularly with veins; the azygos vein, for instance, is a collateral of the inferior vena cava. This vein arises from the iliac vein, ascends in the thoracic cavity along the vertebral column and ends in the superior vena cava.
Arteries also have collaterals. A well-known example is the collateral blood supply of the brain (see Fig. 2.7.45);
- *anastomoses*, junctures between vessels. There are three types:
  • the *venous anastomosis*, in which there is a juncture between two similar veins (see Fig. 2.1.36);
  • the *arterial anastomosis* (Fig. 2.1.44e), in which there is a juncture between two similar arteries (see Fig. 2.1.32) (palmar arcades);
  • the *arterial-venous anastomosis* (Fig. 2.1.44g), in which the juncture forms a 'short-circuit' between an artery and a vein. The capillary network in between is 'omitted'. Such arterial-venous anastomoses are found mainly in the skin (see Fig. 2.5.7);
- *terminal arteries*, the vascular system that until now was taken as the starting point (Fig. 2.1.44). In this system one arteriole takes care of one capillary system. Consequently, the blood supply of the tissue concerned is vulnerable. If the supplying vessel becomes shut, the tissue behind it is threatened with death. This kind of vascular system is – unfortunately!– found in the heart and the brain tissue.

9. Exchange between capillary and m.i.

In a number of functional systems the interchange of substances between capillary networks and m.i. is dominated by homeostasis.

Nutrients are absorbed from the intestines, waste substances are taken out of the blood in the kidneys, gases are exchanged in the lungs, etc. This interchange will be discussed explicitly later.

In this chapter, only the general mechanism of interchange of the substances will be discussed. Apart from the exchange of substances under the influence of *diffusion* (small molecular substances) and *pore-transport* (large molecular transport), there is an interchange on the basis of three connected processes: *filtration, osmosis* and *reabsorption* (Fig. 2.1.45).

At the beginning of the capillary network the blood pressure is about 35 mmHg (4.65 kPa). This pressure has as its purpose the transport of fluids (solutions included) to the m.i. The blood pressure is 'sabotaged' by the difference between the colloid-osmotic value of the blood and the colloid-osmotic value of the m.i. This difference is about 25 mmHg (3.3 kPa).

So the net filtration pressure =
blood pressure minus colloid-osmotic value
= 35 mmHg (4.65 kPa) minus 25 mmHg (3.3 kPa)
= 10 mmHg (1.35 kPa).

In other words: fluids and the substances dissolved in them are pushed into the m.i. through the capillary wall with a pressure of about 10 mmHg (1.35 kPa).
The blood pressure has decreased to about 15 mmHg (1.95 kPa) at the end of the capillary network because of this decrease in blood volume. The colloid-osmotic difference in value is still about 25 mmHg (3.3 kPa).

Therefore:
net reabsorption pressure
= colloid-osmotic value minus blood pressure
= 25 mmHg (3.3 kPa) minus 15 mmHg (1.95 kPa)
= 10 mmHg (1.35 kPa).

In other words: fluid and the substances dissolved in it is absorbed in the blood circulation

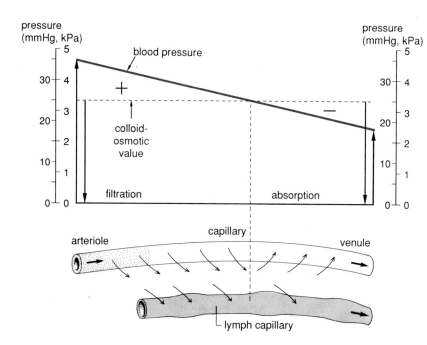

Figure 2.1.45
Filtration, osmosis and
reabsorption

from the m.i. with a pressure of about 10 mmHg (1.35 kPa).

In this way, theoretically, filtration and re-absorption of substances should be equal. In reality, however, the reabsorption is somewhat smaller than the filtration for reasons which will not be discussed here. So there is a constant threat that the volume of tissue fluid will grow. The lymph capillaries in the m.i. absorb and convey this inevitable surplus of fluid.

10. Tissue fluid

Tissue fluid is the internal environment of the cells. In order to be able to function optimally, the cells need a constant composition of the m.i. There is a danger of homeostasis being disturbed by the way the cells are functioning, on the one hand by absorbing nutrients and oxygen from the m.i., on the other hand by giving off waste substances and carbon dioxide. Blood prevents homeostasis from being disturbed by supplying nutrients and oxygen and conveying waste substances and carbon dioxide.

Thus tissue fluid is constantly changed by the blood. The supply and removal of water to and fro the m.i. must also be coordinated. If the supply were larger than the removal, the tissue would swell; were the opposite the case, the tissue would shrink.

Normally the volume of water in the m.i. stays within constant limits as the volume of filtered fluid is equal to the sum of reabsorbed fluid and the fluid carried by the lymph capillaries.

Tissue fluid is colourless and consists almost entirely of water. Electrolytes such as Na^+, Cl^-, K^+, Ca^{2+}, Mg^{2+}, HCO_3^-, are dissolved in it, but it also contains molecularly dissolved substances such as glucose, fatty acids, amino acids and gases. There is a small volume of blood proteins in the tissue fluid. The blood proteins have been pushed into the m.i. through pores in the capillary wall.

As a result of the difference in protein concentration between blood and tissue fluid, the colloid-osmotic value of 25 mmHg (3.3 kPa), mentioned before, exists.

Granulocytes, which have left the capillary by means of leukodiapedesis, can also be found in the tissue fluid (see Fig. 2.1.10). These leukocytes have a phagocytizing function.

Elsewhere, the tissue fluid composition is dependent on the character and activity of the tissue concerned. Intestinal tissue fluid, for instance, will contain a relatively large number of fats, proteins and carbohydrates after a meal.

11. Lymphatic system

The *lymphatic system* comprises two connected parts. Firstly, it comprises a vascular system which supports the circulation in the blood vessels – the lymphatic vessels. It is also a collective term for concentrations of reticular connective tissue in which many lymphocytes are present – *lymphatic tissue* or *lymphoid tissue*. Organs mainly consisting of lymphatic tissue (the lymph nodes, the spleen and the thymus) are called *lymphatic organs*.

Apart from being a support for the blood circulation, the lymphatic system is mainly a resistance system (phagocytosis, humoral and cellular immunity).

11.1. Lymph and lymph vessels

Tissue fluid absorbed in lymph vessels is called *lymph*. The composition of the lymph, therefore, is dependent on the composition of the tissue fluid concerned. The small volume of filtered blood proteins is also conducted away via the lymph vessels. Lymph also contains hormones, enzymes, antibodies and lymphocytes.

Apart from the 'blood capillaries', lymph capillaries are present. They are venules with relatively large pores, 'blindly' beginning in the intercellular spaces of the tissue. The lymph capillaries consist of a layer of endothelial cells covered with a thin fibrous tissue membrane. It is preferable to speak of *lymph capillary networks*, like blood capillary networks (Fig. 2.1.46). The networks conduct the lymph away to small lymph vessels which in their turn join to form larger lymph vessels.

Gradually more layers of smooth muscle tissue appear as the diameters of the lymph vessels become larger. The major lymph vessels in arms and legs have valves.

The major lymph vessels are called *lymphatic trunks* (Fig. 2.1.47).

The lymph from the legs and the pelvic organs is conducted away via the right and the left *lumbar*

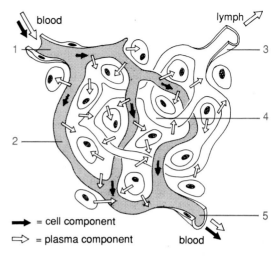

Figure 2.1.46
Circulation in
blood capillaries and
lymph capillaries

1. arteriole
2. blood capillary
3. lymph vessel
4. lymph capillary
5. venule

→ = cell component
⇨ = plasma component

lymphatic trunks. These terminate, together with the *intestinal lymphatic trunk* (with lymph from the abdominal organs, especially the intestines) in a dilatation – the *cisterna chyli*. From the cisterna chyli a large lymph vessel extends upwards: the *thoracic duct*. This vessel goes behind the aorta through the mediastinum and terminates in the left subclavian vein. Just before the terminal thoracic duct, another lymphatic trunk terminates in the thoracic duct. This vessel conducts lymph away from the left arm, the left half of head and neck and from the left lung.

The *right lymphatic duct* terminates in the right subclavian vein. It conducts lymph away from the right arm, the right half of head and neck and from the right lung. In this way, fluid from the blood circulation is added again to the blood circulation via a 'detour'. It could be argued that the lymph vessels form a parallel version of the venous system.

Transport of the lymph is brought about by a number of factors, comparable to the venous back-flow: the presence of valves, the respiration pump, the muscle pump, and the arterial pump.

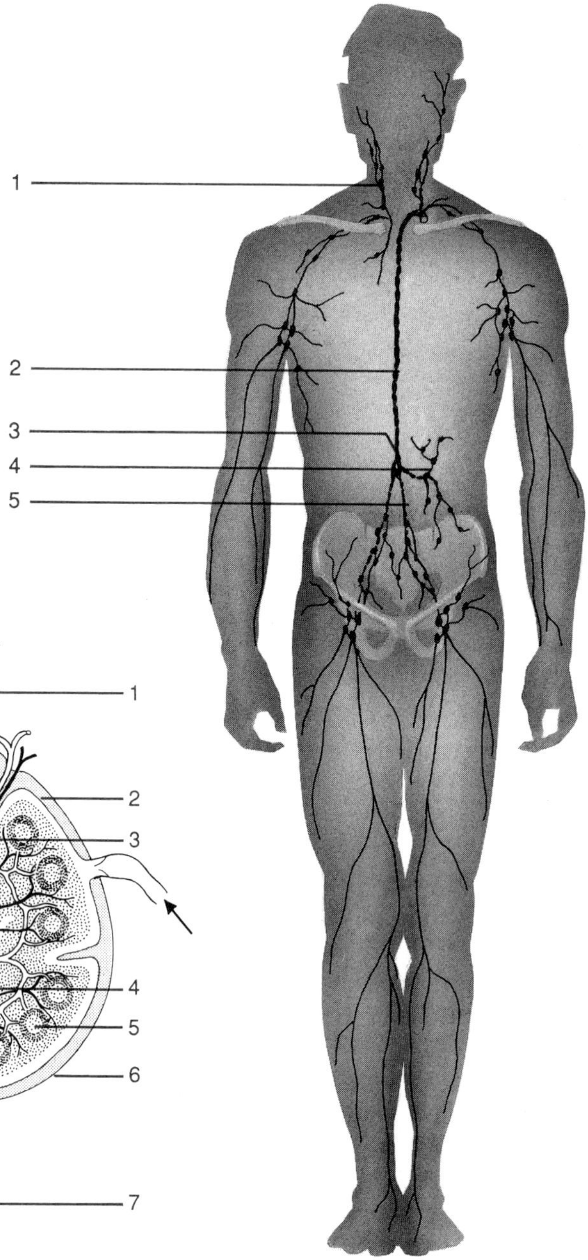

Figure 2.1.47
Topography of the major
lymph vessels

1. *right lymphatic duct*
2. *thoracic duct*
3. *cisterna chyli*
4. *intestinal lymphatic trunk*
5. *lumbar lymphatic trunk*

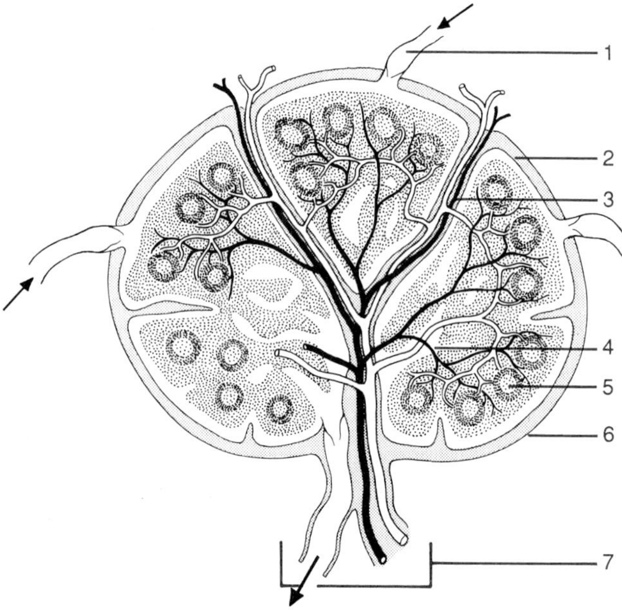

Figure 2.1.48
Structure of a lymph node

1. *afferent vessel*
2. *cortical sinus*
3. *trabecula with small vessels*
4. *medullary sinus*
5. *lymph follicle or nodule*
6. *capsule*
7. *hilum with artery, vein and
   efferent lymph vessel*

11.2. Lymph nodes

Lymph nodes are 'intermediate' stations in the
lymph vessels at sites where a number of smaller
vessels join to form a larger vessel (Fig. 2.1.48).
A concentration of lymph nodes can be found in
the neck, the armpits, the groin, in the mediastinum and the two lung hilums, in the mesenterium and in the pelvis. Such groups are called
regional lymph nodes (Fig. 2.1.49).

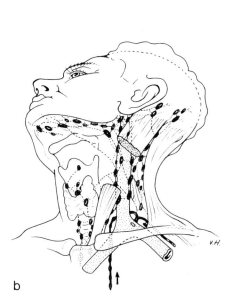

Figure 2.1.49

a. Regional lymph nodes
b. Lymph nodes in the neck
   area

b

Lymph nodes are little bean-shaped organs of different sizes. The largest lymph nodes are about 3 centimetres long. They consist of reticular connective tissue (medulla), divided by fibrous strands (trabeculae) into sections. The trabeculae are connected with the capsule of the lymph node. The capsule consists of collagenous and elastic connective tissue. The small lymph vessels terminate in the medulla of the lymph node via openings in the capsule.

A lymph node has a *hilum*. This is the place where an artery enters the lymph node and where one or two larger lymph vessels leave the node. The larger lymph vessel conducts the lymph away from the medulla. The capillary networks are in the medullary parts, but the larger bifurcations of the blood vessels are in the trabeculae.

The spaces, surrounded by reticular cells inside the capsule and the trabeculae, are called the *lymphatic sinuses*.

The *cortical sinus*, located between the medul-

lary tissue and the connective tissue, and the *medullary sinus*, at all sides surrounded by medullary tissue, are distinguishable. The lymph from the afferent vessels trickles via the sinuses through the lymph node to the efferent vessel. Valves prevent the lymph from flowing in the opposite direction.

The lymph node functions as a filter: the reticular cells have close contact with the passing lymph and make any alien substances harmless. This defence function is supported by the lymph nodes being able to form large numbers of lymphocytes out of reticular cells. This takes place in the lymphatic follicles, spherical germinal centres with many capillaries (Fig. 2.1.48). The lymphocytes produced there are added to the lymph and, after contact with an alien antigen, bring about immunity.

Strongly active lymph nodes swell. This can easily be noticed at some places, for instance at the inner side of the lower jaw, in the armpit, in the groin.

11.3. Spleen

The *spleen*, covered by the peritoneum, is located in the left upper part of the abdominal cavity, below the diaphragm, behind the stomach (see Fig. 1.5.3). The tail of the pancreas touches the hilum of the spleen. The spleen (Fig. 2.1.50), can be considered a large lymph node in the blood circulation. Its structure is comparable to that of a lymph node (capsule, trabeculae and medulla with a great many reticulum cells), as is its function. The *splenic artery* branches from the *celiac trunk* (see Fig. 2.1.35), and enters the spleen via the hilum. The artery branches in the spleen. It is conspicuous that the usual capillary networks are not found in the medulla (spleen pulp). In the medulla tissue the arterioles terminate in the *spleen sinuses*, very porous tube-shaped spaces. The smallest splenic veinlets begin in the medulla tissue in the area of the spleen sinuses.

Figure 2.1.50
Spleen

a. Location of the spleen

    1. *adrenal gland*
    2. *kidney*
    3. *duodenum*
    4. *spleen*
    5. *pancreas*
    6. *abdominal aorta*

b. Cross-section through part of the spleen

    1. *capsule*
    2. *trabecula*
    3. *red pulp*
    4. *splenic lymph follicle*
    5. *white pulp*

c. Detail of red pulp

    1. *efferent vein*
    2. *sinus of spleen*

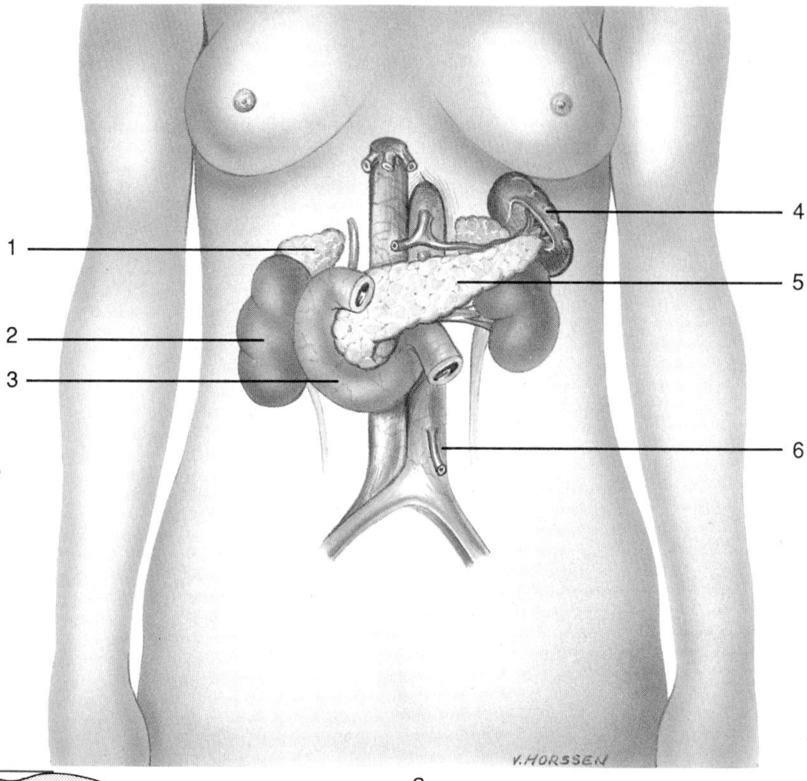

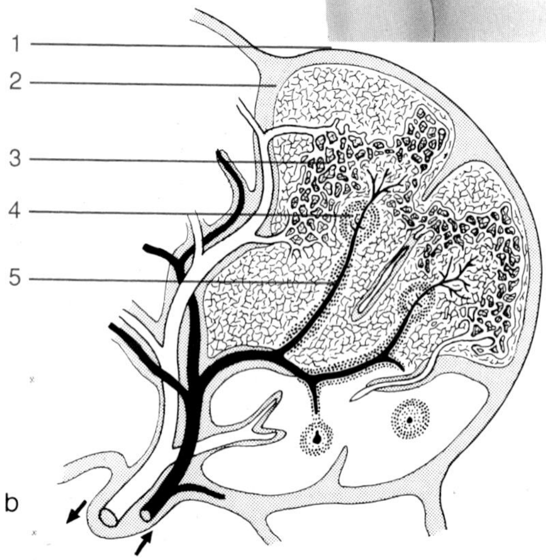

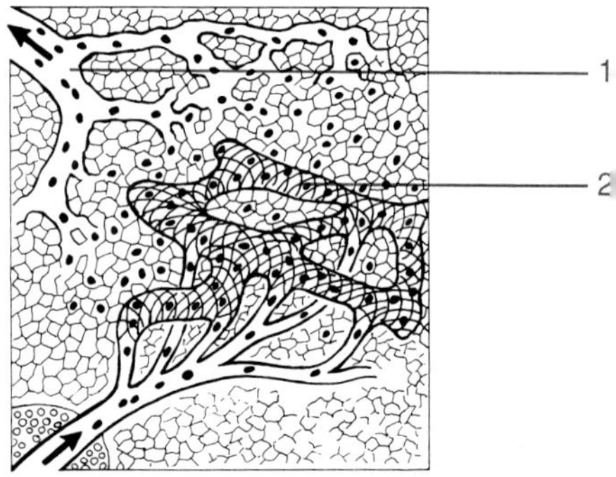

The outgoing *splenic vein* terminates in the portal vein that extends to the liver.

The spleen pulp is divided into *white pulp* (20%) and *red pulp* (80%). It is located around the branches of the splenic arteries and consists of lymphocytes. The white pulp is comparable to the lymph follicles in the lymph nodes. Thicker parts of the white pulp are therefore called splenic lymph follicles. The lymphocytes, made from reticular cells, are added to the blood circulation and cooperate in achieving immunity. The rest of the medulla consists of red pulp: reticular connective tissue, which is red-coloured because it contains blood. The whole of the red pulp, including the spleen sinuses, is comparable to the lymphatic sinuses. The blood is 'filtered' in the spleen; in the spleen sinuses it comes into close contact with the reticulum cells. These cells not only phagocytize alien substances, but also aged and weakened blood cells, especially erythrocytes. It is noticeable that in the decomposition of the erythrocytes the haemoglobin is converted into bilirubin. Connected to albumin, the bilirubin enters the liver via the portal vein. The iron in the haemoglobin is 'recycled' as much as possible. The blood-making tissue is able to build the recycled iron into young erythrocytes. When the iron is not immediately needed, it can be stored in depots.

Apart from these two functions, comparable to those of the lymph nodes, the spleen has a third function – as a blood reservoir. Smooth muscle tissue is found in the capsule. When this contracts, the splenic pulp is squeezed, as if it were a sponge. In this way extra blood can be mobilized. This function of the spleen is, however, not considered very important.

11.4. Lymphoid or tonsillar (Waldeyer's) ring and aggregated lymphatic follicles (Peyer's plaques)

The lymphoid ring is a collective name for scattered areas of lymphatic tissue located at the transition of the oral cavity and the pharynx (Fig. 2.1.51). It comprises the two *palatine tonsils*, the *adenoid* (pharyngeal tonsil) and the lymphoid tissue at the base of the tongue and around the openings of the two auditory tubes.

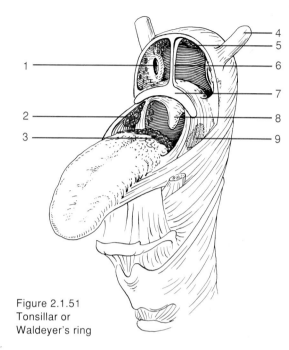

Figure 2.1.51
Tonsillar or
Waldeyer's ring

1. *termination of eustachian tube*
2. *right tonsil*
3. *lingual tonsil*
4. *Eustachian or auditory tube*
5. *pharyngeal tonsil or adenoid*
6. *nasal septum*
7. *palate (section)*
8. *uvula*
9. *left tonsil*

The palatine tonsils are best known. They can be fairly swollen during a cold and are then visible behind the palate. *Aggregated lymphatic follicles* are small scattered areas of lymphoid tissue in the terminal ileum. There are, apart from these relatively large patches, smaller raised nodules of lymphatic tissue everywhere in the wall of the digestive tract (see Fig. 2.2.27b).

Both the lymphoid ring and the lymphatic tissue in the intestines are there as a defence against intruding germs. These defence lines are well placed, as the contents of the digestive tract belong to the m.e.

11.5. Reticuloendothelial system

The RES, *reticuloendothelial system*, comprises all tissues which are able to make blood cells and all cells which have a function in

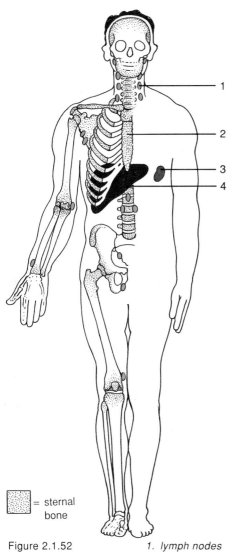

= sternal
bone

Figure 2.1.52
Reticuloendothelial
system

1. *lymph nodes*
2. *bone marrow*
3. *spleen*
4. *liver*

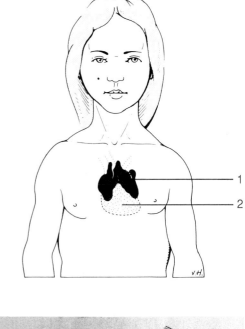

a

b

Figure 2.1.53
Location of the
thymus

a. Projected in the
thoracic cavity

 1. *thymus*
 2. *heart*

b. Location on the heart

 1. *lung*
 2. *diaphragm*
 3. *visceral pleura*
 4. *lobes of thymus*
 5. *thorax wall*
 6. *heart (in heart sac)*

phagocytosis. The red bone marrow, all lymphatic tissue, all reticulum cells and specific cells in the liver are all considered to belong to the RES (Fig. 2.1.52).

11.6. Thymus

The thymus is an organ which is discussed in the chapters on development, Module 3, Chapters 2 and 4. It performs its task till puberty. In adults only a small fat lobe on the heart shows the position of the thymus. In order to present the whole lymphatic system, however, structure and function are discussed in this chapter.

The thymus (Fig. 2.1.53) consists of two, or more, lobes and is located behind the sternum and the heart. As in the structure of the lymph

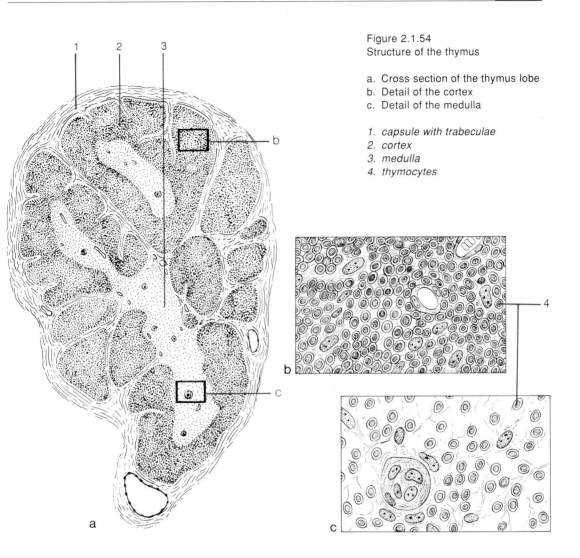

Figure 2.1.54
Structure of the thymus

a. Cross section of the thymus lobe
b. Detail of the cortex
c. Detail of the medulla

1. *capsule with trabeculae*
2. *cortex*
3. *medulla*
4. *thymocytes*

node there is centrally located reticular connective tissue, surrounded by a capsule with trabeculae (Fig. 2.1.54).

Apart from reticulum cells there are vast amounts of *thymocytes* in the reticular connective tissue. Thymocytes are small lymphocytes that eventually develop into the stem cells for T-lymphocytes (*t*hymus lymphocytes). Shortly after birth the thymocytes leave the thymus and settle everywhere in the reticuloendothelial system, where they develop into reticulum cells. These reticulum cells can differentiate into the stem cells of T-lymphocytes as soon as an alien antigen intrudes into the body. A group of lymphocytes that derive from one single stem cell and that is aimed at a specific antigen, is called

a 'clone'. Thereafter, the T-lymphocytes bring about *cell-mediated immunity*. In the production of thymocytes the thymus prevents the production of antibodies that might be reactive to self-antigens. For that reason, all those erythrocytes from which the stem cells of T-lymphocytes – that might be able to produce antibodies against self-antigens – derive, are already eliminated in fetal life. Such unwanted cells are called 'forbidden clones'.

2

Digestive system

Introduction

All the activities of human beings are accomplished by the chemical reactions of *assimilation, dissimilation, anabolism* and *catabolism.*
The main metabolic processes in man are anabolic assimilation and catabolic dissimilation. The former is usually called synthesizing (or building) metabolism, the latter energy releasing metabolism.

So: metabolism = synthesizing metabolism + release of energy.
Or: metabolism = anabolism (building up) + catabolism (breaking down).

The *absorption* of nutrients (food and drink) should correspond with the *consumption* of nutrients. We call this the *energy* and *building material* balance (Fig. 2.2.1).
Supplies are stored when absorption is greater than consumption resulting in an increase in weight. When absorption is smaller than consumption, the supplies are used and there is a decrease in weight.
Absorption and consumption are measured in joules (J).*
An average person uses about 12,500 kJ (3000 kcal) per 24 hours when working moderately, and more than 18,800 kJ (4500kcal) when working intensively.

Figure 2.2.1
Energy and
building materials
balance

a. Consumption
   larger than
   absorption
b. Consumption =
   absorption
c. Consumption
   smaller than
   absorption

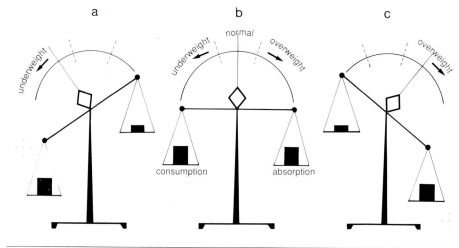

* Formerly, calories (cal) or kilocalories (kcal) were used; 1 kcal = 4.186 kJ, 1 kJ = 0.239 kcal.

Under normal circumstances, the energy required to do physical work far exceeds the energy needed for tissue growth and repair. It therefore makes sense for the obese person to eat less, and also to take more exercise.

The solid and fluid substances, used in the cells during metabolic reactions, are often called nutrients. It is preferable, however, to call them *substrates*, so that the term *nutrients* can be reserved to indicate the elements of our food.

The general function of the digestive system is to make substrates (sources of energy and building materials) available to the circulatory system. Thereafter the circulatory system can carry the substrates to the m.i. of the active tissues.

More specifically, the task of the digestive system is as follows:
- absorption of food from the m.e. (eating);
- mechanical decomposition of the food into smaller parts (chewing);
- transportation of the food through the alimentary tract (swallowing and peristalsis);
- dissection of the food into substrates by enzymes (digesting);
- conveyance of substrates to the circulation (resorption);
- assistance in the removal of waste products (defecating).

This chapter deals with the composition of our food. Thereafter, the structure and function of the digestive system will be discussed – the alimentary tract, the pancreas, the liver and the biliary tracts (gall ducts).

Learning outcomes

After studying this chapter you should have sufficient knowledge and understanding of:
- the different kinds of metabolism in the human body;
- the general function of the digestive system;
- the elements of food and their function;
- the structure and the functions of the alimentary tract and of its various parts;
- the mechanism for the absorption of nutrients into the blood;
- the structure and the functions of the pancreas, the liver and the gall ducts;
- the location of the abdominal organs in relation to the peritoneum.

1. Nutrients

By combining all kinds of foodstuffs and recipes we can put different meals on the table, day after day, but still our food consists of only six elements – *carbohydrates, fats, proteins, minerals, vitamins* and *water*. Carbohydrates and fats are the main sources of *energy* for the human body. Proteins, minerals and water are the main materials. Vitamins help in regulating many metabolic processes. But there is more to the different food elements than their general functions. This will be discussed in the following pages.

Carbohydrates

Carbohydrates (saccharides, sugars) play an important part in the supply of energy.

The average diet in Western Europe consists largely of carbohydrates. The energy value of carbohydrates is relatively low: 17 kJ/g (4.1 kcal/g). This is less than half of the energy value of fats.

Carbohydrates are sub-divided in accordance with the length of their compounds. An elementary counting system is used: one (mono-), two (di-), many (poly-).

The monosaccharides (simple sugars) have a low molecular size so they can diffuse through a semipermeable membrane. The disaccharides (double sugars) must first be separated into two simple sugar molecules. The polysaccharides must also be separated, via disaccharides into monosaccharides, before they can diffuse into the blood. The separation of poly- and disaccharides takes place in the alimentary tract.

The enzymes which influence the carbohydrates are called *carbohydrases*.

– *Monosaccharides*, such as glucose (starch sugar) and fructose, share the molecular formula $C_6H_{12}O_6$. The structure formula of the monosaccharides is always six-ring (Fig. 2.2.2a). The type of monosaccharide is determined by the way in which the corners are arranged.

Glucose is invariably used in the cell metabolism. The other types of monosaccharides are converted into glucose in the liver (Fig. 2.3.36).

– *Disaccharides*, such as maltose, saccharose (beet sugar, cane sugar) and lactose, have the molecular formule $C_{12}H_{22}O_{11}$.

Figure 2.2.2
Molecular formulas and structural formulas of saccharides

a. Glucose (monosaccharide)
b. Maltose (disaccharide)

$C_6H_{12}O_6$
molecular formula

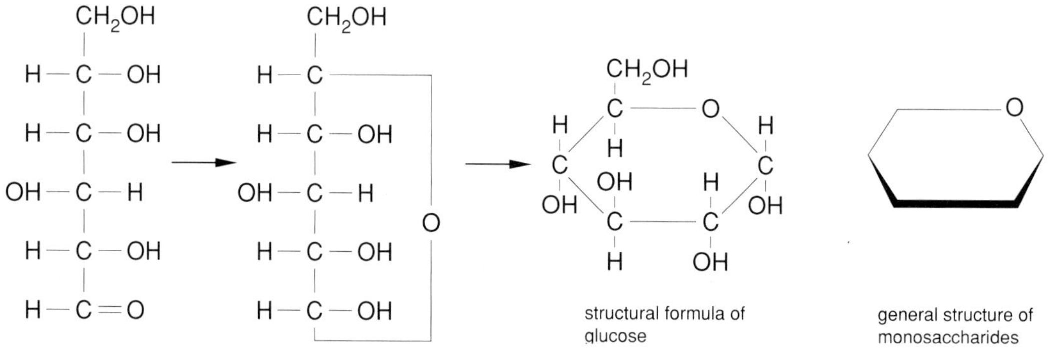

structural formula of glucose

general structure of monosaccharides   a

$C_{12}H_{22}O_{11}$
molecular formula

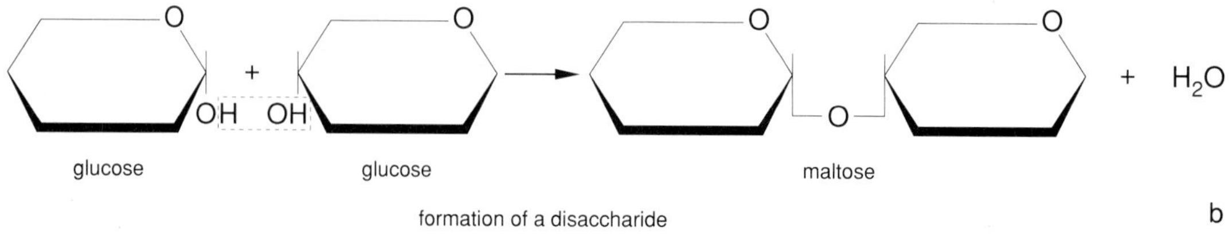

glucose   glucose   maltose

formation of a disaccharide   b

It is evident from the structural formula of the disaccharides that two monosaccharide six-rings are bound, separating a molecule of water (Fig. 2.2.2b). A disaccharide can be separated into two monosaccharides by adding a molecule of water.

One maltose molecule is separated into two glucose molecules.

A saccharose molecule is separated into a glucose molecule and a fructose molecule. A lactose molecule is separated into a glucose molecule and a galactose molecule.

– *Polysaccharides*, such as starch (amylum), cellulose and glycogen, have the formula $(C_6H_{10}O_5)_n$, whereby n indicates the number of glucose molecules from which the polysaccharide molecule is built. The number n is variable and may be very large (greater than 25,000).

When water molecules are added, starch is mainly separated into maltose molecules and thereafter into glucose. Cellulose cannot be broken down in the human body as the body lacks the necessary enzyme (cellulase). However cellulose is important in digestion as it stimulates intestinal peristalsis and secretion of intestinal juices.

Glycogen is formed in liver and muscle cells by linking a large number of glucose molecules. Glycogen is glucose in store, and it is separated into glucose molecules whenever it is needed.

The release of glucose is the objective of the conversion of polysaccharides, disaccharides and monosaccharides (Fig. 2.2.3a), as glucose is one of the main fuels of the mitochondria.

Lipids
The group of fats (lipids) is less homogenous than the group of carbohydrates. The common feature of lipids is that they cannot be dissolved in water. They also have a very high energy rate: 39 kJ/g (9.3 kcal/g).

Triglycerides are the best known lipid molecules. As the name suggests, they can be separated into one glycerol molecule and three

Figure 2.2.3
Digestion and absorption

a. Carbohydrates

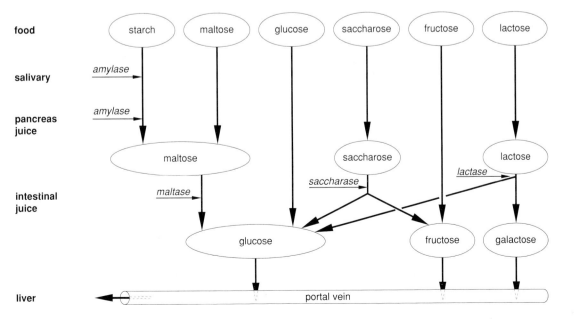

Figure 2.2.3 (cont.)

b. Lipids
c. Proteins

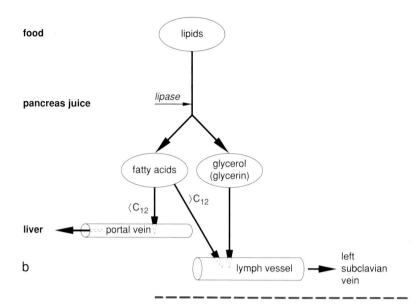

b

fatty acid molecules. The fatty acids can be divided into saturated and unsaturated fatty acids.

The general formula for saturated fatty acids is $C_n H_{2n+1} COOH$.

Some common examples are:
– butyric acid, $C_3 H_7 COOH$,
– palmitic acid, $C_{15} H_{31} COOH$,
– stearic acid, $C_{17} H_{35} COOH$.

The unsaturated fatty acids have an even number of hydrogen atoms less than comparable saturated fatty acids. There is a so-called double bond at the place in the fatty acid molecule where hydrogen atoms could be reabsorbed (Fig. 2.2.4).

Oleic acid, $C_{17} H_{33} COOH$, for instance, is a simple unsaturated fatty acid, with one double bond.

Linolic acid, $C_{17} H_{31} COOH$, is a polyunsaturated fatty acid, with two double bonds.

Phospholipids are fat molecules which have phosphonic acid and a nitrogenous bond as well as glycerol and fatty acids.

Lipids are separated by the enzyme lipase. Fatty acids with chains of less than twelve carbon atoms can be absorbed from the intestines into the blood. Longer chains of fatty acids and glycerol molecules are absorbed into the lymphatic vessels (Fig. 2.2.3b).

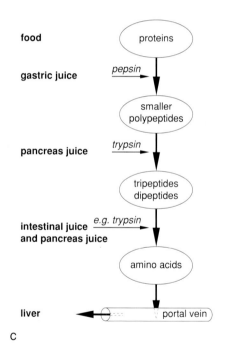

c

Fats are of great importance to the body. They supply much energy, they can be stored for the supply of energy, they provide insulation and mechanical support.

Furthermore they form a building element for the *lipo*protein membranes (especially the phospholipids) and serve as an electrical insulation around the nerve ends.

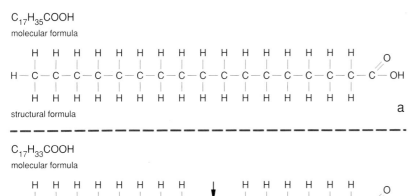

Figure 2.2.4
Fatty acids with (↓) and without double bonds

a. Stearic acid (saturated)
b. Oleic acid (unsaturated, one double bond)
c. Linolic acid (poly-unsaturated, two double bonds)

A number of vitamins (A, D, E and K) can only be absorbed into the blood from the intestinal contents when dissolved in fat.

Proteins

Proteins have a primary structural function, but may also be used as a source of energy.
Proteins are composed of a large number of *amino acids*. An amino acid (*mono*peptide) is a compound in which both an amino-group (NH$_2$) and an acid group (-COOH) are always linked with a C-atom. The (variable) R part in the standard amino acid formula determines the type of amino acid (Fig. 2.2.5a).

In protein synthesis, amino acids link by peptin bonds, separating a water molecule (Fig. 2.2.5b). A *dipeptide* is formed. When more amino acids are linked, the compounds are called – *tri*peptide (three amino acids), *tetra*peptide (four amino acids) and eventually *poly*peptide ('many' amino acids). For this reason, proteins are also called polypeptides.

Figure 2.2.5
Amino acids

a. Standard formula
b. Formation of a dipeptide

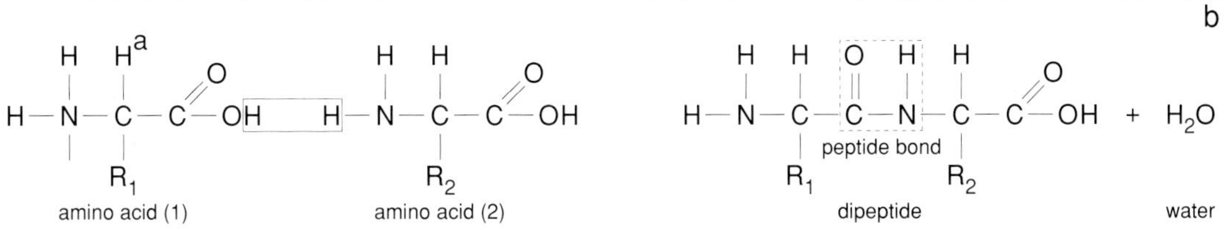

$$NH_2 - \overset{\overset{\displaystyle H}{|}}{\underset{\underset{\displaystyle R}{|}}{C}} - COOH$$

a

b

amino acid (1)   amino acid (2)   dipeptide   water

The enzymes which effect the proteins are called *proteinases* or *proteases*.

The proteolysis in the digestive tract is the opposite of the protein synthesis taking place in the ribosomes. The peptide bond is dissociated by adding a water molecule. Eventually, the polypeptides are separated into amino acids and absorbed from the intestines into the blood (Fig. 2.2.3c).

Twenty different amino acids are needed to build up the tissue proteins. A number of amino acids can be converted into other amino acids by the liver. These are called *non-essential amino acids*, as they need not necessarily occur in food proteins. The amino acids which the body does need but cannot produce itself are called *essential amino acids*, Table 2.2.I, so food must contain these essential amino acids.

Minerals

Minerals are indispensable inorganic elements of food. *Electrolytes and trace-elements* are amongst the minerals. Electrolytes (ions) are salts separated into particles. They form a *buffer* mixture and provide the crystalloid osmotic value.

Table 2.2.I
Amino acids

Non-essential	Essential
alanine	arginine, insufficiently produced
asparagine	phenylalanine
cysteine	histidine (only during growth period)
cystine	isoleucine
glutamine	leucine
glycine	lysine
hydroxyproline	methionine
proline	threonine
serine	tryptophan
tyrosine	valine

A number of minerals are important as *constructive elements*: calcium and magnesium salts (in the bones) and iron (as an element of haemoglobin).

Trace-elements are so called because they need only be absorbed in tiny amounts. They are however indispensable. Fluorine is a trace-element which makes tooth enamel more resistant to the influence of acids. Iodine is an important element of the thyroid hormone thyroxine. Copper, cobalt, aluminium, zinc, chromium and manganese are elements of certain enzymes or hormones.

Vitamins

Vitamins are complicated organic compounds which are indispensable – in fairly small quantities – to the enzyme systems of the cell metabolism. The body itself cannot produce them.

The title 'vitamins' (literally translated: 'aminos of vital importance') reveals that all these compounds were originally assumed to be amino acids. This however is not the case.

Fat soluble vitamins (A, D, E and K) and *water soluble* vitamins (B-complex, C, H) can be identified. Vitamins soluble in fat can only be reabsorbed when the food absorbed contains sufficient fat.

Vitamins have very varied functions in metabolism. Table 2.2.II shows a simple survey.

Water

Water was essential for the origin of life on the planet Earth. Moreover, the greater part of a living organism invariably consists of water.

Water is an important constructive element. More than 75% of cytoplasm consists of water. The m.i. also contains a large volume of water. Furthermore, water is important as a *solvent, medium of transportation* and *heat buffer*.

Every 24 hours a certain volume of water leaves the body in the form of urine and faeces, and through evaporation. This fluid 'loss' balances the fluid 'gain' per 24 hours. Fluid is mainly taken in by drinking, but the fluid in solid food also contributes to the total fluid intake. Finally, there is a production of fluid in the metabolic processes of the mitochondria.

Fluid gain and loss is called the *water balance* (Fig. 2.2.6). This balance should be more or less stable at all times. When there is a considerable discharge of fluids (for instance, during excessive sweating), this loss must be compensated by putting weight on the other scale, i.e. by drinking and/or a decreased production of urine.

Name	Origin	Important for	Table 2.2.II Vitamins
vitamin A	– milk, butter, liver – in carrots as provitamin A (carotene); carotene is converted into vitamin A	– formation of standard epithelium – formation of visual purple in the rods of the retina	a. Fat soluble b. Water soluble
vitamin D	– milk, liver, egg-yolk – provitamin D is converted into vitamin D in the skin under the influence of ultraviolet light	– calcium resorption in the small intestine – residue of calcium salts in bones and dentine	
vitamin E	– vegetable oils, wheat germs, lettuce	– reproduction	
vitamin K	– cabbage, spinach – is produced by the intestinal flora in the colon	– formation of the clotting factor prothrombin	

a

b

Name	Origin	Important for
B-complex: vitamin B_1	– whole wheat bread, potatoes, meat	– carbohydrate metabolism
vitamin B_2	– egg white, corn, milk, liver	– biological oxydation
vitamin B_6	– egg white, corn, liver	– conversion of amino acids into other amino acids (trans-amination)
vitamin B_{12}	– animal proteins	– whole wheat bread, potatoes, meat
vitamin C	– fruit, vegetables	– egg white, corn, milk, liver
vitamin H	– milk, egg yolk, liver	– egg white, corn, liver

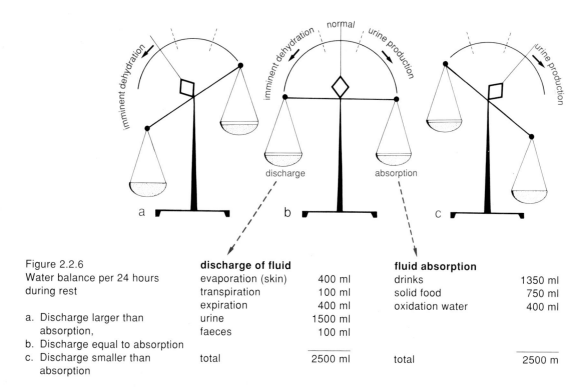

Figure 2.2.6
Water balance per 24 hours
during rest

a. Discharge larger than
   absorption,
b. Discharge equal to absorption
c. Discharge smaller than
   absorption

discharge of fluid		fluid absorption	
evaporation (skin)	400 ml	drinks	1350 ml
transpiration	100 ml	solid food	750 ml
expiration	400 ml	oxidation water	400 ml
urine	1500 ml		
faeces	100 ml		
total	2500 ml	total	2500 m

2. Digestive tract

The digestive tract is the connection between the oral cavity and the anus. The contents of the alimentary tract belong to the m.e.: a part of the outside world absorbed into our body. The entire digestive tract wall is lined with a covering tissue, epithelium. Elements of absorbed food are made suitable for release into the blood (reabsorption) in the digestive canal. The remainder is removed from the body.

The alimentary tract comprises the oral cavity, pharynx, oesophagus, stomach, small intestine and large intestine (Fig. 2.2.7).

The structure of the wall of the digestive tract remains the same over its entire length, but it is different in the oral cavity and pharynx (Fig. 2.2.8). Variations in this general structure have to do with the specific functions of specific areas of the digestive tract.
From inner to outer layer:
– mucosa, the epithelial layer bordering the lumen of the alimentary tract. It has mucus pro-

ducing cells and at some places glandular cells and/or drainage tubes of glands which produce digestive juices. Another word for mucosa is *mucous membrane*. The mucus produced is an excellent lubricant jelly for the transport of food and it also offers the wall of the digestive canal a certain protection from the chemical effect of the intestinal contents (food and digestive juices).
There is a thin layer of smooth muscular tissue directly under the mucosa – the *muscularis mucosae*. The smooth muscle cells aid in pushing the glandular products to the lumen;
– *submucous coat*, the layer of connective tissue around the mucous membrane. It contains blood vessels, lymph vessels, lymphatic tissue and nerve plexuses. The larger glands of the mucosa are located in the submucous coat;
– the *muscularis* is the muscular coat consisting of two layers: a *circular muscular layer*, with a *longitudinal muscular layer* underneath;

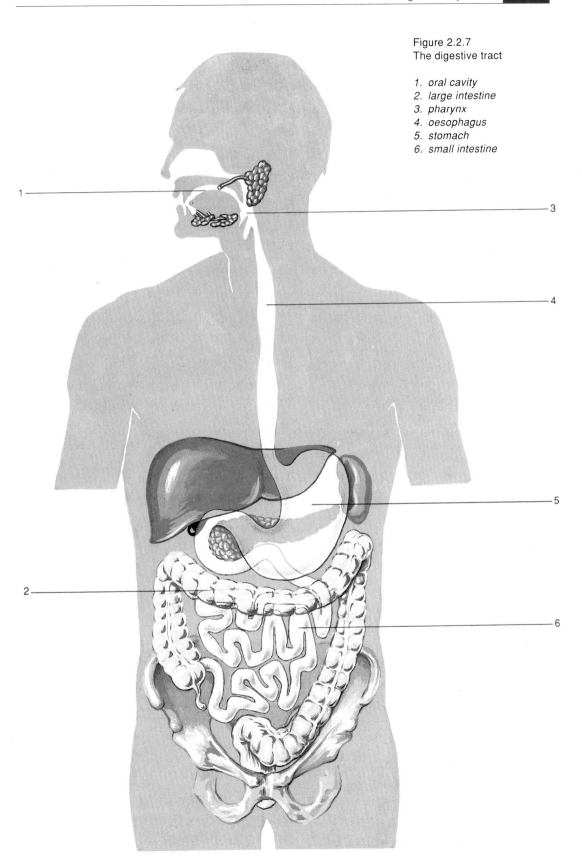

Figure 2.2.7
The digestive tract

1. oral cavity
2. large intestine
3. pharynx
4. oesophagus
5. stomach
6. small intestine

Figure 2.2.8
General structure of the wall
of the digestive tract

1. *mucous membrane*
2. *muscularis mucosae*
3. *submucous coat*
4. *circular layer*
5. *longitudinal layer*
6. *muscularis*
7. *serous coat*

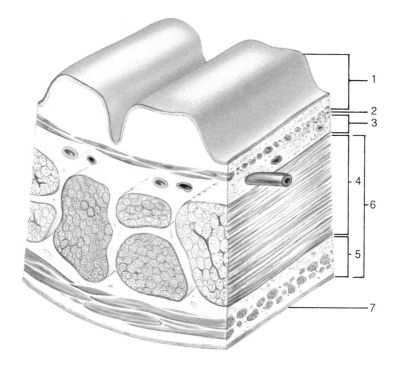

It is worth noting that in a cross-section through the alimentary tract the 'length' fibres are cut, whereas the 'circular' fibres stay intact. In a lengthwise section the opposite is the case;
– *serous coat*, the visceral layer of the peritoneum. There is no serous coat around the oesophagus. It is a thin, smooth membrane consisting of mesothelium on a basic membrane.

2.1. Oral cavity

The oral cavity has a number of functions. It is not only the beginning of the digestive tract but also part of the respiratory system. The oral cavity is important for human speech. The taste function of the mouth, especially that of the tongue, will be discussed in Module 2, Chapter 8.

The mouth is bordered left and right by the *cheeks* containing part of the chewing muscles, on the under side by the *mouth floor* with its muscles, on the front side by the *teeth* and the *lips*. The upper boundary, the roof, is formed by the *palate*.

When the tongue is pressed against the palate, it is noticeable that the anterior part is hard (the hard palate, palatum durum) and that the posterior part is soft (soft palate, palatum molle). The hard palate contains bony tissue that belongs to the bones of the skull. The soft palate contains mainly muscular tissue. The uvula is a pendulous part of the soft palate. The back ridge of the soft palate extends laterally, in folds, downwards to the floor of the mouth. These folds, the *pharyngeal arches*, are the borders between the oral cavity and the pharynx (see Fig. 2.2.16b).

The *anterior and posterior palatine arches* are located parallel to and somewhat ventral of the pharyngeal arches. The *palatine tonsils* are located between the palatine arch and pharyngeal arch, on each side. They are part of the *tonsillar (Waldeyer's) ring*.

The inner side of the mouth is covered with *mucous membrane*, which is a multilayered, squamous epithelium. The lips form the transition to the skin. The epithelium is very thin and has almost no pigmentation. The skin of the lips easily dehydrates; the colour of the lips shows the degree of vascularisation. Normally, the lips are red in colour, due to the blood in the superficial capillaries, but they become pale to purple when it is cold (vasoconstriction).

The inner surfaces of both the lower lip and the upper lip join in the midline to the gums by a *lip bridle*, or *frenulum*. It can easily be felt with the tongue that the upper bridle is larger than the lower bridle.

In the mouth, food is mechanically made smaller by chewing and by mixing it with saliva. This preparatory treatment is a continuation of the preparation that took place before the food was eaten. Both preparations enable a more efficient use of the food.

The jaws, teeth, salivary glands and tongue play an important role in the chewing process.

Jaws and masseters

Part of the upper jaw (maxilla) consists of a horseshoe-shaped bony ridge at the edge of the hard palate. The upper jaw supports the upper teeth (Fig. 2.2.9).

The *lower jaw* (mandible) is a movable bone; it is attached to the skull via a joint located just in front of the external earhole. The part in which the teeth are located has more or less the same horseshoe shape as the upper jaw. As a result, the teeth of the upper and lower jaw usually fit together – occlusion. When the upper incisors overlap the lower one it is known as a *vertical overlap* or *overbite*. The opposite is the case when there is a *horizontal overbite* or *undershot*.

The maxillary joints have a special structure (Fig. 2.2.10). The joint capsule is fairly spacious, the socket is shallow and the articular surface is considerably larger than that of the condyle. An *articular disc* is found between both surfaces of the joint. The articular disc consists of fibrocartilage and is attached to the articular capsule. Consequently, there are two sockets in the maxillary joint. This structure allows more freedom of movement; the lower jaw can go up and down, as when biting, but also from left to right and from front to back. The masseters are structured in such a way that the lower jaw can be drawn

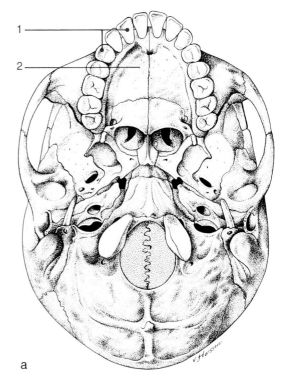

a

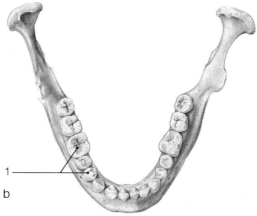

b

Figure 2.2.9
Upper and lower jaw

a. Upper jaw, maxilla
b. Lower jaw, mandible

1. teeth
2. hard palate

strongly towards the upper jaw (Fig. 2.2.11). It is not only tangible but also visible above the ears and above the corners of the lower jaw when the most important masseters are tightened.

Figure 2.2.10
Structure of the
maxillary joint

a. Ligaments

  1. *zygomatic arch*
  2. *upper jaw*
  3. *lower jaw*
  4. *external earhole*
  5. *articular capsule*
     with ligaments

b. Cross section;
   condyle, socket and
   articular disc

  1. *zygomatic arch*
  2. *temporal bone*
  3. *masseters*
  4. *mandible or*
     lower jaw
  5. *articular disk*
  6. *mandibular tip,*
     partly removed
  7. *external earhole*

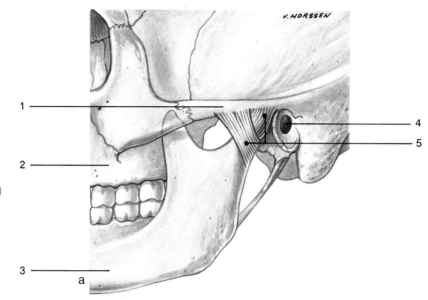

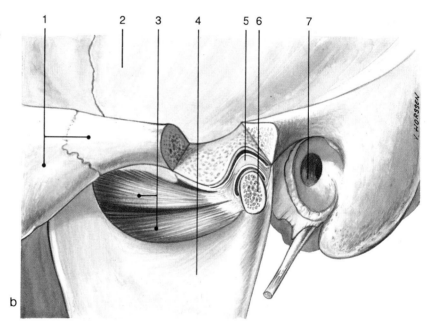

Dentition

A complete adult dentition consists of 32 teeth. There are two mirrored symmetries: upper/lower symmetry and left/right symmetry (Fig. 2.2.9).

There are, in each *quadrant*, two incisors, one canine, two bicuspids, three molars, in that order from the midline. The various kinds of teeth each have their own shapes, suitable to their functions.

The complete dentition can be described by a dental formula (Fig. 2.2.12). The tens indicate the quadrant, the second number indicates the tooth. The 18, 28, 38 and 48 elements are absent or appear later. They are called *wisdom teeth*.

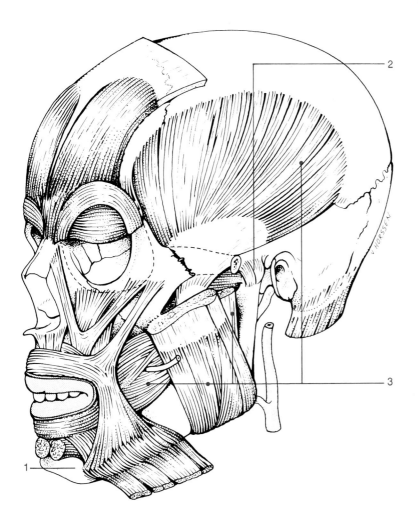

Figure 2.2.11
The masseters

1. *lower jaw*
2. *mandibular joint*
3. *masseters*

right ←								→ left							
18	17	16	*15*	*14*	**13**	12	11	21	22	**23**	*24*	*25*	26	27	28	upper jaw
48	47	46	*45*	*44*	**43**	42	41	31	32	**33**	*34*	*35*	36	37	38	lower jaw

incisors

canines

bicuspids or premolars

molars

Figure 2.2.12
Dental formula of the
permanent dentition

Figure 2.2.13
Cross-section of a tooth

1. crown
2. neck of tooth or cervix
3. root
4. enamel
5. dentine
6. pulp cavity
7. oral mucosa
8. tooth socket
9. cementum
10. root canal

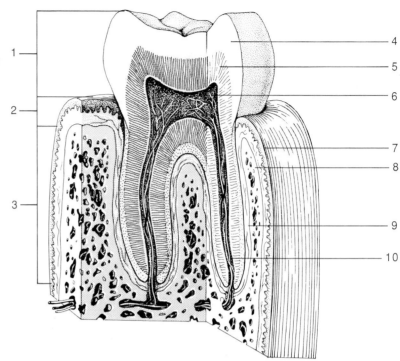

Figure 2.2.14
Embedment of a tooth

1. tooth
2. connective tissue embedment
3. jaw

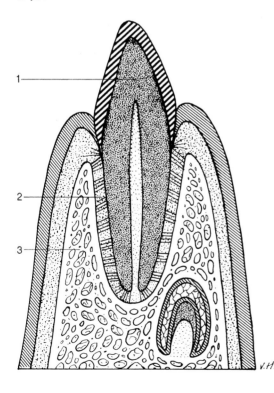

Teeth and molars (Fig. 2.2.13), are made of *dentine*. Dentine is more compact than bone tissue, but at the same time more elastic. The part projecting from the gums is called the crown. It is covered with an extremely hard, shiny substance known as the *enamel*. The root lies entirely in the jawbone. Teeth have one root, molars have two or three roots. By means of strong connections of connective tissue (Fig. 2.2.14), the roots are hung in the jawbone by a spring, but at the same time they are tightly fastened in the jaw as they are covered with a thin layer of 'cementum' (Fig. 2.2.13). Thanks to this structure the teeth can give a little during biting and chewing without becoming loose. The *cervix* is between the crown and the root. At this part, which is covered by the gums, the enamel of the crown changes into the cementum of the roots.

The *dental cavity* and the *root canal* are in the dentine. They are filled with *tooth pulp*, loose connective tissue with small blood vessels and nerve branches. The nerves are connected to a canal in the jaw through the opening at the end of the root canal.

The blood vessels ensure maintenance of the teeth.

The enamel is hard because of a crystal: *calcium apatite*. This crystal is not acid-proof. When exposed to acids, cracks may arise and the acids can affect the underlying dentine. Carbonated drinks are therefore not good for teeth. Sweets, or, in fact, all sweet foodstuffs, can be even worse. Generally this is kept in the mouth for some time and the mouth flora (the microbes that live in the oral cavity) grows very well on it. Microbes produce a considerable amount of acids as waste matter from their metabolism. Such periodical acid eruptions (when sweets are regulary eaten) can lead easily to tooth decay.

Fluorine-apatite is much more acid resistant than calcium apatite. The calcium can be replaced by fluorine by taking fluorine regularly via fluoridated water or via fluoride treatment (tablets, fluoridated toothpaste, gel, etc.).

Salivary glands

Saliva is produced by salivary glands. These are aciniform glands, the ducts of which carry their secretions into the oral cavity. Apart from a large number of minor salivary glands, there are three pairs of relatively large salivary glands (Fig. 2.1.15):

– the *parotid gland* is situated at the outside of the lower jaw below the mandibular joint. Its

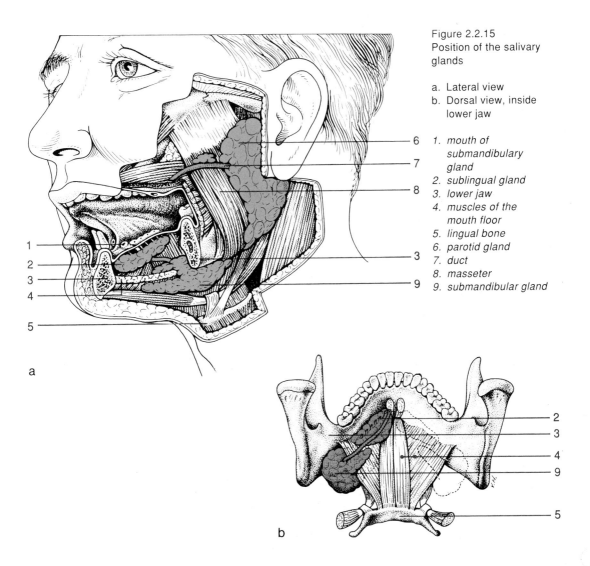

Figure 2.2.15
Position of the salivary glands

a. Lateral view
b. Dorsal view, inside lower jaw

1. mouth of submandibulary gland
2. sublingual gland
3. lower jaw
4. muscles of the mouth floor
5. lingual bone
6. parotid gland
7. duct
8. masseter
9. submandibular gland

a

b

duct is rather long and extends forward in the cheek. It ends in the mucous membrane of the mouth near the second molar of the upper jaw;
- the *submandibular gland* is situated at the inside of the lower jaw near the wisdom tooth. This gland also has a rather long duct which extends forward in the floor of the mouth and ends in a clearly visible papilla on each side of the frenum of the tongue;
- the *sublingual gland* is situated a bit higher in the floor of the mouth, in line with the sub-mandibular gland. This gland has about twenty minor ducts that end lateral of the tongue in the floor of the mouth.

Together, the various salivary glands produce an average of 1–1.5 litres of saliva per 24 hours. The volume and the composition of the saliva produced corresponds with the kind of food consumed. When the food is dry and bitter, a large volume of serous (rich in water) saliva is produced. When food is eaten which is more difficult to make smaller or dilute – such as certain kinds of meat – the saliva contains more mucus, which facilitates swallowing. The parotid glands only secrete serous saliva; the submandibular glands and the sublingual glands secrete a mixture of serous and mucous saliva.

Saliva contains the first digestive enzyme that the food meets on its route through the alimentary tract. This enzyme is called *salivary amylase (ptyalin)*.
Salivary amylase has a pH-optimum of 7. As the environment of the mouth is neutral, the amylum decomposition can begin:

initiated and swallowing is made easier. Moreover, saliva secretion has a function in cleansing the oral cavity. The mouth is continuously rinsed. By swallowing the continuously produced saliva, microbes are also removed from the mouth.
However, saliva also contains a certain amount of inorganic (calcium) salts. This may be deposited on the teeth as tartar. The porous tartar offers an excellent 'hiding place' for microbes in the mouth. So brush your teeth regularly and have them polished once in a while!
The *secretion* of *saliva* is regulated by nerves of the autonomic nervous system.
An increase in secretion of saliva is the result of a reflex and is not solely brought about when there is food in the mouth. When we smell or see food or when we think of something delicious, it 'makes the mouth water'.
A decrease in secretion of saliva is also a result of a reflex and can occur during stress situations. Exam nerves and a dry mouth are a good example of this phenomenon!
The muscles in the area of the salivary glands help the secretion of saliva during chewing.
When you sit behind your textbooks and yawn without keeping your hand in front of your mouth, you may find droplets on your books. This is saliva driven outside by the pressure exerted on the salivary glands during yawning.

Tongue

The tongue is composed of a number of co-operating striated muscles, covered with mucous membrane (Fig. 2.2.16). The tip is the free end of the tongue. On the underside the

$$(C_6H_{10}O_5)_n + H_2O \xrightarrow{\text{salivary amylase}} C_{12}H_{22}O_{11} + C_6H_{12}O_6$$

amylum   maltose   + dextrose
                                    (about 85%) (about 15%)

This reaction is easily noticeable: when a piece of (preferably salt free) bread is chewed on for some time, the taste becomes sweeter.
As a consequence of the secretion of saliva the chewed food becomes thinner, digestion is

tongue is connected with the centre of the inside of the mandible, at the back it is connected to the hyoid bone. Lymphatic tissue of the tonsillar (Waldeyer's) ring is found on the *root of the tongue*, often called the *lingual tonsil*.

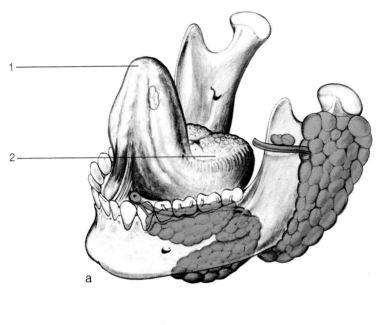

Figure 2.2.16
The tongue

a. Lateral view

   1. *tip of the tongue*
   2. *root of the tongue*

b. View from above
c. Detail

   1. *posterior palatine arch*
   2. *lymphatic tissue on the root of the tongue*
   3. *anterior palatine arch*
   4. *epiglottis*
   5. *palatine tonsil*
   6. *circumvallate papilla*
   7. *filiform papilla*
   8. *serous gland in connective tissue*
   9. *muscular tissue*
 10. *fungiform papilla*

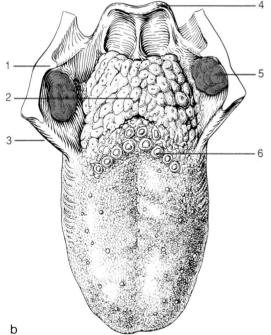

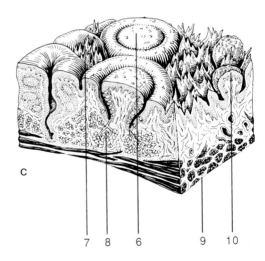

The multilayered squamous epithelium of the mucous membrane of the tongue has many grooves and contains a large number of projections, called *lingual papillae*. Several varieties can be identified with different forms (Fig. 2.2.16b and c). Anterior to the base of the tongue the *circumvallate papillae* can be seen with the naked eye. They lie next to each other in a V-shape. On the tip of the tongue there are chiefly *fungiform papillae*, whereas *filiform papillae* are principally on the sides and the middle part of the tongue. There are *gustatory receptors* or *tastebuds* in the papillae. These tastebuds are sensitive to the various tastes of food. *Sweet* is sensed principally with the tip of the tongue, *bitter* with the root of the tongue, *sour* with the edges and *salt* with the entire surface of the tongue. Each lingual papilla, therefore, has a certain specialism. The receptors give information to the brain via nerve fibres. In fact, taste does not normally occur without smell.

Apart from being a seat for *taste*, the tongue is extremely important for *speech*.
Moreover, the tongue contributes to the cleansing of teeth. The saliva is pushed between the teeth by the tongue and wiped from the enamel. In this way microbes are wiped away and then swallowed.
In addition, the tongue plays an important role in *eating*, *chewing* and *swallowing*.

Eating, masticating, swallowing
Although the incisors have a sharp chisel-shaped upperside, which makes them extremely suitable for biting chunks from food, Western man makes relatively little use of this feature. The food eaten is usually prepared in such a way that it is soft and is taken into the mouth in pieces.
The canines are rather pointed and have a strikingly long root, so that considerable transverse forces can be exerted on these teeth. This is another feature which is not very important for Western man, as he need not tear his food into pieces!
The premolars and molars have more or less square surfaces with a number of cusps. The food is ground between the teeth of the upper- and lower jaw. The masseters give the necessary pressure between the jaws. They also enable the lower jaw to move sideways, forwards and backwards.
During mastication the food is alternately pushed out from between the teeth and put back again by the tongue, the muscles of the cheek and the lips. Moreover, the tongue occasionally pushes the food against the palate. While being ground the food is mixed with saliva. The pulpy mass that results can be swallowed more easily

and can also be affected by the digestive enzymes more effectively compared with unchewed or undiluted food. The duration of chewing depends on the kind of food and the amount of saliva secreted. A piece of bread optimally chewed takes about thirty chewing movements. In reality the duration of the mastication strongly depends on the eating custom.

When swallowing food, there is firstly a *voluntary* phase, followed by the *swallowing reflex*.
The swallowing process begins by closing the mouth; swallowing with an open mouth is difficult. After closing the mouth the bolus of food is moved to the pharynx by means of the tongue pushing the food backwards along the palate (Fig. 2.2.17).
As soon as the bolus of food touches the palatine arches and the wall of the pharynx, the actual *swallowing* reflex begins. This is an *involuntary* activity of the pharyngeal muscles during which the pharyngeal cavity is squeezed shut. During the swallowing reflex, aspiration is reflexively stopped. The passage to the mouth is closed by the tongue. The uvula is pushed upwards, but it is too small to really close the passage to the nasal cavity. This is done in effect by raising the soft palate.
The passage to the open windpipe is closed as the epiglottis is tilted backwards and is pushed over the larynx which is simultaneously raised. The glottal aperture is closed. It is easy to notice that the larynx is raised by putting a finger on the 'Adam's apple' and swallowing.

The oesophagus is slightly opened by the motions of the larynx. The bolus of food is pushed into the oesophagus (Fig. 2.2.17c).

Retching and choking
The opposite of the swallowing reflex occurs when the wall of the throat is tickled roughly, for instance by touching or by something sharp in the bolus of food (fishbone). Such an irritation leads to the *retching reflex*. The food is then moved from the pharyngeal cavity to the oral cavity (from the throat to the mouth).
If food ends up in the larynx and the windpipe,

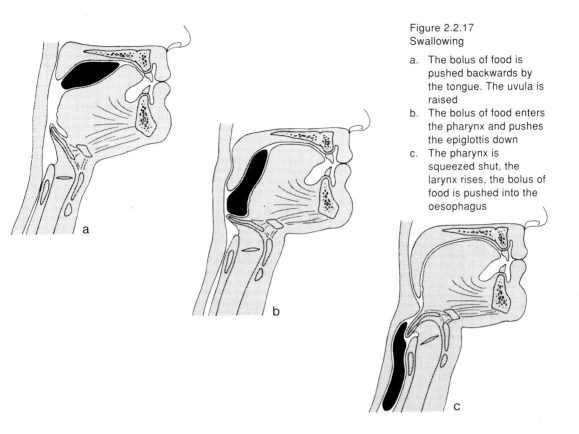

Figure 2.2.17
Swallowing

a. The bolus of food is
    pushed backwards by
    the tongue. The uvula is
    raised
b. The bolus of food enters
    the pharynx and pushes
    the epiglottis down
c. The pharynx is
    squeezed shut, the
    larynx rises, the bolus of
    food is pushed into the
    oesophagus

this is called 'dysdeglutition'. Actually this word does not represent what really happens. Dysdeglutition occurs when there is no deglutition or swallowing. Swallowing always leads to a closure of the larynx. During dysdeglutition food is inhaled, or suddenly enters in some other way, from the mouth into the larynx. The former often happens during talking while eating. At the end of a sentence a person inhales, as a result of which the food in the mouth goes down 'the wrong way'. The latter may happen when food (such as peanuts) is thrown into the mouth. 'Dysdeglutition' (choking) leads to coughing.

2.2. Pharyngeal cavity

The pharynx is a tubular passage behind the nasal cavity. The 'tube' hangs on the lower side of the skull. The pharyngeal cavity extends to the oesophagus and the entrance of the larynx (Fig. 2.2.18).
The pharyngeal cavity is divided into:
– *nasal pharyngeal cavity* (nasopharynx or rhinopharynx), situated behind the nasal cavity. The adenoids, belonging to the tonsillar (Waldeyer's) ring and two exits of auditory (Eustachian) tubes which lead to the middle ear, are situated here. The nasopharyngeal cavity serves respiration.
– *oral pharyngeal cavity* (oropharynx), situated at the level of the palatine arches. It serves both respiration and food transport.
– *laryngopharyngeal cavity*, situated behind the upper part of the larynx. Respiration and food transport should take different routes here.
It is often said that this is the crossroads of food and air passages. Actually this is only the case in nasal respiration. The epiglottis and the triquetral cartilages are linked by a muscular disc. As a result of this structure a tube protrudes into the laryngopharynx (Fig. 2.2.19). The actual entrance of the windpipe is bound by the diagonal upper edge of this tube. The *piriform sinuses* are on each side between the side walls of this tube and the laryngopharynx wall.

Figure 2.2.18
The pharyngeal cavity

1. sinus cavity
2. soft palate
3. palatine tonsil
4. hyoid bone
5. trachea or windpipe
6. sphenoid cavity
7. adenoid or pharyngeal tonsil
8. exit of auditory tube
9. uvula
10. epiglottis
11. fifth cervical vertebra
12. cricoid cartilage
13. oesophagus

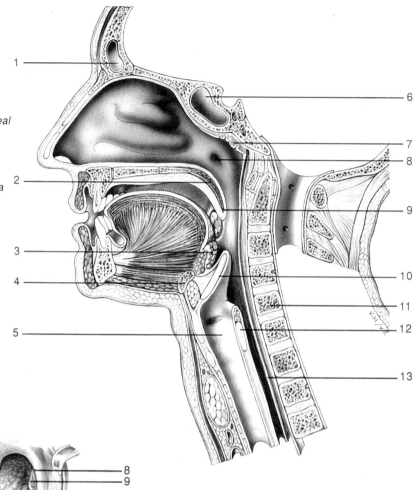

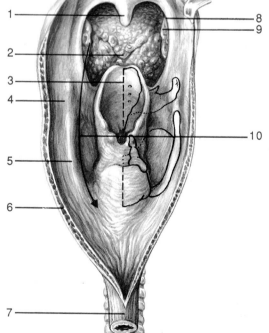

Figure 2.2.19
Open pharyngeal cavity, seen from the back

1. uvula
2. root of the tongue
3. epiglottis
4. hyoid bone
5. dorsal edge of cricoid cartilage
6. cutting edge of pharynx wall
7. oesophagus
8. palatine arch
9. tonsil
10. piriform sinus

Allowing drinks to go from the mouth into the oesophagus via these gutters, without 'deglutition' or 'dysdeglutition', turns out to be possible after some practice.

Small pieces of food can get stuck in the groove in front of the tube, between the epiglottis and palatine tonsil (Fig. 2.2.18). This leads to a 'tickle in the throat'. The inclination to clear the throat is almost uncontrollable. Drinking water usually helps to rinse the groove.

The wall of the pharyngeal cavity consists of a mucous membrane (non-keratinized multilayered squamous epithelium) with striated muscles underneath.

The pharyngeal muscles run more or less horizontally (Fig. 2.2.20). Some of them run vertically to the larynx. During swallowing the horizontal muscles function as circular muscles and squeeze the pharyngeal cavity shut. The vertical muscles lift the larynx, which facilitates closure by the epiglottis and widens the opening of the oesophagus.

2.3. Oesophagus

The oesophagus is a muscular tube, about 30 centimetres long, which connects the pharynx to the stomach. It starts between the cricoid cartilage of the larynx and the fifth cervical vertebra (Fig. 2.2.18). In the throat area it is situated distally between the windpipe and the

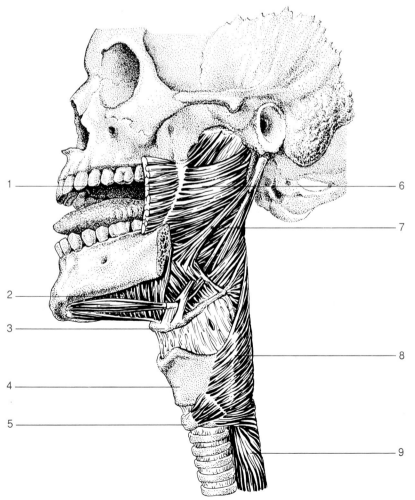

Figure 2.2.20
Outside of the pharynx wall

1. cheek muscle
2. oral diaphragm
3. hyoid bone
4. thyroid cartilage
5. cricoid cartilage
6. upper pharyngeal constrictor muscle
7. longitudinal pharyngeal muscles
8. lower pharyngeal constrictor muscle
9. oesophagus

spine and then extends through the mediastinum of the thoracic cavity. The spine has a dorsal curve in this region.

The oesophagus extends in front of the descending aorta (see Fig. 2.1.33), and enters the abdominal cavity through an opening in the dorsal muscular part of the diaphragm. After another three centimetres it terminates in the stomach. The oesophagus has three constrictions (see Figs. 2.2.7 and 2.1.33). The upper constriction is situated behind the cricoid cartilage, the middle constriction is behind the left principal bronchus and the lower one is where the oesophagus pierces the diaphragm.

The oesphageal arteries are branches of the descending aorta. Most of its veins terminate in the superior vena cava. The lower oesophageal veins form junctures with the veins of the stomach. In this way a small volume of blood from the oesophagus ends up in the portal vein.

Special characteristics of the wall

The oesophagus wall has layers, similar to the ones described in the general structure of the digestive tract. It should be noticed, however, that the emphasis is on muscular tissue (Fig. 2.2.21). There is *striated* muscular tissue at the site of the upper constriction. This part of the wall functions as a valve (upper *sphincter*). This valve prevents air from being sucked from the pharynx. Near the lower constriction there is a lower sphincter. This valve prevents the contents of the stomach from flowing back into the oesophagus. Normally, these two sphincters also retain the unpleasant smell of the stomach contents. The submucous and mucous layers of the oesophagus are poorly developed. The mucosa consists of multi-layered squamous epithelium. There are large numbers of mucus-producing glands in the oesophagus. No enzymes are produced in the oesophagus, nor is there any absorption of nutrients.

Peristalsis

The function of the oesophagus is to transport the food from the pharynx to the stomach. This transport takes place by means of peristalsis: the muscles contract above the chyme, and relax around and directly under the chyme (Fig. 2.2.22). These waves of contraction occur all

Figure 2.2.21
Structure of the oesophagus

1. *mucous gland*
2. *mucosa*
3. *muscular coat*
4. *submucosa*
5. *circular muscular layer*
6. *longitudinal muscular layer*

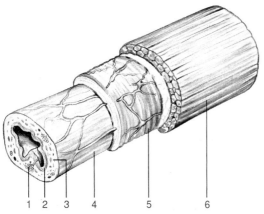

1 2 3   4   5   6

Figure 2.2.22
Peristalsis of the oesophagus
(schematic)

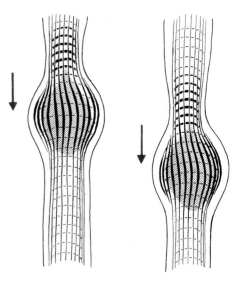

along the stretch of the oesophagus. Consequently, the oesophageal contents are pushed into the stomach. The mucus produced functions as a lubricant. This active transport is very efficient, illustrated by the fact that there is no 'leakage' from the oesophagus when a person drinks a glass of water while lying or hanging upside down.

2.4. The stomach

The stomach looks like an inflated part of the oesophagus with dilatation at the left side. The stomach is situated left of centre in the upper part of the abdominal cavity, resting on the intestines, the upper side bordered by the left part of the diaphragm (see Fig. 2.2.7).
The stomach consists of several parts (Fig. 2.2.23). The oesophagus terminates in the *cardiac stomach*. The *fundus* is above and to the left of the cardiac orifice, lying against the diaphragm. In a vertical position it usually contains 'gas', consisting of swallowed air. When the volume of gas gets larger, for instance by drinking carbonated drinks, the 'gas bubble' can sink below the level of the cardiac sphincter.

Figure 2.2.23
Parts of the stomach

1. *pylorus*
2. *duodenum*
3. *oesophagus*
4. *lesser curvature*
5. *greater curvature*

a = cardiac stomach
b = fundus
c = body of stomach
    (corpus)
d = pyloric antrum

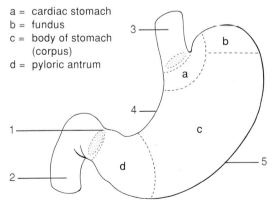

Belching takes place when part of the gas escapes by opening the lower oesophageal sphincter. The *body of the stomach* is the main part of the stomach. It has a more or less vertical position. The *pyloric antrum* is situated between the body of the stomach and the pyloric canal. The pylorus is the sphincter surrounding the orifice which opens into the duodenum. As the stomach is dilated to the left, inner and outer bends have developed, the *lesser curvature* and the *greater curvature*. The stomach-passageway is situated in the lesser curvature. Fluids go almost straight to the exit of the stomach via this route.

The arteries of the stomach are branches of the coeliac trunk (see Fig. 2.1.35). The stomach veins carry blood to the portal veins (see Fig. 2.1.34).

Special characteristics of the wall
Compared with the general structure of the alimentary tract, the wall of the stomach has a number of characteristics of its own (Fig. 2.2.24). The mucous coat consists of a one-layered columnar epithelium that is folded when the stomach is empty. When the stomach is filled, there are fewer folds. At the bottom of the slit-like pits are the openings of the tubular gastric glands. At some places in the mucous coat are concentrations of lymphatic tissue.
The muscular coat consists of three layers (Figs. 2.2.24 and 2.2.25): smooth muscles on the outside, a circular muscular layer underneath, and a diagonal layer on the inside.
Going from the body of the stomach to the centre, the muscular coat gradually gets thicker. The pylorus consists mainly of an extra thickening of the circular muscular tissue.
The outside of the stomach is covered with a serous coat, the *tunica serosa*, the visceral layer of the peritoneum. So the stomach has an intraperitoneal position.

Gastric juice and secretion of gastric juice
About 2.5 litres of gastric juice is produced per 24 hours. It consists largely of water with *pepsin, hydrochloric acid* and the *intrinsic factor*. The stomach also produces mucus:

Figure 2.2.24
Structure of the stomach wall

a. Cross section

1. *fold of mucous membrane*
2. *pit with glandular tissue*
3. *lymphatic tissue*
4. *muscular coat*
5. *submucosa with blood vessels*
6. *muscle layers*
7. *tunica serosa*

b. Detail of gastric glands

1. *parietal cells*
2. *neck cells*
3. *head cells*

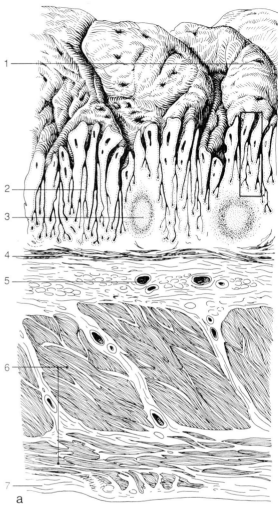

a

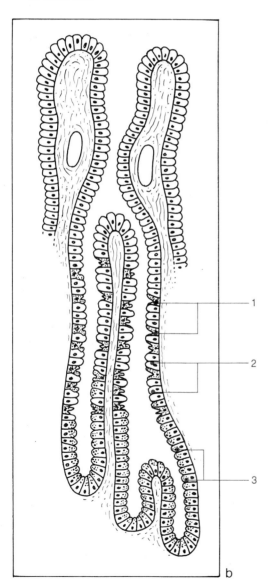

b

– water serves as a solvent and a means of transport;
– pepsin is a protein splitting enzyme. The enzyme itself, however, is not produced in the gastric glands, i.e. in the *head cells*, but the inactive pro-enzyme *pepsinogen*. This prevents the glandular cells, which contain many proteins, from producing their own destroyers;
– hydrochloric acid (HCl) is produced by the *parietal cells* (Fig. 2.2.24b). HCl has a number of functions. It activates the inactive pepsinogen:

$$\text{pepsinogen} \xrightarrow{HCl} \text{pepsin.}$$

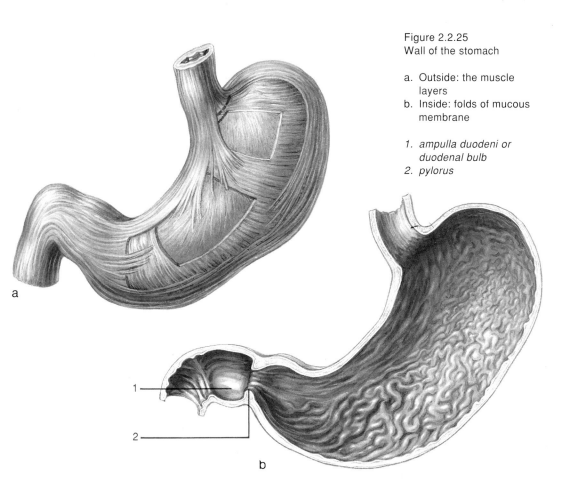

Figure 2.2.25
Wall of the stomach

a. Outside: the muscle layers
b. Inside: folds of mucous membrane

1. ampulla duodeni or duodenal bulb
2. pylorus

Furthermore, it makes the gastric milieu more acid than pH6 (pH 1 to 2), an environment in which the pepsin enzyme functions optimally. HCl also has a decomposing effect on proteins, so pepsin can be more effective. HCl disinfects the stomach contents – micro-organisms, such as microbes and fungi, which may occur in food, are killed;
– intrinsic factor is a substance produced by the parietal cells. The name 'intrinsic factor' (produced by the body itself) is chosen as this substance forms a complex together with an 'extrinsic factor' (a factor absorbed with food), i.e. vitamin B_{12}.
Only when vitamin B_{12} is linked with an intrinsic factor can it pass into the blood via the mucous membrane of the intestines (reabsorption). The intrinsic factor is the 'absorption vehicle' of vitamin B_{12};

– mucus is produced by the mucous membrane of the stomach and also by the *neck cells* (Fig. 2.2.24b) of the gastric glands. In the oesophagus also, the mucus serves as a *lubricant*. Moreover, the mucous layer of about 1 to 2 mm offers an important protection for the stomach wall. On the one hand the layer of mucus is impermeable for pepsin, so this enzyme cannot affect the proteins of the stomach wall itself, on the other hand the gastric mucus is slightly alkaline, so the mucosa is protected against being chemically affected by the hydrochloric acid.
There is a constant production of mucus. The production is somewhat increased by mechanical stimuli, for instance when the stomach is filled.

The production of the gastric juice itself is not only regulated by the direct contact of the stomach contents with the stomach wall (mech-anical), but also by nerves and hormones.

The expression 'stomach rumble' could well be invented for *neuronal regulation*. Seeing, smelling, tasting, or even thinking about food leads reflexively to rapid increased secretion of gastric juice. Branches of the vagus nerve, an important nerve of the parasympathetic (vegeta-tive) nervous system, are responsible for acti-vating the gastric glands.

Nerves of the sympathetic (vegetative) nervous system can make the production of gastric juice slow down. This happens when a person is afraid, suffers pain or is angry.

Hormonal regulation takes place when food comes into contact with the stomach wall. Certain cells of the antrum start producing the hormone *gastrin*. As always, this hormone ends up in the circulatory system (in the portal veins, via capillaries and gastric veins). After a journey through the lesser circulation, the hormones eventually return to the wall of the stomach, via gastric arteries. Here, gastrin causes an increased production of gastric juice.

Storing, churning, digesting and emptying

Psychologically, a hungry feeling corresponds with mealtimes, but it is also linked with con-tractions of the empty stomach. The stomach 'rumbles' because of the shifting of juices in the stomach. The swallowed food is temporarily stored, churned and partly digested in the stomach. Next, it passes to the duodenum, bit by bit. After a meal of average composition, it takes about three hours for the stomach to empty. After a sumptuous meal one or two hours should be added. It is noticeable that, compared to the time used for eating, digestion time is considerably longer.

When the stomach is empty it is rather small. If the stomach becomes filled with food, the muscular coat gradually stretches with the pres-sure. As to the maximal size of the stomach, there are considerable differences between one person and the other. A very full stomach can dilate the abdominal wall at the left side below the ribs.

Stretching the stomach wall contributes to feeling satisfied. About fifteen minutes after the arrival of food in the stomach, the stomach starts making pertistaltic movements. Directly behind the pylorus the muscular wall is interrupted by a ring of connective tissue. As a result, gastric peristalsis is detached from that of the small intestine. The stomach peristalsis goes from fundus to pylorus. By this motion the stomach contents are superficially churned and mixed with gastric juice. Although hydrochloric acid lowers the pH in the stomach, the digestion of carbohydrates continues for some time under the influence of the swallowed salivary amylase. After all, the stomach is filled layer by layer, and only the bolus of food against the wall has any contact with the acid gastric juice. The 'heart' of the stomach contents keeps a pH of 7 for some time, optimal for salivary amylase to be effec-tive. For similar reasons, the digestion of pro-teins takes place mainly at the outside of the bolus.

With the peristaltic wave occurring about every 20 seconds, the thin border layer of the stomach contents is stripped and another layer gets into intensive contact with the gastric juices.

After these processes, the stomach contents are called *chyme*.

As a result of the increasing peristalsis, a small part of the antrum contents is pushed into the first part of the small intestine (the duodenum), after a delay. Next, the pylorus reflexively closes under the influence of the acid stimulus on the intestinal wall. Moreover, stomach peris-talsis is slowed down.

The hydrochloric acid in the chyme serves as a catalyst for the reaction:

$$pro\text{-}secretin \xrightarrow{HCl} secretin$$
$$\text{(from duodenal wall)} \qquad \text{hormone}$$

Secretin is absorbed into the blood and arrives, after the usual detour, at the pancreas, where it stimulates the secretion of pancreatic juice. One of the components of pancreatic juice is sodium bicarbonate ($NaHCO_3$). This substance comes into the duodenum and neutralizes the acidic chyme:

$$NaHCO_3 \text{ (decomposed)} \rightarrow$$

$$HCl \text{ (decomposed)} \rightarrow$$

$$\begin{array}{ccc}
Na^+ & + & HCO_3^- \\
+ & & + \\
Cl^- & + & H^+
\end{array}$$

$$NaCl + H_2CO_3$$
$$\nwarrow\swarrow$$
$$H_2O \quad CO_2$$

In this way the (acid) H^+-ions are neutralized. Sodium chloride (NaCl) and H_2CO_3 are formed. The final products of the reaction can be absorbed into the blood.

When, as a result of this reaction, the pH of the contents of the duodenum has risen to 7 or 8, the pylorus opens again for a short time and so on, and so on, until the stomach is emptied bit by bit. The pylorus rhythm is regulated by the degree of acidity in the duodenum, and this explains why it takes one or two hours more to empty the stomach after a sumptuous meal. Decomposition of fats takes place in the duodenum under the influence of the pancreas enzyme lipase. The fatty acids formed are added to the

hydrochloric acid. The amount of acid to be neutralised is greater, as a consequence of which the pylorus opens after longer periods.

There is hardly any absorption in the stomach. Only fat soluble substances, such as alcohol, are absorbed into the capillary network through the wall of the stomach and reach the liver via the portal veins.

Vomiting

Vomiting can be induced by a strong stimulus of the mucous membrane of the pharynx, the stomach or the duodenum. A well-known cause is an over-full stomach. Vomiting can also be induced by strong stimulus of the olfactory mucosa (smell) or by strong psychic stimulus caused by, for instance, witnessing a terrible accident.

The vomiting reflex begins with a deep breath, the glottal aperture is closed, as is the entrance to the nasal cavity. The gastric muscles and both oesophageal sphincters relax, whereas strong contractions of both diaphragm and abdominal muscles take place.

The pressure in the abdomninal cavity increases as a result of the decrease in volume and consequently the gastric contents are expelled via the relaxed oesophagus.

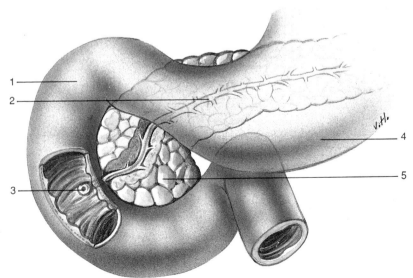

Figure 2.2.26
Position of the duodenum
and the pancreas

1. duodenum
2. pancreatic duct
3. hepatopancreatic
   ampulla (ampulla of
   Vater)
4. stomach
5. tip of pancreas

2.5. Small intestine

The small intestine is the part of the alimentary tract which is responsible for the continued digestion of the chyme and the absorption of the substrates (see Fig. 2.2.7).

Indigestible and undigested remains are carried to the colon. Apart from intestinal juice and pancreatic juice, bile is important for digestion. Intestinal and pancreatic juices contain enzymes to decompose carbohydrates, fats and proteins. Bile contains a substance which helps in the digestion of fats.

The small intestine is slightly more than six metres long. It is divided into duodenum, jejunum and ileum.

The *duodenum* derives its name from its length: twelve finger-widths (20–25 cm.). This was the method of measurement used by the classic anatomist who first described this part of the alimentary tract.

Directly after the pylorus the duodenum has a dilatation: the *duodenal bulb* (Fig. 2.2.25b). The duodenum then arches down to the left. The tip of the pancreas is situated in the 'inner bend' thus created (Fig. 2.2.26). The common mouth of the pancreatic duct and the choledochous duct (bile duct) is situated in the wall of the 'inner bend', the hepatopancreatic ampulla (ampulla of Vater). The ampulla has a valve, the sphincter of the hepatopancreatic

Figure 2.2.27
Structure of the wall of the small intestine

a. Duodenum with the hepatopancreatic ampulla (ampulla of Vater)

1. small intestine (duodenum)
2. muscular coat
3. fold
4. serous coat
5. longitudinal muscle tissue
6. circular muscular tissue
7. gland
8. villi
9. mucous membrane
10. hepatopancreatic ampulla
11. sphincter of the hepatopancreatic ampulla
12. pancreas

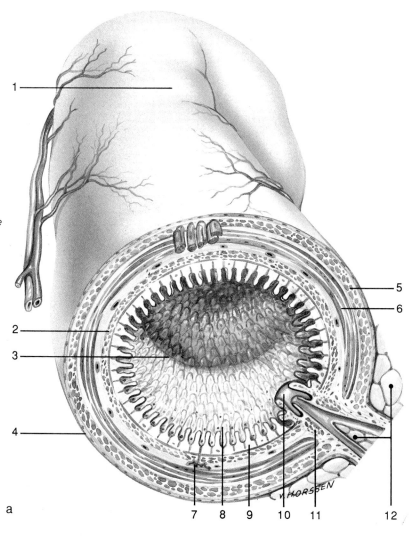

a

ampulla (Oddi sphincter) (Fig. 2.2.27a). This opens when there is chyme in the duodenum. The duodenum ends below the stomach, to the left of the spine. There the jejunum begins, arching ventrally. The jejunum is also called the empty testine, from the belief that this portion of the intestine is always found empty after death. The jejunum has a length of about 2.5 metres and is arranged in coils and loops. Without a clear marking point the jejunum changes into the third part of the small intestine, the ileum. Because of its many coils, this is also called the convoluted intestine. It has a length of about 3.5 m. and is the longest part of the small intestine. It ends in the ascending colon in the lower right section of the abdominal cavity.

The duodenum is supplied with blood by a branch of the coeliac trunk; the jejunum and the ileum are supplied by the superior mesenteric artery (see Fig. 2.1.35). The veins of the small intestine discharge in the portal vein that extends to the liver (see Fig. 2.2.34).

Special characteristics of the wall

The emphasis in structure, in comparison with the general structure of the alimentary tract, is on the mucosa. There are a number of characteristic features. The intestinal wall has a large number of *folds*, parallel to each other, at right

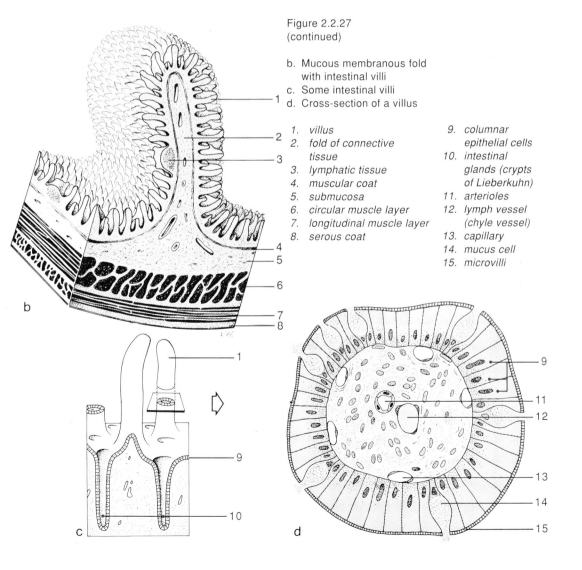

Figure 2.2.27
(continued)

b. Mucous membranous fold
   with intestinal villi
c. Some intestinal villi
d. Cross-section of a villus

1. villus
2. fold of connective
    tissue
3. lymphatic tissue
4. muscular coat
5. submucosa
6. circular muscle layer
7. longitudinal muscle layer
8. serous coat
9. columnar
    epithelial cells
10. intestinal
    glands (crypts
    of Lieberkuhn)
11. arterioles
12. lymph vessel
    (chyle vessel)
13. capillary
14. mucus cell
15. microvilli

angles with the linear direction of the tract (Fig. 2.2.27a).

The circular folds are formed by ridges in the submucosa, covered with mucosa. They are not flattened when the intestine is filled. The presence of the intestinal folds causes the first increase in the surface area and slows down the speed of the chyme, similar to the effect speed ramps have on traffic. The *intestinal villi* cause the second increase in the intestinal surface area. The small finger-like processes of the folds are only just visible to the naked eye, 1 mm high and 0.1 mm wide (Fig. 2.2.27b). The central part of a villus consists of submucosa with capillary network, a lymph capillary and a nerve branch

(Fig. 2.2.27d and e). The intestinal contents is contained by the outside of the villus (internally situated m.e.), which consists of *single layered columnar epithelium*. Between the villi in the mucosa are simple tubular structures or crypts, called the *intestinal glands (crypts of Lieber-kuhn)* (Fig. 2.2.27c). An intensive mitosis of the columnar epithelial cells takes place at the bottom of the crypts.

The newly formed cells push away their neighbouring cells in the direction of the tips of the villi. Here, intestinal cells are continuously replaced (Fig. 2.2.27e). The lifespan of an intestinal cell is about five days.

The epithelial cells, when examined under an

Figure 2.2.27 (continued)

e. Circulation in a villus
f. Intestinal cell with microvilli

1. *columnar epithelial cells*
2. *capillary*
3. *microvilli*
4. *lymph vessel (chyle)*
5. *vein*
6. *artery*
7. *endocytosis*
8. *mitochondria*
9. *endoplasmic reticulum*
10. *exocytosis*
11. *nucleus*
12. *lyosomes*

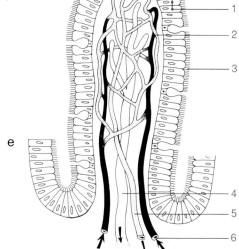

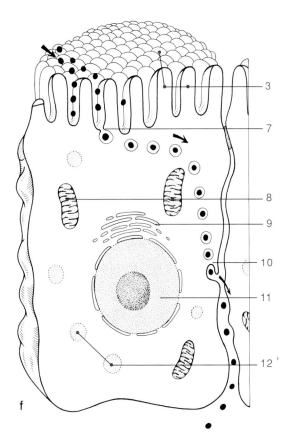

electron microscope, show a great number of membranous processes, the *microvilli* (Fig. 2.2.27d, e and f).

This causes a third increase in surface area. The entire contact surface of the intestine thus becomes about 200 m², Table 2.2.III. This enormous surface contributes to digestion and especially to absorption. This increase in surface area is most evident in the duodenum. It gradually decreases in the direction of the colon.

Table 2.2.III
Increase in surface area of the small intestine

Structure	Surface area (m²)	Enlargement-factor
intestinal tube	0.6	–
folds	1	1.7
villi	8	8
microvilli	200	25

There are a number of mucus-producing cells between the columnar epithelial cells. The mucus serves as a lubricant and as a protective layer on the mucosa.

In the submucosa of the duodenum are the *duodenal glands* or *Brunner's glands*. These produce, besides a quantity of mucus, NaHCO₃. Both protect the duodenal wall against being affected by the acid chyme passing through the pylorus. Furthermore, the mucosa of the duodenal wall produces the enterokinasis enzyme and the hormones *prosecretin* (which is converted into secretin under the influence of HCl) and *cholecystokinin-pancreozymin* (CCK-PZ). Their functions will be discussed later.

Intestinal glands are found throughout the small intestine. They are formed by glandular cells around the crypts of Lieberkuhn (Fig. 2.2.27c).

The glands produce about two litres of intestinal juice per 24 hours. The juice consists mainly of water and contains no enzymes.

Scattered all over the small intestine are small aggregates of lymphoid tissue in the mucosa. In the wall of the ileum these aggregates of lymphoid tissue are arranged in about 25 somewhat larger lymphoid plaques, *aggregated lymphatic follicles* (*Peyer's plaques* or *Peyer's patches*).

The duodenum is situated retroperitoneally. Only the ventral side is covered with serosa. The other part of the small intestine is situated intraperitoneally.

Churning and transport

The peristalsis of the small intestine comprises more than the 'one-way traffic' of the oesophagus. Churning takes place during the digestive process (Fig. 2.2.28). The muscular coat contracts and relaxes at irregular intervals. Consequently, the intestinal contents are again and again divided into portions, so that an intensive mixing with the digestive juices takes place. Simultaneously the substrates can be absorbed through the intestinal wall. While being churned, the chyme is slowly carried in the direction of the ileum. 'One-way peristalsis' takes place here in particular, as a result of which the roughage, the undigested and the non-resorbed remains, is conveyed to the colon.

Figure 2.2.28
Peristalsis in the small intestine (schematic)

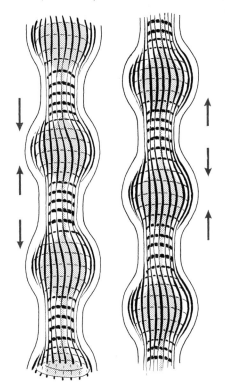

Figure 2.2.29
Emulsification

1. *bile salts*
2. *fat drop*
3. *fat droplets*

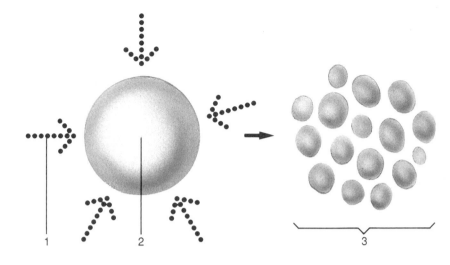

Digestive juices and their secretion

Digestion, which began in the oral cavity and the stomach, is completed in the small intestine. A large amount of enzymes are available for digestion. All these enzymes have a pH-optimum of 7 to 8. Apart from enzymes, digestive juices contain some other important substances. The pancreas is the main producer of digestive juices (more than one litre per day)

Besides *water* and *mucus* the pancreatic juice contains:

– *amylase*, for the conversion of amylum into maltose;
– *lipase*, for the digestion of fats into glycerin and fatty acids. The effect of lipase is considerably enhanced by the influence of choledochous salts on the fats. In a watery environment fats have a rather large *cohesion*, which means that larger drops of fat do not disintegrate into droplets. The surface tension, which keeps the drop of fat together, is reduced by the effect of bile salts (Fig. 2.2.29). As a result, the fat drop disintegrates into a large number of droplets (emulsion). This effect, caused by choledochous acid salts, is called *emulsification*. The sum of all surface areas of the droplets is many times greater than the surface area of the original fat drop. So the field of activity of lipase is expanded;
– trypsinogen, the inactive precursor of the protein splitting enzyme trypsin. If the pancreas produced an active proteinase then the proteins, of which the pancreas itself is made, could be split.

The following reaction is found in the intestinal lumen:

$$\text{trypsinogen} \xrightarrow{\textit{enterokinase}} \text{trypsin}$$

The enterokinase produced by the duodenal lining functions as a catalyst. Once trypsin has been formed, an autocatalytic process takes place:

$$\text{trypsinogen} \xrightarrow{\textit{trypsin}} \text{trypsin}$$

Apart from trypsinogen, the pancreas produces relatively small amounts of precursors of other proteinases. These are only activated in the intestinal lumen. The intestinal lining itself is protected against the effects of proteinases by the mucous layer on the mucosa;

– *sodium bicarbonate* ($NaHCO_3$) neutralizes the acid chyme, which comes from the stomach, in the duodenum. The changes in pH at the beginning of the duodenum regulate the pyloric rhythm. In the second part of the duodenum the pH becomes 7 to 8 by the effect of $NaHCO_3$. This has two advantages:
  • the remainder of the small-intestinal lining, which has a thinner mucous layer than the wall of the stomach and the duodenum, is protected against the chemical effects of acids;

- an optimal environment is created for the enzymes in the digestive juices.

When there is no chyme in the duodenum, the hepatopancreatic ampulla is closed by the sphincter of the hepatopancreatic ampulla. The digestive juices are given *neuronally* or *hormonally* to the intestinal contents, much the same as happens in the stomach. The activity of the vagus nerve causes the secretion of pancreas juice bile and intestinal juice, when there is chyme in the small intestine.

The hormonal regulation is effected by two hormones: *secretin* and *cholecystokinin-pancreozymin* (CCK-PZ).

Prosecretin is produced in the duodenal wall when chyme passes. Under the influence of hydrochloric acid this substance is activated in the chyme, converts into *secretin* and is absorbed into the blood. This hormone stimulates the secretion of pancreatic juice via the bloodstream (and thus indirectly regulates the pylorus rhythm) as well as the secretion of bile by the liver.

The hormone CCK-PZ is also secreted by the duodenal lining when there is chyme in the duodenum. It can be deduced from its name that this hormone, after being absorbed in the blood, stimulates the gall-bladder to contract and the pancreas to secrete pancreatic juice.

Absorption

The processing of the food consumed, as described, is aimed at the absorption of the substrates into the blood. The small intestinal wall is extremely suitable for absorption processes. The total absorption area is large (200 m^2), the columnar epithelium is highly permeable and the submucosa has a good blood supply.

Absorption will be described, using the elements of food as a starting point.

Carbohydrates are broken down into monosaccharides (dextrose, fructose and galactose). They are carried to the capillaries in the villi via active transport, for which the insulin hormone is indispensable.

Proteins are broken down into amino acids which are absorbed actively.

Fats are broken down into glycerin and fatty acids. The fatty acids with a carbon chain of less than twelve atoms diffuse straight into the blood. The major part of the products of fat digestion is converted into triglycerides by the epithelial cells. These triglycerides in their turn form fat droplets. These are discharged to the m.i. of the submucosa by *exocytosis*. Moreover, *pinocytosis* of fat droplets plays a role (Fig. 2.2.27f). Such fat particles cannot be absorbed into the capillaries, but must be absorbed by the more permeable lymphatic capillaries. Such lymph, rich in fat, is called chyle. The lymph vessels concerned are called chyle vessels. The fat particles do not pass through the liver and are discharged into the bloodstream via the thoracic duct.

The *fat soluble vitamins* (A, D, E and K) are absorbed together with the fat particles.

The *water soluble vitamins* diffuse into the bloodstream and the m.i. without any problems. The same is true for *minerals* and *water*. It should be noted that the amount of water and salts (electrolytes) absorbed, does not all come from the food, but also from saliva, gastric juice, pancreatic juice, bile and intestinal juice.

Absorption takes place mainly in the first half of the small intestine. All of the small intestine, however, is used for the absorption of fats. Choledochous acid salts and vitamin B$_{12}$ are only absorbed in the ileum. For the absorption of vitamin B$_{12}$ the intrinsic factor of the gastric juice is necessary, as has been described earlier.

2.6. Large intestine

The large intestine is the last part of the digestive tract (see Fig. 2.2.7). The waste products of digestion are discharged from the ileum into the large intestine, later to be expelled from the body as faeces.

The large intestine is about 1.5 metres long, looks 'indented' and consists of five parts (see Fig. 2.2.7):

- the *blind intestine (caecum)*, situated below the entrance of the terminal ileum. This part of the large intestine is a cul de sac; it has an appendix, the *vermiform appendix* (Fig.

2.2.30). The submucosa of the appendix contains a considerable amount of lymphoid tissue;
– the *ascending colon*, the continuation of the caecum, which extends to the lower part of the liver;
– the *transverse colon*, which arches from the lower part of the liver across the small intestine and the stomach to the left to the spleen;
– the *descending colon*, which extends downwards from the spleen into the pelvis;
– the *sigmoid colon*, which extends to the centre of the pelvic cavity, its bend resembling an 'S';

– the *rectum*, which extends straight down from the pelvis and pierces the pelvic diaphragm and the perineal muscles. The rectum is not crenated. The upper part, the rectal ampulla, is dilated and can easily be stretched. The rectum terminates at the anus.

There is an extra large intestinal fold at the junction of the ileum and the large intestine (Fig. 2.2.30). This fold can serve as a sphincter to prevent reflux of the caecal contents into the ileum.
As this sphincter is situated at the junction of ileum and caecum, it is called ileo-caecal valve.

Figure 2.2.30
Junction of small intestine and large intestine

a. Closed

   1. ascending colon
   2. mesentery
   3. terminal ileum
   4. caecum
   5. (vermiform) appendix

b. Open

   1. teniae coli
   2. ileo-caecal valve

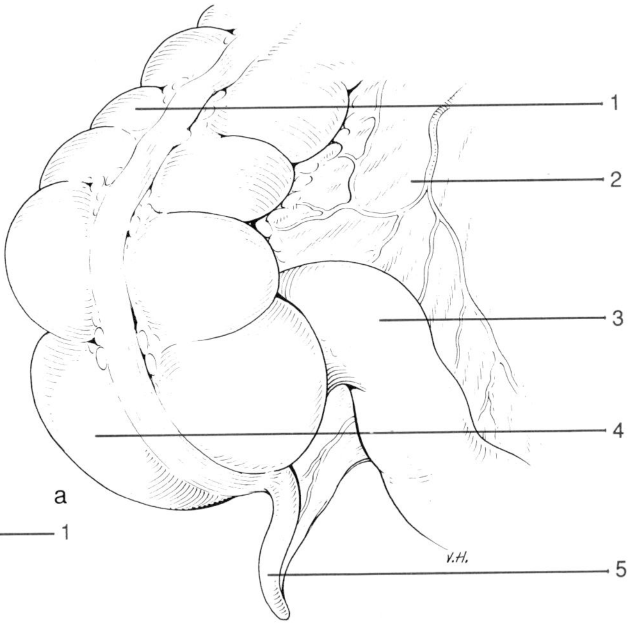

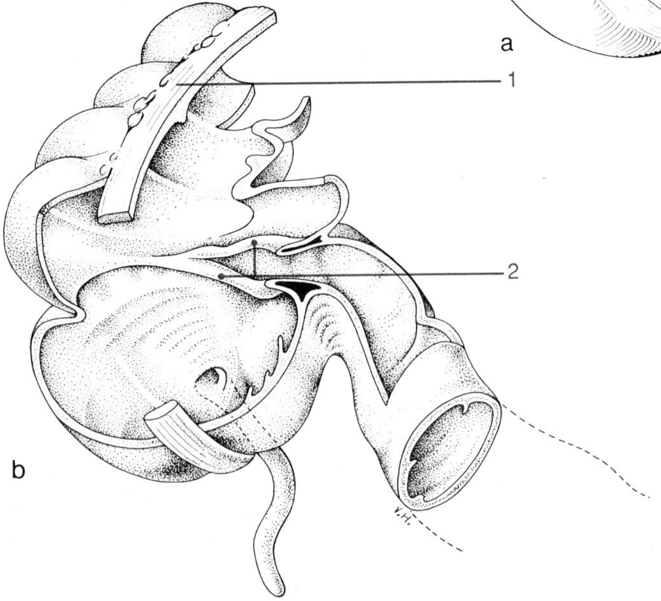

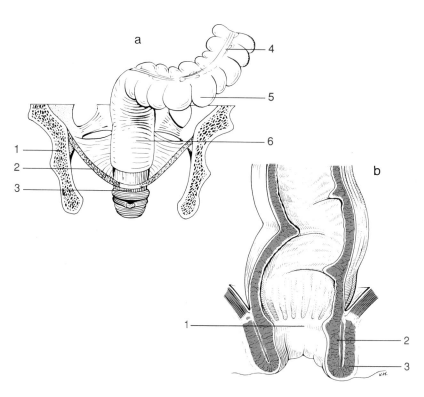

Figure 2.2.31
Rectum and anus

a. Closed

1. pelvic or hip bone
2. perineal muscle
3. anus and
   sphincters
4. teniae coli
5. sigmoid colon
6. rectal ampulla

b. Open

1. transition from
   intestinal mucosa
   to skin
2. internal anal
   sphincter
3. external anal
   sphincter

Special characteristics of the wall

The large intestine has a general structure similar to the one described earlier. The lining is folded but has no villi. The mucosa consists of one layered columnar epithelium with many mucous cells. The muscularis consists of a cylindrical and a longitudinal layer. The circular layer is not entirely covered by the longitudinal layer, however, but by three thickened tape-like bands, the *teniae coli*, 1 cm wide, that extend lengthwise over the entire large intestine (Fig. 2.2.30b). The teniae are somewhat shorter than the rest of the intestinal wall, and cause constrictions at regular intervals, thus producing bulging sacculi.

The anus has two sphincters (Fig. 2.2.31). The internal sphincter is a thickening of the circular part of the muscularis and consequently consists of smooth muscular tissue.
The external sphincter surrounds the longitudinal muscle layer and consists of striated muscular tissue. The lining of the anus shows the transitions from the intestinal mucous membrane (non-keratinized epithelium) to the skin (keratinized epithelium).

The large intestine is situated partly intraperitoneally, partly retroperitoneally. The serosa has a peculiar 'fringe' made of fat tissue and lymphoid tissue. The rectum is situated subperitoneally.

The superior mesenteric artery (see Fig. 2.1.35), supplies the first section of the large intestine; the inferior mesenteric artery supplies the other section. The veins discharge into the portal vein (see Fig. 2.2.34), i.e. into the liver, with the exception of the vein for the lower part of the rectum. This vein discharges straight into the inferior vena cava.

Absorption, transport and storage

No enzymes are secreted in the large intestine. Digestion is complete, the colon contents consist of roughage and indigestible remains. These remains form a culture medium for the microbes that live in the intestines (coli-bacteria), the *intestinal flora*. The bacteria produce vitamin K and vitamins from the B-complex as waste material of their metabolism. Both the coli-bacteria and man profit by this way of 'living together' or *symbiosis*.

Vitamins, some salts and a relatively large volume of water (about half a litre per 24 hours) are absorbed. As a result of this thickening of the colonic contents, the faeces proper are formed. The corresponding meal took place 15 to 20 hours earlier.

Gases (methane, hydrogen sulphide and others) are formed in the large intestine as a result of the activity of the intestinal flora and because of decaying processes. When a person breaks wind these gases escape (flatus).

Due to the peristaltic motions of the large intestine the faeces are carried from the ascending and transverse sections to the descending sections and the sigmoid colon.

The faeces can be stored for varying periods of time.

Defecation

The more the descending colon and the sigmoid fill, the more this part of the colon wall stretches. This causes stronger peristalsis, as a result of which the faeces are pushed into the empty rectum. When the rectal ampulla is filled, stretching sensors in the wall are activated, which in turn activate the defecation reflex. The muscularis of the rectum is stimulated via a synapse with the lower part of the spinal cord, whereas the internal anal sphincter relaxes. Through the process of 'potty-training' we are taught to suppress the defecation reflex and to voluntarily tighten the external anal sphincter.

When the internal sphincter is relaxed and the external sphincter is contracted, we experience the urge to defecate. The urge may disappear for a while, but it returns even stronger. When the external sphincter is relaxed, usually at a suitable time and place, defecation follows: peristaltic movements empty the terminal part of the colon.

By straining, the pressure on the intestine, and therefore on the faeces, increases, which can accelerate defecation.

Defecation usually takes place one to three times a day.

Common elements of faeces are: water and mucus, salts (calcium, magnesium, iron), biliary colours, roughage (for instance cellulose), microbes and mucosal cells. Normally, faeces are sausage-shaped or pulpy, and medium solid to soft in their consistency.

The colour is dependent on the kind of food eaten. The average colour is dark brown, caused by the biliary colours, but after having eaten beetroot, the faeces will be dark red. Vegetarians usually have brownish yellow faeces (vegetable food) and babies have yellow-gold faeces (maternal milk).

3. Pancreas

An obsolete name for the pancreas is 'abdominal salivary gland'. Yet this name describes this organ very well: a gland, situated in the abdominal cavity, which supplies, among other things, water, mucus and enzymes to the digestive tract. The classic anatomist, however, only saw pan-creas, i.e. all flesh, from which it still derives its name.

The pancreas is situated retroperitoneally. It is a mallet-shaped elongated organ (15–20 cm). The tail reaches the spleen, the body is situated behind the stomach and the head, which is somewhat bigger, lies within the loop of the duodenum (Fig. 2.2.32).

The separate 'bunches' of the gland discharge their produce in a common duct, the *pancreatic duct*, which extends through the centre of the pancreas to the duodenum. Just before the terminal opening the pancreatic duct usually unites with the common bile duct. Near the duodenal bend is the terminal in the intestine – the *ampulla of Vater*.

Both the elements and the functions of the pancreatic juice have been discussed, as well as the regulation of the juice secretion.

Pancreatic islets (islets of Langerhans)

The pancreas is clearly an *exocrine* gland. When pancreatic tissue is coloured in a special way (see Fig. 2.6.9), many scattered clusters (islets)

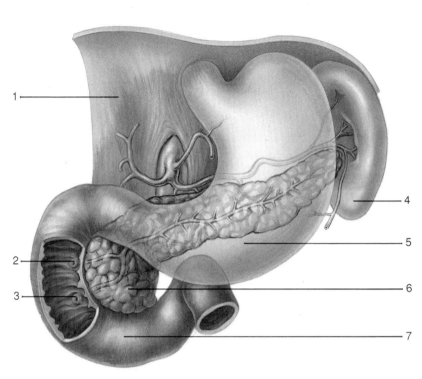

Figure 2.2.32
Location of the pancreas

1. *diaphragm*
2. *papilla (of Santorini)*
3. *ampulla (of Vater)*
4. *spleen*
5. *stomach*
6. *pancreas*
7. *duodenum*

of glandular tissue without excretory ducts can be seen. However, the pancreas is also an endocrine gland – a hormonal gland. The hormones produced find themselves in the m.i. and later in the bloodstream.

The endocrine tissue has two types of hormone producing cell: alpha and beta-cells. These cells produce the hormones *glucagon* and *insulin* respectively.

Insulin is responsible for, among other things, the conversion of glucose into glycogen. This conversion takes place in the liver and the muscles. The hormone is secreted when the volume of glucose in the blood exceeds a certain level. When the blood sugar level threatens to be too low by comparison with the need for energy sources, the adrenalin hormone (from the adrenal gland), supported by the glucagon hormone, reacts in the opposite manner:

$$\text{glucose} \xrightleftharpoons[\substack{adrenalin \\ glucagon}]{insulin} \text{glycogen}$$

The blood sugar level is kept as constant as possible by this so-called *glucose buffer*.

4. Liver

The liver is located in the upper right portion of the abdomen, next to the stomach. It sits on the intestines, and is connected with the right part of the diaphragmatic dome. Although the liver is an abdominal organ, it is protected by the ribs of the thoracic cavity due to the dome-shape of the diaphragm. The liver is pushed down and becomes easily palpable during deep inhalation. The liver is intraperitoneally located (Fig. 2.2.33). On one side it is linked with the lesser curvature of the stomach (lesser omentum) by the peritoneal folds, on the other side with the anterior abdominal wall by the crescent-shaped ligament (falciform ligament of the liver). This ligament (see Fig. 2.2.40), is connected with the round ligament of the liver (ligamentum teres), which is a remnant of uterine life. During that period the umbilical vein extended to the liver, after entering the body via the navel. The liver is divided into a right lobe (2/3 of its total size) and

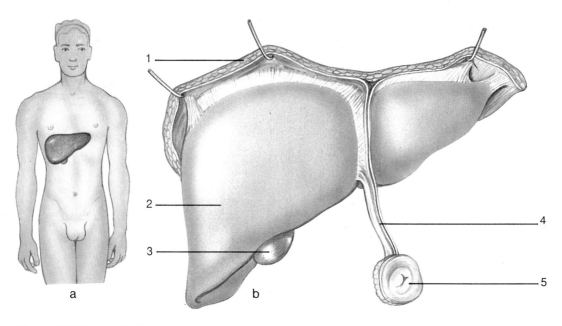

Figure 2.2.33   1. diaphragm
The liver   2. liver
                 3. gall bladder
a. Location   4. round ligament
b. Ventral view 5. navel (umbilicus)

a left lobe (1/3 of the size). The hilum of the liver is on its underside. Here the hepatic artery and the portal vein enter and the hepatic duct and lymph vessels leave the liver (Fig. 2.2.34). Nerves also enter and leave the liver via the hilum. Near to the hilum is the gall bladder. The hepatic veins leave the liver at the upper side and almost immediately discharge into the inferior vena cava.

The liver is the largest organ in the body and weighs an average of 1500 g. It has a key role in the circulatory system. The liver is an intermediate station between the gastrointestinal tract and the spleen on the one side, and the inferior vena cava on the other side (Fig. 2.2.34). All of the blood from the spleen and the gastrointestinal tract, apart from the lower part of the rectum, enters the liver through the portal vein. The liver controls the composition of the blood and is able to influence the concentration of many substances in the blood and maintain a balance by means of a number of chemical processes. The liver has great importance for the homeostasis of the m.i. due to its influence on the composition of blood.

4.1. Structure

The liver is composed of small hexagonal *hepatic lobules* (Fig. 2.2.35). Such a lobule consists of cubic liver cells, surrounded by a connective tissue capsule. At three of the six angles is a space between the lobules, made of connective tissue: the *Kiernan spaces*. These spaces contain fine branches of the portal vein, the hepatic artery and the bile duct.

A number of arterioles and venules branch from the *interlobular arteries and veins*, which enter the tissue of the three bordering lobules at right angles.

The two adducting vessels merge into an integrated vascular network between the cubic liver cells. This network does not normally consist of capillaries, but of somewhat larger vessels (sinusoids), of which the wall consists of endothelium and *stellate reticuloendothelial cells (cells of Kupffer)*. These cells are relatively large, star-shaped phagocytic cells. They belong to the

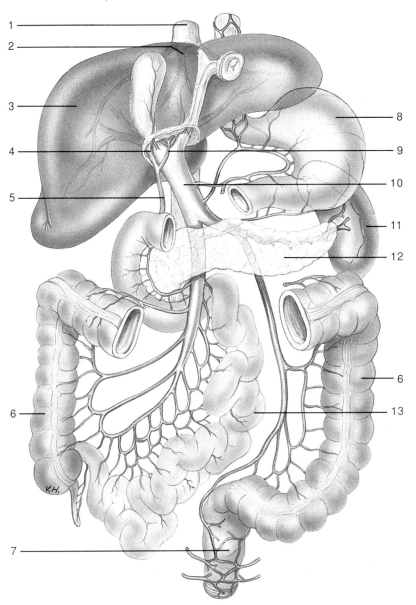

Figure 2.2.34
Portal vein

1. inferior vena cava
2. hepatic vein
3. liver
4. cystic duct
5. common bile duct
   or choledochous
6. large intestine
7. rectum
8. stomach
9. common hepatic
   duct
10. portal vein
11. spleen
12. pancreas
13. small intestine

reticuloendothelial system. The venules which drain the blood from these sinusoids discharge into a vein in the centre of each lobule – the *central vein*. The central veins of the different lobules merge into larger veins. These veins discharge into one of the three hepatic veins that transport the blood to the inferior vena cava.

The blood supply from the hepatic artery is important for the supply of oxygen to the liver cells, as the blood from the portal vein contains many substrates, but no oxygen. The oxygen has already been used by the organs which are drained by the portal vein (Fig. 2.2.34). There is also a biliary capillary network between the liver cells. The bile capillaries drain into the *interlobular bile ducts*. The flow of bile and that of the supplied blood are in opposite directions. The interlobular bile ducts unite into larger bile ducts. All bile ducts from the liver are united into the *common hepatic duct*. This duct leaves the liver through the hilum.

From an anatomical point of view the hexagonal lobule is the functional unit of the liver. From a

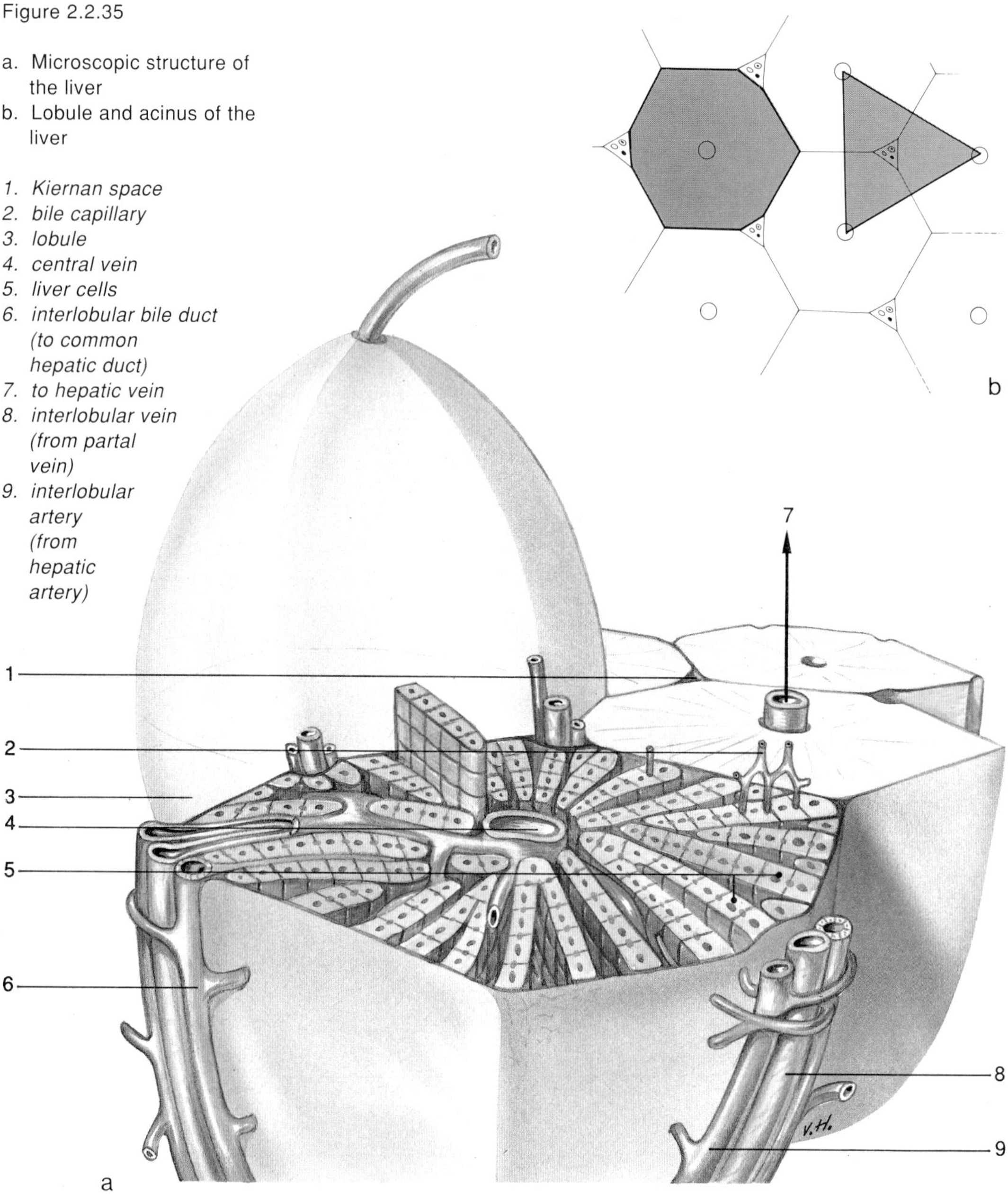

Figure 2.2.35

a. Microscopic structure of
   the liver
b. Lobule and acinus of the
   liver

1. *Kiernan space*
2. *bile capillary*
3. *lobule*
4. *central vein*
5. *liver cells*
6. *interlobular bile duct*
   (to common
   hepatic duct)
7. *to hepatic vein*
8. *interlobular vein*
   (from partal
   vein)
9. *interlobular*
   artery
   (from
   hepatic
   artery)

physiological point of view the *acinus* is the functional unit of the liver (Fig. 2.2.35b).
The acinus is a triangle with a central interlobular space and central veins at the angles. Between the centre and the three central veins is the supply and exchange area of the adducting vessels. From this area the transport of bile to the interlobular bile duct takes place.

4.2. Function

The liver is the main 'metabolic headquarters' of the body. The liver is often compared with a large chemical plant, in which a quantity of substances are converted, broken down, stored or produced. There are an unparalleled number of different enzymes active in the liver.
Of cardinal importance is the role of the liver in

the circulatory system: all nutrients, with the exception of the somewhat larger fat particles, end up in the liver via the portal vein before they are allowed into the greater circulation.

The different functions of the liver are: the metabolism of carbohydrates, proteins and fats, the storage of glycogen, detoxification, the production of bile and the production of heat.

These functions are discussed separately. It will be clear that the liver is able to convert all kinds of substances absorbed by the intestinal lining, before they are passed into the body cells. These processes together are called *intermediary metabolism*. This intermediary metabolism is very important, as not all of the body cells have the enzymes needed to release energy from fatty acids and amino acids, to form fats out of carbohydrates or to synthesize certain amino acids. The liver performs these tasks for the body cells.

Carbohydrate metabolism

Glucose is one of the main sources of energy for the body cells. Nervous tissue uses glucose exclusively for its metabolism. The human body needs at least 150 g of glucose per 24 hours, of which the brains use 120 g. The liver plays an important role in maintaining a constant *blood sugar level*, which is the *volume of glucose per litre of plasma*. To this end the liver has a number of available reactions, in which glucose plays the leading part (Fig. 2.2.36).

The *glucose buffer* is perhaps of the greatest importance. Influenced by the hormone insulin, glucose is converted into glycogen. The proper name for this reaction is *glycogenogenesis*, but it is usually called *glycogenesis*.

When glycogen is converted into glucose under the influence of the hormones adrenaline and glucagon, the reaction is called *glycogenolysis*. The glucose buffer therefore consists of two antagonistic (adverse) reactions. A maximum of 200 g glucose can be stored in or released from the liver. There is a glycogen stock of about 250 g in the skeletal muscles.

Furthermore, glucose is formed in the liver by converting fructose and galactose with the help of the corresponding enzymes, fructase and galactase. These reactions are called *gluco-*

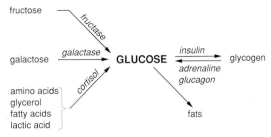

Figure 2.2.36
Carbohydrate metabolism
in the liver

genesis. When the supply of glucose in the liver is larger than can be stored in the glucose depots (hepatic and muscular glycogen), the liver is able to convert glucose into fats. This process is called *lipogenesis*.

Only a small portion of the fat thus formed is stored in the liver itself. The larger part is transported via the bloodstream and stored in the fat depots in the subcutaneous connective tissue.

The liver is also able to form glucose from non-carbohydrate sources. This process is called gluconeogenesis (new formation of glucose) and is very important when the glucose depots are empty, for instance as a result of long exertion or insufficient carbohydrates in food.

Amino acids, glycerol, fatty acids and lactic acid are used in gluconeogenesis. The amino acids may originate from the food absorbed (imbalanced protein-rich food) or from broken down body proteins (especially muscular tissue). Glycerol and fatty acids originate mainly from the fat deposit. Lactic acid is a product of anaerobic glycolysis and is formed when muscles are strenuously exercised.

The cortisol hormone (hydrocortisone) from the adrenal cortex activates the process of gluconeogenesis.

Figure 2.2.37
Fat metabolism in the liver

unsaturated
fatty acids

FATTY ACIDS → body fat

glucose
amino acids

glucose

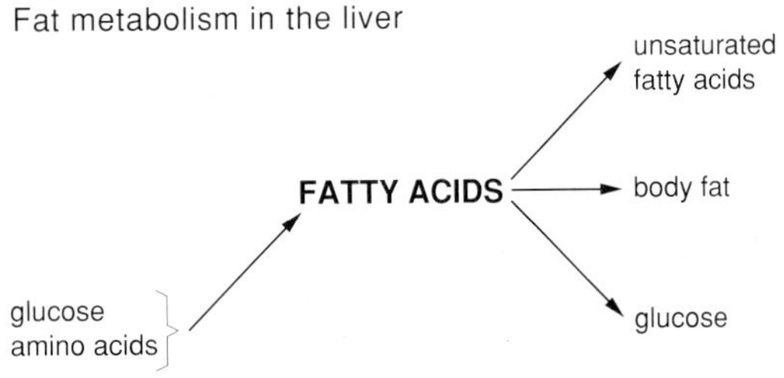

Fat metabolism

The liver has a role in the digestion of fat (Fig. 2.2.37).

Fatty acids become unsaturated in the liver. They can be converted into body fat (fat depots) and they can be burned for the supply of energy. The liver is able to convert glucose, and even amino acids, into fatty acids in the lipogenetic process. Fatty acids and glycerol, in their turn, can be converted into glucose by means of glucogenesis.

Cholesterol is a fatty substance which the liver uses to form bile salts, (elements of bile), and it is very important for the digestion of fat in the small intestine.

Protein metabolism

The amino acids, which enter the liver through the portal vein, are basic elements for all kinds of reactions in the body (Fig. 2.2.38).

The liver synthesizes proteins. This in itself is nothing special, for the synthesis of proteins (for instance for the formation of enzymes or fibres) is within the capability range of most cells. The liver, however, produces practically all *plasma proteins* (albumin, globulins and fibrinogen) and passes them into the bloodstream.

The liver is also capable of converting certain amino acids into others. This process is called *transamination*. Shortages in certain amino acids can be coped with by means of this trans-amination. However, not all 20 amino acids can be formed in this way. The non-synthesizable amino acids must be part of the diet: they are the essential amino acids.

When there is a shortage in fuel or a surplus of amino acids, the liver is able to convert amino acids into glucose by means of *gluconeogenesis*

or into fats by means of lipogenesis. In these reactions the amino groups (NH_2) of the amino acids are separated first. This process is called deamination.

Detoxification

The liver is capable of making certain alien or toxic substances suitable for excretion via the urinary passage or the biliary tract.

The first example is connected with an earlier subject. During the deamination process amino groups are released. The amino groups are converted into the (toxic) substance ammonia (NH_3) and then, with the help of carbon dioxide (CO_2) into *urea*. The liver adds the waste substance, urea, to the blood, the kidneys remove it, with the result that urea is found as an element of urine.

Bilirubin, produced in the spleen when the haemoglobin of the erythrocytes is broken down, enters the liver through the portal vein, linked with albumin. The liver couples (conjugates) the bilirubin with glucuronic acid. *Conjugated bilirubin* is more soluble and can be excreted more easily than bilirubin. The waste is added to the bile. Worn hormones or a surplus of certain hormones (for instance oestrogen, corticol) can also be removed by the liver by coupling them with glucuronic acid. The conjugated hormones end up in the blood, from which they can be removed by the kidneys. The liver is not only responsible for detoxification, necessary because of the normal metabolic processes, but also for the elimination of alien substances, such as drugs or stimulants (alcohol, caffeine, nicotine). The liver renders these substances harmless by changing them chemically and breaking them down, so that they can be excreted via the bile or the urine. This process is called *bioinactivation*.

Storage

It can be concluded from the description above that the liver is a depot for *glycogen*, *fats* and *amino acids*. Together with fats, fat-soluble vitamins (mainly A, D and K) can be stored. Moreover, the liver can form vitamin A out of carotene. The vitamins of the B-complex can also be stored in the liver.

Figure 2.2.38
Protein metabolism in the liver

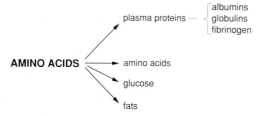

Finally, the liver is an important depot for iron and a number of trace elements, such as cobalt.

Production of heat

In most reactions which take place in the liver, heat is released. The highest temperature in the human body is found in the liver, viz. 39°C (see Fig. 2.5.8). This heat is discharged via the passing bloodstream and spread over the body. The liver plays an important role in keeping the body temperature at 37°C. When a person is active, the muscles, generally speaking, produce a surplus of heat. During rest, however, some heat is produced by the heart and a lot by the liver. This production of heat is normally insufficient to keep the temperature normal, so body heat is enhanced by means of extra insulation (night clothes, blankets and so on).

Production of bile

Apart from water (solvent and means of transport), and mucus (lubricant), bile contains:
- *bile salts* for the emulsification of fats (see Fig. 2.2.29);
- *cholesterol*, which is excreted together with the faeces;
- *conjugated bilirubin*, which is converted by the intestinal bacteria and, after oxidation, causes the faeces to be brown.

Bile can be considered a waste product which is continuously formed by the liver cells during all kind of reactions. At the sime time bile favourably affects digestion as it emulsifies fats. The liver cells discharge bile into the biliary ducts. From this point of view, the liver is an exocrine gland, although its structure is completely different from that of the common exocrine glands.
The liver produces about 700 ml of bile per 24 hours. It is a yellow-greenish, viscid fluid. The gall or bile in the gall bladder is more concentrated.

5. Biliary tracts

The bile produced by the liver cells ends up in the interlobular bile ducts via the bile capillaries (see Fig. 2.2.35). The interlobular bile ducts converge into larger bile ducts and eventually into the *common hepatic duct*. This duct leaves the liver via the hilum. Shortly after it has left the liver, there is a branch to the gall bladder (cholecystis), the *cystic duct*. The bile duct which extends to the duodenum is called the *common bile duct* or *choledochous duct*. Just after the junction with the pancreatic duct it discharges into the inside bend of the duodenum through the hepatopancreatic ampulla with its sphincter.
The gall bladder is located next to the hilum in a fold of the right lobe. It is the size and shape of a small pear.

Special characteristics of the biliary wall

Just like the intestinal contents, the contents of the biliary tracts belong to the m.e., so the lining of the biliary tracts is made of epithelium. Basically, the walls of the bile ducts have a structure similar to that of the intestines: mucosa, submucosa and muscularis. The mucosa and submucosa, however, are poorly developed. A fairly thick layer of mucus lies on the mucosa in order to protect the tissue from being affected by bile.

The mucosa of the gall bladder is made of columnar epithelium. The walls of the empty gall bladder show folds. The muscular coat of the gall bladder is more developed than that of the bile ducts.
Bile ducts and gall bladder are intraperitoneally located.

Storage, concentration and secretion

The continuous flow of bile from the liver is received by the gall bladder. The mucosa of the gall bladder absorbs water from the bile. As a result, the bile becomes thicker and dark green in colour. The maximum capacity of the gall bladder is about 50 ml. The bile is stored in the gall bladder until food (especially fat) passes through the duodenum. The CCK-PZ hormone stimulates the liver to produce bile. The hormone secretin makes the muscularis of the gall bladder contract. The sphincter of the hepatopancreatic ampulla opens and the bile is discharged into the duodenum via peristaltic

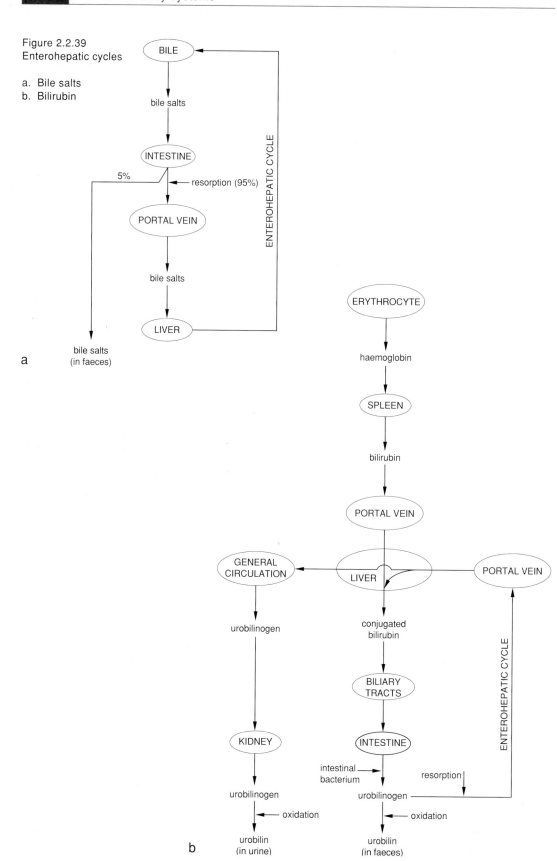

Figure 2.2.39
Enterohepatic cycles

a. Bile salts
b. Bilirubin

movements of the common bile duct. The emulsifying effect of the bile salts is particularly important here.

Enterohepatic cycles

All kinds of substances are discharged into the liver from the intestinal tract via the portal vein. The liver, in its turn, discharges a number of substances into the intestinal tract via the bile ducts. For some substances, therefore, a cycle of being absorbed (intestine) – being discharged (liver) – being absorbed (intestine) – and so on, develops. This is called the *enterohepatic* cycle.

Bile salts follow such a cycle (Fig. 2.2.39a). 95 % of the bile salts, which are excreted via the bile, are absorbed again in the ileum and then discharged into the liver, which adds them to the bile again. The remaining 5% are excreted together with the faeces.

Bilirubin is processed in a somewhat more complicated way (Fig. 2.2.39b). Bilirubin, coupled with albumin, is transported from the spleen to the liver via the portal vein. By conjugating the bilirubin with glucuronic acid the liver makes it more soluble. The *conjugated bilirubin* is discharged into the intestine via the bile ducts. Intestinal bacteria convert the conjugated bilirubin into *urobilinogen*. The larger part of the urobilinogen is then oxydated and converted into *urobilin* and excreted with the faeces. Urobilin makes the faeces brown.

A small part of the urobilinogen, however, is absorbed again and finds its way back to the liver via the portal vein. It is at once partially added to the bile (enterohepatic cycle), and partially remains in the blood, passes through the liver and finds its way into general circulation. The kidneys remove the urobilinogen from the blood; in the urine, with oxygen, it is converted into urobilin. The light yellow colour of urine is caused by the urobilin.

6. Peritoneum

The peritoneum is one of the serous membranes. It is separated into the *visceral peritoneum*, the inside layer that covers the viscera of the abdomen and the pelvis, and the *parietal peritoneum*, the outside layer which lines the inner surface of the abdomen and the pelvis. The peritoneum is very thin, just like the other serous membranes. It is lined by a single layer of flattened mesothelial cells on a thin stretchable base membrane and has many blood vessels. The peritoneal cavity is between the visceral and parietal peritoneum. There is only a thin layer of serous fluid in this cavity. The fluid absorbs the friction between the layers. This friction is caused by the peristalsis of the stomach and the intestines. The location and course of the peritoneum often appear less clear than those of the heart sac and the pulmonary membrane. Yet there is basically no difference. The description of the embryonic development of the alimentary tract and the peritoneum provides an explanation, Module 3, Chapter 2.

The parietal peritoneum lines the abdominal wall and is continuous with the abdominal way. The abdominal organs take up all the space in the abdominal cavity. It is comparable with a balloon, inflated in the abdominal cavity, of which the wall is pushed further and further into ventral direction by the abdominal organs (Fig. 2.2.40). The indented organs are covered by the visceral peritoneum and continuous with the peritoneum.

Intra and extraperitoneal

Some organs are located behind the peritoneum (Fig. 2.2.41). The *retroperitoneal* organs are:
– duodenum;
– pancreas;
– caecum and ascending colon;
– descending colon;
– kidneys and the vertical part of the ureters;
– aorta and inferior vena cava.

Some organs are located below the peritoneum. These *subperitoneal* organs are:
(Fig. 2.2.40b and c.):
– rectum;
– (empty) bladder;
– cervix uteri;
– prostate gland.

When the bladder is filled, its front side rises over the symphysis pubis. The bladder is then

Figure 2.2.40
Peritoneum

a. Location of the
peritoneum in the
abdominal cavity

1. *falciform ligament of
the liver*
2. *stomach*
3. *round ligament of the
liver (free bottom edge
of the falciform
ligament)*
4. *transverse colon
(intraperitoneal)*
5. *greater omentum*
6. *diaphragm*
7. *liver*
8. *lesser omentum (cut)*
9. *pancreas*
10. *omental sac*
11. *duodenum
(retroperitoneal)*
12. *anchoring strap of
duodenum (dorsal
mesentery)*
13. *parietal peritoneum*
14. *visceral peritoneum*

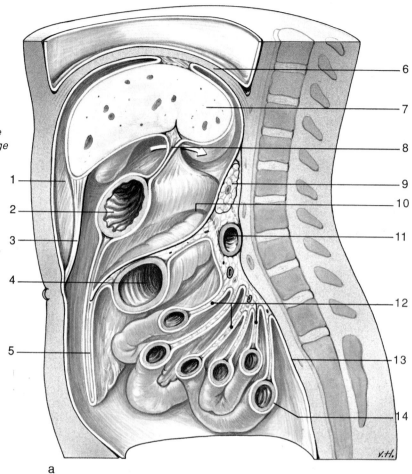

a

partially located *preperitoneally*, in front of the peritoneum.

Retro, sub and preperitoneal locations are all *extraperitoneal* locations, i.e. outside the peritoneum. The other abdominal organs are entirely or partially invested by the serous membrane. They are located inside the peritoneum. The *intraperitoneal* organs are (Fig. 2.2.40a and b):
– stomach;
– jejunum and ileum;
– transverse colon;
– sigmoid colon;
– liver;
– biliary tracts;
– spleen;
– uterus.

It becomes clear from these lists that the portions of the digestive tract from stomach up to and including the rectum are alternatively located introperitoneally and extraperitoneally.

Mesenterium and omentum
The jejunum and the ileum are linked to the dorsal abdominal wall by the *dorsal mesentery* (Fig. 2.2.40a).

There is a space between the peritoneal folds of the mesentery, filled with connective tissue. This is the hilum of jejunum and ileum, in which there are blood vessels, lymphatic vessels and nerves (Fig. 2.2.42).
There is also a *ventral mesentery*. It is a vertical peritoneal fold which extends from the middle of the anterior abdominal wall (between navel

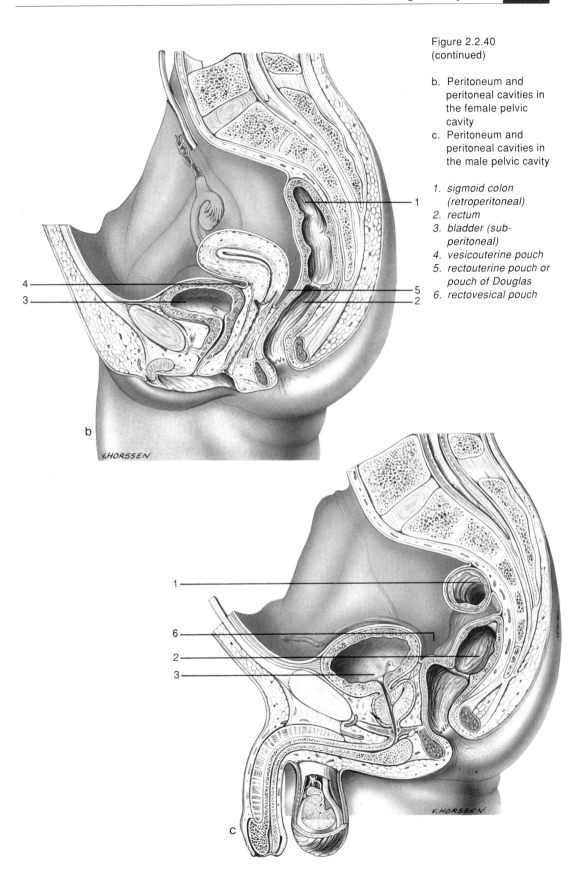

Figure 2.2.40
(continued)

b. Peritoneum and
   peritoneal cavities in
   the female pelvic
   cavity
c. Peritoneum and
   peritoneal cavities in
   the male pelvic cavity

1. *sigmoid colon*
   (retroperitoneal)
2. *rectum*
3. *bladder (sub-*
   peritoneal)
4. *vesicouterine pouch*
5. *rectouterine pouch or*
   pouch of Douglas
6. *rectovesical pouch*

Figure 2.2.41
Parietal peritoneum as
seen from the posterior
abdominal wall

1. inferior vena cava
2. part of the lesser
   omentum
3. attachment of
   mesentery
4. ascending colon
5. ramification of aorta
   (transparent)
6. spleen (intra-
   peritoneal)
7. duodenum
8. left kidney
9. descending colon
10. sigmoid colon

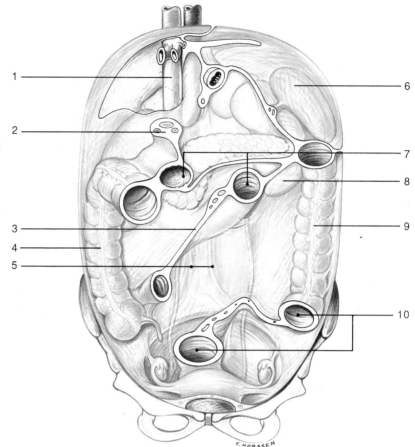

and diaphragm) to the liver and further on to the stomach.

The fold between the anterior abdominal wall and the liver, the falciform ligament, separates the two lobes of the liver (Fig. 2.2.40a). The fold between the liver and the stomach is called the *lesser omentum.*

An extremely large peritoneal fold extends from the stomach like a kind of apron over the intestines, the *greater omentum* (Fig. 2.2.40a). It extends to the bottom of the peritoneal cavity and is often scattered with little fat lobules.

Peritoneal cavities

There is hardly any 'space' in the peritoneal cavity due to all kinds of indented organs. There are, however, clearly defined *peritoneal spaces* between a number of structures. Such a space is

Figure 2.2.42
Mesentery and
intestinal hilum

1. blood vessels
2. intestinal hilum
3. intestine
4. mesentery

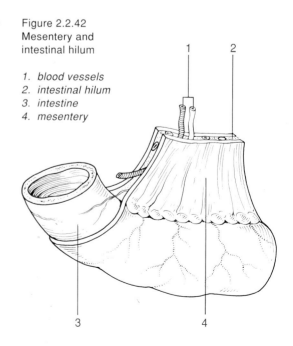

the *omental sac* (Fig. 2.2.40a), behind the stomach. It disappears when the stomach is filled.

Women have peritoneal spaces between the uterus and the bladder (the vesicouterine pouch) and between the uterus and the rectum (recto-uterine pouch). This *rectouterine pouch*, (the *pouch of Douglas*), is the lowest part of the female peritoneal cavity.

In men, the floor of the peritoneal cavity is formed by the space between the bladder and the rectum (the rectovesical pouch).

3

Urinary system

Introduction

The activities of cells and tissue are accompanied by the formation of *waste products*. Waste products are substances which the body cannot use. They are the remains of cell metabolism. These remains are discharged actively or passively into the m.i., which therefore threatens to disturb its homeostasis. The remains, however, are absorbed into and transported by the blood. Although the problem of the homeostasis of the m.i. is initially resolved in this way, the problem is then transferred to the blood. In order to be able to guarantee the homeostasis of the m.i., there must be a regulation of the composition of the blood. Not only do waste products affect the balanced composition of the blood, substances such as water, salts and vitamins, which are absorbed from the alimentary tract in considerable quantities and which cannot be stored in depots in these quantities, also exert their influence.

The role of the urinary system is to help regulate the composition and volume of the blood.
This role is separated into three components:
– regulating the *water and salt metabolism* and consequently the *blood pressure* is kept under control;
– maintaining the *balance between salts and bases* and consequently keeping the *acidity of the blood* under control;
– *secretion of waste materials of metabolism.*

Urine is produced as a result of the function of the urinary system.
The structure and function of the parts of the urinary system (including the kidneys, the ureters, the bladder and the urethra) are discussed in this chapter.

Learning outcomes

After studying this chapter you should have sufficient knowledge and understanding of:
– the general function of the urinary system and its three main components;
– the structure and the functions of the kidneys and the urinary passages;
– the mechanism of urine discharge;
– the composition of urine.

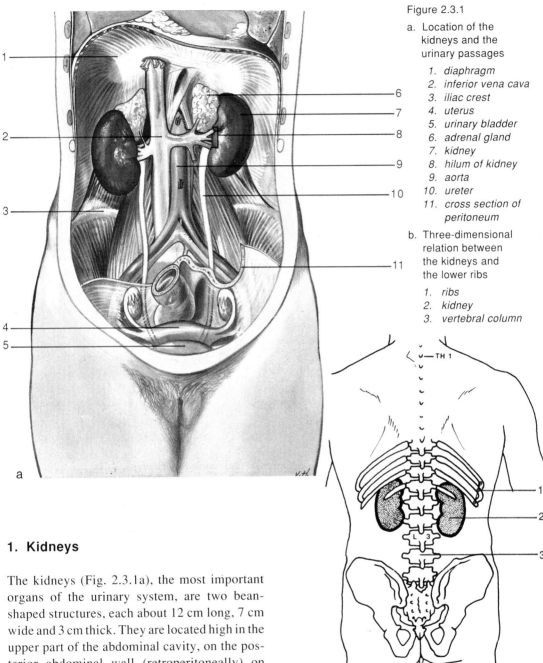

Figure 2.3.1

a. Location of the kidneys and the urinary passages

1. diaphragm
2. inferior vena cava
3. iliac crest
4. uterus
5. urinary bladder
6. adrenal gland
7. kidney
8. hilum of kidney
9. aorta
10. ureter
11. cross section of peritoneum

b. Three-dimensional relation between the kidneys and the lower ribs

1. ribs
2. kidney
3. vertebral column

1. Kidneys

The kidneys (Fig. 2.3.1a), the most important organs of the urinary system, are two bean-shaped structures, each about 12 cm long, 7 cm wide and 3 cm thick. They are located high in the upper part of the abdominal cavity, on the posterior abdominal wall (retroperitoneally) on either side of the vertebral column. Although the kidneys are abdominal organs, they are protected by the thorax, due to the diaphragm being dome-shaped (Fig. 2.3.1b). The left kidney borders the diaphragm. It extends from the level of the border of the twelfth thoracic vertebra (Th_{12}) to the third lumbar vertebra (L_3). The right kidney is slightly lower.

The concave sides of the kidneys face each other. The centre of the concave side is the *hilum*, through which blood vessels, nerves, lymph vessels and the ureter enter and leave the kidney. On top of the kidneys, as if they were 'dollops' of cream, are the adrenal glands. These are hormonal glands.

Figure 2.3.2
Embedment of the
kidney in peri-renal fat

1. *peri-renal fat*
2. *inferior vena cava*
3. *adrenal gland*
4. *kidney*
5. *ureter*
6. *aorta*

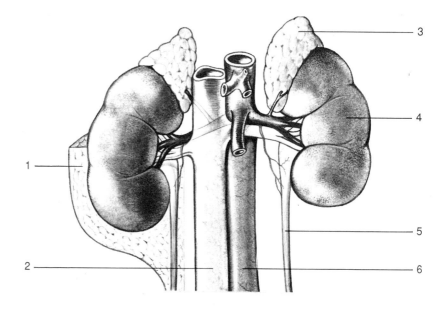

The adrenal glands and kidneys are surrounded by *supportive fat* (Fig. 2.3.2). The kidney and the peri-renal fat are surrounded by a sheath of condensed connective tissue, the fibrous sheath of kidney or renal fascia, bordering the peritoneum ventrally and the posterior abdominal wall dorsally. The fascia leaves room for ingoing and outgoing structures at the medial side, and is open at the bottom. The ureter extends through this opening. The supportive fat and the fascia have a securing and shock absorbing function.

Figure 2.3.3.
Internal structure
of the kidney

1. *capsule*
2. *papilla*
3. *renal pelvis*
4. *renal artery*
5. *renal vein*
6. *ureter*
7. *renal pyramid /*
   cortical rays
8. *interlobular artery*
9. *cortex*
10. *renal calix (pl. calices)*

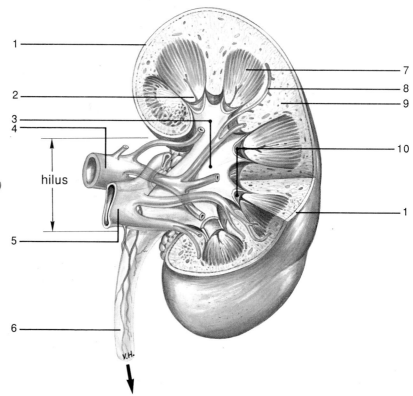

hilus

1.1. Structure
Cortex, medulla and renal pelvis

The outside layer of the kidneys consists of a thin, strong connective tissue capsule (Fig. 2.3.3).

Two different areas of tissue can be seen in a half-section of the kidney.

The *cortex* is the spotted area which lies just inside the connective tissue capsule. In some places the cortex extends between darker coloured striated fields, the *medulla*, to the centre of the kidney.

The medullary inner part comprises a number of conical-shaped masses, the *renal pyramids*. The striae of the medullary masses extend from the tops of the renal pyramids to the bases and are called the *renal tubules*.

The bases of the pyramids are more or less parallel to the capsule. The apices are directed to the centre of the kidney and form the *renal papillae* projecting into small cavities, the *minor calices*. Three to six renal pyramids terminate at each papilla. The number of papillae varies from five to eleven per kidney. The papillae form the transition to the m.e. The calices terminate in a cavity at the level of the hilum, the *renal pelvis*.

The calices and renal pelvis are covered with transitional epithelium.

Vascularization

The *renal artery* is a short branch of the aorta. When the kidneys are at rest they receive about 20% of the cardiac output. With a cardiac output of five litres, each kidney must therefore process half a litre per minute. This is a large amount, in comparison to other organs.

At the level of the renal hilum the renal artery branches into smaller arteries which enter the renal tissue at the renal pelvis and the calices. Branches, the *interlobular arteries*, extend along the edges of the renal pyramids to the outer cortical layer. The interlobular arteries then split into branches which extend over the bases of the pyramids. From these branches arterioles first penetrate into the cortical tissue and then into the medullar tissue. They form a special vascular system, the *portal circulation*. The venules and veins, which carry blood from the kidneys, run parallel to the arterioles and arteries described earlier. At the level of the renal hilum a number of veins unite into the renal vein, which discharges directly into the inferior vena cava.

Functional renal units

Each kidney has approximately one million microscopic units that each contribute to the total urine production. Such a functional unit is called a *nephron*. A nephron consists of a glomerular (Bowman's) capsule, a proximal tubule, a loop of the nephron (loop of Henle), a distal tubule and a collecting tube (Fig. 2.3.4). Each nephron has trombone-like loops. A *glomerular capsule* is a double-layered sac of single-layered squamous epithelium. Its shape is comparable to a brandy glass: the diameter of the capsular opening is relatively small. There is a jumble of capillary vessels in a glomerular capsule, called a *glomerulus*.

The afferent arteriole is a branch of the interlobular artery, but the efferent vessel is also an arteriole. The glomerulus is thus a capillary network between two arterioles.

A glomerular capsule containing the glomerulus is called a *renal pyramid (pyramid of Malpighi)*. This is the filtering unit of the kidney. All filtering units are located in the cortex.

A glomerular capsule is hollow. There are slits between the cells of the squamous epithelium, which is very near to the capillaries of the glomerulus. This structure makes the inner lining of a glomerular capsule readily permeable.

At the base of the renal pyramid, the capsular cavity changes into a number of canals. Like the wall of the capsule, the walls of these canals consist of single-layered epithelium. The cells, however, are not flattened, but cubic or cylindrical in shape.

The base of a glomerular capsule is connected with the proximal tubule, which is a small strongly convoluted canal. This first convoluted tubule is wholly located in the cortex.

The proximal tubule becomes straight when it enters the cortex layer and changes into the *loop of the nephron* (loop of Henle). This consists of a narrow, descending part extending in the

Figure 2.3.4
The nephron

a. Parts of the nephron
   located in the cortex
   and the medulla

   1. renal pyramid
   2. proximal tubule
   3. distal tubule
   4. collecting tubule
   5. loop of the nephron
   6. artery
   7. vein

b. Functional scheme
   of a nephron

   1. interlobular artery
   2. efferent vessel
   3. glomerulus
   4. glomerular capsule
   5. convoluted tubule I
   6. loop of the nephron
   7. interlobular vein
   8. juxtaglomerular
      cells (near a
      glomerulus)
   9. afferent vessel
   10. convoluted tubule II
   11. collecting tubule
   12. renal pelvis

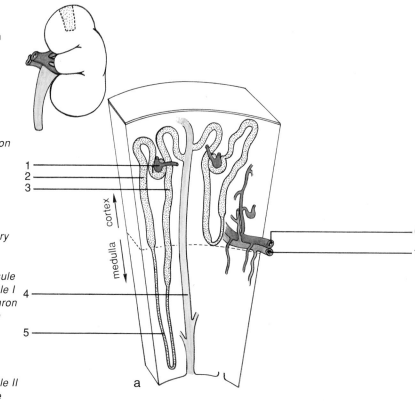

a

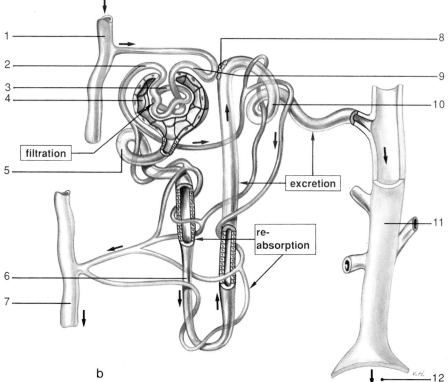

b

direction of the papilla, a 180° bend and an initially narrow, but later thicker, ascending part, extending parallel to the descending part, and then back to the cortical layer. At the level of the corresponding renal pyramid (in the cortical layer) is the transition to the second *convoluted tubule* or *distal tubule*. Before the afferent vessel enters a glomerular capsule, it comes into contact with the distal tubule. At this point of contact, hormones that are needed for the regulation of the blood pressure are produced in *juxtaglomerular cells*. The distal tubule terminates in a collecting tubule into which several nephrons discharge. The collecting tubules pass through the cortex and terminate at the tip of the papilla in the corresponding calix. The collecting tubules, the medullary or cortical rays, extend parallel to the loops of the nephrons and give the medulla its striated look.

The lumen of the efferent vessel of the glomerulus is smaller than that of the afferent vessel. The efferent vessel branches into a loop-shaped second capillary network around the tubule and the loop of the nephron. The transition from arterial to venous takes place in this network. The afferent vessels discharge their blood into the *interlobular veins,* from which the blood eventually enters the renal vein.

1.2. Function

Connecting mechanisms

Each nephron contributes to the renal function. Three linked mechanisms play a role in the production of urine: filtration, reabsorption and excretion.

– *Filtration.* During rest approximately 1 litre of blood enters the nephrons. Fluid is pushed from the glomeruli through the pores of the inner coat of the glomerular capsule into the capsular cavity. This process of filtration through a membrane whose pores are so small that they are impervious to large molecules is called *ultrafiltration.* The ultra-filtrate has almost the same composition as blood serum: blood cells and blood proteins are too large to pass through the pores.

The pressure with which ultrafiltration takes place can be calculated:

net filtration pressure =
blood pressure – opposing pressure

Blood pressure in the glomerulus is 45 mmHg (6 kPa). This pressure is 10 mmHg (1.35 kPa) higher than the pressure usually found at the arterial side of a capillary network. This is due to the fact that the distance to the abdominal aorta is relatively short and the drop comparatively small.

The blood pressure is 'opposed' by two forces, firstly by the difference in colloid-osmotic value of the blood compared to that of the capsular fluid. As the blood proteins are too large to be able to leave the capillaries, the colloid-osmotic value of the blood is larger than that of the capsular fluid. This difference in colloid-osmotic value is 25 mmHg (3.3 kPa), just like that in the exchange mechanism between capillaries and m.i. Secondly, the fluid in a glomerular capsule exerts an opposing pressure. This pressure is 10 mmHg (1.35 kPa), so:

net filtration
pressure $= 45 - (25 + 10)$
             $= 45 - 35$
             $= 10$ mmHg (1.35 kPa)

From this calculation it can be derived that a lowering of the blood pressure rapidly minimizes the filtration pressure. Consequently, a well regulated blood pressure is of paramount importance for the function of the kidneys. We shall see later that the kidney itself keeps the blood pressure under control.

The glomerular capsules of both kidneys produce 125 ml/minute of ultrafiltrate or primary urine as a result of filtration. Approximately 1 litre of blood per minute goes to the glomeruli, in other words 550 ml of plasma (55% × 1 litre). This means that a little less than 25% of the plasma ends up in the glomerular capsules as ultra-substrate 7.5 litres per hour (60 × 125 ml), 180 litres per 24 hours (24 × 7.5). Fortunately, the volume of the urine proper is only a fraction of the amount of primary urine. If it were different, one would continuously pass urine, and one would have to drink copious amounts of fluid.

– *Reabsorption.* The primary urine is concentrated into urine by the mechanism of reabsorption. Reabsorption takes place from the proximal tubule, the loop of the nephron, the distal tubule and the collecting tubule. The renal process of reabsorption is different from that taking place in the exchange between capillary and m.i.

There is not only a passive component (reabsorption as a result of the osmotic value), but also an active component (reabsorption under the influence of hormones and enzymatic pumps). As is the case in every active process, energy (ATPs) is used.

Approximately 80% of the primary urine volume is reabsorbed from the proximal tubule, 6% from the loop of the nephron and about 13% from the distal and the collecting tubules, so the entire reabsorption is 99%. This is more than 178 litres, out of 180 litres of primary urine per 24 hours. This reabsorption is accompanied by a precise selection of substances which will or will not be reabsorbed.

The objective of this *selective reabsorption* is the:

- reabsorption of useful substances into the blood;
- regulation of the water and salt content of the blood;
- regulation of the acid/base balance in the blood.

These subjects will be discussed separately.

– *Excretion.* The primary urine is concentrated into proper (secondary) urine after the selective reabsorption. Urine from several nephrons is discharged into a collecting tubule. Although the contents of the small canals of the nephrons, strictly speaking, belongs to the m.e. (they even have epithelium), the urine really enters the m.e. at the renal papilla, where the urine trickles into the calices from a number of collecting tubules (excretion). An average of approximately 1.5 litres is excreted every 24 hours. The volume produced, of course, corresponds with the volume drunk. At a party where the bar is popular, the toilets are popular as well!

Reabsorption of useful substances

The filtering mechanism is relatively non-selective. The energy needed is supplied by the heart. All small molecular and useful substances enter the capsular cavity together with the enormous volume of filtered water.

Active reabsorption of *glucose* and almost all *amino acids* takes place in the proximal tubule, via the mechanism of the enzymatic pump (see Fig. 1.3.5).

Normally all glucose is reabsorbed, but when the volume of glucose in the blood is greater than 160 mg per 100 ml of blood (for instance after a meal with an excessive amount of carbohydrates), the capacity of the enzymatic glucose reabsorption pump is too small. The *renal threshold* value for glucose (160 mg%) is then exceeded and glucose is found in the urine (glucosuria).

The parathyroid hormone, produced by the parathyroid gland, increases the reabsorption of calcium and magnesium ions from the distal tubules. Phosphate ions, on the contrary, are selectively excreted.

Water and salt balance: blood pressure control

The composition of the blood is influenced by the absorption of water and salts from food and by the production of water in certain metabolic reactions. As the kidneys regulate the content of water and salts in the blood, they keep the blood pressure under control. At first large volumes of water and salts are non-selectively absorbed from the blood. Reabsorption puts these substances back into the blood, also by selection.

There is active reabsorption of Na^+ and K^+ ions from the proximal tubule. Negative ions (Cl^- and HCO_3^-) and water go along passively, as a result of the high crystalloid-osmotic value of the medulla.

The hormone *aldosterone,* produced by the outer layer of the adrenal cortex, regulates the relationship between sodium and potassium (Na/K balance). This takes place mainly in the ascending part of the loop of the nephron and the distal tubule.

The reabsorption of Na^+ ions is stimulated

(sodium retention) and is accompanied by the excretion of K^+ ions (potassium depletion).

As a result of the exchange of ions and water in the renal medulla, the salt concentration (and therefore the crystalloid-osmotic value) in the area just above the renal papilla is relatively high. The result is that the urine, trickling down in the collecting tubule, is again concentrated, and the 'sucked in' water is then absorbed from the m.i. of the medullar tissue into the capillaries that accompany the loop of the nephron.

Aldosterone indirectly influences the water content of urine. The ADH (anti-diuretic hormone) or vasopressin, discharged into the blood from the hypothalamus, influences it directly. It increases the water permeability in the epithelium of the distal and collecting tubules. As a result, the 'sucking' activity of the salt-rich medullar tissue directly above the papilla is strengthened and the urine becomes more concentrated.

Aldosterone and ADH each influence in a different way the water and salt content of urine. Both hormones also influence blood pressure.

Through the selective exchange of Na^+ against K^+ in the medulla, under the influence of aldosterone, the Na^+ concentration of the m.i. is greater and more Na^+ ions than K^+ ions diffuse from there into the blood. As the water binding cap-acity of Na^+ ions is larger than that of K^+ ions, an increase in aldosterone results in an increase in blood pressure (see Fig. 2.1.3).

As a result of the increased permeability of the distal and the collecting tubules and the influence of ADH, more water is absorbed into the m.i. of the medullar tissue and from there into the blood. An increase in the volume of ADH results in an increase in blood pressure (and vice versa).

Aldosterone and ADH are *blood pressure controlling* hormones.

Renin is the third blood pressure controlling hormone. It is produced by the juxtaglomerular cells (Fig. 2.3.4b). When the blood pressure in the afferent vessel becomes lower, these cells discharge more renin into the blood (and vice versa). This hormone converts the protein *angiotensinogen,* present in the plasma, into *angiotensin.* Angiotensin directly and indirectly influences blood pressure (Fig. 2.3.5). It causes vasoconstriction in the arterioles and an increased production of aldosterone in the adrenal cortex.

These three blood pressure controlling hormones together form the *RAA-system (renin-angiotensin-aldosterone-system).*

Acid/base balance; control of blood acidity

The pH-value is a measure for the concentration of H^+ ions. A *low pH* value indicates a *large concentration of H^+ ions* and consequently *high acidity.* The average blood pH is 7.4, which means that blood is slightly basic (a pH of 7

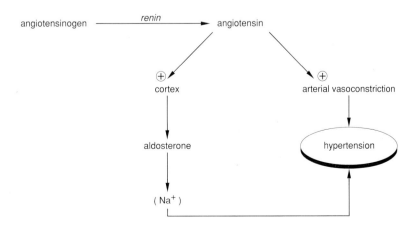

Figure 2.3.5
Hypertensive action
of angiotensin

angiotensinogen ——*renin*—→ angiotensin

cortex ⊕

aldosterone

(Na^+)

arterial vasoconstriction ⊕

hypertension

indicates neutrality). It is very important that the pH-value should be maintained, as the optimal activity of enzymes is particularly dependent on the pH-value. When the pH-value of blood is lower than 7.0 or higher than 7.8, serious metabolic disorders arise.

The most important reaction that influences the pH of the blood is:

$$CO_2 + H_2O \leftrightarrows H_2CO_3 \leftrightarrows H^+ + HCO_3^- \quad (1)$$

The blood becomes more acid, so the pH goes down, when the reaction goes from left to right. The blood becomes more alkaline, and the pH goes up, when the reaction goes from right to left.

Together with acid metabolic products, such as lactic acid (anaerobic metabolism), acetic acid, ketone acids, H^+ ions can be added to the blood. This, in turn, reduces the pH of the blood.

During the aerobic metabolism in the mitochondria carbon dioxide and water are produced. The imminent lowering of the pH in the blood (reaction 1) is opposed by the haemoglobin buffer, the respiration and the kidneys.

The haemoglobin buffer is a system in which haemoglobin can absorb H^+ ions while releasing oxygen (O_2), and can absorb oxygen while releasing hydrogen (H^+):

$$H^+ + HbO_2^- \leftrightarrows HHb + O_2 \quad (2)$$

Thus, in active tissues two birds are killed with one stone. Reaction (2) can absorb H^+ ions from reaction (1). The imminent surplus of HCO_3^- ions can be excreted by the kidneys.

Respiration is responsible for the removal of carbon dioxide from the blood: reaction (1) goes to the left, resulting in a decrease in the amount of H^+ ions. Due to the fact that reaction (2) in the lungs moves to the left as well (Module 2, Chapter 4), an increase in the pH of the blood is prevented. As the relationship between the volume of H^+ ions and the volume of Cl^- ions, on the one hand, and that between the H^+ ions and the HCO_3^- ions, on the other hand, is constant, a surplus of Cl^- ions and a shortage of HCO_3^- is imminent, because of the shift of reaction (1) to the left.

The *kidneys* offer a solution to the impending surplus of Cl^-.

The surplus of Cl^- ions could be linked with Na^+. But then useful salt would be lost. In order to prevent this, the cells of the distal tubule *synthesize* NH_3. The Na^+ ions from the primary urine are exchanged against H^+ ions from the blood. This does not make any difference to the buffer reaction (1) and the Na^+ ions are preserved.

In the tubule the H^+ ions react with NH_3 and form positive NH_4^+ ions, which are then excreted together with the Cl^- ions. A surplus of Cl^- ions is thus removed together with the urine, without the urine becoming too acid.

The imminent shortage of HCO_3^- ions is compensated by reabsorption of all HCO_3^- ions from the primary urine and by formation of HCO_3^- ions in the tubular cells.

Control of the capacity to carry oxygen

The volume of blood offered to the kidneys remains more or less constant. For that reason the kidneys are extremely suitable as blood monitors. They not only influence blood pressure and blood acidity, but also the production of erythrocytes and thereby the oxygen carrying capacity of the blood.

When the blood which enters the kidney contains too little oxygen, the renal tissue reacts by discharging the *renal erythropoietic factor* (REF) into the blood.

The liver reacts to the shortage of oxygen by discharging a certain kind of globulin into the plasma. This globulin and the erythropoietic factor together form the hormone *erythropoietin*, which stimulates the production of erythrocytes in the red bone marrow. When the shortage of oxygen is rectified by the now enlarged number of erythrocytes, the production of REF and globulin, and therefore that of erythropoietin, stops.

As the oxygen pressure becomes lower, when, for instance, a person stays for some time at a higher altitude, erythropoietin is responsible for the increase in the number of erythrocytes.

2. Urine

The regulating processes in the nephrons eventually lead to the production of urine. It could be said that the urine which trickles from the papillae into the calices is the result of the pluses and the minuses in the filtration and reabsorption processes. The composition of the urine produced is not constant, as the composition of the blood, passing through the kidneys, is not constant.

The volume of urine produced during 24 hours is approximately 1.5 litres. This production, however, is heavily dependent on the fluid intake. The more one drinks, the more urine is produced. The volume of urine produced also depends on the amount of sweat secreted. The production of urine can go down to less than half a litre per 24 hours, resulting from heavy perspiration, when insufficient drinks are taken.

In order to accurately assess the composition of urine, it is necessary to monitor it over a 24 hour period. The normal constituents are:
- *water*, about 95%;
- *salts*, mainly sodium chloride (NaCl);
- *waste products of proteins*: urea is formed when amino acids are broken down in the liver, creatinine is formed when certain amino acids are decomposed in the muscles, uric acids are formed when nucleoproteins are broken down;
- *urobilin*. A small part of the absorbed urobilinogen passes the liver in the enterohepatic cycle and enters the greater circulation. The urobilinogen is discharged by the kidneys into the urine, where it is converted into urobilin by oxidation (see Fig. 2.2.39b). Urobilin gives the typically yellow colour to urine;
- *cell casts*, rejected epithelial cells of the nephrons and the urinary passages.

Furthermore, vitamin C can be present when more is absorbed than used.
After excessive consumption of carbohydrates the corresponding threshold value can be passed and glucose is present.

The amount of gonadotropic hormones is considerably larger during pregnancy. Some of it is present in the urine (pregnancy test).

3. Urinary passages

The urine produced by the nephrons continuously trickles from the renal papillae. This is the starting point of the *urinary passages:* calices and renal pelvis, ureter, urinary bladder and urethra. The urinary passages, apart from the last part of the urethra, are lined with transitional epithelium. On this is a mucous layer which protects it against the effects of urine.

Renal calices and pelvis, ureter

The calices are cup-like processes of the renal pelvis around the papillae (see Fig. 2.3.3). The renal pelvis is a centrally located cavity, of which the smaller, funnel-shaped tip extends through the hilum to continue as the ureter.

The ureters are 25 to 30 cm long and extend from the hilum to the urinary bladder, at the left and right sides of the spine. They are initially located in the retroperitoneal connective tissue, but at the level of the sacral bone they arch ventrally and enter the subperitoneal area above the pelvic floor. There they extend diagonally and open up into the bottom of the urinary bladder. The ureters pass diagonally through the smooth muscle tissue of the bladder wall.

Around the transitional epithelium of the renal pelvis, and especially around that of the ureter, smooth muscular tissue is found, circularly and longitudinally arranged. This enables a peristaltic motion. A peristaltic wave, which pushes the urine from the renal pelvis into the bladder, takes place approximately once every three minutes. In relation to the force of gravity this may appear unnecessary. Yet this peristalsis is essential, not only because urine is also produced and must be carried to the bladder when a person is in a supine position (during sleep, for instance), but mainly because the mouth of the ureter at the bladder is normally closed. The tension in the muscular tissue of the wall of the bladder affects the mouth of the ureter in much the same way as the valve rubber of a bicycle valve affects the air outlet. This structure prevents the urine in the bladder from flowing back to the ureters, even when their orifices (mouths) are below the urine level.

Figure 2.3.6
Location of urinary bladder
and urethra

a. In the female

1. ureters
2. inguinal ligament
3. uterus
4. ovary and uterine
   tube (fallopian tube)
5. urinary bladder
6. symphysis pubis
7. vagina
8. urethra
9. clitoris

b. In the male

1. ligament
   (inguinal ligament)
2. symphysis pubis
3. deferent or
   spermatic duct
4. urethra
5. penis
6. ureters
7. urinary bladder
8. prostate
9. testicle

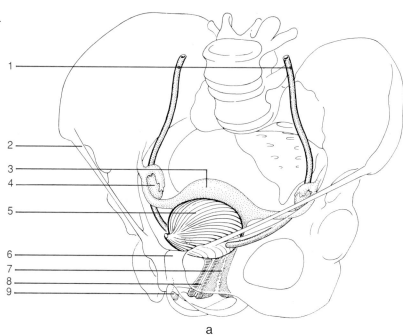

a

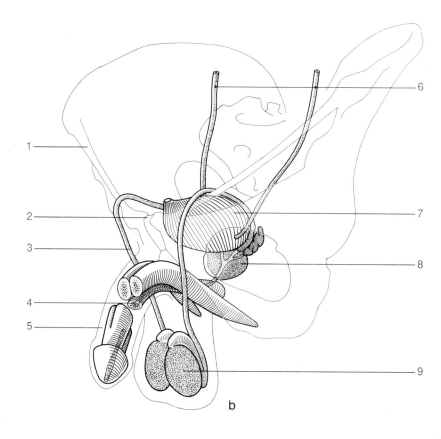

b

Urinary bladder (vesica urinaria)

The urinary bladder is located on the floor of the pelvic cavity, behind the symphysis pubis (Figs. 2.3.6a and b). When it is empty it is located subperitoneally, but when it fills, its anterior side expands over the symphysis. The filled bladder is then pre-peritoneal.

The flexibility of transitional epithelium is paricularly useful in the bladder (see Fig. 1.4.1f). When the bladder fills, the two layers of high narrow cells stay close to each other and the epithelium will not 'leak'.

Multilayered smooth muscular tissue encircles the transitional epithelium. Compared to, for instance, the ureters and the intestines, there is less arrangement of circular and longitudinal muscular tissue, but there is an outer arrangement. There, most fibres are parallel to each other from the bottom of the bladder, along the tip, to the exit (Fig. 2.3.7). The tip of the bladder is a ligament which connects the bladder with the inner side of the ventral abdominal wall.

When the bladder is empty, the inside is ribbed. There are folds in the mucous membrane, but not in the area of the vesical triangle or *trigone of the urinary bladder*. This part of the bladder, between the ureters and the exit of the bladder, does not stretch, so that the orifices of the ureters cannot be opened.

There are two sphincters around the bladder exit, where the urethra begins. The internal sphincter (sphincter muscle of urinary bladder) is part of the bladder wall and therefore consists of 'involuntary' smooth muscle fibres. The external sphincter (sphincter muscle of urethra) is part of the muscles of the pelvic floor and consists of 'voluntary' striated muscle fibres.

Urethra

Urine leaves the body via the urethra. The structure of its wall is similar to that of the ureters, but in the last part of the urethra, columnar epithelium is found and no transitional epithelium. There are differences between the urethra in the male and the female (Figs. 2.3.6a and b).

In the female the urethra extends throughout its length ventrally of the vagina. It is a tube of

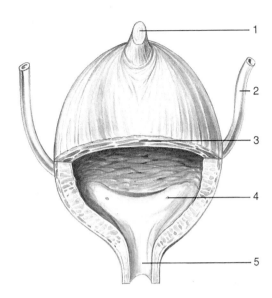

Figure 2.3.7.
Structure of the wall of the bladder

1. tip of the bladder (apex)
2. ureter
3. muscular layers
4. orifice (mouth) of ureter
5. urethra

about 3 cm which opens up into the vestibule of the vagina between the labia minora.

In the male the urethra extends through the prostate gland which is situated directly under the orifice of the bladder. This part of the urethra is called the *prostatic urethra*. Directly below the pelvic floor the urethra enters the corpus spongiosum of the penis. This part is called the *spongy urethra*. The male urethra is approximately 20 cm long and ends in the head of the penis (glans penis).

4. Urination

Urine is removed from the body by urination. This process does not take place continuously, but about five to six times every 24 hours. The urine is stored temporarily in the bladder until a certain level of filling is reached.

The empty bladder is gradually filled as urine is continuously supplied. Initially the wall of the bladder stretches under the influence of the filling pressure. The muscular tissue stretches and the pressure in the bladder grows. When the bladder contains about 350 ml of urine, stretching sensors in the bladder wall become active and the *micturition reflex* comes into action.

The sensors are connected to the sacral part of the spinal cord. There the transmission to the parasympathetic nervous system takes place, which makes the bladder wall contract rhythmically and makes the sphincter relax. An *urge* is felt. One becomes conscious of this urge because the brain gets information from the sacral part of the spinal cord. During severe cold or strongly felt emotions, this urge can be felt while the bladder is only partly filled. Just before an examination one often experiences a desire to pass urine. When a person is concentrating hard on something, the urge may not be felt until the bladder is very full.

Quite often the urge is suppressed for some time, until the bladder contents are about 500 ml. At a suitable place the external sphincter is also relaxed, after which the urine can flow through the urethra. The presence of urine in the urethra reinforces the micturition reflex. The bladder wall contracts and urination gradually becomes stronger.

Controlled urination, in other words the command of the the external sphincter, must be learned. In children urination is automatic until the moment that they are 'potty trained'.

Urination can be stimulated by *pressing*. The abdominal cavity is a closed space. When the volume decreases, the pressure increases. The abdominal cavity can be made smaller by tightening the abdominal muscles and by making the diaphragm go down. The increase in pressure also affects the bladder: the bladder volume becomes smaller, the pressure rises and consequently urination is stimulated.

Respiratory system

Introduction

Combustion of nutritional substances involving oxygen (aerobic combustion) yields 18 times the number of 'energy packages' (ATPs – see also pages 53–55) yielded in combustion without oxygen (anaerobic combustion).
Respiration is the process by which oxygen is withdrawn from the atmosphere and conveyed to the blood. In human beings this takes place in the lungs. The oxygen can then be carried and transferred to the m.i. of the active tissues by means of the circulatory system. The cells can then use the oxygen from the m.i. in the citric acid cycle (Krebs cycle). Very simply, the citric acid cycle involves the following chemical reactions:

- glucose + oxygen → carbon dioxide + water + ATPs (glycolysis)
- fatty acids + oxygen → carbon dioxide + water + ATPs + waste products (lipolysis)

In these reactions water is produced and can be discharged – if necessary – via the kidneys. Carbon dioxide is also released. It diffuses to the m.i., is absorbed within the bloodstream (mainly by diffusion) and is then carried to the lungs, where it is transferred to the atmospheric air. In this chapter we will discuss the structure and function of the respiratory tract, the air tracts, ventilation, the exchanges of gases, and the regulation of respiration.
The respiratory system not only plays a part in the exchange of gases, but it is also important in a number of other functions such as giving off heat, discarding water (expired air is warmer and more saturated with water vapour than inhaled air), as well as in the capacity to express oneself and communicate with others (speaking, singing, whistling, sighing) and in smelling.

Learning outcomes

After studying this chapter you should have sufficient knowledge and understanding of:
- the general function of the respiratory system;
- the structure of the air passages and the functions performed by the various parts of these air passages;
- the structure and function of the pulmonary tissue and membranes;
- the movements involved in respiration;
- the performance of the lungs and the importance of quantitative data on respiratory efficiency;
- the exchange of gases via the blood between lungs and active tissues;
- the regulation of respiration.

1. Air passages

Air passages constitute the communication between the atmospheric air and the lung tissue.
Oxygen can be transported to the lung tissue and
carbon dioxide out of the lungs to the external
world via the air passages. The contents of the
air passages belong to the m.e., so it is only
natural that the air passages are lined with epithelium.
We can identify the following parts of the air
passage (Fig. 2.4.1): *nose* and *mouth cavity*,
pharynx, larynx, windpipe (i.e. trachea) and
the *bronchi* which subdivide extensively and
eventually end in the *alveoli*.

Nose cavity

The nose in humans comes in many shapes and
sizes. This diversity is caused by differences
in nasal bones (Fig. 2.9.11), by differences in
nasal cartilage and also by differences in

enveloping skin. The *base* of the *nose* is
located between the two eye sockets, the tip of
the nose is the most anterior part and at the sides
we find the *nostrils*.
The *nasal cavity* is considerably greater than we
would expect judging by the nose itself. It is a
deep hollow cavity, stretching from over the
hard palate at the back to between the eye sockets (Fig. 2.4.2). The nose cavity is divided into
two symmetrical spaces by the *nasal septum*
which runs medio-sagittally, and of which the
back part consists of bony tissue and the front
part of flexible cartilage. The upper border, the
roof of each nose cavity, is formed by the *ethmoid bone plate*, a part of the base of the skull in
which we see a large number of holes which let
through nerve fibres of the olfactory nerve. The
front border of the nasal cavity runs obliquely
from the base of the nose downward and forward. The first few centimetres consist of bony
tissue, rather than of flexible cartilage.

Figure 2.4.1
The air passages

1. nose cavity
2. mouth cavity
3. right lung
4. pharynx
5. larynx
6. trachea
7. left main bronchus
8. major bronchi

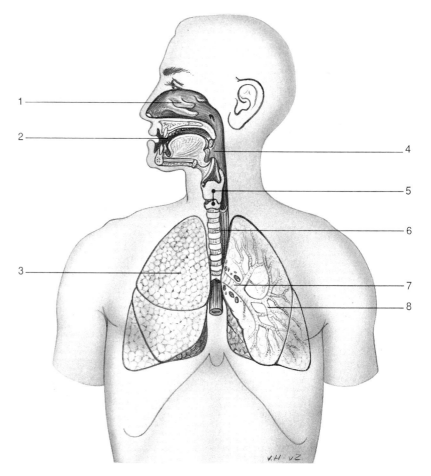

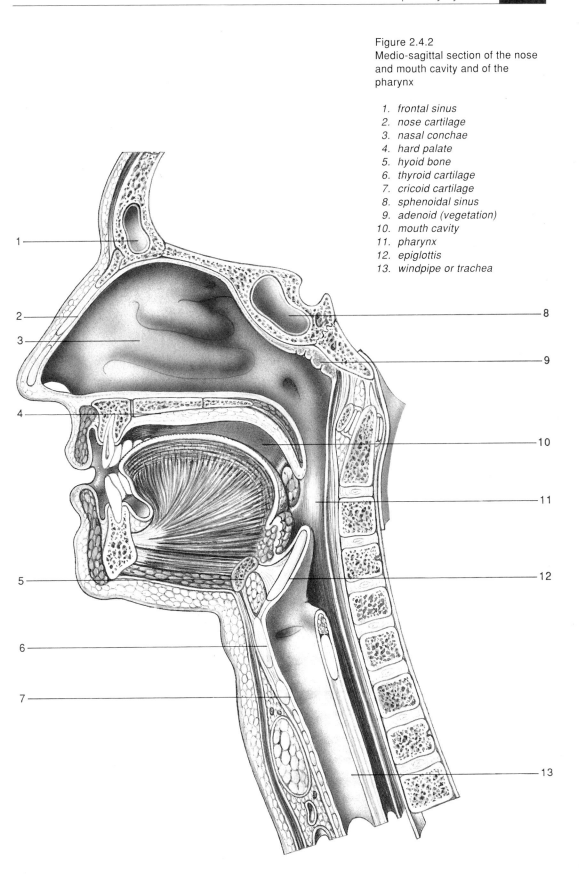

Figure 2.4.2
Medio-sagittal section of the nose
and mouth cavity and of the
pharynx

1. frontal sinus
2. nose cartilage
3. nasal conchae
4. hard palate
5. hyoid bone
6. thyroid cartilage
7. cricoid cartilage
8. sphenoidal sinus
9. adenoid (vegetation)
10. mouth cavity
11. pharynx
12. epiglottis
13. windpipe or trachea

Figure 2.4.3
Nasal and paranasal
cavities

a. Nasal conchae
b. A frontal section of
   nasal cavity and
   paranasal sinuses

1. frontal sinus
2. opening to the tear
   duct (nasolacrimal
   duct)
3. hard palate
4. olfactory duct
5. ethmoid bone
6. sphenoidal sinus
7. nasal conchae
8. uvula
9. eye socket, orbit
10. ethmoidal sinus
11. inferior meatus
12. upper jaw cavity or
    maxillary sinus

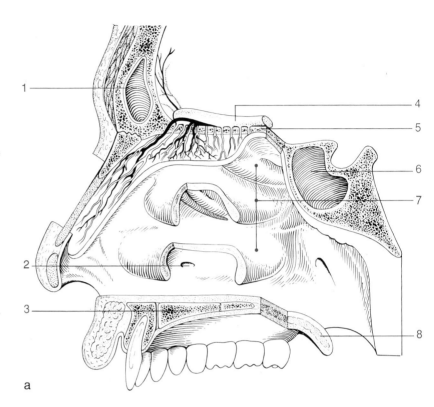

a

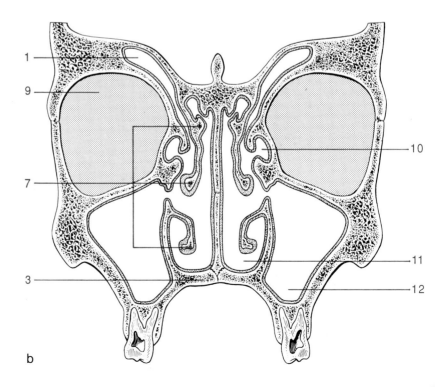

b

The base of the nose consists of bony tissue, which transforms, near the nostrils, into flexible cartilage, ending in a fold of skin. The base of the nose is therefore very solid and the front part is very flexible. The anterior openings of the nose are more or less directed downwards and constitute the entrances to the nasal cavity. The posterior *nasal apertures* (choanae) constitute the connection with the nasal and pharyngeal cavities.

The nasal septum is a flat sheet of bone, forming the medial wall of each half of the nasal cavity. The lateral wall of each of these cavities is far more irregular. In the side wall there are three bony projections, which to a large extent narrow down the air passage (Figs. 2.4.2 and 2.4.3). Beneath each concha we find a groove, or *meatus*, which opens to the nasal cavity. The lowest one, meatus inferior, is the biggest; when breathing through the nose by far the largest amount of air flows through this meatus. The nose cavity is lined with a *mucous membrane*, ciliated epithelium, in which there are many little serous glands and mucus cells which together excrete a thin, watery, mucus-like liquid. Beneath the flimsy layer of epithelium we find an extremely dense capillary network.

In the upper part of the nose cavity we find *olfactory epithelium* as the lining of the bony plate of the ethmoid. The sensors for the perception of smell lie in the olfactory epithelium between the epithelial cells (see Fig. 2.8.6).

When breathing through the nose there is an intensive contact established between the air and the olfactory epithelium of the nasal cavity, for the passageways are relatively narrow and the intricacy of the walls gives the air a great deal of turbulence.

Breathing in through the nose has a number of advantages over breathing in through the mouth cavity:

- the air can be *cleansed*: generally air taken in (as inhaled air) contains quite a lot of dust particles. The cilia with the sticky mucus all over them catch many of these pollutants out of this air. In a more dusty environment more mucus is produced. That this barrier is effective is clearly demonstrated if we look at our handkerchief when we blow our nose after having been in dusty surroundings for some time. Because of the movements of the ciliae the layer of mucus in the nose is shifted towards the nasopharyngeal cavity. The polluted mucus can then be swallowed after which the gastric juices and acids can destroy any of the disease (pathogenic) germs it may contain. Under normal circumstances a part of this polluted mucus is dried up within the nose. Picking or blowing your nose will dispose of this dried up mucus;

- the air can be *warmed*: atmospheric air is in general colder than the body temperature. The capillary network of blood vessels in the nose – positioned close under the surface – gives off the heat of the blood to the air which is inhaled. As a result of this quick transfer of heat the temperature of the lung tissue is prevented from being considerably lower than the temperature of e.g. the heart. Even when the temperature of the atmospheric air is some degrees below zero, the air in the pharynx has already been warmed to 32° or 33°C;

- the air can be *moistened*: moistening of the inhaled air prevents the lung tissue from becoming parched. Moist lung tissue benefits the exchange of gases. In this respect we can compare lung tissue with a sponge: when a sponge is dried up water poured onto it simply runs down the outside of the sponge; whereas an already wet sponge absorbs water much easier. The mucous membrane is constantly producing moisture which is passed on to the inhaled air. In addition, the air is moistened by evaporation of tear fluid for the lacrimal duct flows into the lowest meatus, the one through which most air passes (Figs. 2.4.3a and 2.4.4). The lacrimal gland, which lies lateral to the eyeball, continually produces tear fluid to keep the conjunctiva of the eye moist. The surplus of this moisture is led away via two *lacrimal canals* in the medial corner of the eye to the tear duct. When there is a lot of tear fluid produced (e.g. when weeping), the tear tubules cannot cope with so much tear fluid, so the tears will roll over the edge of the lower eyelid and then down the cheeks;

Figure 2.4.4
The lacrimal apparatus

1. *lacrimal gland*
2. *lacrimal canal of the upper eye lid*
3. *lacrimal sac*
4. *lacrimal canal of the lower eye lid*
5. *nasolacrimal duct, flowing out into the lowest meatus*

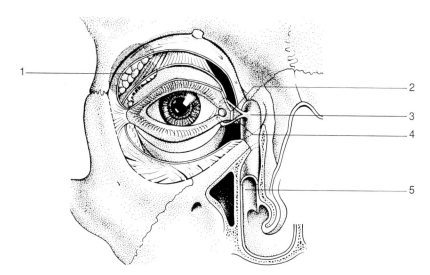

– the air can be *sampled:* the olfactory epithelium alerts us to the quality of the inhaled air by smelling.

Taking in air via the nose is not possible when large amounts of air must be transported, e.g. during certain exertions. The narrow passageways of the nasal cavity do not allow the flow of air in such quantities in a short time; the person concerned must switch over to breathing through the mouth.

Mouth cavity

Breathing in through the mouth has the disadvantage that the air becomes less 'compatible', than air inhaled via the nose. The contact of the air with the mucous membrane of the mouth is less intensive than it is in the nose: in the mouth there are hardly any turbulences, no ciliated epithelium and no sticky mucus, no capillary network situated near the surface, no sufficient moistening, no sampling of the air.

Breathing out through the mouth makes *sound production* possible (singing, speaking, whistling). The exhaled air flows past the vocal cords, which start to vibrate so that sound is produced. The sound produced receives a certain sonorous quality because of the shape of the mouth and nasal cavity, and because of the position of and/or the movements of tongue and lips.

It is remarkable that inhaling air during speech almost always takes place through the mouth.

That is why a glass of water is often provided for a person who is to deliver a speech.

Paranasal sinuses

In the neighbourhood of the nasal cavity we find a number of *paranasal sinuses.* These are in open contact with the nasal cavity via small apertures and are lined with the same type of ciliated epithelium as in the nasal cavity. The paranasal sinuses make the skull less massive and they constitute with the nasal and mouth cavity the resonance box of the voice. If one pinches the nose closed, this resonance box is changed – at least in its role in voice production – so that the quality of the sounds which are made, i.e. the timbre of the voice, completely changes.

We distinguish (Fig. 2.4.3):

– the *maxillary sinus,* the open hollow space of the upper jaw ($\times$ 2; i.e. left and right);
– the *ethmoidal sinus,* the open hollow space in the ethmoid bone in the wall between the nose cavity and the eye socket ($\times$ 2);
– the *frontal sinus,* the hollow space in the frontal bone over the eye sockets ($\times$ 2);
– the *sphenoidal sinus,* the hollow space in the sphenoid in the back wall of the nasal cavity ($\times$ 1).

Pharynx

The *pharynx* lies behind the nose and mouth cavities and belongs to both the respiratory and digestive systems. This is so because at the

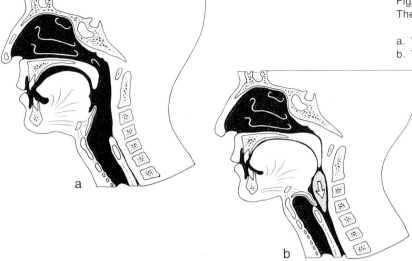

Figure 2.4.5
The pharynx

a. When breathing
b. When swallowing (food)

height of the mouth-pharyngeal cavity we find the crossing and therefore the division of the food tract and the air tract. Air flows into the ventrally positioned *larynx* and food goes into the dorsally positioned *oesophagus*. At a moment when food is passing, the *epiglottis* covers the entrance to the larynx completely. The glottis, i.e. gap between the vocal cords, is also closed. It is then, of course, impossible to breathe.

When at rest and during breathing the epiglottis stands up, almost vertically, and the oesophagus is closed; because of the elasticity of the oesophageal tissues the air is therefore free to pass through into the respiratory tract (Fig. 2.4.5).

The lymphatic ring (of Waldeyer) is also in a strategic position for respiratory activities (see Fig. 2.1.51).

Larynx

The *larynx* lies in the cervical area ventral to the oesophagus. The construction of the larynx consists of a number of cartilages, connected by ligaments and striated muscles (Fig. 2.4.6).

– In relation to the skull the larynx is connected to the hyoid bone by means of a plate of connective tissue. The hyoid in its turn is connected by means of ligaments and muscles to the lower jaw, the base of the skull and

the breast-bone (sternum). In relation to this connection we speak of the 'reins', the bands of muscular tissue by which the hyoid is controlled (Fig. 2.4.7). The hyoid bone supports the tongue and provides attachment for some of its muscles, and a number of muscles forming the wall of the pharynx (cavity). The hyoid is a bone in the shape of a horse-shoe. Its ventral part can easily be felt if you palpate directly above the Adam's apple when your head is slightly bent backwards. In skeletons which have been prepared for educational purposes you will hardly ever find the hyoid bone. This is probably due to the fact that it does not have any articulation at all with other parts of the skeleton in an intact body.

– The largest body of cartilage in the larynx is the *thyroid cartilage*. The side walls of the thyroid cartilage form a rather sharp angle to one another. The upper edges slope forward in the anterior region – particularly in the Adam's apple in males. To the back the thyroid cartilage is open; the side walls taper out upwards into rather long tips – the topmost upper horns or superior cornu. At its lower end, too, we find similar tips, though shorter – the inferior cornu.

To the front the thyroid cartilage can easily be palpated as it is only covered by skin.

– The *epiglottis* is a little elastic plate of cartilage, the shape of which is something half

Figure 2.4.6
The larynx

a. Seen from the front
   and the side
b. Seen from the back
c. Inner wall

  1. hyoid bone
  2. thyroid cartilage
  3. thyroid gland
  4. vertebral column
  5. first rib
  6. clavicle
  7. epiglottis
  8. arytenoid cartilages
  9. cricoid cartilage
10. trachea (windpipe)

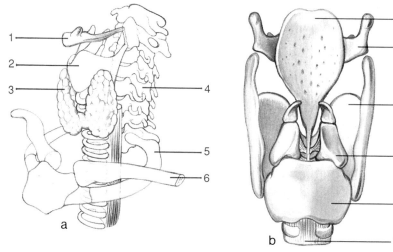

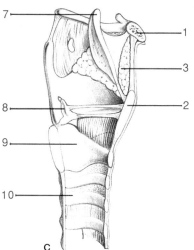

way between the end of a shoehorn and a tablespoon without a handle. The narrow end is attached to the inside of the thyroid cartilage by means of ligaments. In its vertical position the epiglottis reaches as far up as just above the hyoid bone. During the action of swallowing, it is the base of the tongue which presses the epiglottis across and over the opening of the larynx (Fig. 2.4.5). In this way food is prevented from entering into the trachea.

– The *cricoid cartilage* is in the shape of a signet-ring. The high back part, the 'signet part' of this ring points up between the back edges of the thyroid cartilage; the narrow arch of the ring is to the front. It can easily be palpated ventrally, a little lower than the thyroid cartilage.

To the sides the cricoid cartilage is connected to the thyroid cartilage by means of ligaments. This construction allows for a slight frontal and dorsal incline.

– The two *arytenoid cartilages* stand with their base on top of the upper rim of the 'signet-ring cartilage' (the cricoid). Their shape is like that of a pyramid. From the inside of the thyroid cartilage, directly beneath the attachment of the epiglottis the vocal cords run in pairs, parallel to the bases (median ridge) of each of the arytenoid cartilages.
At the back of the arytenoid cartilages there are a number of little muscles attached which run to the cricoid cartilage. As a result of contractions of these muscular bands the arytenoid cartilages can make the following movements on to the edge of the cricoid

cartilage (Fig. 2.4.8): they can shift to and fro, they can turn round their longitudinal axes (rotate), and they can tilt back and forward. These movements influence the position of and tension in the vocal cords.

The oral and the pharyngeal cavities are coated with a type of pavement epithelium which is not subject to keratosis. This is also true for the larger part of the inside of the larynx.

In the wall of the lowest part of the pharynx and in the wall of the larynx we find muscular tissue. From the top and the side edges of the arytenoid cartilages stretches a muscular sheet-plate, coated with pavement epithelium to the sides of the epiglottis, one on the left-hand side and one on the right-hand side. There is also a similar tissue connection between the two arytenoid cartilages (Fig. 2.4.9). In summary there is a kind of cylinder standing on the larynx cavity, which is more prominent at the front than at the back.

In between the walls of this cylinder and the side plates of the thyroid cartilage we see two grooves, one to the left and one to the right. These notches run from the base of the tongue to the back of the cricoid cartilage. After some practice it is possible to let drinks flow to the oesophagus via these notches without swallowing.

The posterior edges of the thyroid cartilages and the end tips of the hyoid bone are inte-

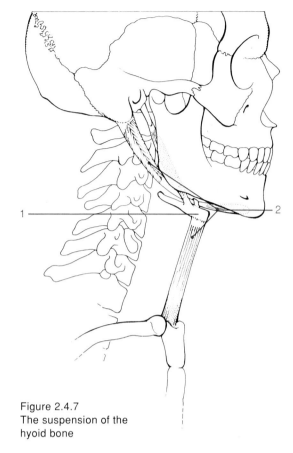

Figure 2.4.7
The suspension of the hyoid bone

1. hyoid bone
2. the muscles, controlling reins

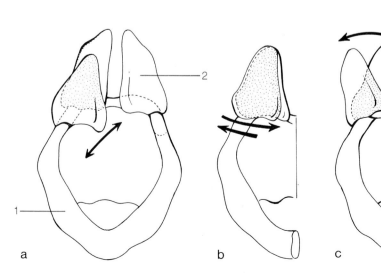

Figure 2.4.8
Movements of the arytenoid cartilages

a. Shifting
b. Rotation
c. Tilting

1. cricoid cartilage
2. arytenoid cartilage

Figure 2.4.9

a. Pharyngeal cavity folded open, seen from the back
b. View inside of the larynx

1. uvula
2. arch of the pharynx
3. tonsil
4. base of the tongue
5. epiglottis
6. hyoid bone
7. back rims of the thyroid cartilage
8. pharyngeal wall, cut and folded open
9. oesophagus
10. false vocal cord
11. real vocal cord
12. arytenoid cartilage
13. part of the cricoid cartilage

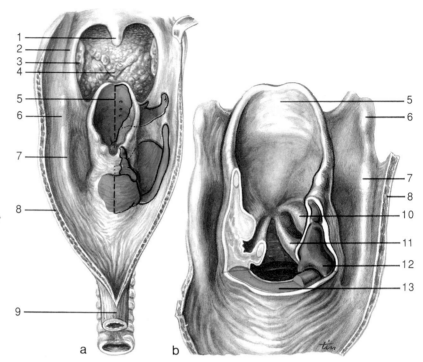

grated in the muscular wall of the pharyngeal cavity.

The cricoid cartilage marks precisely the separation of the air tract and the food tract: its inner side belongs exclusively to the air tract and is covered with ciliated epithelium; the dorsal part is also a part of the wall of the oesophagus. This side of the cricoid is covered with pavement epithelium and belongs exclusively to the food tract.

Vocal cords and vocalization

The vocal cords (Fig. 2.4.9), are two pairs of folds stretching between the arytenoid cartilages and the middle of the thyroid cartilage. The upper pair of vocal cords are called the *false* vocal cords. They derive their name not from any false tones they would produce, but from the fact that they play no part at all in the actual production of sound. They consist of connective tissue coated with pavement epi-thelium in which there are many glands. These glands keep the lower pair of folds, the *real vocal cords*, moist and supple.

The real vocal cords consist of muscular tissue: the vocal muscle. This is coated with pavement epithelium. The opening slit between the vocal

cords is called the *glottis*. Because of the muscles, as mentioned earlier, the arytenoid cartilage can execute different movements which can narrow or widen the glottis and tighten or slacken the vocal cords. Also the tilting of the thyroid cartilage in relation to the cricoid cartilage has an influence on the tension in the vocal cords, as have the contractions of the vocal muscle itself.

Innervation of the muscles of the larynx is achieved by a nerve which branches from the left and right vagus nerve at the level of the lung tips, and after that takes up its course again – the *recurrent laryngeal nerve*.

When the vocal cords are in their widest, most open position the exhaled air cannot make the glottis vibrate. This is the case in normal breathing; it is easy for air to pass the glottis. When the glottis is made narrower the exhaled air can make the vocal cords vibrate. A 'sound' arises. The pitch of the tone depends mainly on the tension of the vocal cords, and the higher this tension, the higher the tone. In addition to this the length, the elasticity and the mass of the vocal cords play a part. In this respect it is possible to compare them with the strings of a harp.

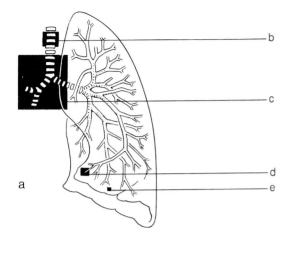

a

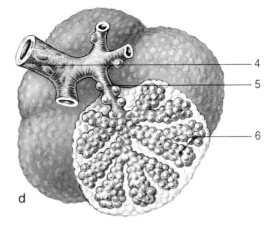

d

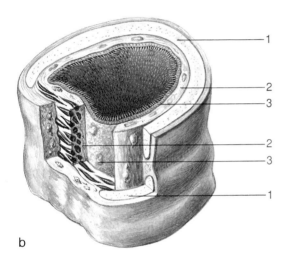

b

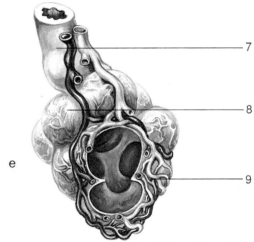

e

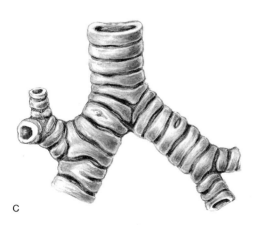

c

Figure 2.4.10
Structure of the air
passages

a. Divisions and
    structures of the
    trachea in the left
    lung
    The boxes refer to
    the detailed
    drawings
b. Structure of the
    trachea, seen
    obliquely from the
    back
c. Dividing the trachea
    into the two main
    bronchi, seen from
    the front
d. Final division of a
    small bronchus,
    bronchiole, half
    schematically

e. Detail of an atrium, half
    schematically
    It can clearly be seen
    that the lung sacs
    (alveoli) are bulges of
    the atrium

1. *a ring of cartilage*
2. *layer of connective
    tissue with muscle
    fibres and mucus
    pituitary glands*
3. *mucous membrane*
4. *cartilagenous scales,
    chips and muscular
    tissue in the wall of the
    bronchiole*
5. *smooth muscular tissue
    in the wall of the
    bronchiolus*
6. *lung atrium*
7. *bronchiole*
8. *alveolus*
9. *capillary network
    around the alveolus*

Generally the vocal cords of males are longer and thicker than of females. In a choir the bass section mainly consists of men and the soprano section of women and children.

The volume of the tone depends on the force with which expiration is executed.

The tones formed in this way are transformed into sounds, and thus into voice sounds, by resonances in the nasopharyngeal cavity, the paranasal cavities and the mouth cavity and especially by movements of the tongue, the lips and the lower jaw.

Trachea and main bronchi

The *windpipe* (i.e. trachea) links up with the cricoid cartilage of the larynx. Ventrally of the oesophagus it descends vertically from the cervical area (6th cervical vertebra) between the two lungs in the mediastinum till the upper border of the 5th thoracic vertebra (see Fig. 2.4.1). There it divides (bifurcation) into left and right main *(pulmonary) bronchi*. The angle made by the left main bronchus with the trachea is sharper than the same angle with the right main bronchus.

Because of the shape of the left half of the heart the left main bronchus *lies* a little more horizontally. This is the reason why small objects (such as a peanut) which accidentally get in the trachea, usually end up in the right lung through the somewhat more vertical right main bronchus.

In the cervical area there is a small part of the trachea palpable. Immediately below the cricoid cartilage we notice the narrow part of the thyroid gland (see Fig. 2.6.5).

Within the thorax the arch of the aorta runs just above the bifurcation (of the trachea) and then along the side of the trachea. The aorta descends then takes its route along the back of the left main bronchus (see Fig. 2.1.33). The lung (pulmonary) arteries and the large coronary veins lie in front of the main bronchi.

The trachea (and the main bronchi) has a framework of incomplete rings (horseshoe shaped) of hyalin cartilage which keep the 'lumen' (more or less moon-shaped) open. These 16–20 rings – placed at regular intervals – are united by fibrous collagen connective tissue (Fig. 2.4.10b). The inner walls of these parts of the air tract are lined by a mucous membrane: ciliated epithelium with numerous pituitary and serous glands. Between this mucosa and the collagen connective tissue we find a layer of connective tissue interwoven with a large number of horizontal fibres of smooth muscle tissue.

In the back wall of the trachea and the extrapulmonary bronchi there is no cartilage; there the incomplete rings of cartilage have their openings. The wall here consists of fibro-elastic tissue and not striated muscle. The oesophagus is situated at the back of this wall.

The mucous membrane in the trachea and the main bronchi has a similar function to the mucosa in the nose cavity. As a result of the movements of the ciliae the thin layer of mucus in which dust particles and other pollutants have been caught is shifted in the direction of the larynx and the pharyngeal cavity. Having arrived there the mucus is swallowed and bacteria etc, if any, are made harmless by gastric acids.

Large bronchi and lung lobes, bronchi and bronchioles

The two main bronchi divide further and further. From this point onwards the branches of the air tract are completely enveloped by lung tissue. The right main bronchus divides into three 'large' bronchi, the left one into two (see Fig. 2.4.1). Each large bronchus is responsible for one lobe of the lung. The left lung consists of two lobes and is smaller than the right lung which has three lobes. This difference is related to the position of the heart.

The construction of the large bronchi is equal to that of the trachea and the main bronchi, the only difference being that the incomplete rings of cartilage get slightly more incomplete and irregular.

The large bronchi divide into ever smaller bronchi. Bronchi supply *lung segments*. There are ten of these segments per lobe. Just like the lobes these segments are functionally and anatomically separate lung units.

The inner coating of the bronchi also consists of a mucous membrane; instead of cartilage horse shoes here we only have small *cartilage scales*,

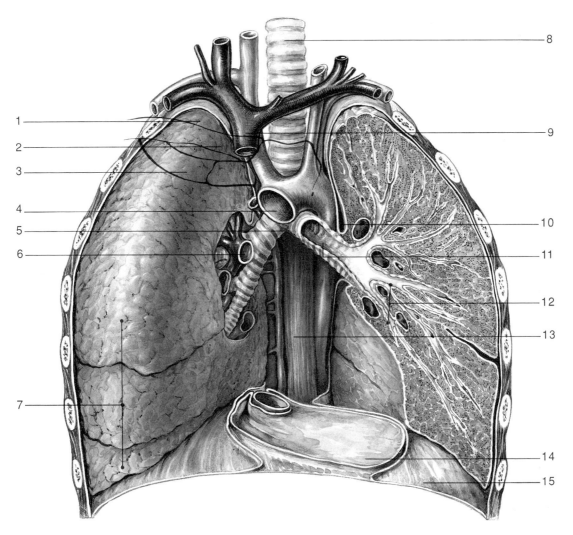

Figure 2.4.11
Overall view of the lungs
and the mediastinum; the
heart has been removed

1. apex of the lung
2. clavicle
3. first rib
4. aorta
5. bifurcation
6. right pulmonary artery

7. lung lobules
8. trachea
9. venae cavae
10. folding edge of
    visceral pleura to
    parietal pleura

11. left main bronchus
12. large bronchi
13. oesophagus
14. bottom of the cardiac
    sac
15. diaphragm

(Fig. 2.4.10d). In the smallest branches, called
bronchioles, even these are absent. All of their
wall is connective tissue containing fibres of
smooth muscle. The muscle fibres relax a little
when we breathe in, and contract a little again
when we breathe out. So during this last action
the already narrow lumen of the bronchiole
(approx. 1 mm) becomes even narrower.

Alveoli

The bronchioles open into one or more atria
(Fig. 2.4.10). The diameter of such an *atrium* is

greater than the diameter of the feeding tubule:
the bronchiole. Remarkable and important in the
walls of the atria are the numerous bubble-like
bulgings known as the *alveoli*. The actual lung
tissue consists of these alveoli which are closely
packed and piled together. It forms a sponge-
like entity with a pale pinkish colour.

The largest part by far of the thorax is filled with
the lungs. They lie against the inner wall of the
thorax. The *top of the lung* (apex) reaches up to
behind the *clavicle*, above the first rib (Fig.

Figure 2.4.12
Relationship between the
alveolus and the capillary
network of the lung
(schematic)

1. *pavement epithelium*
   (plate-epithelium) of the
   alveolus
2. *basal membrane*
3. *endothelium of the*
   capillary
4. *erythrocyte*

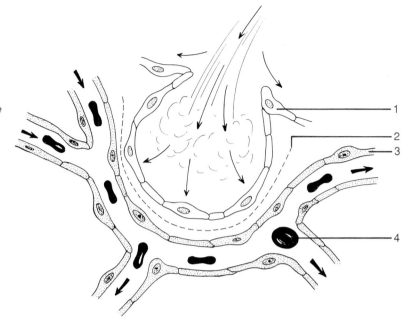

Figure 2.4.13
Ratio of surface : volume =
$L^2 : L^3$

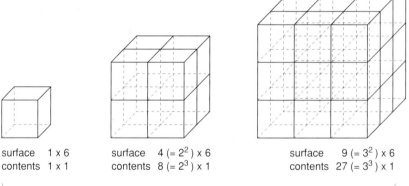

surface 1 x 6	surface 4 (= 2^2) x 6	surface 9 (= 3^2) x 6
contents 1 x 1	contents 8 (= 2^3) x 1	contents 27 (= 3^3) x 1

ratio of surface: volume = $L^2 : L^3$

2.4.11). The *base of the lung* rests on the diaphragm and follows its dome-like shape. Because of this the lung base tapers out at the sides which border the thoracic wall.

The medial side of the lungs borders on the mediastinum (Fig. 2.4.11). At its superior end the mediastinum is bordered by the connective tissue of the neck, at the inferior end by the diaphragm, anteriorly by the breastbone (sternum) and posteriorly by the vertebral column. Within the mediastinum we find the heart, the major blood vessels, the oesophagus, the trachea and the main bronchi, important nerves and lymph vessels.

The alveoli constitute the functional units of the lung. They consist of *one-layered pavement epithelium* (plate-epithelium, sheet-epithelium) on a basal membrane. They are surrounded by a capillary network (see Figs. 2.4.10 and 2.4.12). Between the m.e. in the alveoli and the intra-vascular milieu in the capillary networks an exchange of oxygen and carbon dioxide takes place.

The total surface of all the alveoli is called the *respiratory surface*. Due to the three-dimensional structure of the alveoli this is simply enormous, at rest ca. 70 m² (ca. 765 ft²). During effort the respiratory surface increases consid-

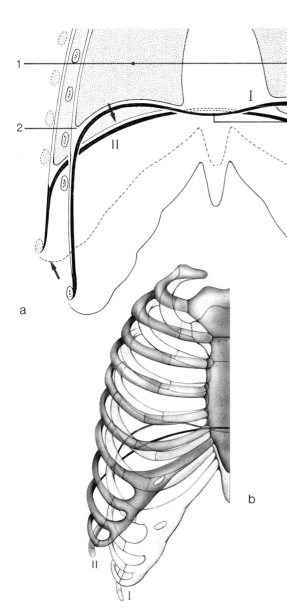

Figure 2.4.14
Respiratory movements;
I exhaling, II inhaling

a. Movements of the
   diaphragm

   1. *lungs*
   2. *pleural sinus*
   3. *ribs*
   4. *diaphragm*

b. Movements of the ribs

the cell wall is – relatively – large enough for a sufficient supply of oxygen via diffusion through this wall. In the case of a human being the surface of its skin (ca. 1.65 m²) is far too small. The respiratory surface increases to the third power. The fact that the respiratory surface lies completely within the body has the additional advantage that the respiratory surface is prevented from drying out.

2. The pleura

The pleura belongs to the serous membranes. The 'inner leaf' is called the pulmonary pleura; the 'outer leaf' is called the visceral pleura (Fig. 2.4.14). The visceral pleura exists of mesothelium on to a subjacent thin, elastic, basal-membrane.

The rim where the visceral pleura is turned down is folded double, and this becomes the parietal pleura. This gateway, the hilum of the lung, is also where the pulmonary bronchus, the blood vessels, the lymph vessels and the nerves enter and leave the lungs.

The parietal pleura has developed together with the thoracic wall, the diaphragm and the adjacent structures in the mediastinum. The walls and surfaces of the visceral and parietal pleura which are turned towards each other are extremely smooth; they are lying in close contact with one another, divided only by a mere film of

erably, to well over 100 m² (ca. 1100 ft²) because the lung tissue is then stretched (i.e. lengthened), and a number of unopened alveoli are unfolded and put to action.

The size of the respiratory surface is one of the important factors for the diffusion speed of gases in the lungs. When we compare the measurements of a human being with those of a unicellular creature it is striking that the surface of the body has increased to the second power and the contents to the third power (Fig. 2.4.13).

The need for oxygen also increases to the third power. In a unicellular creature the surface of

serous moisture. This construction neutralises any friction that would otherwise result from the 'inner and outer leaf' rubbing over each other during respiratory movements.

The potential space between them is the *pleural cavity*, it is airtight and is no thicker than the film of fluid between them. Though the leaves can slide over one another, they cannot be pulled away from each other. We can best illustrate this situation by two sheets of glass with a film of water in between. The pressure in the pleural cavity (intrapleural pressure) appears to be even lower than the atmospheric pressure. The reason for this is that the elastic lung tissue in the thorax has been attached to the walls by means of the pleura in a somewhat elongated state. Were any air to enter the pleural cavity the lungs would be pulled together because of the elasticity of their tissue.

The diaphragm dome rises very steeply from its attachment to the thoracic wall (Fig. 2.4.14), the recessus diaphragmaticus. The lung tissue itself does not extend deep into this sharp notch, but the parietal pleura does. Thus there are two parts of the pleura lying against one another, the small part of the visceral pleura lying against the parietal pleura, and also against lung tissue in the recessus diaphragmaticus. We call this the pleural sinus. When the pleural sinus is in this position, and the recessus diaphragmaticus is percussed, we hear a clear tone.

This clear tone indicates 'solid mass'. This tone can be heard especially on the right side, because of the proximity of the liver.

3. Respiratory movements

There is a constant exchange (diffusion) of gases to and from the alveoli and the surrounding capillary networks. Oxygen is transferred from the alveolus to the passing blood, carbon dioxide from the passing blood into the alveolus.

If we did not breathe, after some time this diffusion would halt. Then, at a certain moment, the difference in the concentration of the gases in the alveolus and the gases in the blood would no longer exist.

The alveolar air is, however, continuously replaced. This is known as *ventilation*. Ventilation is the supply of 'fresh' air to the alveolus and the

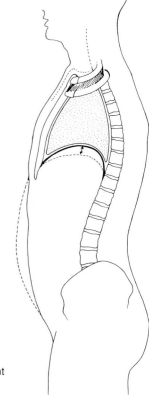

Figure 2.4.15
Movement of the abdominal wall as a result of the movement of the midriff

removal of 'used' air to the outside world. Ventilation occurs by means of *respiratory movements* (Fig. 2.4.14): alternating between inspiration or inhalation (breathing in) and expiration or exhalation (breathing out).

Breathing in

Breathing in (inspiration) is achieved by enlarging the volume of the chest cavity. The thoracic volume can be enlarged by:

- the flattening of the diaphragm: when the muscular part of the diaphragm contracts, the diaphragm dome is flattened. As a result of this flattening of the diaphragm the pleural sinus is widened, so that the lung tissue can take up the enlarged space (Fig. 2.4.14a). At percussion of the recessus diaphragmaticus during inspiration, the tone gradually sounds increasingly dull and hollow. The lung pushes itself into the space, which is now becoming available between the diaphragm and chest wall; the liver is pushed downwards.

Enlarging the thoracic volume by flattening of the diaphragm is accompanied by an imminent reduction in the volume of the abdomen. Since in a closed space (such as the abdominal cavity) the product of the volume (V) and the pressure (P), shows a constant (c) outcome (formula: $V \times P = c$), the pressure in the abdominal cavity should increase during inspiration. Most of the time, however, we see that the abdominal wall gives way under the influence of this force (Fig. 2.4.15) and therefore the volume of the abdominal cavity remains more or less the same in size, though its shape changes. The movement of the abdominal wall in the ventral direction (i.e. forward, away from the spine) has led to the title 'abdominal breathing' for this type of respiration. The term *diaphragmatic breathing* is more appropriate;

– lifting (or drawing up) the ribs: in the position of rest the ribs run forward and downward at an angle (Fig. 2.4.14b). When the ribs are brought into a more horizontal position, both the sagittal and the transverse section of the thoracic cavity increases, in turn increasing its volume (since volume = length × width × height). This raising of the ribs is achieved by contracting the *external intercostal muscles* (see Fig. 2.9.37).

As the thoracic wall moves forward so distinctly, this type of breathing is called chest breathing (thoracic respiration, pectoral respiration).

When we breathe lightly, we use almost exclusively diaphragmatic respiration. When exerting ourselves this type of breathing becomes increasingly combined with the thoracic type of breathing. When extremely deep inhalations are necessary the raising of the ribs can be helped by auxiliary respiratory muscles – the muscles of neck and shoulder.

Since the parietal pleura has developed with the thorax wall and the diaphragm, this membrane moves along with every enlargment of the thoracic cavity. The intrapleural pressure decreases further – in relation to the atmospheric pressure – therefore the visceral pleura is sucked to the outside with the parietal pleura. Within the lung the higher air pressure pushes the inner membrane out towards the outer membrane until the intrapleural pressure has become as high as the atmospheric pressure, i.e. the pressure in the lungs.

The visceral pleura in its turn has developed with the elastic lung tissue and because of its elasticity, the lung tissue then expands.

The lung volume (i.e. the volume of all the lung lobules and alveoli) increases. This would have caused a lowering of the pressure in the lung ($V \times P = c$), if the lung were a closed space. The impending lowering of the pressure is prevented by an influx of atmospheric air via the open respiratory tract, nose, mouth and bronchi.

Now we understand that breathing in is an active process (consumption of energy); muscular activity causes a gradual increase of lung volume. The then higher pressure of the atmospheric air results in a replenishing of the impending lower pressure in the lung alveoli.

Figure 2.4.16
Volume of air

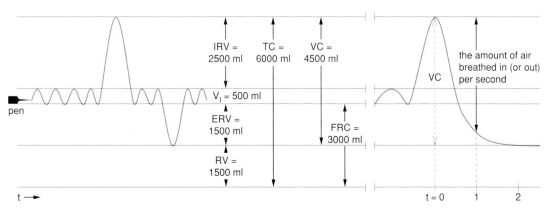

pen

IRV = 2500 ml TC = 6000 ml VC = 4500 ml

V₁ = 500 ml

ERV = 1500 ml

RV = 1500 ml

FRC = 3000 ml

VC

the amount of air breathed in (or out) per second

t →   t = 0   1   2

Breathing out

The process of breathing out (expiration) is effected by reducing the volume of the chest cavity.

When the muscles which have brought about the inspiration start to relax the ribs will lower. The diaphragm dome returns to its position of rest, i.e. it rises under the influence of the pressure in the abdominal cavity.

This results in a reduced thoracic volume; and the lung tissue – because of its elasticity – returns from its extra stretched state. The lung volume is gradually reduced, which is why the pressure in the alveoli is somewhat higher than the pressure of atmospheric air. This force pushes the air from the lungs via the open respiratory tract to the outside world.

You will now understand that breathing out is in principle a passive process (no energy consumption); gravity takes care of the lowering of the raised ribs and the elasticity of the lung tissue gives it back its normal, reduced shape in the position of rest, as it is no longer stretched.

It is possible to breathe out actively (using force) by contracting the interior intercostal muscles (see Fig. 2.9.37). This will reinforce the lowering of the ribs. In addition to that we can contract the abdominal muscles which will push up ($P \times V = c$) the diaphragm dome higher. We can see this active, forceful breathing out with long distance runners, players of wind instruments and opera singers.

4. Lung functioning

When someone is connected to a spirograph it is possible to record (on paper) the amount of air that this person breathes in and out (Fig. 2.4.16). The registered data contains information on the functioning of the lungs.

Respiratory volumes

In the different volumes of air the following can be distinguished (Fig. 2.4.16):
– the volume of air per inhalation (V_I): this is the amount of air inhaled (and therefore exhaled) per 'respiratory cycle'.
  When at rest this is 500 ml: V_I = 500 ml. 0 V_c (per cycle)

During exercise, the volume per breath becomes greater;
– *inspired reserve volume* (IRV): this is the amount of air that could be breathed in after a normal inhalation.
  When at rest IRV = 2.5 litres.
  During exercise, the inspired volume of air becomes smaller;
– *expired reserve volume* (ERV): this is the amount of extra air that can be breathed out, after a normal exhalation.
  When at rest: ERV = 1.5 litres.
  During exercise, the expired reserve volume decreases;
– *residual volume* (RV): this is the quantity of air that remains in the lungs after a maximum exhalation.
  When at rest or during exercise this holds good: RV = 1.5 litres;
– *total capacity* (TC): this is the quantity of air contained by the lungs after a maximum inhalation.
  At rest or during exercise, this is 6 litres.
  In a formula:
  TC = (IRV + V_I + ERV + RV) = 6 litres;
– *vital capacity* (VC): this is the maximum amount of air that can be breathed out after a maximum inhalation.
  Formula:
  VC = (IRV + V_I + ERV) = 4.5 litres.
  TC = VC + RV;
– *functional residual capacity* (FRC): this is the amount of air that has remained in the lungs after a normal expiration.
  FRC = ERV + RV.
  When at rest the FRC = 3 litres.
  During exertion the functional residual capacity diminishes.

The figures presented are rounded approximations for the average person. For women these volumes average about 30% lower.

Functional residual capacity and the gas pressure

From Figure 2.4.16 we can deduce that during breathing – when at rest – the lungs are neither being filled to capacity nor being emptied completely.

Even after a maximum exhalation there is still air left in the lungs. The air left in the lungs after exhalation, however, is again and again subject to gaseous exchange in the alveolar wall and thus deteriorates in quality.

The inhaled 'fresh air' is again and again mixed with the 'old' air of the functional residual capacity. This mixing makes the gas pressures in the alveolus (P_A) change less forecfully than when there was no functional residual capacity. ($P_A O_2 = 100$ mmHg (13.35 kPa) and $P_A CO_2 = 40$ mmHg (5.35 kPa)).

The residual volume is caused by the fact that the lungs are 'hung' in the thorax in an out-stretched form. When the intrapleural cavity is filled with air, the lung collapses due to the elasticity of the lung tissue itself and then the residual volume is to a very large extent driven out of the alveoli. A relatively small part of the residual volume remains in the cartilaginous (therefore open) bronchi, trachea and larynx.

Vital capacity and the one-second-value

The vital capacity provides information on the motility of the thorax. If the deflection on the distance moved by the ribs decreases, then the thorax volume will become smaller and accordingly the difference between maximum inhalation and maximum exhalation will be smaller. The value attached to the vital capacity, however, does not reveal enough information on the functional state of the lungs. So we must also mention, as an important factor, the *speed* with which someone can transport a certain amount of air. For this factor the *one-second-value* is used. This one-second-value gives the percentage of the vital capacity that one can breathe out within one second after a maximum inhalation (Fig. 2.4.16). The one-second-value averages 83% (in our example: 3.7 litres). The size of the vital capacity also provides infor-mation about the elasticity of the lung tis-sues. When this elasticity diminishes (through age or disease) the residual volume is bound to increase. Since the total capacity always remains the same in this case so the vital ca-pacity will decrease.

Respiratory frequency and breath/minute-volume

The respiratory frequency (f_R) is the number of inspirations or expirations per minute.
The breath/minute-volume (BVM or AMV) is the amount of air breathed in (or out) per minute.

Represented in a formula:
$$AMV = f_R \times V_I$$

For the standard person when at rest science has found:

$\left. \begin{array}{l} f_R = 15 \text{ times per minute} \\ V_I = 500 \text{ ml} \end{array} \right\}$ $\begin{array}{l} AMV = \\ (15 \times 0.5) = \\ 7.5 \text{ litres} \end{array}$

During effort not only the volume of air per inhalation rises (a greater part of the inspiratory reserve volume is now used and a greater part of the expiratory reserve volume), but also the respiratory frequency.
The minute volume of breathing can easily reach more than 100 litres (for example: $f_R = 30$ times per minute, or 30/minute, and a V_I of 3.5 litres gives an A/MV of $30 \times 3.5 = 105$ litres). It is worthy of note that the respiratory frequency and the volume of air per inhalation are to a certain degree related. A very fast respiration is a very superficial respiration. Very deep respir-ation goes with a low respiratory frequency.

Dead space

A part of the total pulmonary volume is known as *dead space*. This term indicates those places where there is *no* exchange of gases. This ab-sence of gaseous exchange can result from ana-tomical or physiological reasons.
The *anatomical dead space* is that space in the air tract which is *not* lined with pavement epi-thelium, namely nose or mouth cavity (depend-ing on the route followed), pharyngeal cavity, larynx, trachea, bronchi and bronchioles. The volume of the anatomical dead space amounts to about 150 ml.
The physiological dead space is that space where there is indeed a lining of pavement epi-thelium, but where the capillary networks are (temporarily) not supplied with blood. When at rest the lung tissue in the apexes is unfolded and

not supplied with blood. The physiological dead space represents a reserve capacity.

The effect of the anatomical dead space is (as in the case of the residual capacity) that the gas pressures in the alveoli cannot show dramatic variations. After expiration the air in the dead spaces will contain 'polluted' alveolar air. However, during inspiration this 'polluted' air is mixed with 'fresh' atmospheric air resulting in less pollution. As this mixture of air flows into the alveoli it is further polluted.

Gas analysis

As a result of gas exchanges and of the influence of the various respiratory volumes, there are apparent differences in the composition of inhaled air, alveolar air and exhaled air (Table 2.4.I).

Atmospheric air (and so alveolar air) averages a gas pressure of 760 mmHg (= 101.3 kPa = 1013 mbar). This air pressure is called one atmosphere. The air pressure is derived from the different components of air. The concentration of the component gases is called partial pressure.

- Nitrogen constitutes the major component in the mix of the inhaled air (i.e. atmospheric air). Nitrogen has no role in the gas exchanges in our lungs or our body. This does not mean that there is no nitrogen in our blood or tissues. The gas pressure of nitrogen ($P_A N_2$) in the body equals the nitrogen pressure in the alveolar air (i.e. atmospheric air), namely about 570 mmHg (76 kPa). Though nitrogen does not take part in the gas exchange, the percentage of nitrogen in the three columns is not equal. This is caused – in simple terms – by the changes occurring in the other component parts. The most important change here is the amount of hydrogen added to the air in the

alveoli and so to the exhaled air: this influences the percentages of the other gases in the total mix.

- The oxygen pressure in the atmospheric air ($P_I O_2$) is about 160 mmHg (21 kPa). The $P_A O_2$ is about 100 mmHg (13 kPa) lower because of the incessant diffusion of oxygen into the blood. The oxygen pressure of exhaled air is higher than in alveolar air, because this is being mixed during exhalation with the air in the anatomical dead spaces, where there is no diffusion, so no loss of oxygen takes place.
  The air in the anatomical dead space contains, relatively, the highest amount of oxygen and is the first to be exhaled. This is advantageous in mouth-to-mouth-resuscitation.
- The $P_I O_2$ is extremely low because of the activity of chlorophyll which binds carbon dioxide.
  The $P_A CO_2$ is highest as a result of the continuous diffusion of carbon dioxide from the blood into the alveolar air ($P_A CO_2$ = 40 mmHg (5.3 kPa)).
  The $P_E CO_2$, the CO_2 level in the exhaled air, is lower than in alveolar air through the influence of the anatomical dead space.
- Atmospheric air contains very little hydrogen. It is almost impossible to pinpoint this level since the air humidity varies with all kinds of climatic circumstances.
  During inhalation the air is heavily moistened when passing over the mucosa. Alveolar air is therefore saturated with hydrogen. This has a positive effect on gas diffusion.

A certain amount of water is taken away with the hydrogen in the exhaled air. This is accompanied by a loss of heat: exhaled air is warmer than inhaled air and calories are dispelled with the exhaled hydrogen (as with other

	Inhaled air % P_I	Alveolar air % P_A	Exhaled air % P_E
Nitrogen (N_2)	79	75	77
Oxygen (O_2)	21	13	16
Carbon dioxide (CO_2)	0	5	4
Hydrogen	0	7	3

Table 2.4.I

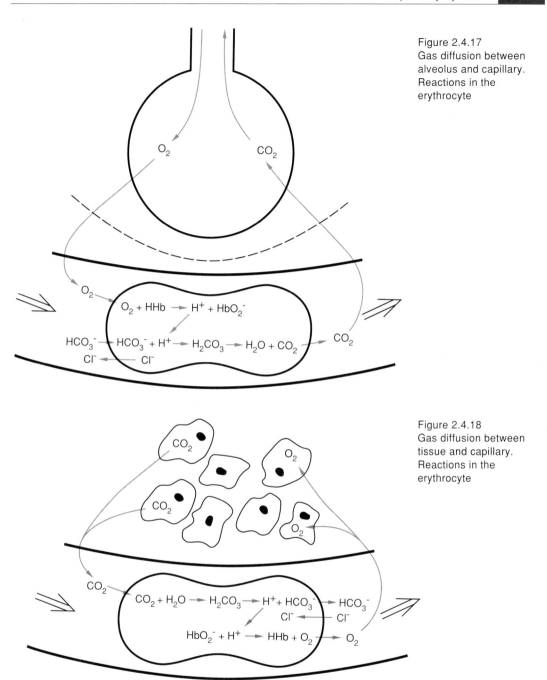

Figure 2.4.17
Gas diffusion between
alveolus and capillary.
Reactions in the
erythrocyte

Figure 2.4.18
Gas diffusion between
tissue and capillary.
Reactions in the
erythrocyte

exhaled substances). However, the heat regulatory function of breathing is insignificant
when compared to that of the skin.

5. Circulation and gas exchange

The heart pumps blood into the lesser circulation via the pulmonary artery. This blood has a
high concentration of carbon dioxide and a low

level of oxygen. When this blood arrives in the
lungs, in the alveoli, gas exchange takes place
through the alveolar walls: oxygen diffuses out
of the alveolar air into the blood, carbon dioxide
diffuses out of the blood into the alveolar air.
This blood, now rich in oxygen and low in
carbon dioxide, flows from the lungs into the
heart via the pulmonary veins. The heart then
pumps it into the greater circulation. In the

active tissues gas exchange again takes place: oxygen diffuses out of the blood via the m.i. to the cells, carbon dioxide diffuses out of the cells via the m.i. into the blood. Now the blood is low in oxygen and rich in carbon dioxide. This blood returns to the heart via the venae cavae. The cycle is complete.

We shall now concern ourselves in greater detail with the two points of contact at which the gas exchange takes place (Figs. 2.4.17 and 2.4.18).

Gas diffusion between alveolar air and blood
The values for blood pressure in the lesser circulation are considerably lower than in the greater circulation. In the capillary networks round the alveoli the blood pressure is about 10 mmHg (1.3 kPa), whereas the colloid-osmotic pressure of the blood is about 25 mmHg (3.35 kPa). This means that the net filtration pressure along the total length of the alveolar capillary network is negative (–15 mmHg (–2 kPa)).

In contrast to the exchange between capillary and the m.i. in the greater circulation (see Fig. 2.1.45), there is no fluid given up in the lung capillary. This explains the fact that the alveoli do not contain any fluid, though they are only three unicellular membrane-layers away from the 100% moist milieu of the blood.

The gaseous exchange between alveolar air and blood (Fig. 2.4.17), depends on the difference in the concentrations of these gases in the two compartments. The blood arriving in the lungs shows a venous oxygen pressure/level (P_vO_2) of about 40 mmHg (5.3 kPa). The alveolar oxygen-pressure in the alveolar air (P_AO_2) amounts to 100 mmHg (13.3 kPa). This vast difference in pressure leads to a fast diffusion of oxygen to the blood at first. Later on it lessens, as the difference in pressure diminishes.

The oxygen pressure of the arterial blood leaving the lung (P_aO_2) amounts to about 90 mmHg (12 kPa).

There is very little oxygen dissolved in the blood plasma (0.3 ml O_2 in 100 ml of blood). The erythrocytes, however, are capable of combining with large amounts of oxygen. The red blood pigment *haemoglobin* forms a loose,

easily reversed bond with the oxygen molecules. Simplified this reaction reads:

$$HHb + O_2 \rightarrow HbO_2 + H^+$$

Reduced haemoglobin (dark red in colour) reacts with oxygen to form oxyhaemoglobin (bright red in colour) and hydrogen (ion).

About 20 ml of oxygen is bound to haemoglobin per 100 ml of blood. This quantity of oxygen is generally quite sufficient for the tissues.
The blood arriving in the lungs out of the active tissues shows a venous carbon dioxide pressure (P_vCO_2) of 45 mmHg (6.0 kPa). As a result of constant ventilation, refreshing of the alveolar air, the CO_2 level here (P_ACO_2) is around 40 mmHg (5.3 kPa). In spite of this small difference in pressure the CO_2 passes quickly into the alveolar air for the CO_2 level in the arterial lung blood (P_aCO_2) is already equal to that of the alveolar air (P_ACO_2): both 40 mmHg (5.3 kPa). Of all the carbon dioxide transported about 10% is dissolved in the blood; about 70% is present in the blood in the form of HCO_3 and the remaining 20% is bound on to the haemoglobin. The HCO_3 forms a bond (loose and easily reversible) with the H^+ ion in the erythrocyte, which is set free in the oxidation of haemoglobin (see above). The formula for this is:

$$HCO_3 + H^+ \rightarrow H_2CO_3 \rightarrow H_2O + CO_2$$

Note that the binding of oxygen (H^+ ion set free) facilitates the discarding of carbon dioxide. The water formed in this reaction can be expelled via the kidneys; the carbon dioxide diffuses to the alveolar air and is exhaled.

Gas diffusion between blood and active tissue
The gas diffusion between blood and the m.i. of active tissue also depends on the difference in gas pressure on both sides of the capillary wall (Fig. 2.4.18). The blood which arrives in the active tissue has, of course, a P_aO_2 of 90 mmHg (12 kPa); the P_aCO_2 is 40 mmHg (5.3 kPa).
Due to the metabolic activities in the cells the CO_2 level in the cells and so in the m.i. is constantly higher than in the blood which flows past. Along the entire length of the capillary network, CO_2 diffuses from the m.i. into the

blood. This is accompanied by this chemical reaction in the erythrocyte:

$$CO_2 + H_2O \rightarrow H_2CO_3 \rightarrow H^+ + HCO_3$$

This causes the P_vCO_2 to rise to 45 mmHg (6 kPa). This reaction also informs us that the increase in the amount of CO_2 threatens to create an increase in the H^+-ions and thus a lowering of the pH-level.

The H^+ formed in this reaction now reacts with the oxyhaemoglobin. In a (simplified) formula:

$$H^+ + HbO_2^- \rightarrow HHb + O_2$$

Oxygen (O_2) is set free so the increased amount of CO_2 here stimulates the freeing of oxygen out of the erythrocyte. Because of the consumption of oxygen in the active cells, the oxygen pressure in the cells, and therefore in the m.i., is constantly lower than that in the blood which flows past. Along the full length of the capillary network, this oxygen diffuses from the blood into the m.i., and from there into the cells. This results in a lowering of the P_vO_2 to the level of 40 mmHg (5.3 kPa).

Comparison of Figs. 2.4.17 and 2.4.18 shows that the occurrences in the active tissues exactly mirror the occurrences in the lungs. The processes which involve haemoglobin follow an opposite course; we speak of a *haemoglobin-buffer*.

The bronchial artery

From the discussions so far you could conclude that the lung tissue itself gets a poor deal; it seems to be offered only blood with little oxygen. The pavement epithelium of the alveoli is a tissue that would be able to withdraw oxygen for its metabolism from the alveolar air since it is in direct contact with it. All the other tissues of the lungs and the lower branches of the bronchi have no direct access to this source of oxygen. These tissues are supplied with oxygen via a branch of the aorta (greater circulation): the *bronchial artery* (see Fig. 2.1.22). This artery divides itself into arterioles and after that it forms alongside the capillaries of the small circulation an integrated network of capillaries.

6. Regulation of respiration

Though the respiratory muscles are striated and thus according to definition belong to the voluntary muscular system, breathing is generally a process which takes place completely without our volition. In spite of the fact that we can voluntarily control the respiratory frequency and the volume of air per inhalation, it is correct to refer to respiratory automatism.

Yet this is an automatism of a quite different nature than the heart automatism or the automatism of the leg muscles when walking. Heart automatism cannot be stopped by the individual by any means whatsoever; breathing automatism indeed can be stopped, but only temporarily. At a certain moment the urge to breathe becomes stronger than our will. The breathing-muscles are innervated animally and the striated breathing-muscles belong to the animal motor system.

Respiratory centre

In the medulla oblongata and the pons of the brain stem (see Fig. 2.7.25), there is a large number of neurones which form a functional centre for the regulation of our breathing: *the respiratory centre*. In this breathing centre we can distinguish a number of inseparable parts which work together in such a finely tuned way that any inspiration is followed by an appropriate expiration in a smooth and harmonious manner, and then again by an inspiration, and so on throughout a lifetime.

Not only is this regular alternation controlled and directed, but also the frequency and the volume of air per inspiration is regulated.

The muscles involved in respiration are innervated by the respiratory centre; the muscular part of the diaphragm is innervated by the *phrenic nerve* (Fig. 2.7.21); and the intercostal muscles by the *intercostal nerve*. In order to be capable of regulating breathing in an efficient manner the respiratory centre must be informed of the respiratory movements and the gas pressures at various places in the blood.

In the wall of the bronchi we can find interoceptors; they are the *Elftmannian stretch receptors*. During inspiration the elastic lung tissue is being increasingly stretched and the stretch receptors are stimulated, so they give impulses to the respiratory centre with an increasing frequency via the vagus nerve. As a result of this, inspiration firstly slows down, then stops completely, after which expiration follows. During expiration the frequency of the impulses from the stretch receptors diminishes. After some time the frequency becomes so low that an impulse to start an inhalation is sent to the centre and after a while there again follows an inhalation.

The continuous alternation from inhaling to exhaling under the influence of receptors in the lung tissue is under the control of *the Hering-Breuer reflex*, or *inflation reflex*.

The need for ventilation

The respiratory centre is informed of the need for ventilation in the body mainly by chemoreceptors which are sensitive to the P_aCO_2 or the pH of the blood. Such chemoreceptors which receive chemical data can be found in the aortic arch (see Fig. 2.1.33), and in the common carotid artery (see Fig. 2.1.34).

A rise in the P_aCO_2 and/or a drop in the pH of the blood bring about a very strong stimulus for the respiratory centre to make respiration deeper and quicker. As a result of this more CO_2 can be expelled out of the blood; which leads to the P_aCO_2 coming down to normal values and breathing returns to 'normal'.

The respiratory centre itself is excited mainly by changes in the pH-level of the blood. It has specific receptors sensitive to this type of change.

On the one hand the acid concentration in the blood is directly influenced by an increase of carbon dioxide, but on the other hand it can also be caused by an increase in the amounts of other acids in the blood, such as lactic acid which is formed in anaerobic oxidation.

A lowering of the amount of oxygen in the blood is a far weaker respiration stimulant than a lowering of the pH or a raising of the carbon dioxide pressure. Under normal circumstances the oxygen level plays hardly any role. When we stay at great altitudes (e.g. when climbing mountains) this is different. The pressure of the atmospheric air decreases with increasing altitude, and along with it the P_1O_2 decreases. When the pressure of the atmospheric air is 500 mmHg (66.6 kPa) the P_1O_2 is about 105 mmHg (14 kPa).

The lower P_aO_2 leads to hyperventilation (quickened and deeper breathing). This brings down the P_aCO_2 and makes the pH rise. These changes slow down the respiratory centre, but the rise of the pH is negated in and by the kidneys. The lowering of the pH reinstates hyperventilation. This adaptation improves the oxygen supply at higher altitudes. After some time at a higher altitude more erythrocytes are formed under the influence of the hormone erythropoetin, as the erythrocytes are no longer completely saturated with oxygen, because of the lower pressure. Larger numbers of erythrocytes compensate to a certain extent for this decrease in oxygen transport per erythrocyte.

7. Interruptions of the respiratory automatism

The regular alternation of inspiration and expiration can be interrupted with or without our will.

We normally regulate the flow of expiratory air in the voluntary interruptions of this automatism. We do this when speaking, but in an even more explicit way in singing, or when playing a wind-instrument or blowing glass. Whilst giving birth a number of respiratory techniques are used in order to facilitate the delivery. A stuffy environment 'takes your breath away'; a cold shower has an undoubted effect on breathing; when frightened you 'catch your breath'. The glottis can be closed almost instantaneously because of reflexes, when irritating vapours threaten to become inhaled. In such cases the glottis functions as a safety valve. Apart from these irregularly occurring phenomena there are a number of frequently occurring interruptions of respiration.

Sighing

In normal quiet breathing there is a deep inhalation at rather regular intervals. During this deep inhalation those alveoli which remained closed flat during the shallow, quiet ventilation are fully unfolded again. This *sighing reflex* is effected via specific receptors, sensors in the lung tissue and/or a local shortage of oxygen.

Yawning

Yawning is a spasmodic breathing movement along with the tightening of the muscles in the jaw and neck. It occurs especially in the morning and evening, and after awakening it is often accompanied by stretching the entire body.

The stimulus which initiates yawning is not known. It is noteworthy that the phenomenon also occurs at the perception of hunger, when you feel bored or when seeing other people yawn. Yawning is thought to establish a better oxygen supply to the brain. The deep inhalation brings more oxygen into the blood and the stretching of the muscles which have been at rest brings about a redistribution of the blood to the benefit of the brain.

Swallowing

When we swallow, respiration is halted by a forceful brake-signal from the respiratory centre. The flow of air comes to a standstill, the epiglottis can close the air tract at the very moment that the lump of food reaches the entrance to the trachea. Thus swallowing into the respiratory tract is prevented.

Sneezing

When the mucosa of the nasal cavity is irritated by e.g. inhaled dust particles, a powerful reflex occurs: the *sneeze reflex*. This reflex brings about a deep inhalation followed by a violent contraction of the exhalation muscles so that the air is exhaled explosively. As the mouth cavity is automatically closed (the reflex being instigated from the nose area) the explosive airstream passes through the nose cavity, and the irritating mucous or dust particles are exploded out.

Coughing

The *cough reflex* comes into action when a mucous membrane of the deeper lying parts of the air tract (larynx, trachea, bronchi) is irritated. After a deep inhalation the glottis is closed. Then follows a forceful contraction of the exhalation muscles which causes a high pressure in the thorax ($P \times V = c$). When the glottis is then suddenly opened the air shoots out through the mouth cavity and sweeps the cause of the irri-tation away in its stride. The nose cavity is generally closed during this action.

Hiccup

Having hiccups is an extraordinary interruption of the breathing rhythm caused by a sudden and forceful contraction of the diaphragm muscles, which occurs from time to time. This results in a forceful influx of air, during which the glottis is suddenly closed; the inhalation is suddenly stopped by the vocal cords which causes the characteristic 'hiccup' sound.

Eating or drinking hastily often leads to a fit of the hiccups. The reason for hiccups is not known nor is the purpose.

We can put an end to the hiccups by holding our breath for a long time after a deep inhalation. The diaphragm is then maximally tightened for a longer period of time.

Vomiting

The *vomiting reflex* also interrupts the breathing rhythm. The contents of the stomach are returned to the pharyngeal cavity. These stomach contents must be prevented from entering into the air tract when passing the larynx. That is why the vomiting reflex starts by causing a deep inhalation, after which the glottis and the entrance to the nose cavity close. Thus the contents of the stomach are led to the pharyngeal and oral cavity.

Pushing

When a person pushes – usually after a (deep) breath – the glottis becomes closed. Then the abdominal muscles are tightened and/or the diaphragm is lowered in order to reduce the abdominal cavity. The enhanced pressure which is now executed onto the abdominal organs has a positive effect on e.g. the evacuation of urine and on defecation.

5

Skin

Introduction

The skin is an indicator of temperature, emotion, age and race, as well as being the contact point between the individual and the external environment.

The skin has a wide range of functions.

Firstly, *protection* against certain outside influences. These influences can be mechanical in nature, for instance a punch, a superficial cut, or pressure, e.g. on the sole of the foot. They can also be chemical in nature, for instance the cleansing of the hands with white spirit. The skin also protects against biological influences, for instance against invading microbes. It protects against radiation, and against loss of water and heat.

Secondly, *excretion*. Water is evaporated and salts are excreted via the skin. Salt is easily perceptible when sweat is tasted.

Thirdly, *heat control*. The skin (and the subcutaneous connective tissue) gives a certain degree of insulation; the body temperature is controlled by adapting the amount of perspiration and the volume of blood going to the skin.

Fourthly, *synthesis of vitamin D*. Under the influence of energy from UV-radiation, ergosterol, present in the skin cells, is converted into vitamin D.

Fifthly, skin has an important *sensory* function. There are a great number of sensory organs in the skin which respond to pressure, touch, pain and changes in temperature.

These and other functions of the skin will be discussed in this chapter. The skin (cutis) consists of two layers: the epidermis and the dermis or cutis vera. The subcutaneous connective tissue, the movable connection between the skin proper and the structures underneath (general fascia and muscles), will also be discussed.

Learning outcomes

After studying this chapter you should have sufficient knowledge and understanding of:

- the general functions of the skin;
- the structure and function of the epidermis, the hair, the nails and the glands of the skin;
- the structure and function of the dermis and the subcutaneous connective tissue;
- the vascularization of the skin and the meaning and functioning of temperature control by the skin.

1. Epidermis

The *epidermis* (Fig. 2.5.1), consists of multilayered squamous epithelium. The appearances of the cells in the different layers reflect the stage of life of the cells. The cells gradually move from the deeper layers, in which mitosis takes place, towards the surface. During this process they gradually die, wear off at the surface and are discarded as flakes.

The lower layer is called the *stratum basale*. Its cells are cuboidal in shape and rest on the basement membrane. Epidermis and dermis are linked here by connective tissue elevations of the dermis which interdigitate with the epidermis. These elevations are called the *skin papillae* (Fig. 2.5.1).

The cells in the stratum basale contain pigment grains, melanin, which determine the colour of the skin and protect the skin against being adversely affected by the rays of the sun.

In the stratum basale mitosis takes place. A kind of epithelial stem cell continuously produces daughter cells, which are divided at high speed and pushed to the surface. Because of lack of room the basal layer becomes undulated.

On their way to the surface the cells become more and more flattened, they lose their pigment and the possibilty of mitosis diminishes.

In the layer on the stratum basale the cells make linkages with neighbouring cells via processes or spina. This layer is called the *stratum spinosum*. The skin gains a large part of its firmness from this layer.

The next layer is the *stratum granulosum*. Its cells contain grains (granulae) which are a pre-stage of keratin. The cells of the next layer are almost flat, have hardly any nucleus and are translucent due to the colourless pre-stage of keratin. The layer is called the *stratum lucidum*

or *clear layer of skin*. The most superficial layer is the *stratum corneum*. The cells are dead and form a layer of flakes, which contain the corny substance keratin.

The first three layers discussed are also called *stratum germinativum* or germ layer (Fig. 2.5.1b), as the cells in this area are alive and divide. In this process the stratum basale is always active, whereas the stratum spinosum and stratum granulosum, when the skin is damaged, can be activated by the production of the hormone histamine.

The stratum lucidum and stratum corneum together are called the horny layer (i.e. stratum corneum). As division is no longer possible in these cells the horny substance is an important element. Unfortunately the term 'stratum corneum' is therefore ambiguous.

There is no vascularization in the epidermis and it has no lymphatic vessels. The m.i. of the living stratum germinativum is changed via diffusion from the underlying connective tissue. There is no nervous tissue in it either, apart from free-ending fine branches for the sense of pain. The sensation of pain takes place when these nerve endings are stimulated by substances which are released from damaged cells. There is itching when these substances are released in small quantities, and the cells remain intact.

Callus or callosity is a thickening of the stratum corneum in response to repeated friction or pressure.

The skin of the palmar surface of the hand and the fingers always has a somewhat thicker horny layer than the other skin areas. The epidermis has many ridges and grooves, and produces, among other things, the characteristic and unique finger print. The same is true of the feet. These ridges and grooves are called *dermatoglyphics* (Fig. 2.5.2). The dermatoglyphics are formed by the skin papillae being

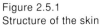

Figure 2.5.1
Structure of the skin

a. Three-dimensional
   picture of the skin
   and the
   subcutaneous
   connective tissue

   1. epidermis
   2. dermis
   3. subcutis
   4. sweat gland
   5. stratum corneum
   6. stratum lucidum
   7. stratum
      granulosum
   8. stratum
      spinosum
   9. stratum basale
   10. skin papillae

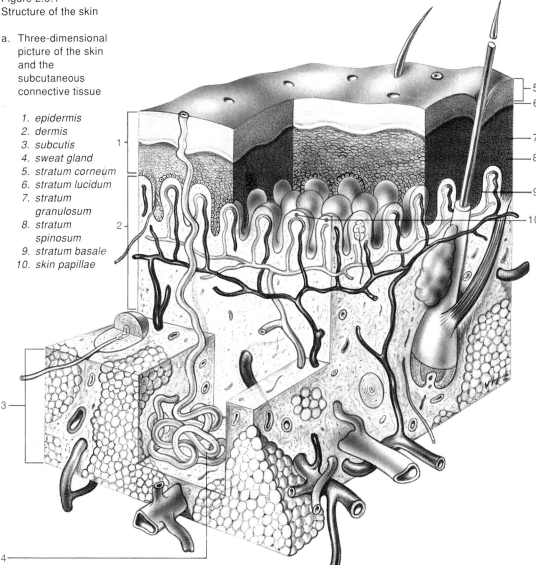

b. The skin layers,
   arranged

   1. horny layer
   2. germ layer
   3. dermis or cutis
      vera

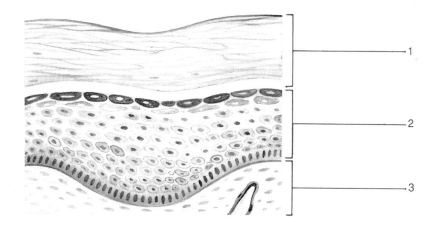

Figure 2.5.2
Dermatoglyphics on the tip of the finger

arranged in rows. This arrangement is genetically determined, so each individual has unique finger prints.

Skin colour

The colour of the skin is dependent upon the amount of melanin which is formed by cells in the stratum basale. The formation of pigment is dependent upon the amount of sunshine (especially UV-radiation, harmful to tissue) to which the skin is exposed. Melanin prevents UV-radiation from penetrating deep into the tissue.

At some places in the skin the quantity of melanin producing cells is greater than at other places, as it is the case around the anus, in the area around the nipples, in the skin of the scrotum and the labia majora and minora. Moreover, the skin of the face and the limbs is more pigmented than that of the abdomen and the back. A good illustration is the difference in burning after a day of excessive sunbathing.
Genetically darker races produce more melanin. Most probably this has to do with the climate in which races originally lived.
Everybody has more or less birth marks. They are local clusters of cells which produce extraordinary amounts of melanin.
In a freckled person, the pigment producing cells are not evenly spread in the basal layer, but clustered together. When exposed to sunshine the freckled effect becomes more apparent because of the contrast. The usually reddish colour of freckles can be attributed to the presence of another pigment known as *carotene*.

Hair

Hairs are extraordinary epidermal formations. They are implanted obliquely in the tissue and reach as far as the subcutis. A hair consists of several elements (Fig. 2.5.3). The part protruding from the skin is the *hair shaft*. In the skin the shaft changes into the *hair root* and then into a thickened end, the *hair bulb*. Mitosis takes place here, as a result of which the hair, which actually consists of stratum corneum, gradually shifts upwards. The hair sits in the hair follicle, which only consists of local epidermal layers. In the hair bulb the hair and the epidermal layers actually merge. The bulb has a cavity, the hair papilla, in which nerve branches and a number of capillaries are located. The hair itself consists of several layers and is normally solid.
At the side of the smallest angle with the skin is a small muscle (smooth muscular tissue), at the junction of hair follicle and bulb, extending upward in an oblique direction and connected with the subcutaneous connective tissue known as the arrector pili muscle. Contraction of such hair muscles makes the hairs stand upright and pulls the follicles upward, as a result of which the so-called 'goose flesh' occurs. Strong haired animals, such as dogs and cats, can keep a thick layer of air 'trapped' on the skin through this mechanism, to prevent cooling. Air is an excellent insulator, as it is a poor conductor of heat.
The hair growth on man does not have an insulatory function. Goose flesh is probably nothing more than an evolutionary rudiment. Clothing can take over the original function of hairs. Woollen clothing is known for its 'warmth'. This is not surprising as wool is sheep hair and can retain heat extremely well.
Apart from an aesthetic function, hair has a sensory function. A number of small nerve branches circle the lower parts of the follicles. These branches are stimulated, together with the branches in the bulbs, when the hair changes its position. These nerve ends are extremely

Figure 2.5.3

Longitudinal section through
a hair follicle

1. *hair shaft*
2. *hair root*
3. *sebaceous gland*
4. *hair muscle (arrector pili muscle)*
5. *hair follicle*
6. *hair bulb*
7. *hair papilla*

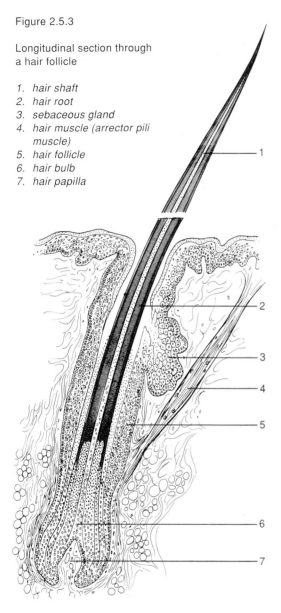

sensitive. This is easily noticed when any one hair is touched.

It is known that human hair is not evenly spread and not equally thick everywhere. In adults the only hairless parts are the palms of the hands and the soles of the feet, the lips, the navel and parts of the external genital organs.

Some hair grows continuously, whereas other hair hardly grows or does not grow at all, or only when it is cut or shaved off. On the skull there are 100,000 to 200,000 hairs, which grow an average of 2 mm a week and have an estimated life span of approximately four years. After the hair has fallen out, a new one is formed in the bulbus.

The hairs of the eyebrows are thick and do not grow. The way they are implanted largely prevents sweat from streaming straight from the brow into the eyes. The eyelashes grow up to a certain length, characteristic for the individual; when this length is reached, the growth stops. Eyelashes regularly fall out and are replaced. The sensory function of the eyelashes is developed further and the eyes are shut reflexively before a dust particle can reach them.

The hair in the armpit is relatively soft and does not grow, nor does the hair on the arms and the legs. Pubic hair is usually thicker and harder and does not grow.

Generally speaking, men have more hair growth than women. The growth on the face, moustache and beard is especially conspicuous. This secondary sexual characteristic is related to the influence of the different sex hormones.

In addition there is a race-linked difference in hair growths (curly hair, straight hair, etc.).

Like the colour of skin, the colour of hair is also dependent upon the quantity of pigment. After all, hairs are epithelial formations. A large amount of 231 pigment causes the hairs to be black, little or no pigment makes them white. In a person with red hair the pigment carotene is present.

Nails

The nails, located on the distal phalanges of the fingers and the toes, are hard, but flexible, horny plates (Fig. 2.5.4). They are transparent due to the large amount of a colourless pre-stage of keratin present in them. At its base the nail is embedded in a small fold of the skin, the nail skin which has a small fold on the nail, the *cuticle*. The nail skin appears to be the place where the nail ends, but the skin folds on the inside and folds again after about 5 mm. The nail is, apart from its tip, embedded into its underlying layer, the *nail bed*. This consists only of the stratum germinativum.

The growth of the nail starts from the second fold. This area is called the *nail root* or *matrix*. From the double stratum germinativum located here, the cells slide distally over the nailbed.

Part of the matrix is usually visible below the cuticle as a white coloured half moon, the *lunula*.

Sometimes there are white spots in the nails. These are air cells, caught in the horny layer. The fingernails are useful for scratching and for increasing the accuracy with which the fingers can grasp.

Fingernails grow an average of 0.5 mm per week, toenails grow more slowly.

Perspiratory glands

Perspiratory or sweat glands are epidermal formations of which the sweat producing part is located in the dermis. A sweat gland is a convoluted tubule-shaped gland of one-layered epithelium, encircled by smooth muscle fibres. These fibres push the product of the sweat gland to the surface of the skin via the conducting tube. Sweat glands are well vascularized, which is understandable given the volume of fluid excreted by the cells. Sweat consists of 99% water, the rest consists of dissolved salts (NaCl) and acids, but also of waste products of nutrients. This is noticeable in a person who has eaten a lot of garlic.

The number of sweat glands is estimated at 2 to 3 million. There are more of them in the palm of the hand, the sole of the foot, the forehead and the back than on other parts of the body.

The sweat glands continuously produce fluid. This evaporates immediately via the skin. During evaporation a large amount of heat can be lost from the body. When a large volume of sweat is produced in a short time, this cannot evaporate and drips from the skin onto the clothes or is wiped away. It then has very little function in the carriage of heat.

Evaporation cannot take place unless the surrounding air is able to take up vapour. 'Sultry summer nights' and tropical sea climates are characterized by high humidity, making a person sweat easily.

The body not only discharges fluid by means of sweat glands. There is also a continuous diffusion of water from the lower skin layers through the epidermal surface, where it evaporates. As this process is not discernible, it is called *insensible sweating*.

The loss of weight caused by this evaporation can be measured by using a very accurate pair of scales.

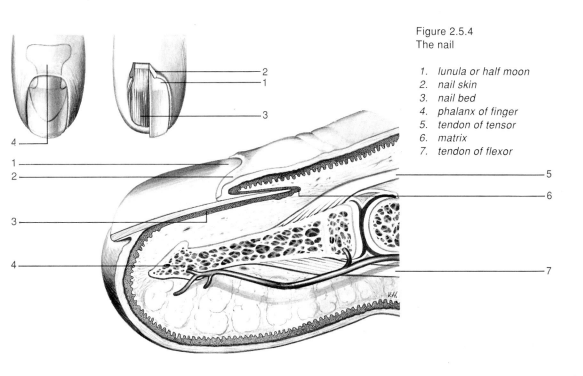

Figure 2.5.4
The nail

1. lunula or half moon
2. nail skin
3. nail bed
4. phalanx of finger
5. tendon of tensor
6. matrix
7. tendon of flexor

Specific sweat glands, those in the armpits, around the nipples and in the area of the anus and the sex organs, secrete not only sweat but also aromatic substances, which are specific for each individual. A tracker dog uses this characteristic, but men can also recognize certain body odours. Some mothers – and fathers – can pick out their child's T-shirt, worn during one day, from a pile of other T-shirts.

Women have more aromatic glands than men. There are also race-linked differences. It is not known whether these aromatic substances have any sexual arousal function. Microbes on the surface of the skin decompose these glandular products, as a result of which a smell develops which is often experienced as unpleasant. Deodorant producers make a living out of counteracting this smell.

In the external auditory meatus sweat glands are located which produce the waxy substance *cerumen*. This ear wax keeps the skin and the eardrum supple and water resistant, which assists hearing.

Sebaceous glands

Sebaceous glands are lobulated glands which produce the greasy substance *sebum* (Fig. 2.5.3). They mainly end in pairs in a hair follicle. The sebum secreted spreads and keeps skin and hair greasy and therefore supple. It protects against dehydration and intruding microbes.

Mammary glands

Both men and women have mammary glands, but in the female they develop into breasts and only start functioning during pregnancy. In the male they remain rudimentary.

The breasts lie on the left and the right pectoral muscles.

A small part of the breast consists of glandular tissue, the larger part consists of subcutaneous connective tissue and fat tissue (Fig. 2.5.5). The size of the breasts is determined by the amount of fat tissue, the firmness is determined by the amount of connective tissue.

The glandular tissue comprises approximately twenty glands composed of lobules which are drained by ducts which join together so that each gland has a single duct opening on the surface of the nipple. The glandular lobules are surrounded by firm connective tissue plates (Fig. 2.5.5). The lobules, especially the ducts, have smooth muscle cells, that push the milk outwards when the time is right. The glandular tissue is surrounded by an extensive capillary network. The nipple (papilla mammae) is on the top of the breast and encircled by well pigmented *areola* of the nipple, comprising many small protuberances. Just under the skin of the nipple smooth muscular tissue is located which contracts reflexively when the nipple is stimulated (for instance by touching or sucking). This can also be effected by hormonal or neuronal impulses.

As a result of contraction the nipple *erects.* The breasts have an evident function as milk secreting glands for babies, but they also have a sexual function.

2. Dermis

The *dermis* consists of connective tissue (see Fig. 2.5.1). The border with the epidermis is not smooth, but undulated. The dermis has elevations which interdigitate with the overlying epi-dermis. They are called the *dermal* or *skin papillae*. The connective tissue of the papillae and the dermis underneath contain fine collagen and elastin fibres. This is called the *papillary layer.*

Going deeper, the amount of collagen fibres grows. The skin gets its toughness and elasticity from the fibres in the dermis. The collagen fibres form a rather regular network of longitudinal reticula (Fig. 2.5.6.a). This is the *reticular layer* of the dermis. The reticular layer determines, by the course of the collagen fibres, the *splitting direction* of the skin. When the skin is cut in longitudinal direction, few collagen fibres are cut through and the edges of the wound stay more or less together due to the pressure on the other fibres. The result after recovery is a small scar.

An incision at a right angle to the longitudinal direction of the reticular network will cut through many collagen fibres. The edges of

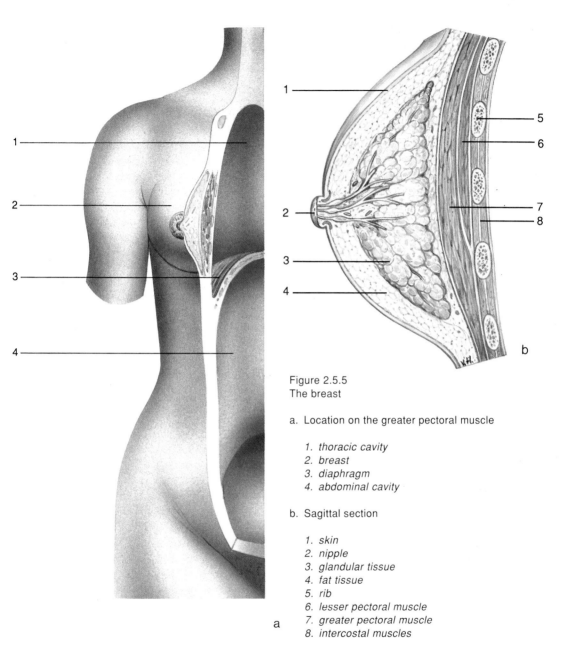

Figure 2.5.5
The breast

a. Location on the greater pectoral muscle

    1. thoracic cavity
    2. breast
    3. diaphragm
    4. abdominal cavity

b. Sagittal section

    1. skin
    2. nipple
    3. glandular tissue
    4. fat tissue
    5. rib
    6. lesser pectoral muscle
    7. greater pectoral muscle
    8. intercostal muscles

the wound will gape, because of the tension on the cut fibres, as a result of which there may be a broad scar. The splitting direction of the skin differs from area to area. However, the directions of the splitting lines in different individuals are very similar (Fig. 2.5.6b). Surgeons, and especially plastic surgeons, will take into account as far as possible the splitting direction of the area in which they intend to make an incision.

Apart from the glandular tissue discussed earlier, the dermis contains plexuses of blood and lymphatic vessels, nerve fibres and sensory nerve endings. The sensibility of the skin is transmitted to the brain via the skin sensors (pressure, temperature, pain, sense of touch).

The sensory function of the skin will be discussed in Chapter 8 of Module 2.

Figure 2.5.6
Splitting direction of the skin

a. structure of the reticular
   layer of collagen fibres.
   A cut made in the skin in
   situation (I) will cause a
   narrow wound and a
   narrow scar, in situation
   (II) a gaping wound and,
   generally speaking, a
   broader scar

b. Splitting lines in the
   different areas of the
   skin

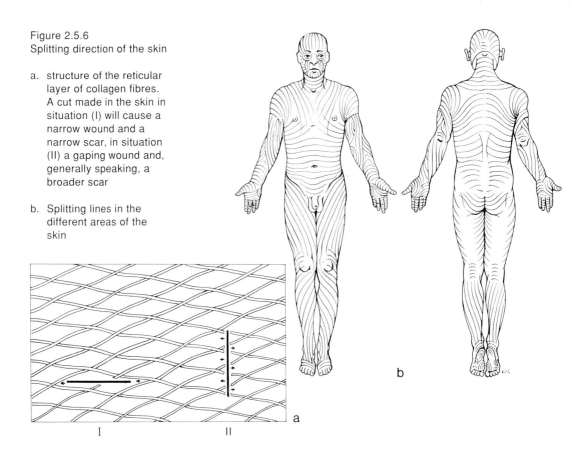

3. Subcutaneous connective tissue

The *subcutaneous connective tissue* does not belong to the skin. However, in view of its structure and function, it makes sense to discuss this layer of the body together with the skin.
The subcutaneous layer consists of loose connective tissue in which a large number of connective tissue cells have developed into fat cells. The connective tissue is continuous with the underlying common fascia. Whether it is easy to shift the skin, as on the back of the hand, or difficult, as on the palm of the hand, depends on the extent to which the tissue and the fascia are continuous.
In some areas of the body there is more subcutaneous fat tissue than in other areas. These are the *preferential areas*. They are located in the back, the buttocks, above the iliac crest and in the upper legs. The spread and amount of fat tissue is also dependent on sex. Generally speaking, women have more fat tis-

sue than men, at different places (breasts, shoulders, hips). The quantity of subcutaneous adipose tissue is also, of course, strongly related to individual eating habits.

The subcutaneous fat tissue functions as a *thermal insulator* (fat is a poor heat conductor), as a *food stock* and as a buffer. The *buffer* function is evident in the palms of the hands and the soles of the feet, where the layer of fat is not only thick, but also provided with and surrounded by collagen fibres, due to which these structures are firm.
Apart from blood and lymphatic vessels there are skin sensors and nerve fibres in the subcutaneous layer.

4. Vascularization

As mentioned earlier, only the dermis and the subcutaneous layer are vascular. There are three layers of capillary networks (Fig. 2.5.7). The first layer is located immediately between the fascia and the subcutaneous tissue. The arteries come from the underlying muscles and form an extensive arterial network on the superficial fascia. Branches of this first network form an arterial network in the border area between the subcutaneous tissue and dermis. From this network extend branches, which form a network of arterioles, next to the papillary layer of the dermis. From these arterioles the capillaries arch to the higher network of venules known as the *sub-epithelial venous plexuses*. The combined cross-sections of the vessels of these plexuses are immense, in order that this network can contain a large volume of blood. In two steps it discharges into venous networks which are at the same level as the two arterial networks mentioned earlier. From the lower venous network the blood is transported via veins. The three networks are called (from inside to outside); the *fascial network*, the *cutaneous network*, and the *sub-papillary network*. The venous and arterial capillary networks join in the sub-papillary network. There are also connections between the arterial and venous networks of the two lower networks. Such a network is called an *arteriovenous anastomosis*. Anastomotic networks are very significant components in temperature regulation.

The rest of the skin is relatively poor in capillary networks, apart from the areas around the glands, hair roots, skin sensors and in the adipose tissue.

5. Temperature regulation

Man is warm-blooded, which means that the body temperature is relatively constant even when the temperature of the surroundings changes. The human organism is able to generate sufficient heat to keep the body warm in an

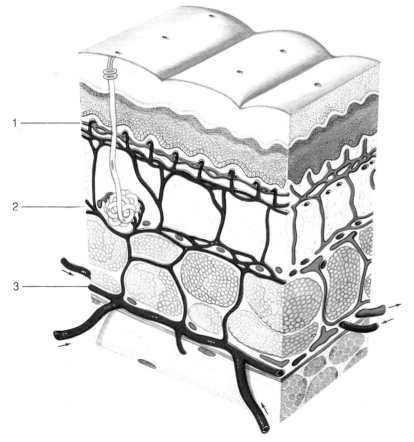

Figure 2.5.7
Blood supply of the skin

1. sub-papillary network
2. cutaneous network
3. fascial network

environment with a temperature that is lower than that of the body. This is made possible by a precise temperature regulating mechanism, which is particularity active via the skin.

Body temperature

A constant body temperature does not exist. The body temperature of an adult under conditions of physical and mental rest, in an environment with a temperature of 20°C, varies from 28°C (skin of hands and feet) to 39°C (liver) (Fig. 2.5.8). After some exercise or at a higher environmental temperature (for instance 35°C), there is a different distribution of heat (Fig. 2.5.8b). In both situations, however, there is a centrally located area with temperatures of 37°C. The temperature of this area is the *core temperature*. The temperature of the areas around this core is called the *'peel temperature'*. When a person speaks of *the* body temperature (37°C) this refers to the core temperature. At this temperature the metabolic reactions in the body are at their fastest as most enzymes have their optimum temperature around 37°C. Evidently, the core temperature can be best measured in the 'core', which can be reached via the anus (rectal temperature). Measuring under the tongue or in the armpit is a less valid indication of core temperature.

Heat production

During anaerobic metabolism of nutrients, especially of glucose and fats, approximately 40% of the energy present in those nutrients is used for the formation of 'energy packages' (ATP) and 60% is released as heat. When energy is released from ATP, another 40% of the energy contents of ATP is converted into heat. Consequently, a little more than 75% of the entire energy contents of nutrients is converted into heat and no more than 25% is readily available as energy.

When no external work is done (for instance, no muscles are contracted), the entire energy content of the food is eventually released as heat.

Figure 2.5.8
Body temperatures

a. in rest
b. during exercise

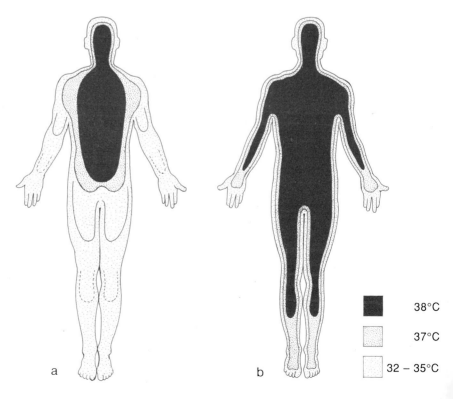

a   b

■ 38°C

  37°C

  32 – 35°C

Mechanism of heat loss

The metabolic reactions in the cells of the organs (especially in those of the muscles and the liver) are accompanied by the production of heat. However, the average core temperature, should not rise much higher than 37°C. Heat is therefore carried to the periphery (Fig. 2.5.9). This takes place to some extent by *conduction*. Conduction is the immediate transfer of heat, comparable to diffusion. Warmer parts give off heat to colder parts and eventually to the periphery, which can transfer heat to the environment. This mechanism, however, is absolutely insufficient and might have the risk of certain (active) organs becoming overheated. Moreover, subcutaneous fat is a poor conductor, therefore a good insulator.

Transportation of heat to the periphery takes place mainly via the circulatory system. The blood as it flows absorbs heat out of the active (core) parts through conduction and evenly distributes the heat over the colder (peripheral) parts.

The skin, with its average surface area of 1.65 m², is the major cooling organ.

The skin makes use of different mechanisms:

– *heat radiation*. Warmer objects lose heat to a colder environment via direct radiation. About 60% of the heat loss takes place via radiation, in a comfortably warm environment. When the environmental temperature is higher than the body temperature, the body absorbs heat from the environment.

– *thermal conduction*. Here also, two-way traffic is possible. Air is a bad conductor, water is a very good conductor. As the air immed-iately bordering the skin is continuously 'changed' because it flows, the possibility of heat loss increases. Heat loss via conduction is also a significant activity in water.

– *evaporation* of water. This has an enormous cooling effect, even if little water is used. After all, water has a high evaporation heat coefficient.

The two mechanisms first mentioned become virtually inactive at temperatures over 36°C and are insufficient during strong exertion. At that moment evaporation comes into action. The water, needed for evaporation, ends up on the surface of the skin by diffusion from lower skin layers, especially from the sweat glands. In extreme situations approximately 6 litres of sweat (insensible perspiration) can be produced per 24 hours.

Regulation

In order to be able to regulate, the brain must be informed about the situation. There are temperature sensitive sensors in the skin, which give information about the environmental temperature to the central nervous system.

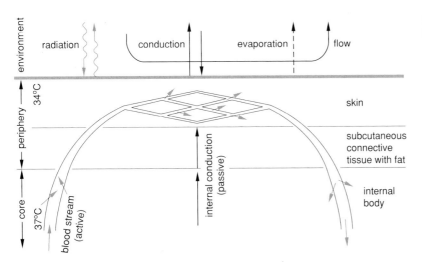

Figure 2.5.9
Mechanisms of heat-loss

Moreover, the temperature regulating centres in the medulla oblongata and the hypothalamus are themselves sensitive to changes in temperature of the blood which flows past.

From these centres impulses are sent to the vessels in the sub-papillary network and the cutaneous network, which then contract or dilate, according to need. When considerable heat must be discharged the skin is red and the secretion of sweat is superfluous.

When less heat must be carried away, the blood stream through the sub-papillary vessels becomes smaller. When heat must be retained, the sub-papillary bloodstream stops and the skin is pale. The arteriovenous anastomoses of the cutaneous layer take over the bloodstream's function. Goose flesh is formed and the tissue in the skeletal muscles is stimulated to make short uncoordinated contractions known as *shivering*. The masseters make one's teeth chatter.

Moreover, there is also an inclination to increase the production of heat by voluntary muscular activity (clapping the arms). In extreme cold the cutaneous layer can even be shut off. The blood does not go any further than the fascial layer, which is deep in the adipose tissue, an excellent heat insulator. In this case, non-vascularized structures may die (frostbite).

An impending extremely low body temperature activates a number of other reactions. The thyroid is, via hypothalamus and hypophysis, stimulated to produce the thyroid hormone. This hormone stimulates cell metabolism. Also, the nervous system increases the blood sugar level. There can be an increased metabolism in the mitochondria, which is accompanied by the production of heat.

Integration

Introduction

In the previous five chapters we have discussed the following vegetative systems: circulation, digestion, urinary tract, respiration and the skin. These systems perform the vegetative tasks of life; they are the functions which serve to maintain the life of the cells.
In the next five chapters we shall concentrate on the *integration* of these systems.

Learning outcomes

After studying this section you should have sufficient knowledge and understanding of:
- the importance of integration;
- the differences between vegetative and animal integration;
- the three components of the process of integration;
- the body systems which are concerned with integration.

1. Self preservation and preservation of the species

Integration means *to make into an entity*. The vegetative systems must be integrated in order to achieve efficient functioning of the organism. The interaction between the body's vegetative organ systems is constantly being regulated; the aim is to keep the composition of the m.i. constant – *homeostatis*. The optimal composition of the m.i. is constantly under threat of disruption as a result of changes in internal and external factors. If keeping the optimal composition fails, in other words if the vegetative integration fails, the human being will die.
However, vegetative systems and vegetative integration by themselves are not sufficient to make human life possible. The interaction *between the individual person and the environment* is also essential for man. A human being obtains information from his environment and is himself actively involved with that same environment and so changes it. In order to have this interaction between man and his environment running smoothly man has animal organ systems at his disposal and there is an animal integration in order to have an efficient functioning of the entity.
If this animal integration fails the individual person will certainly eventually die.

So vegetative and animal integration are of vital importance. The combined purpose of the vegetative and animal integration is the *preservation of the individual*.

A very special form of integration is *procreation*. The purpose of procreation is the preservation of the species and for a community, for mankind in its totality, it is indeed of vital importance. Procreation is also undoubtedly a part of the interaction between the individual person and his environment. Yet, all in all, procreation in itself does not have the same direct and vital significance for the individual human being.

2. Inadequate labels

The terms vegetative and animal are often used in functional anatomy. However, it is extremely difficult to offer precise definitions of these terms. Literally vegetative means 'of or concerned with growth in plants' and animal is self explanatory.

It is often said of someone who is in a coma that he or she 'is vegetating'. In view of the fact that there is no communication between this person and the environment this is a correct statement. Yet there is *some* interaction with the environment; a drop in temperature, for example, can result in a pale skin and goose flesh for someone in a coma.

For the notion *vegetative* we often use synonyms such as *autonomous* or *involuntary*. These synonyms are inadequate for, although the regulation of heart frequency and blood pressure are vegetative (involuntary) processes, it turns out that most people under certain circumstances are capable of voluntarily bringing about certain changes in their heart beat and blood pressure. In a similar way, breathing, which is a vegetative function, is maintained by striated 'voluntary' muscles.

For the term 'animal' we often use the synonym *voluntary*. In relation to the functioning of the organs of the five senses (defined as an 'animal sensor system'), the notion 'voluntary' is meaningless.
There are all kinds of animal reflexes which often have very little 'voluntary' elements in them. Strolling down a shopping street is an animal activity. However, putting one foot before the other does not seem to be 'voluntary' rather it appears much more like an automatism at work.

Using terms like 'unconscious' and 'conscious' respectively for vegetative and animal, brings no more clarity in the matter either. Most regulatory processes take their course without us being aware of them; we can become aware of a number of these processes if need be, such as a change in the heart beat.

In this book – keeping in mind all the restrictions identified above – we shall use the term *vegetative*, as far as possible, only for everything that is of or related to the *interaction within the individual person*, and the term *animal*, as far as possible, only for everything that is of or related to the *interaction between the individual and the external environment*.

3. Information, processing, action

The process of integration has three components: information, circulating or dealing with information, and action undertaken on the basis of the processed information:
- *Information*: in order to be capable of regulating, coordinating and directing there must be data available on the state of affairs and the changes which take place. The perception of information is called sensorics. In many, of even the smallest parts of all the body systems there are little organs sensitive to certain changes, which act as a kind of antennae, *the sensors*.
- *Processing*: the information is passed on to the two particularly integrating body systems: the hormone system (Chapter 6 of Module 2) and nervous system (Chapter 7 of Module 2). Between these two regulating systems there are very many links and connections. These systems support, facilitate and intensify each other's function. They influence one another, and complement one another.
There are a number of striking differences between the regulations of these two systems:
- *hormonal regulation* starts with endocrine glands. These glands produce on the basis of the presented information a smaller or greater amount of *chemical* 'messenger-molecules' known as the hormones. These

endocrine hormones are transported via the blood vessels (rather than via specialized ducts) to the organs which must be regulated;

- *neuronal regulation* takes place via nerve tissue. Originating from the sensors there are *electric impulses* sent and processed via *nerve fibres*. These kind of electric impulses are called *action-potentials*. After being processed there are action-potentials conducted via other nerve fibres towards the organs which require regulation.

Hormonal regulation is slower than neuronal regulation. Generally the effects of hormonal regulation last longer than the effects of neuronal regulation. The hormone system is mainly a vegetative integration system; whereas the nerve system shows more or less clearly distinguishable vegetative integrating and animal integrating parts.

– *Action:* the target organs are called *effectors*. In the case of neuronal regulation these can be muscles (smooth or striated) or glands (endo- or exocrine). The controlling or regulatory action is called *motor*. In the case of hormonal regulation the target organs are widely varied; they can be cells in the kidneys, in the bone tissue, in other hormone glands, in the liver, etc.

In both the sensory and motor systems we can again distinguish between vegetative and animal. The vegetative sensory and the vegetative motor systems have been covered when discussing the vegetative systems in the previous chapters.

The animal sensory and the animal motor systems will be discussed in Chapter 8 of Module 2 (Sensory system) and Chapter 9 of Module 2 (Motor system).

4. Reproduction

Reproduction comprises the entire process of finding a heterosexual partner, the mating, pregnancy, parturition and birth and finally the care and education of children. In Chapter 10 of Module 2 (the Reproductive System), the structure and the function of the reproductive organs of man and woman will be discussed.

In the process of mating – just as in many other human activities – many vegetative and animal processes are integrated.

In the majority of instances human mating does not result in procreation. Human sexual intercourse is often mainly aimed at experience and pleasure.

6

Hormonal system

Introduction

Hormones are substances which act upon one or more specific target organs or upon the regulation of metabolic processes. They are produced in glands which have no ducts, *ductless* or *hormonal* glands (see Fig. 1.4.3b). The hormone molecules are synthesised in the separate glandular cells. They are mostly proteins. They are discharged via active transport into the m.i. and from there into the bloodstream. For this reason hormonal glands are called *endocrine glands*.

Firstly, the general effects of hormones will be discussed in this chapter. Next the different endocrine glands and the functions of the hormones produced will be discussed (Fig. 2.6.1). The hormone producers can be divided into three groups:
- endocrine tissue, arranged into a wholly separate organ:
  pituitary gland, pineal gland, thyroid gland, parathyroid glands and adrenal glands;
- endocrine tissue, embedded into another organ: the pancreatic islets (Islets of Langerhans) in the pancreas, the endocrine tissue in the sexual glands, the endocrine tissue in the kidneys;
- hormone producing cells, scattered over specific tissues, such as the renal tissue, the gastric wall and the duodenal wall.

Learning outcomes

After studying this chapter you should have sufficient knowledge and understanding of:
- the general effect of hormones;
- the elements of the hormonal system and the linkage with the nervous system;
- the structure of the hormone producers: pituitary gland, thyroid, parathyroid glands, adrenal glands, pancreatic islets, sexual glands, and specific tissues in the kidneys, the gastrointestinal tract and the skin;
- the functions of the different hormones produced.

1. Effects of hormones

There are more than ten hormone producers in our body. Several hormonal glands also produce more than one hormone. Dozens of hormones enter the bloodstream and are distributed throughout the body. The thyroid hormone, for instance, is able to regulate the metabolic level of all cells. Adrenaline can have its effect in several places. Apart from these examples of hormones with an extended field of activity, there are a number of hormones which effect the function of only one organ. The antidiuretic hormone (ADH) from the pituitary gland exclu-

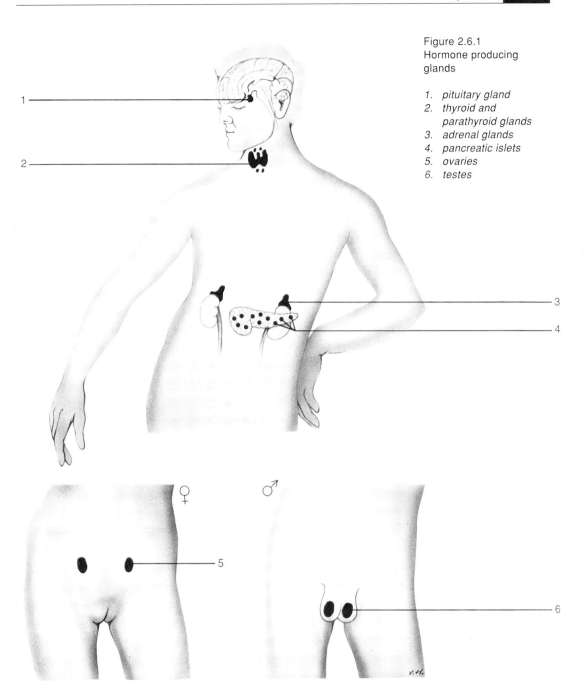

Figure 2.6.1
Hormone producing
glands

1. pituitary gland
2. thyroid and
   parathyroid glands
3. adrenal glands
4. pancreatic islets
5. ovaries
6. testes

sively effects the permeability of the tubule cells of the renal nephrons and gastrin from the gastric wall effects exclusively the production of gastric juice.

The chemical structure of each type of hormone represents, as it were, a message. Hence, hormones are also called *messenger molecules*. On its way through the bloodstream the message concerned is continuously transmitted, but it is only understood by cells that are sensitive to the specific chemical structure of the hormone. In this context the name target-cells is often used. This name, however, misleadingly suggests that the hormones are specifically searching for their

Figure 2.6.2
Hormonal gland,
bloodstream and target
organ

1. hormone producer
2. bloodstream
3. target organ for
   hormone ▲; the organ
   is activated
4. target organ for
   hormone ■; the organ
   is on the verge of
   being activated
5. target organ for
   hormone ●; the organ
   has not been
   activated

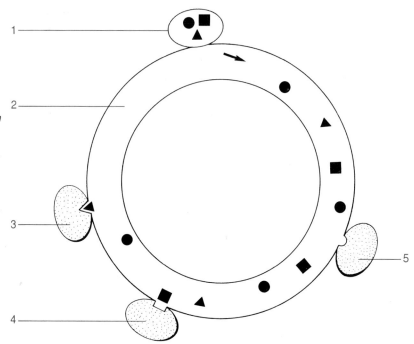

targets. It is probably more accurate to refer to specifically sensitive *tissue receptors* in the cell membrane. The shape of the receptor is such that only the matching hormone fits (Fig. 2.6.2). As soon as the hormone attaches itself to the matching receptor, a chain of reactions starts. In this process *cyclic adenosinemonophosphate* is formed in most cells. This substance regulates the activity of the cell, which usually means producing specific enzymes.

Negative feedback

The concentration of each hormone in the blood plasma is dependent upon the relationship between production and decomposition. Production is by a number of hormonal glands, decomposition of hormones is mainly in the liver. If the production is greater than the decomposition, the concentration, and usually the effect, of the hormone increases. If the decomposition is greater than the production, concentration and effect decrease. Generally speaking, there will be a subtle fluctuation around the equilibrium, as the hormones are subject to a regulating cycle in which the *feedback mechanism* plays an important role. The regulating cycle for the maintenance of a constant concentration of thyroid

hormone in the blood plasma is shown in a simplified way in Figure 2.6.3. The pituitary gland produces thyroid stimulating hormone (TSH). This hormone stimulates (+) the thyroid gland to produce thyroid hormone. Thyroid hormone, in its turn, slows down the production of TSH in the pituitary gland (–). This process is called the *negative feedback*. An increase in TSH hormones is relatively quickly reversed. The thyroid hormone keeps is own concentration within narrow limits.
When the hormones are discussed, the matching regulatory cycles will also be discussed.

Master gland

The pituitary gland (hypophysis) is considered the master gland of the hormone system. Many other hormone producers are controlled by the pituitary.
The pituitary in its turn is directly influenced by the nervous system via the *hypothalamus*. The pituitary gland may direct affairs, but the gland itself is also tied into the regulatory mechanism.

This is a good illustration of an important functional and structural relationship between the hormonal system and the nervous sytem.

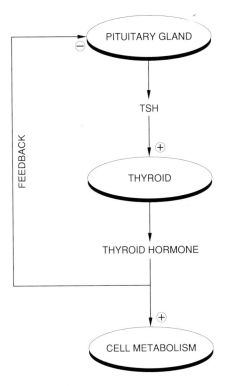

Figure 2.6.3
Regulating cycle in order to
maintain a constant thyroid
hormone concentration

2. Pituitary gland

The pituitary literally hangs from the floor of the
hypothalamus (Fig. 2.6.4). It is approximately
as large as a pea and and sits on the *sella turcica*.
This is a saddle-shaped depression of the
sphenoid bone at the base of the skull (see
Fig. 2.9.12). It is attached to the *hypothalamus*
by a rather thin stalk. The hypothalamus is the
nerve tissue which forms the floor of the third
ventricle (see Fig. 2.7.44). The crossing of the
optic nerves (optic chiasma) is found ventral to
the pituitary stalk.

The pituitary consists of two main lobes, clearly
discernible as to structure and function: the
posterior lobe or *neurohypophysis* and the
anterior lobe or *adenohypophysis*.

2.1. Neurohypophysis

The neurohypophysis differs considerably from
all other endocrine glands. It arises in embryo-
genesis from developing cerebellar tissue and is
not made of glandular cells, but of a special kind
of glial cell. Many branches of nerve cells which
are situated in the hypothalamus terminate between these glial cells. The cell bodies produce
the hormones which are then carried in the axons
via the pituitary stalk to the fibre ends in the
pituitary (Fig. 2.6.4c). The hormones are secreted (*neurosecretion*) and are immediately
absorbed in the capillary network. The neurohypophysis itself is not really a hormone producer. It is evident that there is not only a connection between the nervous system and the
hormonal system, but that there are also certain
nerve cells that can even produce hormones.

The neurohypophysis produces two hormones:
- the *antidiuretic hormone (ADH)*, which is
  produced when specific receptors in the hypothalamus indicate that the colloid osmotic
  value of the blood has risen, in other words,
  that there is a shortage of fluid in the circulatory system. ADH is discharged from the
  pituitary into the bloodstream and becomes
  active in the kidneys. It increases the permeability of the epithelium of the distal tubuli
  and the collecting tubuli. The result is that
  considerably more water is reabsorbed into
  the bloodstream and that the urine becomes
  more concentrated. When the colloid osmotic
  value returns to normal, the production of
  ADH gradually slows down and eventually
  stops altogether.
  ADH is sometimes also called vasopressin.
  This name misleadingly suggests that the rise
  in blood pressure which is effected by the
  hormone is a result of activity of the blood
  vessels;
- *oxytocin*, a hormone which is increasingly
  produced at the end of pregnancy. It stimulates the contraction of smooth muscles,
  especially of the uterine muscles. The sex
  hormone progesterone, however, makes
  the smooth muscle tissue insensitive to
  oxytocin. The delivery begins when the concentration of progesterone falls. Then oxytocin becomes active and labour contractions
  start.

Oxytocin also stimulates the contraction of
the smooth muscle cells in the glandular lobules and the ducts of the mammary glands.

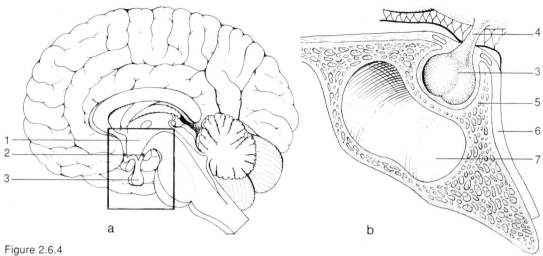

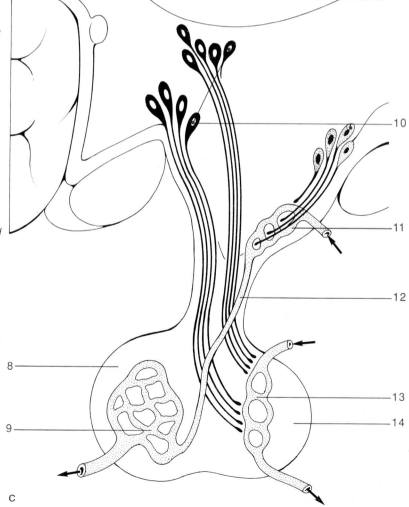

Figure 2.6.4

Hypophysis or pituitary
gland

a. General view, median
    cross-section
b. Location of the pituitary
    gland on the sella
    turcica
c. Adenohypophysis and
    neurohypophysis and
    their relation with the
    hypothalamus

1. *third ventricle*
2. *hypothalamus*
3. *hypophysis or pituitary
    gland*
4. *pituitary stalk*
5. *sella turcica*
6. *dura mater*
7. *depression of sphenoid
    bone*
8. *adenohypophysis*
9. *venous capillary
    network in the
    hypophysis*
10. *hormone producing
    nerve cells*
11. *capillary network in the
    hypothalamus*
12. *portal venule*
13. *capillary network in the
    neurohypophysis*
14. *neurohypophysis*

Stimulation of the nipples, such as sucking of the newborn baby, leads to an increase in oxytocin production via the nervous system. The ejection of milk caused by oxytocin increases the flow of milk caused by the sucking of the baby.

It often happens that mothers who breast-feed their babies find the milk 'rushes in' around feeding time. Thinking of the baby can increase the secretion of oxytocin, resulting in milk trickling out of full breasts. The increased oxytocin production during lactation can lead to renewed uterine contractions. In many cases no sensation is felt in the abdomen, but sometimes the contractions are painful. These post-partum contractions are important during the period immediately after delivery, as they help the stretched uterus to return to its original proportions.

2.2. Adenohypophysis

In the anterior lobe of the pituitary gland, the adenohypophysis, hormone production is, in effect, undertaken by glandular cells. Here, also, a relationship with the hypothalamus is indispensable for production. Nerve cells in the hypothalamus are attached to a capillary network in the pituitary stalk via nerve fibres (Fig. 2.6.4c). This capillary network converges at first, but branches again later between the glandular cells of the adenohypophysis into a venous capillary network. Such a vascular system is called a *portal circulation.*

The hormone production of the adenohypophysis begins and develops from the hypothalamus. Its nerve cells, mentioned earlier, produce a message which is specific for each hormone of the adenohypophysis, a *releasing hormone* (RH). After neurosecretion, these releasing hormones enter the capillary network of the pituitary stalk and are carried to the second capillary network through the portal veins (Fig. 2.6.4c). There, the releasing hormones leave the bloodstream and stimulate the glandular cells of the adenohypophysis to produce a specific hormone. The hormones produced are absorbed into the secondary network and then set off on their journey through the body.

The hypothalamus is able to discharge stimulating release hormones into the bloodstream and some inhibitory substances as well, known as the *inhibitory hormones* (IH). Releasing and inhibiting hormones are, together with the regulating cycles, responsible for a subtle hormonal production in the adenohypophysis.

Some seven hormones, which can be divided into two groups, are produced by the adenohypophysis:
– hormones which stimulate or regulate the function of other endocrine glands, the *trophic hormones*;
– hormones which immediately influence certain body functions, the *effector hormones.*

Trophic hormones

– The *thyroid stimulating hormone* (TSH), stimulates the thyroid to produce thyroid hormones. The thyroid hormones, in their turn, slow down the TSH-production (negative feedback).
TSH-synonyms are: *thyrotropic hormone, thyrotropin.*
– *Adrenocorticotropic hormone* (ACTH), literally; adrenal cortex stimulating hormone. The production of corticoids (cortical hormones) is stimulated. Some corticoids, in their turn, slow down the ACTH-production. The ACTH-concentration in the blood plasma shows striking fluctuations over 24 hours. The hormone is important for the regulation of day and night rhythms.
– *Follicle stimulating hormone* (FSH), which in the female stimulates the ovarian follicles to change into (Graafian) follicles, and in the male stimulates testicular spermatogenesis. Certain sex hormones slow down the FSH-production.
– *Luteinising hormone/interstitial cell-stimulating hormone* (LH / ICSH) which are chemically identical.
In the female it is called luteinising hormone, as it stimulates the ripening of follicle and later ovulation. After ovulation, the yellow body (corpus luteum) is formed at the site of the 'empty shell' of the Graafian follicle. The corpus luteum produces pro-

gesterone (see Fig. 2.6.10). The luteinising hormone stimulates the formation of the corpus luteum.

In the male this hormone is called interstitial cell-stimulating hormone. The ICSH hormone stimulates the interstitial cells (Leydig cells), which are located between the seminiferous tubules in the testes, to produce the male sex hormone testosterone (Fig. 2.6.11).

Within the group of trophic hormones, the FSH and the LH / ICSH belong to the *gonadotropic hormones*. This name clearly shows the kind of gland which is stimulated, as ovaries and testicles are the gonads, or sex glands.

It should be noted that the concentrations of all gonadotropic hormones are kept within defined limits by means of relatively simple regulating cycles with negative feedback mechanisms.

Effector hormones

- *Somatotrophic hormone* (STH), usually known as *growth hormone*. It not only stimulates the synthesis of proteins, but also influences the metabolic processes in which carbohydrates, lipids and minerals are involved. The hormone has a stimulating effect on the growth of all tissues; it stimulates both cell division (mitosis) and the growth in size of the cells. This effect is obvious in children, especially in the growth in length of the bones, notably that of the long bones of the arms and the legs. Growth hormones are however produced and secreted throughout life. The growth in length stops when the growth plates or epiphysial discs are ossified. Synonyms of STH are: human growth hormone (HGH) and somatotrophin.

- *Melanocyte stimulating hormone* (MSH), a hormone which is probably responsible for the normal intra-uterine pigmentation of the skin.

- *Mammary stimulating hormone*, often called *prolactin releasing hormone* (PRH). It stimulates, together with STH, the development of the breasts and lactation after childbirth.

- *Luteotrophic hormone* (LTH), has the same chemical formula as PRH. Its name indicates a second function. LTH is not only an effector hormone but also a trophic hormone in the beginning of pregnancy. LTH is then responsible for preservation of and secretion from the corpus luteum, so it is more specifically a gonadotropic hormone.

3. Epiphysis or pineal gland

The epiphysis is a small, conical gland, posterior to the third cerebellar ventricle (see Fig. 2.7.25). Its significance is not precisely known. It probably releases a hormone, melatonin, which holds back the secretion of gonadotropic hormones and thus regulates the time of onset of puberty.

4. Thyroid

The thyroid is named after the thyroid cartilage, below which it is located (Fig. 2.6.5a). Its shape is like a butterfly. The 'wings' are called *lobes*. The lobes are joined at the front by an isthmus. The gland often has a protuberance on the isthmus that reaches as far as the thyroid cartilage. The frontal and lateral sides of the lobes are located on the borderline of larynx and trachea. The back of the 'wings' are in apposition to the beginning of the oesophagus. The branch of the vagus nerve which innervates the larynx and the vocal chords, the recurrent laryngeal nerve, extends to this point.

The glandular tissue consists of cuboid epithelial cells arranged around a large number of large or small cavities (Fig. 2.6.6). This is called a follicular structure. Between the follicles is an extremely extensive capillary network. In the cavities is a protein-like substance known as *colloid*.

Two types of hormones are produced by the thyroid: *thyronine* and *calcitonin*.

- *Thyronine* is produced by the epithelial cells and temporarily stored in the colloid. Under the influence of TSH the hormone is released from the colloid and discharged into the capillary network.

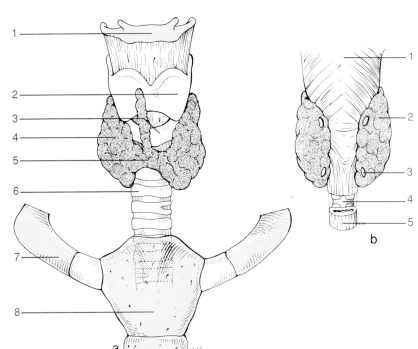

a

b

Figure 2.6.5

Location of thyroid and parathyroid glands

a. Front

1. hyoid bone
2. thyroid cartilage
3. cricoid cartilage
4. lobe of thyroid
5. isthmus
6. trachea
7. first rib
8. sternum

b. Back

1. pharyngeal wall, from the back
2. lobe of thyroid
3. parathyroid
4. trachea, anterior to oesophagus
5. oesophagus

There are two manifestations of thyronine: *tri-iodothyronine* (T_3) and *tetra-iodothyronine* (T_4). The number indicates the number of iodine atoms in the molecule.

T_4 is known as *thyroxine*. T_3 is approximately five times as effective as T_4 and works faster. The volume of T_4, however, is at least ten times larger and is often regarded as stock, as T_3 is formed when an iodine atom is removed. The production of enzymes which are important in the metabolic processes of carbohydrates and fats is stimulated by thyronine (T_3 and T_4). The synthesis of proteins is also stimulated and consequently hormones influence the growth.

To summarise, thyronine regulates the metabolic rate in the cells.

– *Calcitonin* is produced by the C-cells, which are located on the outside of the follicles and discharge their product directly into the bloodstream. Calcitonin is formed when the concentration of calcium ions (Ca^{2+}) in the blood plasma is increased. Calcitonin decreases the Ca^{2+} concentration by depositing calcium in the bones, by decreasing the reabsorption of calcium from the renal tubules and finally by decreasing the calcium absorp-

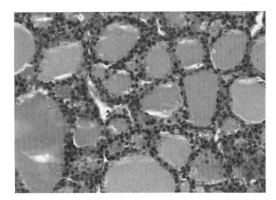

Figure 2.6.6
Thyroid follicles filled with colloid and surrounded by glandular cells

tion from the intestines. When the Ca^{2+} concentration in the blood plasma decreases, the calcitonin production also decreases.

5. Parathyroid glands

The parathyroid glands are located on the four 'wing tips' of the thyroid lobes (Fig. 2.6.5b). They are the size of a lentil and are not discernible from the thyroid tissue with the naked eye, though Figure 2.6.5 suggests otherwise. The parathyroid glands consist of small groups of epithelial cells, surrounded by capillaries and connective tissue.

The hormone of the parathyroid glands is called *parathyroid hormone* or *parathyrin* (PTH). Its effect is opposite to that of calcitonin.

PTH is produced when the Ca^{2+} concentration of the blood plasma decreases. The hormone mobilises calcium from the bones into the blood, stimulates the reabsorption of calcium from the renal tubules and stimulates calcium absorption from the intestines. When the Ca^{2+} concentration in the blood plasma increases, the PTH production gradually decreases.

Antagonists and synergists

Calcitonin and PTH are called antagonists (Fig. 2.6.7). By their opposite effects in the regulating cycle they not only keep each other's concentration in balance, but they also keep the Ca^{2+} concentration in the blood plasma within defined limits. A good calcium balance is important for metabolism in the bones and a well-functioning muscle contraction mechanism.

The calcium balance is linked with phosphate balance, as these two substances often form a compound, calcium phosphate, which is found especially in bones.

Just like PTH, vitamin D generates an increased absorption of calcium from the intestine and a decreased calcium excretion by the kidneys. In this respect these two substances are synergists, i.e. they work together. However vitamin D also generates the deposit of calcium in the bones (mineralisation); in this case vitamin D and PTH are functioning as antagonists.

6. Adrenal glands

On each kidney sits an adrenal gland, like a dollop of cream (Fig. 2.6.8a). The adrenal glands are enclosed by the perirenal adipose tissue (see Fig. 2.3.2). The name 'adrenal' only indicates the location; the adrenal gland is not a kind of 'auxiliary kidney', it is a hormonal gland. It is more correct to characterize each

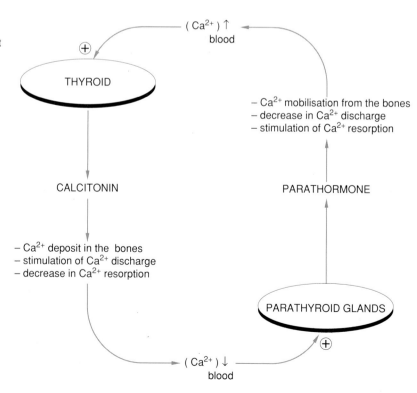

Figure 2.6.7
Regulating cycle for the preservation of a constant concentration of calcium ions in the blood plasma

$(Ca^{2+}) \uparrow$ blood

THYROID

$(+)$

– Ca^{2+} mobilisation from the bones
– decrease in Ca^{2+} discharge
– stimulation of Ca^{2+} resorption

CALCITONIN

PARATHORMONE

– Ca^{2+} deposit in the bones
– stimulation of Ca^{2+} discharge
– decrease in Ca^{2+} resorption

PARATHYROID GLANDS

$(+)$

$(Ca^{2+}) \downarrow$ blood

adrenal gland as two separate hormonal glands. The adrenal glands comprise, in structure and function, two clearly separate parts; the *cortex* and the *medulla* (Fig. 2.6.8b). The cortex is affected by the adenohypophysis (ACTH), the medulla is affected by the sympathetic nervous system.

The adrenal gland is supplied with blood mainly by a branch of the renal artery, and has many capillaries, especially in the cortex.

6.1. Cortex

The adrenal cortex, surrounded by a connective tissue capsule, produces *corticoids* or *corticosteroids*.

There are three layers in the epithelium of the cortex, each of which produces its own group of hormones (Fig. 2.6.8c).

Mineralocorticoids

The outer layer, the zona glomerulosa, is the thinnest and produces the *mineralocorticoids*. The hormones of this group control the mineral balance. They regulate the concentrations of sodium and potassium in the blood plasma. Of the mineralocorticoids, 95% consist of aldosterone, the production of which is controlled by ACTH and angiotensin. The production depends on the Na/K balance and the volume of water in the blood. The aldosterone production increases when the Na^+ concentration is too low, the K^+ concentration is too high, and when the volume of blood plasma has reduced. Abundant perspiration activates the production of this hormone.

Aldosterone stimulates the reabsorption of Na^+ ions (sodium retention) and the excretion of K^+

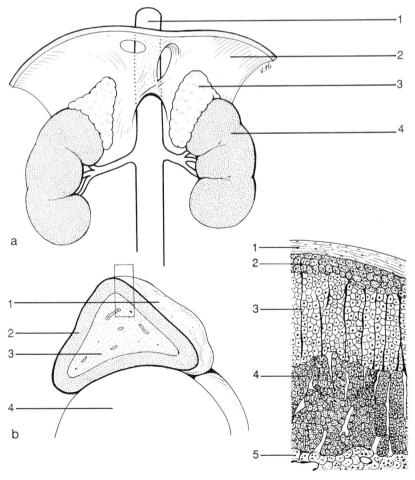

Figure 2.6.8
The adrenal glands

a. Location of the
    adrenal glands

    1. aorta
    2. diaphragm
    3. adrenal gland
    4. kidney

b. Cross-section of
    adrenal gland

    1. capsule
    2. cortex
    3. medulla
    4. superior fold of
        the kidney

c. The layers of the
    cortex

    1. capsule
    2. outer layer of
        cortex
    3. middle layer of
        cortex
    4. inner layer of
        cortex
    5. medulla

ions (potassium depletion) in the kidneys. This regulation of the Na/K balance is accompanied by the reabsorption of Cl⁻ ions and the excretion of H⁺ ions. Aldosterone indirectly controls blood pressure, as the selective Na⁺ reabsorption increases the crystalloid-osmotic value in the area directly superior to the renal papilla. As a result of this, an extra volume of water is 'sucked' from the primary urine and discharged into the bloodstream.

Glucocorticoids

The middle layer (zona fasciculata) of the cortex produces the *glucocorticoids*, hormones specifically controlling the glucose balance. More than 95% of all glucocorticoids consist of the *cortisol* hormone (hydrocortisone). It is produced under the influence of ACTH and controls in its turn ACTH-production in the pituitary (negative feedback).
Cortisol stimulates *gluconeogenesis*; glucose is formed in the liver and muscles, out of amino acids, glycerol, fatty acids and lactic acid. The absorption of glucose by the cells is also reduced. As a result of these processes, the blood sugar level rises. For that reason cortisol and insulin are called antagonists (see Fig. 2.2.36). Cortisol slows down the protein synthesis and stimulates the decomposition of proteins in the tissues, notably in the muscles. As a result, amino acids for gluconeogenesis are released.

Cortisol has a restraining influence on common inflammatory reactions and on the production of antibodies. This seemingly unfavourable effect can be life saving under certain circumstances.
In some allergic reactions, in serious infections, at extreme temperatures or in serious wounds, an accompanying strong inflammatory reaction could make things worse. In situations of physical and/or mental stress the production of cortisol stays higher.

Sex hormones

The inner layer of the cortex (zona reticularis) produces both androgenic hormones, *androgens,* and oestrogenic hormones, *oestrogens.*

As the amounts are minimal compared with the amount of gonadotropic hormones, their influence on the development of the primary and secondary sexual characteristics is minor.

6.2. Medulla

The medulla is centrally located in the adrenal gland (Fig. 2.6.8b). Like the neurohypophysis the adrenal medulla originated from developing nerve tissue. It is innervated by a part of the vegetative nervous system, i.e. by the sympathetic system. The epithelial cells of the medulla produce two related hormones: *adrenaline* and *noradrenaline*, in the proportion of three to one, during rest.

Basically, adrenaline effects the organs via the bloodstream in much the same way as the sympathetic system does via the nerve fibres. It prepares the body for an assault, defence or flight. During strong emotions, such as fear or sudden fright, and in emergency situations, large quantities of the hormone are produced. Adrenaline may justly be characterised as an 'action hormone'. Adrenaline stimulates the conversion of glycogen into glucose in the liver and the muscles. As a result, the blood sugar level rises, to the advantage of the active cells. Adrenaline and insulin are antagonists (see Fig. 2.2.36). Adrenaline stimulates heart activity, and cardiac output is increased when the heart frequency increases. In this way adrenaline indirectly causes an increase in blood pressure. In the skeletal muscles vasodilation takes place, muscular tension grows, and muscles prepare themselves for swift action. The increased tension can lead to small movements. In most other vessels there is vasoconstriction, especially in the gastrointestinal tract and in the skin. After a while, adrenaline can effect vasodilation in the skin and stimulate the sweat glands. The intestinal peristalsis and the secretion of digestive juices decrease. The hormone makes the sphincters of anus and urinary bladder relax. In some animals, this release may be the difference between a life or death escape. For man this effect of adrenaline can be rather unpleasant. And last but not least, adrenaline dilates the pupils of the eyes.

Noradrenaline is always produced together with adrenaline. It has the same effects, in such a way however, that it does not bring about vasodilation in the skeletal muscles. Consequently, its effect is greater than that of adrenaline.

7. Pancreatic islets (Islets of Langerhans)

Although the pancreatic islets are embedded in the pancreas tissue (Fig. 2.6.9), the functions of the pancreas, as an exocrine gland, and those of the pancreatic islets, as endocrine glands, are totally independent of each other.
An 'islet' consists of two kinds of hormone producing cells; alpha-cells and beta-cells, in the proportion of one to four. The alpha-cells produce the *glucagon hormone*, which stimulates the conversion of glycogen into glucose, similar to adrenaline. Glucagon and adrenaline are synergists.

The beta-cells produce the hormone *insulin*. Insulin stimulates the conversion of glucose into glycogen in the liver and the muscles. Moreover, it accelerates the transport of glucose across the cell membranes. Consequently, insulin has a strong effect on the lowering of blood sugar level. It is the antagonist of glucagon and adrenaline. These three hormones play a role as the *glucose buffer*, which keeps the blood sugar level as far as possible constant. The production of glucagon and insulin is linked with the glucose concentration in the blood plasma by means of negative feedback.

8. Production of hormones in the sexual glands

The sexual glands (gonads) produce reproductive cells (gametes). The structure of the gonads in relation to this function will be discussed in Module 2, Chapter 10. Endocrine

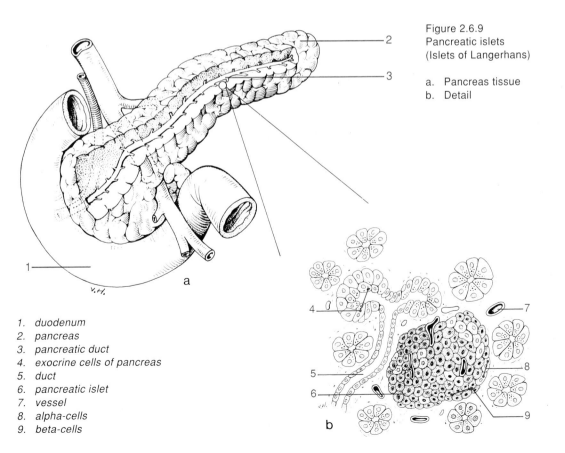

Figure 2.6.9
Pancreatic islets
(Islets of Langerhans)

a. Pancreas tissue
b. Detail

1. duodenum
2. pancreas
3. pancreatic duct
4. exocrine cells of pancreas
5. duct
6. pancreatic islet
7. vessel
8. alpha-cells
9. beta-cells

tissue is found both in the female gonads, (ovaries), and the male gonads (testes).

Oestrogen and *progesterone* are produced in the ovaries. Both hormones are female sex hormones. *Androgens*, male sex hormones, are produced in the testes.

8.1. Follicular cells and corpus luteum

During its first stage of development in the ovary, the egg cell is surrounded by a layer of epithelial cells, the *follicular cells*, (Fig. 2.6.10). The follicular cells produce *oestrogen* in response to FSH from the adenohypophysis. In negative feedback, oestrogen in its turn slows down the FSH production and stimulates the adenohypophysis to produce LH. In response to oestrogen, the female primary and secondary sexual characteristics develop. It stimulates (in girls) growth in height. The hormone is best known for its part in the menstrual cycle.

The pre-menstrual stage is controlled by oestrogen. The uterine endometrium (mucosa of the uterus) grows (see Fig. 2.10.12). The glandular ducts become wider and longer; there is good vascularisation. This stage of growth, the *proliferation phase*, lasts until ovulation. The oestrogen concentration then temporarily decreases because of the follicles transforming in response to LH. The epithelial cells of the 'empty shell', which is left behind in the ovary after ovulation, divide and fill up the cavity within a few hours. The cells themselves swell and contain a yellowish granular substance. This fatty substance is called *lutein*. The little organ which has developed is called the *corpus luteum*. It is an endocrine gland which continues to produce oestrogen, and it produces *progesterone* as well. In response to progesterone the endometrium enters the *secretion phase* after ovulation. The glands swell even more and start secreting mucus. Capillarization increases further and glycogen is stored in the uterine wall. In this stage the endometrium is prepared for a possible implantation of a fertilised ovum.

Progesterone also stimulates the glandular lobules in the breasts. The cells grow a little, as a result of which the breasts may swell. A number of women experience a somewhat painful tension in the breasts during the second half of the menstrual cycle. Progesterone restrains the LH production (negative feedback).

When no fertilization takes place, the corpus luteum degenerates and the production of progesterone stops. As a result, the *menstruation phase* starts.

A number of hormones which control menstruation are linked in this way:

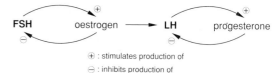

FSH   oestrogen ⟶ LH   progesterone

⊕ : stimulates production of
⊖ : inhibits production of

8.2. Interstitial or Leydig cells

The spermatic ducts of the testes have continuously produced sperm cells since puberty. Between the spermatic ducts, in the testicular con-

Figuur 2.6.10

Cross-section of ovary; hormone producers

1. *follicle cells, produce oestrogen*
2. *graafian follicle*
3. *corpus luteum, produces oestrogen and progesterone*

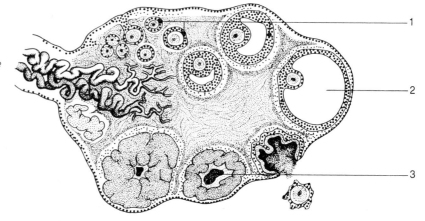

nective tissue, the *interstitial cells* or *Leydig cells* are located (Fig. 2.6.11). These cells produce the male sex hormone testosterone in response to ICSH from the adenohypophysis. Testosterone in its turn restrains the ICSH production (negative feedback).

Testosterone promotes the development of male primary and secondary sexual characteristics. It stimulates the growth of the penis and prostate, it promotes the production of sperm by stimulating both the formation of sperm cells and the activity of spermatocytes and prostate. Testosterone influences hair growth in the male, axillary and pubic hair develops. In the male the pubic hair is usually more extensive than in the female and it often extends in a triangular shape up to the navel. Beard and moustache grow in

response to the hormone, as does the hair on the body, which is often thicker and more dense than in the female. Testosterone stimulates the growth in height in the male. It causes the cartilage of the larynx to grow, as a result of which the vocal chords grow, causing the male voice to be lower than the female voice.

In response to testosterone the skin becomes thicker all over the body. Oestrogen does not have this effect, giving the skin a smoother and softer appearance. A general action of testosterone is to promote protein synthesis (anabolic activity). The effects described earlier are linked with protein production. Comparing the body composition of both sexes, it is notable that the male body contains more protein and less fat, by percentage, than the female body.

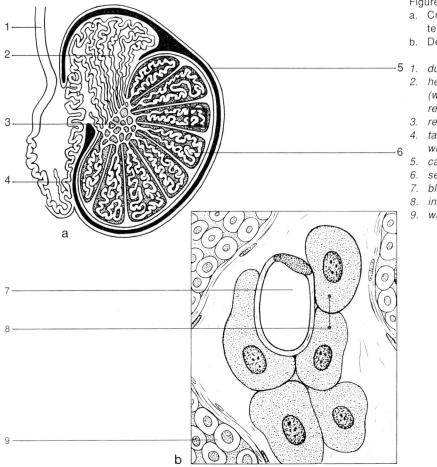

Figure 2.6.11
a. Cross-section of the testis; interstitial cells
b. Detail

1. *ductus deferens*
2. *head of epididymis (with ducts from the rete testis)*
3. *rete testis*
4. *tail of the epididymis with convoluted tubule*
5. *capsule*
6. *seminiferous tubule*
7. *blood vessel*
8. *interstitial cells*
9. *wall of spermatic duct*

9. Juxtaglomerular cells

The nephrons are the filtering units of the kidneys (see Fig. 2.3.4).

Opposite to the wall of the vas afferens, at the site where this vessel enters the glomerular capsule, is a group of hormone producing epithelial cells, the juxtaglomerular cells (Fig. 2.6.12). As soon as the blood pressure in the vas afferens decreases, these cells begin to produce the hormone *renin*. When the blood pressure increases, renin production is restrained. Like all hormones, renin is discharged into the bloodstream. It then causes the conversion of the plasma-protein *angiotensinogen* into *angiotensin* (see Fig. 2.3.5). Angiotensin is a vasoconstrictor and has an immediate hypertensive effect. It stimulates the production of aldosterone in the adrenal cortex and thus indirectly influences the blood pressure.

10. Tissue hormones

So far, only glandular tissue which is clearly arranged has been discussed. However, hormones are not only produced by this endocrine tissue, but also by cells scattered all through the tissue. Hormones, produced by these cells, are called *tissue hormones*.

Renal erythropoietic factor and erythropoietin

Renal tissue produces a hormone called *renal erythropoietic factor* (REF). This hormone is discharged into the bloodstream when the blood that enters the kidney has too little oxygen. The REF reacts with a specific plasma globulin in the blood to form the hormone *erythropoietin*, which stimulates the production of erythrocytes in the red bone marrow.

When a person stays at a high altitude for a long period of time, as a result of erythropoietin a larger number of erythrocytes circulate in the bloodstream, in this way compensating for the lower oxygen pressure.

Gastrin

When food comes into contact with the gastric wall, certain cells in the gastric antrum react by producing *gastrin*. This hormone is absorbed into the bloodstream and eventually ends up in the gastric wall again, via the portal vein, the inferior vena cava and a journey through the lesser circulation. Gastrin stimulates the production of gastric juices.

Secretin and CCK-PZ

As soon as a chyme passes through the duodenum, certain cells in the duodenal wall react by producing *pro-secretin*. This substance is converted into the *secretin hormone*, with the help of hydrochloric acid, and absorbed into the bloodstream.

After a 'detour', similar to the one described for gastrin, secretin stimulates both the secretion of pancreatic juice and the secretion of bile in the liver.

Figure 2.6.12
Juxtaglomerular cells

1. afferent tubule
2. glomerular capsule
3. juxtaglomerular cells
4. efferent tubule
5. glomerulus

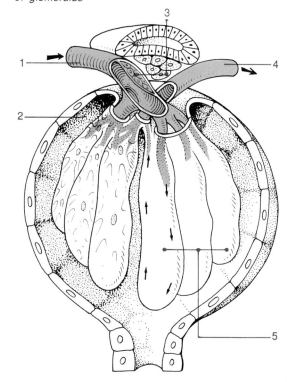

When the chyme passes, the *cholecystokinin-pancreozymin* (CCK-PZ) hormone is produced. As its name implies this hormone stimulates the gall bladder to contract, and the discharge of pancreatic juice.

Histamine
Many tissue cells all over the body are able, after being damaged for instance, to produce the hormone *histamine*. This tissue hormone stimulates local vascular dilatation and increases capillary permeability. When histamine is produced in the skin, it stimulates mitosis in the stratum spinosum and the stratum granulosum.
Repair or maintenance of the tissue concerned is promoted by histamine.

7

Nervous system

Introduction

The functional anatomy of the nervous system appeals to the imagination of many people. This is not just because the nervous system is responsible for the integration of the vegetative functions in man, but in addition the structure and function of the nervous system are thought to have, and indeed do have, characteristics which man considers of fundamental value to life. These characteristics comprise awareness and self-awareness, the ability to learn and remember, moods and emotions, dreams and fantasy, passions and self-control, logic and creativity. Most people credit the nervous system with their personality, with all its characteristics.

We can only touch on these characteristics in this book as we must focus upon the relatively simple features of the structure and function of the nervous system.

After dividing the nervous system according to structure and location (anatomical criteria), and function (physiological criteria), the general functional anatomy will be discussed. Thereafter the different parts, classified by structure and function, will be described. And last but not least, the cerebral membranes, the cerebral fluid and the vascularization of the brain will be discussed.

Learning outcomes

After studying this chapter you should have sufficient knowledge and understanding of:
- the general function of the nervous system;
- the anatomy and the physiology of the nervous system;
- the structure and function of the nerve tissue;
- the structure and function of the central nervous system: spinal cord, brain stem, cerebellum, diencephalon and cerebrum;
- the structure and function of the vegetative nervous system;
- the different cerebral membranes and the circulation of cerebral fluid;
- the blood supply of the brain.

1. Division

Although functional anatomy always focuses on describing structures and functions simultaneously, it makes sense to divide the nervous system into separate parts for study purposes.

The *anatomical division* starts from the *structure* and *location* of the elements of the nervous system; the *physiological division* is made with *functional* criteria as the starting point.

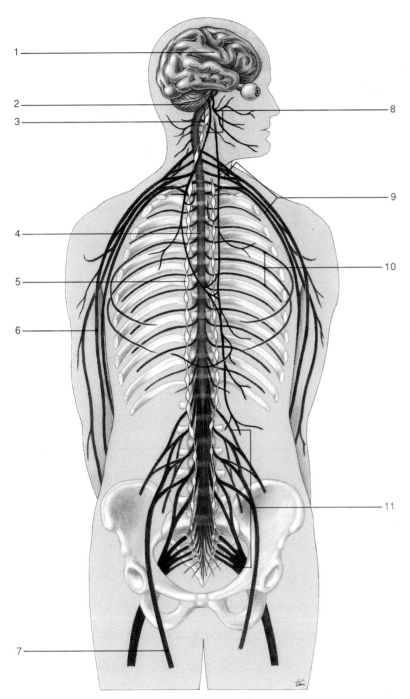

Figure 2.7.1

Elements of the nervous system

1. *cerebrum*
2. *cerebellum*
3. *brain stem*
4. *spinal cord*
5. *truncus sympathicus*
6. *peripheral nerves (arm)*
7. *peripheral nerves (leg)*
8. *cranial nerves*
9. *cervico-brachial plexus*
10. *peripheral nerves (thoracic nerves)*
11. *lumbo-sacral plexus*

Anatomical division

The structure and location of the nervous system can be divided into two parts:

– *central nervous system* (CNS). This is the part which is wholly enveloped in a bony covering, the skull and the vertebral canal. The CNS comprises, in sequence from caudal to cranial: *spinal cord or medulla spinalis, brain stem, cerebellum* or little brain, *diencephalon* or midbrain, and *cerebrum* or large brain (Fig. 2.7.1);

– *peripheral nervous system* (PNS). This part is not or only partially enveloped in a bony covering. The PNS is responsible for connections between the central nervous system and all systems in the body, the connections between the systems, and the connections within the systems. The peripheral nervous system includes 32 pairs of *spinal nerves*, 12 pairs of *cranial nerves* and two *ganglionated nerve trunks*, the *sympathetic trunks*, lateral to the vertebral column, one on each side (Fig. 2.7.1).

Physiological division

From a functional starting-point there are three aspects:
– integration:
  • the vegetative part of the nervous system integrates the vegetative systems. This part is called the *vegetative system*, with as its synonyms *autonomic* or *involuntary* nervous system. Vegetative integration amounts to *regulation* of the separate systems and *coordination* between the systems. It is called autonomic or involuntary as the regulation and coordination are not under voluntary control. Examples are: regulation of blood pressure, gastrointestinal activity and respiration. The vegetative nervous system comprises two systems with antagonistic effects: the *sympathetic system* and the *parasympathetic system*. The sympathetic system is active when man is outwardly active: it speeds up heart activity and respiration, it raises blood sugar level and the tension of skeletal muscles. It relaxes the longitudinal musculature, and consequently slows the digestive process.
  The parasympathetic system is active when a person is outwardly calm and passive. It promotes peristalsis, it slows the activities of the heart and the respiratory organs, it slows muscular activity, etc. The sympathetic system and parasympathetic system are tuned exactly to each other (Fig. 2.7.2);
  • the animal part of the nervous system, the *animal nervous system*, is responsible for the integration of man and his surroundings. Sometimes this is called the *voluntary* nervous system.

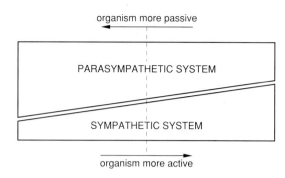

Figure 2.7.2
Tuning of parasympathetic/ sympathetic nervous system. When the organism becomes more active outwardly, the parasympathetic control decreases and the sympathetic control increases

The animal nervous system controls the interaction between the individual and his surroundings. One of its important aspects is *communication*;
– hierarchy:
  The nervous system can be seen as the control station of the body. Different levels for different functions, which are hierarchically related, are distinguishable in this control station. Higher levels have priority over lower levels. One example: When somebody's finger is unexpectedly pricked on a needle, the person draws back his finger reflexly. This movement, brought about by a lower level, can be suppressed, to a certain extent, by a higher level, for instance, when somebody's finger is pricked in order to take a blood sample. Generally speaking, the power in the sequence of the elements of the central nervous system grows, from spinal cord to cerebrum;
– direction of the signal:
  • when there is a *peripheral-to-central* impulse, there is *supply of information*. This is called the *afferent* direction;
  • when there is a *central-to-peripheral impulse*, there is an *initial impulse for action*. This is called the *efferent* direction;
  • within the central nervous system there are two main directions.
    Signals which go from a higher level to a lower level take *descending tracts*; signals

that go from a lower level to a higher level, take *ascending tracts*.

Descending tracts are often called efferent tracts; ascending tracts are often called afferent tracts.

Categories of neurones

Classified according to the directions of their signals, there are three categories of neurones (Fig. 2.7.3):

– *sensory neurones*, conducting signals from peripheral to central and from low to high. Sensory neurones are afferent neurones, supplying information from the somatic nervous system;
– *motor neurones*, conducting signals from high to low and from central to peripheral. Motor neurones are efferent neurones, stimulating muscular and glandular activity;
– *connector* or *interneurones*, transmitting signals from one neurone to another. The terms afferent and efferent are not applicable. Con-

nector neurones are more normally found in the central nervous system.

In attempting to discuss the nervous system it is impossible to stay within the limits set as dividing criteria. The anatomical division will be the guideline, but it must be complemented with physiological data. The vegetative system will be discussed separately.

2. Nervous system

The nervous system consists of arranged nerve tissue. There are two categories of cells in the nerve tissue, the *neurones* or conducting cells, and the *neuroglial cells* or supporting cells, in the proportion of approximately one to ten.

2.1. Neurones

Nerve cells are as different, in size and type, as most other cells. They have, however, a common structural plan (Fig. 2.7.3).

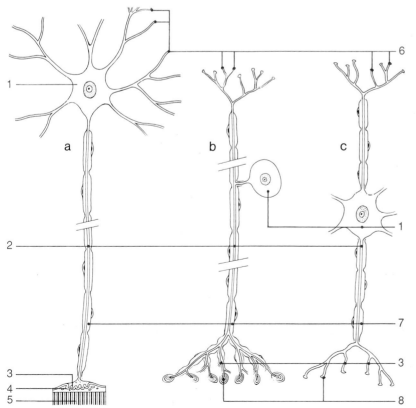

Figure 2.7.3
Categories of neurones; structures

a. Motor neurone
b. Sensory neurone
c. Connector neurone

1. cell body with nucleus
2. axon
3. telodendria
4. neuromuscular synapse
5. muscle
6. dendrites
7. myelin sheath
8. synapse

The cell body is comparatively large. It contains the nucleus and nucleoli, an extensive endoplasmic reticulum (Nissl substance) and a number of Golgi apparatuses. The cell membrane has one or more radiated processes, which are hollow and contain cytoplasm. These processes can range from being short or very long, and are often ramified.

A neurone has one, usually long, thin process, a *neurite* or *axon*, which normally has few branches.

Directly on the cell membrane are a large number of shorter extensions, which can have many branches, the *dendrites*.

The neurites are usually covered by a layer of *myelin*. This fatty white-yellowish substance forms the *myelin sheath* or *medullary sheath* (Fig. 2.7.4).

The myelin sheath has an insulating function. The speed of conduction of neurones is directly proportional to the thickness of the myelin sheath. Fibres without myelin conduct considerably slower than fibres with myelin.

At regular intervals (of about 1 mm) the myelin sheath is interrupted. These interruptions are called the *neurofibral nodes (nodes of Ranvier)*. It will be seen later that these interruptions make the signals pass along the fibre at an even faster rate.

Surrounding the myelin sheath in the peripheral nervous system are *the neurolemmocytes (Schwann cells)*. These form the *neurolemma (Schwann membrane)*, that envelopes the neurite (Fig. 2.7.4b). The neurolemma and the myelin sheath prevent the signal in a neurite from jumping from one neurite to another.

In the central nervous system the neurite is not encircled by neurolemmocytes but by a special type of neuroglial cells. The neurolemmocytes and the neuroglial cells have similar functions: insulating and nourishing the neurite that they encircle.

At its end the neurite can have branches: the *telodendria*. The thickened tips of the telodendria are called the *terminal buttons* or *synaptic buttons*. They play a role in the transmission of signals. A neurite without telodendria is called an *axon*.

Dendrites have neither myelin sheath nor supporting cells. Numerous signals from other nerve cells are transmitted on dendrites, as well as on the cell body itself (Fig. 2.7.8). The dendrites conduct the impulse to the cell body. The neurites of motor neurones and con-

Figure 2.7.4
Structure of a nerve fibre

a. Longitudinal section

1. *neurite*
2. *myelin sheath*
3. *neurofibral node (node of Ranvier)*
4. *nucleus of a neurolemmocyte (Schwann cell)*
5. *neurolemma (Schwann membrane)*

b. Cross section

1. *nucleus*
2. *neurolemmocyte (Schwann cell)*
3. *neurite*

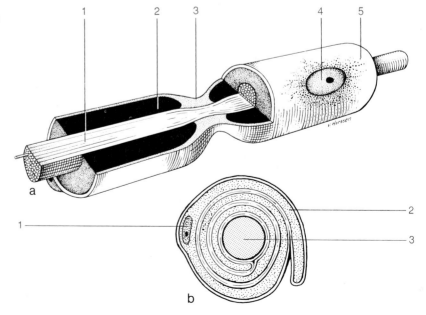

nector neurones conduct the impulse away from the cell body (Figs. 2.7.3a and c). The nerve fibre of a sensory neurone should be called a dendrite, since the fibre conducts the impulse towards or past the cell body (T-cell). But because of its length and its having myelin, it is called *a neurite*.

From the above it can be inferred that there is one-way traffic in the nerve fibres. The impulse always goes from the dendrite to the cell body and via the axon to the telodendria.

Stimuli and action potentials

In functional anatomy, the concept of stimulus denotes every change in the surroundings of the organism or part of it (organ, tissue, cell) which leads to a reaction.

More specifically at cellular level, a stimulus is a short-lasting change in the m.i. of the cell resulting in an electrical action of the cell membrane.

In response to the stimulus an *action potential* arises. An action potential is an electrical current which is transmitted through the cell membrane. Almost all categories of cells, and especially nerve cells, are sensitive to stimulation. Due to their having long processes (fibres), nerve cells are extremely suitable to transmit stimuli. The nerve fibres could be seen as the electrical wires of the body.

Na⁺ influx, K⁺ efflux

The propagation of electrical signals is a result of changes in ion concentrations inside and outside the cell membrane. When a nerve cell is at rest (polarized neurone), the concentration of K^+ ions in the intracellular milieu is greater than in the m.i. The same is true for the Na^+ ions: there are more Na^+ ions in the m.i. than inside the cell membrane (Fig. 2.7.5).

This difference could be overcome by diffusion, if the membrane were equally permeable to sodium and potassium. In fact, an enzymatic pump (see Fig. 1.3.5) maintains the Na/K proportion. The Na/K pump actively removes the Na^+ ions from the cell and absorbs smaller volumes of K^+ ions. All this creates a difference in charge between the inside of the cell and the m.i. The inside of the cell has a negative charge compared to the outside (Fig. 2.7.5). This is called the *resting potential* of the nerve cell.

Stimulating the membrane causes a change in permeability and consequently in ion concentration. Suddenly Na^+ ions pass through the membrane to the inside of the cell (Na^+ influx), resulting in a positive charge on the inside of the cell in proportion to the m.i. (depolarization). The depolarization of a part of the cell membrane thereafter generates the Na^+ influx into the bordering part of the mem-

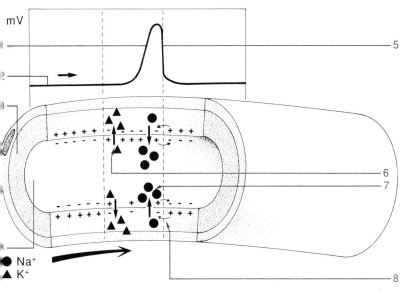

Figure 2.7.5

Depolarization and impulse conductivity in the nerve fibre. The sodium influx causes the inside of the cell to become positive in proportion to the bordering part of the cell membrane. This causes a net transfer of the stimulus to the right

1. *depolarization*
2. *resting potential*
3. *extracellular*
4. *intracellular*
5. *repolarization*
6. *K⁺ efflux*
7. *Na⁺ influx*
8. *transfer of the charge along the membrane*

mV

● Na⁺
▲ K⁺

brane as the positive charge acts as a stimulus. This process causes a net transfer of the charge over the membrane of the cell and the nerve cell, *the action potential.*

The electrical situation of rest (repolarization) is caused by an exit of K$^+$ ions (K efflux). Thereafter the Na/K pump restores the correct proportion of Na$^+$ and K$^+$ ions inside and outside the nerve cell.

In the period between the beginning of the depolarization and the end of the repolarization, the ion concentrations are not the same as during a period of rest. The cell is not sensitive to stimulation and cannot immediately conduct another stimulus. This is the *refractory period.*

The speed at which the action potentials are propagated along the nerve fibres is dependent on the thickness of the fibres and the corresponding thickness of the myelin sheath. The thinnest fibres have no myelin and conduct slowly. The thicker the fibre, the thicker the myelin layer becomes and the larger the intervals between the neurofibral nodes. Due to the insulating capacity of the myelin sheath, depolarization can only take place at the level of the neurofibral nodes. A flow of ions from the one depolarized node to the next one arises in the cytoplasm of the axon. The same happens in the m.i. outside the axon in the opposite direction. This leads to depolarization at the level of that node, etc. The action potential jumps, as it were, from node to node. This is called *saltatory conduction.*

The jumping time is always the same. When the intervals between the nodes become greater, as is the case when fibre and myelin sheath become thicker, the speed of conduction increases. Thin fibres without myelin have a conduction speed of 1.5 to 2 m/sec. The thickest myelinated fibres have a conduction speed of 90 m/sec.

Impulse transmission

The impulse goes via the axon to the telodendria, where it must be transmitted. There are two possibilities:
– transmission of the impulse from one neurone to another, or *neuro-neuronal transmission;*

– transmission of the impulse from a nerve fibre to a muscle fibre, or *neuromuscular transmission.*

In sensory neurones and connector neurones there is always neuro-neuronal transmission. In motor neurones there is always neuromuscular transmission.

Transmission of the impulse takes place at the terminal buttons of the telodendria. The action potentials, propagated along the telodendria, release *transmitters* from the terminal buttons. Such *neurotransmitters* are hormone-like substances which stimulate the next neurone (in neuro-neuronal transmission) or the muscle fibre (in neuromuscular transmission).

– In neuromuscular transmission the junction betweeen telodendrite and muscle fibre is called a *neuromuscular synapse* (Fig. 2.7.6). The end of the telodendrite becomes broader and is embedded in the wall of the muscle fibre. The junction is covered by neurolemmocytes. A small cleft remains between the membrane of the muscle fibre and the membrane of the nerve fibre.

There is a great number of vacuoles in the terminal button of the telodendrite, apart from the mitochondria. These vacuoles contain the neurotransmitter *acetylcholine* and the enzyme *choline-acetyl-transferase*, which is needed to build acetylcholine (Fig. 2.7.6b).

As soon as action potentials reach the neuromuscular synapse via the telodendria, acetylcholine is released from the vacuoles and diffuses into the cleft between nerve fibre and muscle fibre membranes. Acetylcholine thereafter stimulates depolarization of the muscle fibres. The *cholinesterase* from the muscle fibre swiftly decomposes the acetylcholine into acetic acid and choline. The stimulating activity of the transmitter stops. The membrane of the muscle fibre is ready for the next impulse. For muscle fibre contractions there are 10 to 100 depolarizations per second.

Acetic acid and choline are taken up again by

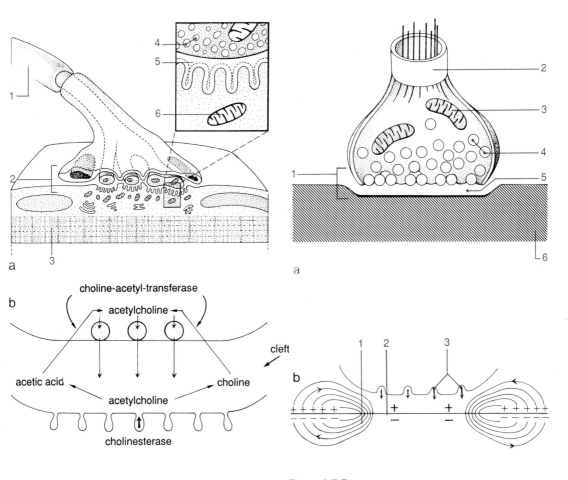

Figure 2.7.6

The neuromuscular synapse

a. Structure

1. telodendrite
2. neuromuscular synapse
3. muscle tissue
4. vacuoles with
   neurotransmitter
5. cleft
6. mitochondrium

b. Schematic diagram
of reactions involved
in impulse
transmission

Figure 2.7.7

The synapse

a. Structure

1. synapse
2. telodendrite
3. mitochondrium
4. vacuoles
5. synapse cleft
6. membrane of, for
   instance, next
   neurone

b. Schematic diagram of
impulse transmission

1. transfer of the charge
2. depolarizations
3. transmitters are released
   in the synapse cleft

the terminal button. The choline-acetyl-transferase enzyme realizes the resynthesis to acetylcholine in the vacuoles.

– In neuro-neuronal transmission the junction of telodendrite and the membrane of the next neurone is called a *synapse* (Fig. 2.7.7). Between the terminal button and the membrane of the dendrite or cell body is the *synaptic cleft*. The terminal button contains a large number of synaptic vacuoles filled with transmitter substances and enzymes.

As soon as action potentials reach the terminal buttons via the telodendria, these neuro-transmitters are released from the vacuoles and diffuse into the synaptic cleft, where they stimulate the next neurone.

Figure 2.7.8

The contacts of a
nerve cell

1. neurites of other
   nerve cells
2. synapse
3. dendrite
4. cell body
5. nucleus
6. axon
7. myelin sheath

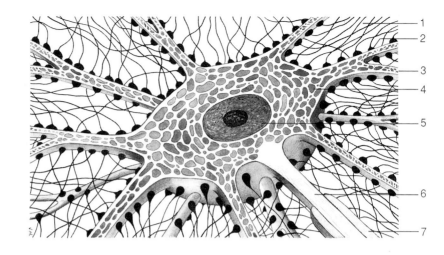

There are *excitatory* and *inhibitory* neuro-transmitters. Excitatory transmitters cause depolarization of the neurone: the impulse continues. Inhibitory transmitters make the neurone less excitable. *Acetylcholine* is a well-known excitatory transmitter substance. It is, among other things, found in the neuro-muscular synapses and in the synapses of the parasympathetic system.

Furthermore there is the group of *catecho-lamines*, comprising *adrenaline, noradrena-line (hormones)* and *dopamine*. These belong to a group of neurotransmitters, of which little is known at present. There are also amino acids, for instance, gamma-amino fatty acid, an inhibitory transmitter, and *serotinine*.

Most neurones have hundreds of synapses on their dendrites and cell bodies (Fig. 2.7.8). Moreover, many terminal buttons are further influenced by other synapses. It could be said that the function of the nervous system amounts to the result of pluses and minuses in the innumerable synapses.

The synapse guarantees *one-way traffic* in the nervous system. Although the action poten-tials can be propagated, basically, along the neurite in both directions, the transmitter and the enzymes in the synapse permit trans-mission in one direction only.

The synapse causes impulse transmission to slow down. The processes in the synaptic cleft take some time. The more synapses an impulse has to pass, the more the impulse slows down.

Description of functions

In order to understand the mutual co-operation between neurones, a description of the func-tional parts is sufficient (Fig. 2.7.9).

A neurone consists of:

– a *receptive part*, containing the cell body and the dendrites, where the impulses are re-ceived;
– a *conductive part*, in which the impulse is propagated along the cell membrane, and the neurite (i.e. axon + telodendria) conducts impulses away from the cell body;
– a *transmission part*, in which the impulse is transmitted to another neurone or a muscle fibre through a synapse or a neuromuscular synapse.

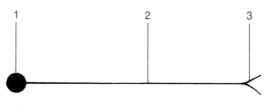

Figure 2.7.9   1. receptive part
The functional parts of a   2. conductive part
neuron   3. transmission part

2.2. Neuroglial cells

Neurones appeal to the imagination, it often being said that they determine the function of the nervous system. However, the importance of the supporting cells, the neuroglial cells, should not be underestimated. Their number is ten times that of the nerve cells and their function is indispensable to the neurones. They are helping, rather than supporting, cells.

It is not illogical to designate the neuro-lemmocytes (Schwann cells), that form the myelin sheath in the peripheral nervous system, as supportive, and therefore call them neuroglia. The term neuroglia, however, is reserved for the supporting cells in the central nervous system.

There are three categories of neuroglial cells (Fig. 2.7.10):
- *astrocytes*, which are about the same size as neurones. They have a large number of radiating, branched extensions. The ends of the extensions are a little broader. With these 'feet' they have contact with the membranes of the neurones on the one hand, and with the membranes of the capillaries on the other hand. Astrocytes are important for the exchange of substances between nerve cell and bloodstream;
- *oligodendrocytes*, which are smaller than the astrocytes. They have relatively few, scarcely ramified, short extensions. The oligodendrocytes have a function in the formation of the myelin sheath;
- *microglial* cells, which are the smallest of the three types of glia. They have short, ramified extensions. They behave as if they were spiders; they move through the nerve tissue by moving their extensions. When tissue is damaged, the microglia phagocytize the remains of the neurones.

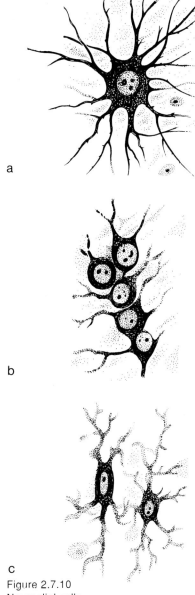

a

b

c

Figure 2.7.10
Neuroglial cells

a. Astrocytes
b. Oligodendrocytes
c. Microglia

2.3. Grey and white substance

Neurites are myelinated, dendrites and cell bodies are not. When studying the nervous system, it is obvious that there is a difference in colour between the one substance and the other.

When *grey substance* is found, mainly cell bodies and dendrites are present. When, however, *white substance* is found, myelinated neurites are present. From a functional point of view, grey substance can be characterized as the *transmission centre*, and *white substance* as the *conduction tract*.

Figure 2.7.11
The structure of a nerve

1. nerve
2. bundle of nerve fibres
3. epineurium
4. perineurium
5. endoneurium
6. neurolemmocyte
   (Schwann cell)
7. axon
8. dendrite
9. cell body
10. nucleus
11. myelin sheath
12. neurofibral node
    (node of Ranvier)

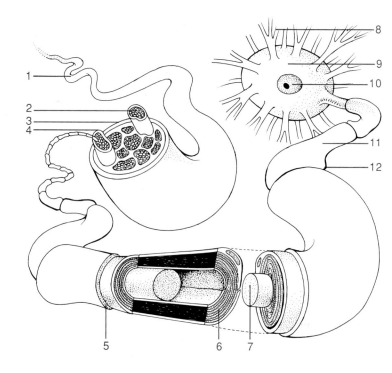

Nerve

A nerve consists of hundreds of axons (Fig. 2.7.11). Around the neurolemma (Schwann's membrane) of each axon is a thin layer of connective tissue, the *endoneurium*. A *nerve bundle* is a large number of axons surrounded by the *perineurium*, a layer of connective tissue. Several nerve bundles are surrounded, in their turn, by the strong connective tissue layer of the nerve itself, the *epineurium*.

The inside of a nerve is white, because of the prevailing myelin.
Nerves belong to the peripheral nervous system. Generally speaking, a nerve contains both efferent and afferent nerve fibres, and is therefore called a *mixed nerve*.

Tract

A tract is also a bundle of myelinated axons, but *inside* the central nervous system. There are tracts with bundles which course from lower to higher centres, the ascending tracts, and tracts with bundles which course from higher to lower centres, *descending tracts*. There are *short tracts* in the spinal cord, and *association tracts*

and *commissural tracts* in the brain. Consequently, where there are tracts there is white substance.

Nucleus

A nucleus is an aggregate of cell bodies and dendrites with a common function inside the central nervous system, for instance the red nucleus, the olivary nucleus and the caudate nucleus. A nucleus is recognizable by its grey colour (see Fig. 2.7.30(6)).

Ganglion

A ganglion can be seen as a nucleus outside the central nervous system. Its inside, therefore, is grey.
The above terminology is not always used consistently. Certain nuclei in the brain are often called basal ganglia.
The difference between central and peripheral can also be confusing. Some cerebral nerves are almost entirely located inside the skull and yet belong to the peripheral nervous system. The name *spinal ganglions* indicates that they belong to the peripheral nervous system (Fig. 2.7.13), even though they are inside the

vertebral canal. This is the accepted termin-ology.

3. Spinal cord or Medulla spinalis

3.1. External characteristics and connections

The spinal cord is the part of the central nervous system which is located in the vertebral canal. It extends from the foramen magnum of the skull to L_1 or L_2, the first or the second lumbar vertebra (Fig. 2.7.12). The spinal cord is thinner than is usually supposed. Its cross-section mid-way is the size of a pound coin. It has a cervical fusiform enlargement (intumescentia cervica-lis) and a lumbar fusiform enlargement (intumescentia lumbalis) which are involved in the innervation of the arms and the legs, respect-ively. The spinal cord has a dorsal and a ventral median groove; the ventral groove is a little wider than the dorsal one. The spinal cord has a left/right symmetry.

The vertebral canal is not filled to the end with spinal marrow, due to the fact that the vertebral canal and the spinal cord did not develop at the same pace. It is as if the spinal cord has ascended in the vertebral canal (see Fig. 3.2.18).

Throughout the whole of its length, there are straight lines of nerve bundles, to left and right of both the ventral and the dorsal fissures, which are connected with the central part of the spinal cord (Fig. 2.7.13).

At regular intervals these fibres join to form the *roots*, both ventrally and dorsally, and left and right symmetrically. In their turn, a *ventral root* and a *dorsal root* join to form a *spinal nerve* which leaves the vertebral canal through the *intervertebral foramen*, an opening formed by the apposition of the notches of two adjacent vertebrae.

32 pairs of spinal nerves leave the vertebral canal, left and right, between every two verte-brae. Apart from the cervical nerves, they are always designated by the number of the vertebra beneath which they appear: Th(oracic)$_1$ up to Th$_{12}$ inclusive, L(umbar)$_1$ up to L$_5$ inclusive, S(acral)$_1$ up to S$_5$ inclusive. C(ervical)$_1$ leaves between the skull and the first cervical vertebra,

Figure 2.7.12
The spinal cord; location in the vertebral canal

1. spinal ganglion
2. dorsal root (radix dorsalis)
3. spinal process of first thoracic vertebra
4. edge of dissected dura mater
5. cross-section through posterior arch of vertebra
6. spinal process of the first lumbar vertebra
7. cauda equina
8. lower tip of dural sac

Figure 2.7.13
A segment of the spinal
cord

1. lateral column
2. anterior or ventral
   column
3. membranes
4. posterior or dorsal
   column
5. posterior or dorsal
   horn
6. lateral horn
7. anterior or ventral horn
8. posterior or dorsal root
9. anterior or ventral root
10. spinal ganglion

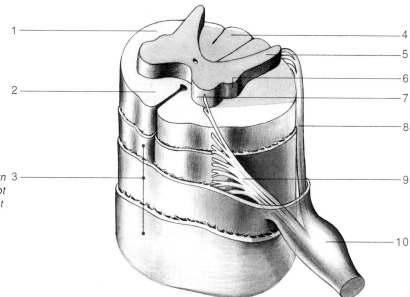

C_8 between the seventh cervical and the first thoracic vertebra.

Although the spinal cord is a continuous structure, it is thought of as being divided into as many segments as there are pairs of nerves which leave it: i.e. 32. The segments are named after the spinal nerves with which they have a connection: segment C_1 up to C_8 inclusive, Th_1 up to Th_{12} inclusive, L_1 up to L_5 inclusive, S_1 up to S_5 inclusive and one or two caudal vertebral segments.

The phrase 'spinal nerves leave' suggests that there is an efferent one-way traffic. This is not the case. The spinal nerves contain both motor and sensory nerve bundles; they are *mixed nerves*. However, there is one-way traffic in the anterior roots of the spinal nerves, as they almost entirely contain motor (and therefore efferent) nerve bundles. They are called *motor roots*.

In the posterior roots there is one-way traffic as well. They only contain sensory and therefore afferent nerve bundles. They carry sensations from the skin, the intestines, and the motor system and are called *sensory roots*.

Immediately behind the bifurcation into an anterior and a posterior root of the spinal nerves is a thickening in each posterior root, the *spinal ganglion*. The spinal ganglion (grey substance) consists of the cell bodies of the sensory neurones.

Such neurones have earlier been referred to as T-cells.

Cauda equina

Due to the difference in length of spinal cord and vertebral canal, the spinal segments from distal to cranial are less and less located opposite their synonymous vertebrae (Fig. 2.7.14a). This means that the spinal nerves of the first cervical elements leave the vertebral canal more or less horizontally, but that they have to descend more and more in order to reach their exit point from the vertebral canal, with a result that below the end of the spinal cord, at the level of L_1/L_2, only a 'brush' of spinal nerves is found: the horsetail or *cauda equina* (Fig. 2.7.14).

Segmentation

The spinal segments and the vertebrae represent the rudiments of a certain articulation in the embryonic development of the body; comparable components are connected in series. This segmented structure is found not only in the spinal nerves and the vertebrae,

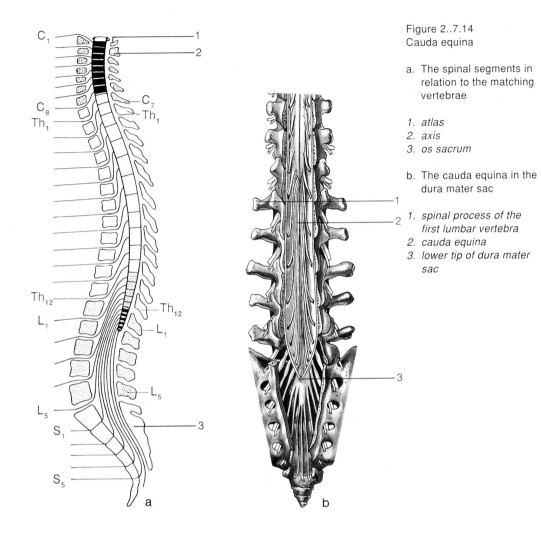

Figure 2..7.14
Cauda equina

a. The spinal segments in relation to the matching vertebrae

1. *atlas*
2. *axis*
3. *os sacrum*

b. The cauda equina in the dura mater sac

1. *spinal process of the first lumbar vertebra*
2. *cauda equina*
3. *lower tip of dura mater sac*

but also in the structure of the vertebral muscles and the connective tissue of the skin.

The elements of the skin that belong to a segment are called *dermatomes*. Each spinal nerve supplies the afferent innervation of a specific dermatome (Fig. 2.7.15). In the head and trunk the dermatomes are arranged in an orderly manner. It is remarkable that the dermatomes in the arms and the legs seem to have been moved from the trunk to the extremities.

The muscles, also, have rudimentary segmentation, the *myotomes*. In the development of the musculature, however, there are many shifts. Muscular elements from different myotomes merge into one muscle. Moreover, a myotome can take part in the formation of several muscles.

However a muscle originated, it can be said that the innervation of a muscle can be derived from the segment or segments from which the muscle arose. This explains why most muscles can be innervated from more than one spinal segment and why one spinal segment innervates several muscles.

In the vertebral and intercostal muscles the segmentations can easily be recognized, as innervation here is relatively simple (see Fig. 2.7.1). The innervation of the abdominal muscles is more complicated. The muscles originate from several segments. The segmental origin of the muscles of arms and legs seems to have been lost entirely. Although there is an enormous shift of muscular elements and the muscles are innervated from three or more segments, usually only one nerve reaches the

Figure 2.7.15
Dermatomes

a. Ventral view
b. Dorsal view

C = cervical or neck
      segments
T = thoracic or chest
      segments
L = lumbar or loin
      segments
S = sacral or sacrum

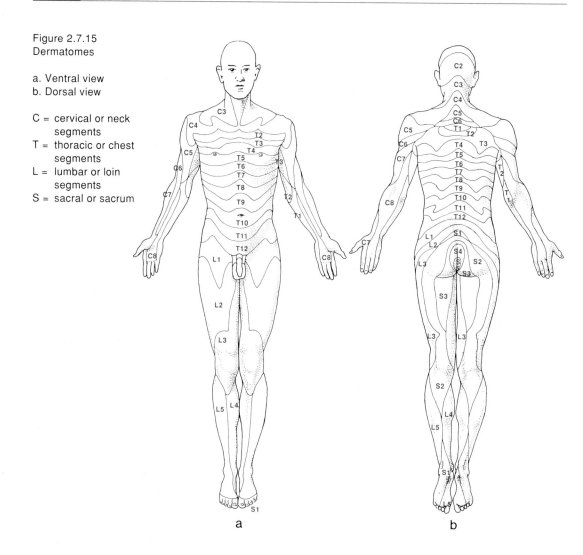

a   b

muscles. This means that there must be a rearrangement of nerve fibres outside the central nervous system. Such a network, in which fibres for the same muscles are re-arranged, is called a *plexus*.

The muscles in the head/neck area are supplied by the *cervical plexus*, the muscles in the arms by the *brachial plexus*. These two networks are somewhat related, so together they are often called the *cervico-brachial plexus* (see Fig. 2.7.1).

A similar plexus is found in the loin/sacral bone area for the supply of the muscles in the legs, the *lumbo-sacral plexus* (see Fig. 2.7.1).

3.2. Internal characteristics and connections

In a cross-section through the spinal cord a 'butterfly'-shaped figure, surrounded by white substance, is seen (Fig. 2.7.13).
The grey substance consists of cell bodies and dendrites, whereas the white substance consists of mainly myelinated fibres.

Grey substance

In the centre of the H-shaped figure is a small hole, filled with cerebral fluid called the *central canal* (Fig. 2.7.17). It extends centrally through-out the length of the spinal cord and is a contin-uation of the ventricles. The grey substance has a number of *horns* (see Fig. 2.7.13). The

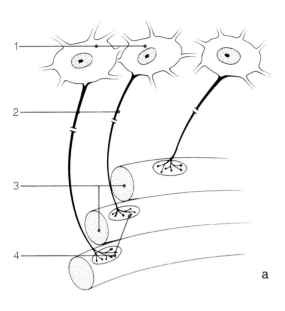

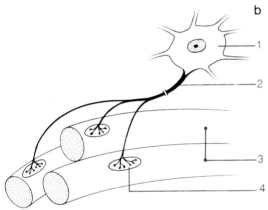

Figure 2.7.16
The motor unit

a. Small motor unit; three
   neurones innervate three
   muscle parts
b. Large motor unit; one
   neuron innervates three
   muscle parts

1. *motoneurone*
2. *axon*
3. *muscle fibres*
4. *neuromuscular synapse*

like the muscles in the back for instance, have relatively few motoneurones in the anterior horn.

A motoneurone, together with the muscle fibres it controls, is called a *motor unit* (Fig. 2.7.16). Small motor units are characterized by little ramification of the neurite. They guarantee minute control and the opposite holds true, as to structure and function, for large motor units.

The fibres of the sensory neurones enter the *posterior horn*. These fibres synapse with the receptor neurones in the posterior horn.
The development cells of the sympathetic system of the vegetative nervous system are located in the *lateral horns*.
The parts of the grey butterfly-shape not named contain mainly large numbers of connector neurones.

Organization in the butterfly is mainly horizontal. Upward and downward connections are made only over short distances.

White substance
There are a number of *columns* in the white substance (see Fig. 2.7.13), distinguishable in *anterior, posterior* and *lateral columns*. These columns contain myelinated axons in bundles, called *tracts*. The tracts extend vertically.

Immediately beside the butterfly figure are the *short tracts* (Fig. 2.7.17), in which extend the

anterior horns are a little broader and duller than the posterior horns. The segments C_7 up to I_2 inclusive are located in the lateral horns and in the sacral segments.

The *anterior horn* contains the cell bodies and dendrites of connector cells, but also the relatively large cell bodies of the motor cells, of which the neurites pass through the anterior horns. That is why they are often called *motor horns*. Muscles which can be minutely controlled such as, for instance, the muscles of the hand, need extensive 'wiring' (i.e. innervation). Consequently, such muscles must have a large number of motoneurones in the anterior horn. Muscles which can only be slightly controlled,

Figure 2.7.17
Tracts in the white
substance

red: descending tracts
blue: ascending tracts
grey: short tracts

1. *central canal*
2. *posterior horn*
3. *pyramidal tract*
4. *lateral horn*
5. *anterior hom*

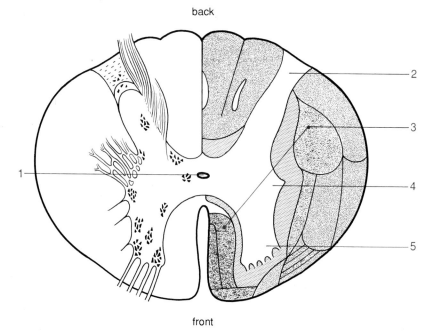

fibres of connector neurones joining the spinal segments which are closely above or below each other. For instance, a cell body and its dendrites is located in the grey substance of the L_3 segment. Via a short tract the neurite extends distally and thereafter gives off telodendria to the grey substance of L_5. Spinal segments have the capability to co-operate via the short tracts. Information is transmitted, activity is co-ordinated.

The *descending tracts* pass through the anterior column and the lateral column beside the short tracts. The descending tracts comprise the axons of the *pyramidal* and *extra-pyramidal tracts* which extend from the cerebrum. Impulses to the cells of the motor horns are transmitted via these motor (efferent) tracts. For instance, when we want to jump and contract our calf muscles, impulses go from the cerebrum, cerebellum, mesencephalon (midbrain) and brain stem, via the descending tracts, through the spinal cord to the level of the L_5 up and S_3 inclusive. There, the connection with the motoneurones which belong to the motor units of the calf muscles, takes place. A muscle which can be painstakingly controlled, not only has a large number of motor units, but also, of course, an equally large number of descending fibres as its 'wiring'.

Caudally, the number of descending tracts decreases. Tracts which supply areas located higher in the body with efferent innervation, drop out all the time. This is shown by the average (decreasing) size of the distal spinal cord. The fusiform enlargements in neck and loin areas are caused by the presence of large numbers of connector neurones in the segments of these parts of the spinal cord. This is related to the innervation of the arms and the legs, respectively.

The *ascending* tracts extend in the posterior column and in the lateral column, where they appose the descending tracts. The ascending tracts contain the *sensory tracts* which come from the posterior horns.

Stimuli from the skin, the intestines and the motor system are transmitted to the cerebrum, the cerebellum, the diencephalon and the brain stem via these sensory (afferent) tracts.

Body areas with high sensitivity or discrimination (see Fig. 2.8.4) need an extensive 'wiring system'. Consequently, a relatively large amount of space in the posterior roots and the posterior horns is taken up by the neurites which innervate these areas, and for the

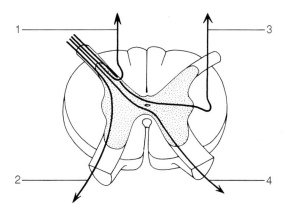

Figure 2.7.18
The four main
afferent routes

1. homolaterally ascending
2. homolaterally horizontal
3. contralaterally ascending
4. contralaterally horizontal

connector neurones with which they synapse. It is evident that an equal number of ascending fibres extends from the posterior horns as the 'wiring system' for the higher levels.

Towards the cranium, the number of ascending tracts increases, as the tracts from higher levels are added to the tracts from lower levels, with a resulting growth in the average cross-sectional size of the spinal cord.

Homolateral and contralateral
It seems as if everything is connected with everything inside the grey butterfly figure (horizontally) and in the short tracts (vertically) via connector neurones. There are connector neurones which make connections inside one half of a vertebral segment. Other neurones connect two apposing segments at the same side of the body (homolaterally), while others again omit one or two segments before they synapse, etc. Apart from these homolateral connections there are just as many connector neurones crossing medially. They make connections with the other side of the spinal cord (contralaterally).

With the afferent (sensory) fibres entering the spinal cord via the posterior horn as a starting point, at least four main routes are discernible (Fig. 2.7.18):
– *homolaterally horizontal*: via the grey substance of the posterior horn on the motoneurones in the homolateral anterior horn.

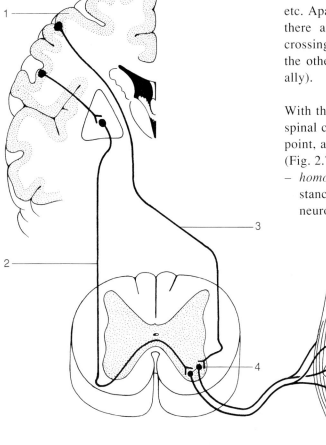

Figure 2.7.19
The two main efferent routes

1. cerebral cortex
2. contralaterally descending (extrapyramidal tract)
3. homolaterally descending (pyramidal tract)
4. motoneurones

None, one or more connector neurones can be involved;

- *contralaterally horizontal*: via the grey substance of the posterior horn on the motoneurones of the contralateral anterior horn. Connector neurones are involved.

These two routes are of great importance in reflexes;

- *homolaterally ascending*: the fibres are immediately connected, via the posterior horn, with the homolateral posterior columns and conduct the information to the brain via ascending tracts. There is a median crossing, but it takes place at a higher level.

This route is often taken by a specific part of the sensory system, namely by the senses of touch, pressure, vibration and by information from the motor system (proprioception, the awareness of position, movement and balance).

- *contralaterally ascending*: the fibres, coming from the posterior horn, cross medially and conduct information to the brain via the ascending tracts in the lateral column. This route is followed by another part of the sensory system, namely the senses of pain and temperature.

With the descending (motor) tracts as a starting point, at least two main paths are discernible (Fig. 2.7.19):

- *homolaterally descending*: the fibres have crossed medially at the level of the brain stem and travel in the descending tracts. Upon

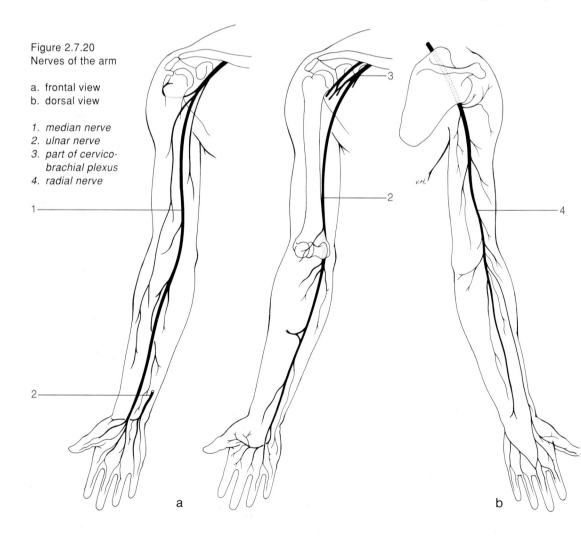

Figure 2.7.20
Nerves of the arm

a. frontal view
b. dorsal view

1. median nerve
2. ulnar nerve
3. part of cervico-
   brachial plexus
4. radial nerve

arrival at the segment concerned, they are homolaterally connected with the motoneurones. These tracts are called the *pyramidal tracts*. They supply the fine motor function and cortical control of voluntary movements.
– *contralaterally descending*: the fibres travel in the descending tracts. Upon arrival in the segment concerned, they first cross medially and then connect with the contralateral motoneurones. These paths are called the *extra-pyramidal tracts*. They subserve aspects of control of the locomotor apparatus.

It is a general rule for both ascending and descending tracts that the left half of the brain supplies the right half of the body and the right half of the brain supplies the left half of the body. When the median line is not crossed at the level of the segment of entering or leaving, this crossing takes place at supra-spinal level.
Obviously it is essential to know the impulse paths in order to be able to diagnose disorders of the nervous system.

3.3. Spinal nerves

As mentioned above, the spinal nerves are connected to the spinal cord. In the peripheral areas, notably in the arms and the legs, the spinal nerves are extensively ramified. The main branches will be discussed.
There are three thick nerves in the arm area,

Figure 2.7.21
The phrenic nerve

1. *phrenic nerve*
2. *diaphragm*
3. *ribs*
4. *heart sac*

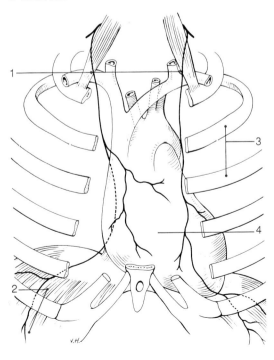

Figure 2.7.22
Femoral nerves

a. frontal view
b. dorsal view

1. *part of lumbo-sacral plexus*
2. *femoral nerve*
3. *ischial nerve*

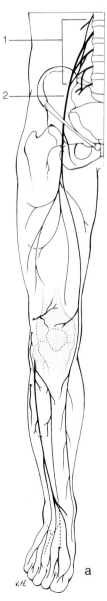

a   b

(Fig. 2.7.20), which originate from the cervico-brachial plexus: the *radial nerve*, the *median nerve* and the *ulnar nerve*. Their branches almost entirely supply the motor and somatic nervous systems of the arm and hand.

The nerves in the trunk which should be mentioned are the *phrenic nerve*, which innervates the diaphragm (Fig. 2.7.21), and the twelve *thoracic nerves*, and their branches that supply the skin and the thoracic muscles (see Fig. 2.7.1). The lumbo-sacral plexus gives rise to, among other things, the very thick *ischial nerve* and the *femoral nerve* (Fig. 2.7.22). These nerves branch in the leg. They largely supply the motor and the somatic nervous systems of the leg and foot.

3.4. List of functions

The spinal cord constitutes the lowest level of the central nervous system.

A number of functions have been explicitly discussed earlier, such as:
– the connections between the different spinal segments via the short tracts, by which related information, which arrives at different segments, can be connected;
– the capability to co-ordinate related motor activity starting from different segments;
– the connection with higher centres, via the ascending and descending pathways in the spinal cord;
– the connections with the periphery, both afferent (sensory) and efferent (motor), via the spinal nerves.

4. Reflexes

When a person comes out of a dark room into the daylight, his pupil constricts. When a person pricks his foot on a drawing pin, his leg is swiftly withdrawn. When dust particles enter the nose, one sneezes, often more than once. When in standing position a person lifts a leg to scratch his ankle, there are immediate corrections in the other leg to prevent him from falling. As soon as one smells delicious food, one's mouth begins to water.

These are all examples of *reflexes*. The complete list of all reflexes is very long.

4.1. General principles of reflexes

It can be seen from the examples that a reflex is always an involuntary (motor) activity of an organ, induced by some peripheral stimulus (sensory). A part of the central nervous system supplies the connection between sensation and motor activity.

Depending on the part of the central nervous system where the reflex connects, there are:
– spinal reflexes (via the spinal cord)
– brain stem reflexes
– cortical reflexes (via the cerebrum).

Automatic, involuntary but not always unaware

Reflexes are automatic; we cannot control them. Often the stimuli which generate the reaction and its effects, are not registered in our consciousness. Just think of the salivary reflex, the pyloric reflex, the accommodation reflex, the blink reflex, the breathing reflex, etc.

This does not mean that the consciousness is always uninvolved. For instance, we are aware of a number of reflexes: the vomiting reflex, the choking reflex, the sneezing reflex and the coughing reflex, etc.

Even so, awareness does not come into being until after the motor effect. Awareness is dependent on relevant information which travels in the ascending tracts to the cerebrum.

Reflex arc

A reflex arc is the pathway of the impulse between the sensor (site of stimulus) and the effector (site of reaction).

The sequence of the distinguishable parts is: sensor, sensory neurone, connector neurone (none, one or more), motoneurone, muscular fibres or glandular tissue.

As usual, more than one 'wire' is involved in most reflexes and it is better to distinguish: *sensor* (or sense), *afferent nerve, reflex centre* (in the spinal cord, brain stem or cerebral cortex), *efferent nerve, effector* (muscle or gland).

In the animal nervous system, the effectors are always striated muscles. In the vegetative

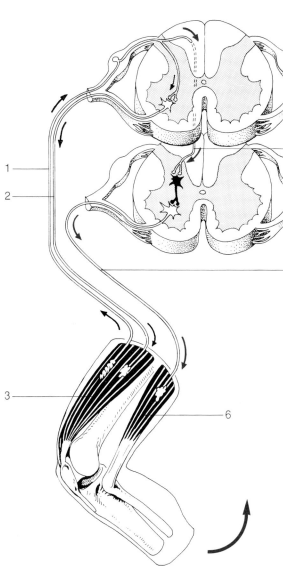

Figure 2.7.23

Reciprocal inhibition. The red arrows indicate the stimulated neurones, the black arrows indicate the inhibited neurones

1. sensory neurone of the muscle
2. motoneurone extending to the same muscle
3. extensor of the leg
4. branch of the sensory neurone
5. motoneurone (inhibited) extending to the flexor of the leg
6. flexor of the leg

nervous system, the effectors can be smooth muscles, the heart, and exocrine and endocrine glands.

When a person steps on a drawing pin, the flexors of the leg are activated reflexly (Fig. 2.7.23). At the same time the extensors must relax. However, in order to achieve this, the withdrawal reflex activates the motoneurones of the flexors (excitation) and simultaneously inhibits the motoneurones of the extensors (inhibition). The latter is called *reciprocal inhibition*.

Reflexes therefore respond to both excitatory and inhibitory factors.

There are several categories of reflexes distinguished by the type of the reflex arc:

– *monosynaptic* and *multisynaptic reflexes*.

Monosynaptic reflexes have no connector neurone, and therefore only one synapse. The afferent fibre immediately connects with the motoneurone. The period of time between stimulation and reaction (reflex time) is very short, as only one synapse needs to be bridged. Monosynaptic reflexes are sometimes simply called *single reflexes*. An example is the patellar or knee jerk reflex, induced by tapping the patellar tendon (Fig. 2.7.24a). Multisynaptic reflexes are typified by one or more connector neurones between afferent and efferent fibres, and therefore by two or more synapses. The reflex time is proportionate to the number of synapses. These reflexes are called *compound reflexes*. An example is the withdrawal reflex (Fig. 2.7.24b).

– *monosegmental* and *multisegmental reflexes*. In monosegmental reflexes the reflex arcs course in one spinal segment, in multisegmental reflexes the reflex arcs extend over more than one segment. Multisegmental reflexes are sometimes called *spread* reflexes. The stronger the stimulus, the more an originally monosegmental reflex will have the characteristics of a spread reflex.

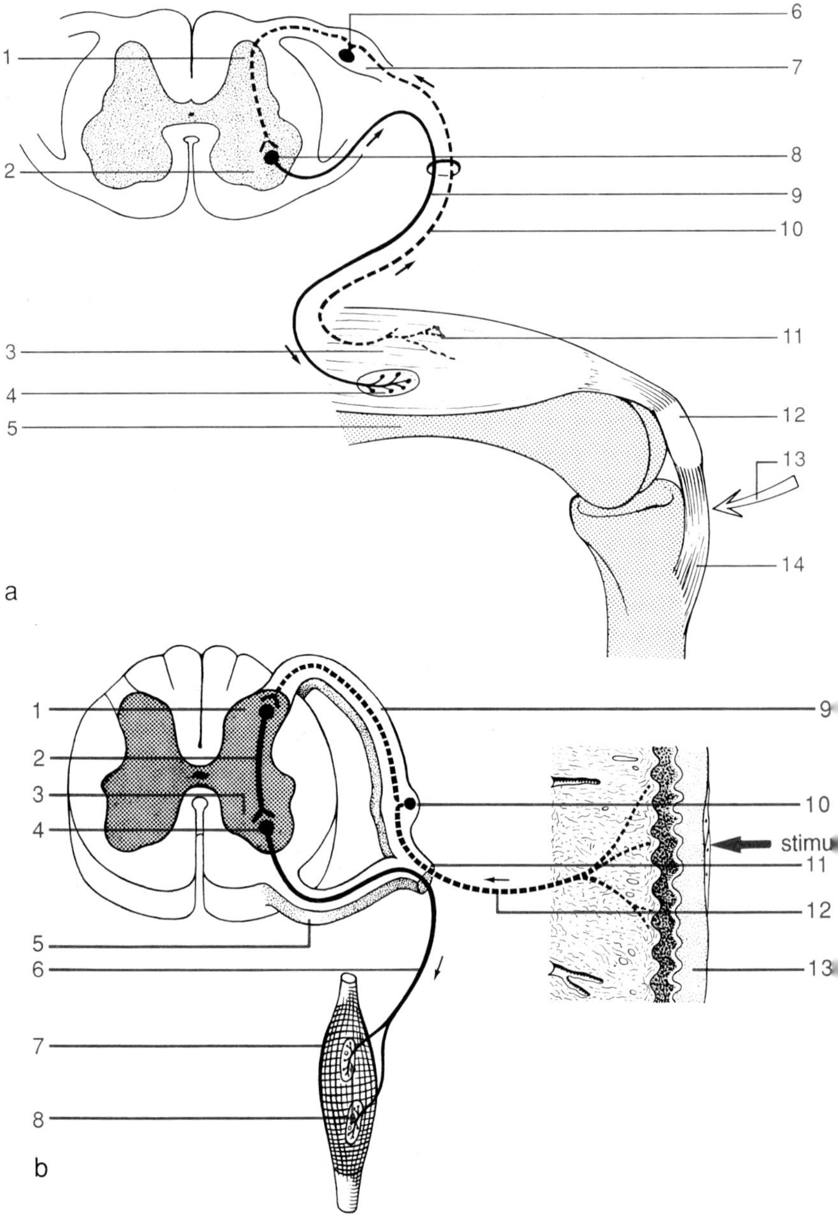

Figure 2.7.24

a. The muscle spindle stretch reflex, better known as the knee jerk reflex

1. posterior horn
2. anterior horn
3. femoral muscle
4. neuromuscular synapse
6. cell body of sensory neurone
7. spinal ganglion
8. cell body of motoneurone
9. efferent fibre of motoneurone
10. afferent fibre of the sensory neurone
11. sensors sensitive to stretching (muscle spindles)
12. patella
13. stretching influence on the extensor
14. knee tendon

b. The withdrawal reflex

1. posterior horn
2. connector neurone
3. anterior horn
4. motor anterior horn cell
5. anterior root
6. motor nerve fibre
7. muscle
8. neuromuscular synapse
9. posterior root
10. spinal ganglion
11. mixed nerve
12. sensory nerve fibre
13. skin

– crossed and uncrossed reflexes.

The reflex arc crosses or does not cross medially. An example of a reflex that crosses the median line has been described earlier. When a person, standing on both legs, bends one leg, this action is accompanied by the stretching of the other leg; a crossed extensor reflex.

Conditioned and unconditioned reflexes

A new-born baby has a range of inborn reflexes, such as the sucking reflex, the gripping reflex, the withdrawal reflex, the cough reflex, etc. Such reflexes, intrinsic to the nervous system, are called *unconditioned reflexes*. They are seen in every healthy individual; they travel through the spinal cord or the brain stem. An unconditioned reflex is induced by a non-changing specific stimulus.

Under certain conditions certain unconditioned reflexes may occur in response to changed, acquired stimuli. Such reflexes, dependent upon condition, are generated via the cerebral cortex.

They are called *acquired* or *conditioned re-flexes*. The stimulus which generates such reflexes is individually different and depends on experience.

When two people walk in a street lined with restaurants, one person's mouth may start watering (a conditioned reflex) when seeing restaurant A, whereas the other person may experience this reflex when seeing restaurant B. After all, tastes differ!

4.2. Spinal reflexes

Reflexes in which the spinal cord is the reflex centre, are called *spinal reflexes*. A number of them will be discussed:

- the *flexor reflex* or *withdrawal reflex*. Following a painful (noxious) stimulus to one of the limbs, the limb is usually swiftly withdrawn. The flexors are activated, the extensors are inhibited (Fig. 2.7.4b).
- the *crossed extensor reflex*. When a person, standing on both legs, bends one leg, the body weight must suddenly be borne by the other leg. In this leg the extensors are increasingly activated.
- the *muscle spindle stretch reflex*. When a muscle is suddenly stretched, there is an almost immediate contraction of the same muscle. The muscle resists the stretching, but later 'exaggerates'. The reflex is generated by the sensors in the muscles, the *muscle spindles*, which are sensitive to stretching.
  The best known muscle spindle stretch reflex is the *patellar tendon* or *knee jerk reflex* (Fig. 2.7.24a). In response to tapping on the patellar tendon, the quadriceps femoral muscle is suddenly stretched, resulting in the well-known kick. The same kind of reflex can be generated in other muscles, for instance in the calf muscles, by tapping on the corresponding tendons.
- the *plantar reflex*. On mechanical stimulation (by a pointed object) of the outer edge of the sole of the foot from the heel to the little toe, the toes are flexed in the direction of the sole (plantiflexion), the *negative Babinski sign*.
- *superficial abdominal reflex*. Upon stroking the skin of the abdomen with a sharp object,

in a transverse direction, the local abdominal wall musculature contracts.
- the *cremasteric reflex*. When, in the male, the inner surface of the thigh is scratched or stroked, the homolateral testis is lifted due to contraction of the cremaster muscle.
- the *defecation reflex*. The sphincters relax and peristalsis is stimulated in response to the filling of the rectum.
- the *micturition reflex* or *bladder reflex*. The external sphincter relaxes and the bladder wall contracts in response to the level of urine in the bladder.

5. Brain stem

At the level of the foramen magnum the spinal cord becomes the brain stem. This part of the central nervous system is located inside the skull. From caudal to cranial it is divided into the *medulla oblongata* or *spinal bulb*, the *pons cerebelli* or *pons varolii* and the *mesencephalon* or *midbrain*. On the terminal part of the pons is the cerebellum. The mesencephalon connects with the diencephalon. There are ascending and descending tracts, connections with the cerebellum and a number of functional centres (nuclei) in the brain stem. Twelve pairs of *cranial nerves* arise from the brain stem.

The brain stem controls functions which are literally of vital importance for the maintenance of life, such as circulation and respiration.

5.1. Medulla oblongata

The *medulla oblongata* or *spinal bulb* is the continuation of the spinal cord in the skull. As to its structure, there are similarities with the spinal cord (Fig. 2.7.26). It has a grey 'butterfly' figure (connector neurones) encircled by white substance (tracts). There are lines of nerve bundles connected with the medulla oblongata. They do not, however, form anterior and posterior horns and afterwards mixed nerves, but form separate *cranial nerves*. It should be noted that a number of ventrally located lines of nerve bundles – and consequently also the cranial nerves which they form – contain both sensory and motor fibres. The ventral bundles of the spinal cord have only sensory fibres. There is an

Figure 2.7.25
Sagittal section through
the brain

1. thalamus
2. hypophysis
3. hypothalamus
4. third ventricle
5. pons
6. reticular formation
7. cerebrum
8. corpus callosum
9. epiphysis or pineal
   gland
10. cerebellum
11. cerebral aqueduct
12. fourth ventricle
13. mesencephalon or
    midbrain
14. medulla oblongata or
    spinous bulb

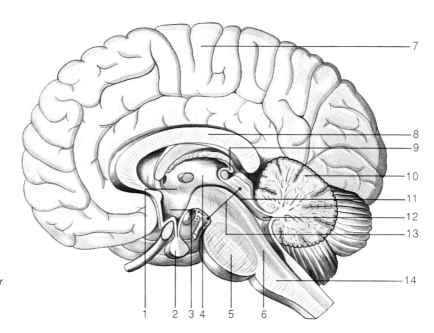

enlargement at both sides of the spinous bulb, between the ventral and dorsal lines of nerve bundles. These enlargements are called the *olivary nuclei* (Fig. 2.7.26). They are the connector centres for balance in the body.

The central canal in the medulla oblongata becomes wider at its top end. It changes into the fourth ventricle, which is located between the brain stem and the cerebellum (Fig. 2.7.25). It is one of the cerebral cavities, which is filled with fluid.

Apart from the nuclei of the cranial nerves and the olivary nuclei, there are functional centres in the medulla oblongata which are anatomically harder to distinguish. They are located in the floor of the fourth ventricle, therefore partly in the pons. They are the *cardiac centre*, the *vasomotor centre*, the *respiratory centre*, the *thermoregulatory centre* and a *vomiting centre*. A *cough centre* is also often thought to be located in the medulla oblongata, but it is not certain whether such a centre does exist separate from the respiratory centre. It is absolutely certain, however, that all stimuli which generate coughing and sneezing are integrated and co-ordinated in the medulla oblongata, in close co-operation with the respiratory centre.

The functional centres, mentioned above, belong to the vegetative nervous system. They are involved in vegetative integration and regulation, and are influenced by superior cerebral centres, notably by the *limbic system*, controlling human emotions (see Fig. 2.7.34). This is the reason why strong emotions, such as anger, excitement, fear, fright, tension, etc. can be accompanied by heart palpitations, a rise in blood pressure (flushing), and even vomiting and fainting. The coughing often encountered when people are nervous, is probably also a response to these connections.

Tracts, similar to those in the inferior spinal cord, are seen in the white substance. The pyramidal tracts show something special. The *pyramidal tracts* are two thick bundles in which the fibres cross at the site where the medulla oblongata changes into the spinal cord. After crossing they travel homolaterally to the corresponding motor cells of the anterior horns. This crossing is called *pyramidal decussation* (Fig. 2.7.26).

5.2. Pons (cerebelli) or pons varolii

The pons or pons varolii is the part of the brain stem which is characterized by a clear anterior enlargement (Figs. 2.7.25 and 2.7.26). This

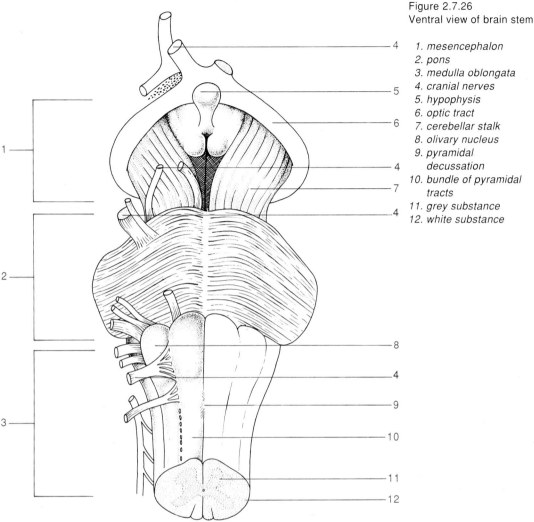

Figure 2.7.26
Ventral view of brain stem

1. *mesencephalon*
2. *pons*
3. *medulla oblongata*
4. *cranial nerves*
5. *hypophysis*
6. *optic tract*
7. *cerebellar stalk*
8. *olivary nucleus*
9. *pyramidal*
   decussation
10. *bundle of pyramidal*
    tracts
11. *grey substance*
12. *white substance*

enlargement is caused by a large amount of transverse fibres around the column of vertical fibres. These transverse fibres supply the connections between both halves of the cerebellum.

Apart from a number of nuclei of cranial nerves, the so-called *pontine nuclei* are located in the pons. They form the connector centres of the left/right connections. The widest part of the fourth ventricle is found between the pons and the cerebellum.

5.3. Mesencephalon or midbrain

The relatively small part of the brain stem situated between the pons and the diencephalon is called the *mesencephalon* or *midbrain*. In the mesencephalon the fourth ventricle narrows to the *cerebral aqueduct* (aqueductus Sylvii), a narrow canal which forms the connection with the third ventricle.

Below the midbrain are two thick nerve bundles emerging from beneath the transverse fibres of the pons and thereafter forming the connection, travelling underneath the optic tracts, with the other parts of the brain (Fig. 2.7.26). They are called the *cerebellar stalks* (pedunculi cerebri). They contain the fibres of the pyramidal tracts and also sensory fibres on their way to specific parts of the cerebrum.

The mesencephalon contains the nuclei of the cranial nerves, which are involved in vision.

At the bottom of the anterior part are the *nucleus*

Figure 2.7.27
The brain stem, dorsal view

1. mesencephalon
2. pons
3. medulla oblongata
4. thalamus
5. fourth ventricle
6. connection with the
   cerebellum
7. cut edge of the roof of
   the fourth ventricle

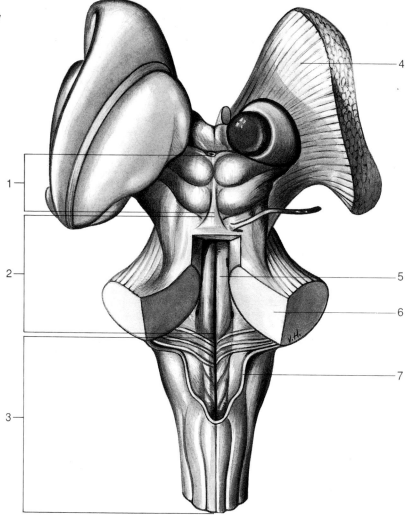

ruber (red nucleus) and the *substantia nigra* (black substance). These nuclei constitute connector centres in the extra-pyramidal tracts.

At the top of the posterior part, the connector centres for the fibres from the eyes and the ears are located.

These centres are connected with the thalamus and the optic and auditory parts of the cerebral cortex, respectively. They also supply the connections for the eye and ear reflexes.

5.4. Reticular formation

Throughout the length of the brain stem there is a diffuse network of short neurones forming a functional nucleus, the *reticular formation* (Fig.

2.7.25). This centre controls the functional level of the central nervous system, the 'state of alertness'. For this purpose it receives information (afferently) from the somatic nervous system, the motor system and other parts of the nervous system, such as the red and black nuclei, the diencephalon (notably from the thalamus) and the cerebrum. At the same time, the reticular formation is connected (efferently) with the cerebrum, the diencephalon (especially the thalamus) and the spinal cord.

The reticular formation is often characterized as a co-operating *sleep and wakefulness centre*. The interaction of both centres could determine the extent of general alertness.

When a person is awake, the reticular formation stimulates all parts of the brain to higher activity and the muscles are also activated to tighten. This, in its turn, stimulates the reticular formation. This explains why an individual can stay awake for a long time during captivating events or interesting work.

During sleep the reticular formation inhibits the activity of the brain and the muscle cells. In its turn, the reticular formation is inhibited. This explains why an individual can fall asleep during a boring TV programme.

During sleep, stimuli also continuously reach the reticular formation, but they will not wake the individual unless the stimulus is very strong (say, a noisy thunderstorm), or unusual (he is touched during his sleep), or very significant (the baby cries).

The reticular formation itself shows a rhythmic activity, more or less coinciding with the diurual rhythm, known as the sleep-alert rhythm. This is sometimes called the biological clock or biorhythm.

The transitional circumstances in this rhythm can be described as follows: after a day of activity the reticular formation gradually becomes less active, probably due to tiredness of the synapses. This can be seen as a decreased capacity in the metabolism of neurotransmitters. Due to this decreasing activity of the reticular formation, the nucleus cells become less activated and muscular tension decreases, which in its turn deactivates the reticular formation. Due to this downward spiral a person falls asleep. When the synaptic system has recovered and the preparedness of the reticular formation for stimulation increases, sleep becomes less deep. An alarm clock, a touch, or an impulse from the cerebral cortex, leads to awakening. After awakening the reticular formation, and with it the central nervous system, is more and more activated in an upward spiral.

5.5. Cranial nerves

Nerve bundles arise from the brain stem in lines, in the same way as they do from the spinal cord (Fig. 2.7.26). They constitute the twelve pairs of cranial nerves. These nerves are designated by their names, but also by Roman numerals, I up to XII inclusive. In order to distinguish the cranial nerves from the other peripheral nerves, N. is used for the first group and n. for the second one in abbreviations.

Some cranial nerves branch after they have been formed from bundles (Fig. 2.7.28). The branches of the cranial nerves supply the innervation of the senses and the muscles in head and neck. The vagus nerve (N. X) is an exception. This nerve is extensively branched and innervates many organs in the trunk. Some cranial nerves contain sensory fibres only, others have motor fibres only, and others are mixed nerves. Most cranial nerves have clearly distinguishable nuclei located in the white substance of the brain stem.

The cranial nerves are (Fig. 2.7.28):

I   the *olfactory nerve*. This is a sensory nerve consisting of approximately 20 thin bundles of fibres (fila olfactoria) which enter the cranial cavity from the olfactory epithelium via the ethmoid bone. It then extends, apposing the frontal lobe, to the brain stem, among other things (see Fig. 2.4.3a).

II the *optic nerve*. This is a sensory nerve which carries the impulses from the retina of the eye to the brain. The optic nerves cross at the level of the pituitary gland (optic chiasma) and then become the optic tract (see Fig. 2.7.26).

III the *oculomotor nerve*. This is a motor nerve. It originates just above the pons, ventral to the mesencephalon. It innervates the muscles of the eyelid and all external eye muscles, except the oblique superior muscle and the exterior straight muscle. This nerve also has parasympathetic motor fibres for the internal eye muscles.

IV the *trochlear nerve*. This motor nerve originates at the dorsal side of the mesencephalon and innervates one of the external eye muscles (the oblique superior muscle).

Figure 2.7.28
Basal view of the brain;
origin of cranial nerves

1. *frontal lobe of the*
   cerebrum
2. *temporal lobe of the*
   cerebrum
3. *pituitary gland*
4. *pons*
5. *cerebellum*
6. *first spinal root*

The cranial nerves as
designated by both their
names (in the text) and
by the Roman numerals
I to XII (on this diagram)

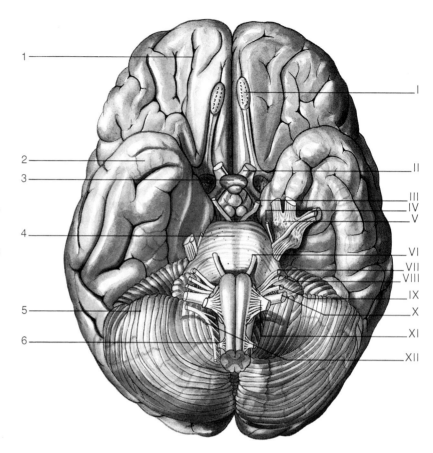

V   the *trigeminal nerve*. This is a mixed nerve emerging from the lateral side of the pons and dividing into three separate nerves. The sensory components supply part of the face, the oral cavity and the tongue. The motor fibres innervate the masseters.

VI   the *abducens nerve*. This motor nerve originates just below the pons, at the ventral side of the olivary nucleus, and innervates one of the external eye muscles (the exterior straight muscle).

VII the *facial nerve*. This mixed nerve originates underneath the connections of the pons with the cerebellum. It carries the taste stimuli of the anterior two-thirds of the tongue and contains motor fibres for the muscles of facial expression. Together with this nerve, parasympathetic motor fibres travel for the innervation of the lacrimal gland and two of the salivary glands (submandibular and sublingual glands).

VIII the *auditory nerve*. This is a sensory nerve composed of fibres from the organ of balance and the auditory organ. Functionally, this nerve is often called the N. stato-acusticus. Anatomically it is called the N. vestibulo-cochlearis. It connects with the cerebral cortex near the facial nerve.

IX   the *glossopharyngeal nerve*. This mixed nerve connects with the brain stem, just below the pons, dorsal to the olivary nucleus. The sensory branches carry stimuli from the posterior third of the tongue, from the laryngeal cavity and the tympanic cavity, from the sensors in the wall of the aortic arc and the carotid arteries which register blood pressure and blood acidity.
     The motor branches innervate the

muscles of the pharyngeal wall and the larynx. Moreover, the nerve has a parasympathetic motor branch that innervates the parotid gland.

X   the *vagus nerve*. This is a large, mixed, mainly parasympathetic, nerve which emerges in a number of bundles from the medulla oblongata medially behind the olivary nucleus. It supplies sensory fibres from the larynx, the air passages, the lungs and the mucosa of the alimentary tract. The parasympathetic motor fibres innervate the heart, the lungs, and the muscles of the roof of the mouth and the pharyngeal wall that make the swallowing movements possible. The motor branch supplies the vocal chords.

XI   The *accessory nerve*. This motor nerve arises from the medulla oblongata, immediately below the N. X. It innervates the trapezius muscle and the sternocleidomastoid muscle.

XII   The *hypoglossal nerve*. This motor nerve arises from the medulla oblongata, ventrally of the olivary nucleus. It provides innervation for the tongue muscles.

5.6. Brain stem reflexes

The cranial nerves provide the afferent and efferent pathways for the cranial reflexes, just like the spinal nerves, being part of the reflex arc of the spinal nerves. The reflex centres are in the brain stem. Of the large number of brain stem reflexes a few are mentioned:

– the *salivation reflex* is induced by smelling, seeing or thinking of food. The salivary glands are stimulated to increase the secretion of saliva.

– The ocular *fixation reflex* makes both eyes focus on one point, projected on the retina, when this point moves within the visual field.

– The *pupillary reflex* controls the amount of light admitted to the eye.

– The *accommodation reflex* changes the shape of the eye lens in order to maintain a clearly focused retinal image of objects at varying distances.

– The *corneal or blink reflex* takes place in response to irritation of the cornea by, for instance, a dust particle. The same stimulus can lead to the *lacrimal reflex*.

– The *sneezing reflex* or *nasofacial reflex* occurs when the nasal mucosa is irritated.

– The *cough reflex* is induced when the epithelium of the windpipe is stimulated.

– The *swallowing, vomiting*, and *choking* or *gag reflexes* are induced by specific stimulation or irritation of mouth and pharyngeal wall.

The brain stem also plays an important role in reflexes which maintain body posture. These reflexes are not discussed here.

6. Cerebellum

The cerebellum is situated at the posterior part of the pons, under the dome of the posterior part of the cerebrum (see Fig. 2.7.25). It is connected with the pons by the *cerebellar peduncles* (stalks) (Fig. 2.7.27). The cerebellum bridges the fourth ventricle. Its two halves, (hemispheres) reach as far as the sides of the pons. The cerebellar surface shows folds, in which a large number of more or less parallel fissures are visible (Figs. 2.7.28 and 2.7.25). This folded outer layer consists of grey substance (cell bodies) and is called the *cortex*. The white substance (myelinated axons) is underneath and is called *medulla*. In a sagittal section through the cerebellum (see Fig. 2.7.25), a kind of 'tree', of which the branches are formed by white substance and the 'leaves' by grey substance, is visible. This figure is called the '*arbor vitae*'.

It is noticeable that, unlike the organization in spinal cord and brain stem, the grey substance of the cerebellum – and also of the cerebrum – is located on the outside and the white substance is situated more centrally.

The peduncles are located at either side of the fourth ventricle and consist of nerve bundles which can be divided into three parts according to their direction. The superior bundle consists of fibres which emerge from one of the cerebellar hemispheres, travel to the mesencephalon, ramify there and give off branches to the red nucleus, thereafter crossing and terminating in the thalamus.

The middle bundle consists of fibres which emerge from the cerebrum, synapse in the pons, cross and terminate in the cerebellar hemisphere.

The lower bundle consists of fibres connecting the spinal cord with specific nuclei of the brain stem, on the one hand, and with the cerebellum on the other hand.

The function of the cerebellum is the *coordination of the motor system*. The cerebellum has connections with the cerebrum, the brain stem and the spinal cord. The cerebellum informs the cerebrum about the character of intended movements. Information from the organs of balance is processed in the brain stem. Thereafter the cerebellum is informed about posture and changes in posture. Information on articulatory position, muscular length and tension is supplied via ascending tracts. In this way the cerebellum is informed via the spinal cord, of the interrelation of the positions of several parts of the body.

The cerebellum, in its turn, influences the anterior horn motor cells via pathways to the red nucleus, from which efferent tracts leave.

The left and right hemispheres have contact via internal left-right connections and also via pons fibres.

How does the co-ordination of the motor system take place?

A voluntary movement, for instance lifting a telephone receiver, is initiated from the cerebrum. The anterior horn motor cells of the muscles involved in this movement are stimulated via descending tracts. Simultaneously, the cerebellum receives a copy of the movement from the cerebrum. As soon as the movement has started, the cerebellum controls and co-ordinates the course of the movement. The hand which is going to lift the receiver must be activated in time; after being picked up the receiver must gradually be raised to the ear, etc. The movement made is continuously compared with the 'copy' which was sent to the cerebellum, so that adjustments can be made. The

cerebellum warns the cerebrum, via the thalamus. There is therefore more than one form of feedback.

7. Diencephalon

The diencephalon is located between the brain stem and the cerebrum. It is the continuation of the brain stalk and consequently consists of ascending and descending fibres, notably from two important functional nuclei: the *thalamus* and the *hypothalamus* (see Fig. 2.7.25). The third ventricle is located in the diencephalon. This is a narrow, vertical cavity, of which the floor is formed by the hypothalamus and the sides by the medial sides of the left and right thalamus.

The thalamus: several connecting stations

Both thalami are made of a number of nuclei (grey substance) with related functions. They are the connecting stations of most sensory tracts. These ascending tracts have crossed the median immediately above the mesencephalon, before they synapse in the thalamus.

The thalamus is not only a connecting station for sensory tracts (senses of touch, hearing and seeing, and sensory impulses from the motor system), but also for the contacts between cerebellum and cerebrum, and between hypothalamus and cerebrum. The thalamus functions as a selective filter. It prevents the cerebrum from becoming overloaded with information from the periphery. In order to function properly, the cerebrum must receive only part of the immense flood of continuous sensory impulses. For instance, a person intently watching a movie, 'is in another world', due to the thalamus. It could be said that the extent of concentration on a specific job is provided for in the thalamus. In order to bring about this phenomenon, impulses which can stimulate or inhibit its synapses travel from the cerebral cortex to the thalamus.

Apart from providing specific attention, the thalamus also has a role in controlling the general level of cortical functioning. Some parts of the thalamus are considered to belong to the reticular formation.

The thalamus connection between cerebrum and cerebellum enables more delicate motor coordination as information from the motor system and the skin reaches the thalamus.

Information on the course of movements in the muscles, tendons and joints, and, for instance, from pressure sensors in the soles of the feet, is combined with information on balance and the movement which was originally planned.

The connection between the hypothalamus and the cerebrum belongs to the limbic system (see Fig. 2.7.34). This system is involved in controlling emotions and their autonomic effects, and is therefore the foundation of our moods, feelings, etc.

The hypothalamus: vegetative control centres

The hypothalamus comprises the floor of the third ventricle and consists of a number of nuclei. Part of the hypothalamus has close contacts with the main producer of hormones, the pituitary gland, or hypophysis.

Another part controls a number of vegetative functions, such as:

- the *centre for the production of hormones*. On the one hand, specific nerve cells in the hypothalamus produce hormones which the axons carry to the neurohypophysis via the pituitary stalk; on the other hand, specific cells in the hypothalamus produce the 'releasing hormones' ending up, via the bloodstream, in the adenohypophysis causing it to produce hormones.
- the *thermoregulatory centre*. The body temperature is controlled by specific centres in the medulla oblongata and the hypothalamus. When the temperature of the blood flowing through these centres rises or falls, impulses are sent to the vessels in order to bring about either vasoconstriction or vasodilation. When there is a threat of the body temperature becoming too high, the thermoregulatory centre induces 'panting' in the respiratory system. When there is a threat of the body temperature becoming too low, the thyroid is activated, via the hypothalamus, resulting in cell metabolism intensifying, accompanied by the production of heat.
- the *thirst centre*. A feeling of thirst and the urge to drink arise when the crystalloid-osmotic value in the cells of the thirst 'centre increases. Drinking and the decreased excretion of water as a result of ADH-production make the water concentration in the thirst centre rise and the crystalloid-osmotic value drop. The sensation of thirst disappears.
- the *appetite centre*. The sensation of hunger grows stronger when the appetite centre receives stimuli resulting from a decrease in the volume of insulin, on the one hand, and a decrease in the volume of glucose in the blood plasma, on the other. When the glucose level is raised by the intake of food, the production of insulin increases (positive feedback) and the sensation of hunger lessens.

8. Cerebrum

Of all parts of the nervous system the *large brain* or cerebrum appeals most to the imagination. Functions which are considered typically human, such as thinking, self-confidence, creativity, etc, are attributed to the cerebrum. At the same time, however, from a functional-anatomical point of view, it is notable that relatively little is known about these functions.

8.1. Structure

The cerebrum consists of two mirrored halves, *the hemispheres*, and their connections (Fig. 2.7.29). One hemisphere is more or less the shape of a quarter of a ball. The diencephalon has the hemispheres as a roof, especially the main connection between the two halves, the corpus callosum or great commissure.

Sulci and gyri

There is a separating fissure between the hemispheres, the *longitudinal fissure*. The surface of the cerebral hemispheres is folded, although not as strongly as that of the cerebellum, as a result of which there are *furrows* and convolutions. A furrow is called a sulcus, a convolution a gyrus. Due to the folding, the surface of the cerebrum, the cortex, is grooved. This allows room for very many cell bodies (grey substance)

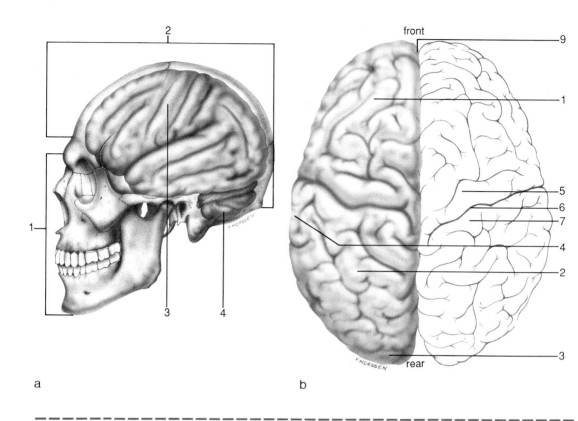

a   b

of neurones. When spread out, the surface is 50 × 50 cms on which approximately 10 billion cell bodies are located.

All sulci and gyri are designated by names. A few of them will be discussed (Fig. 2.7.29).

On the lateral surface, at about one third from the bottom, a virtually horizontal groove is located, the *lateral sulcus* (Sylvii). At the base of this sulcus is another cortex area. This part of the cortex is the insula (Fig. 2.7.30). Approximately halfway on each hemisphere is a vertical sulcus, the *central sulcus* (Rolandi).

The gyrus in front of the central gyrus is called the *precentral gyrus* and the one behind it is the *postcentral gyrus*.

Lobes

The hemispheres are divided into lobes (Fig. 2.7.29). The *frontal lobe* is located in front of the central sulcus. Behind the central sulcus is the *parietal lobe*, bordered at its base by the (notionally elongated) lateral sulcus. The temporal lobe is located below the lateral

sulcus. The remaining part is called the *occipital lobe*.

Connections

Underneath the cortex lies the white substance (myelinated axons) of the medulla. The white substance is divided according to the interconnections. Interconnections within a hemisphere are called *association tracts*. They interconnect different areas of the cortex without crossing the median line, as a result of which information between parts of the cortex with different functions can be exchanged.

– *Commissural tracts* are interconnections between the hemispheres. They cross the median line.

The main commissure, the *corpus callosum*, is situated between the floor of the longitudinal fissure and the roofs of the first and the second ventricles.

The commissures enable the one hemisphere to know what is going on in the other hemisphere. Information between the

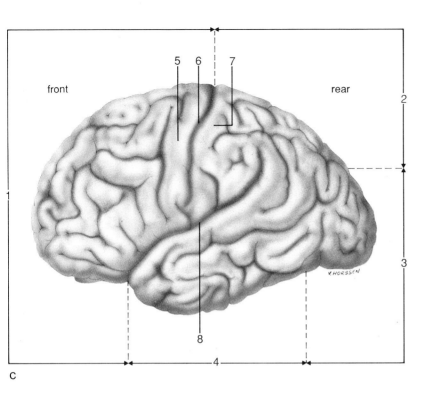

Figure 2.7.29
The cerebrum

a. Position of the
   cerebrum in the skull

   1. *facial bones or*
   visceral cranium
   2. *cerebral cranium*
   3. *cerebrum*
   4. *cerebellum*

b. View from above
c. Lateral view of
   cerebrum

   1. *frontal lobe*
   2. *parietal lobe*
   3. *occipital lobe*
   4. *temporal lobe*
   5. *precentral gyrus*
   6. *central sulcus*
   7. *postcentral gyrus*
   8. *lateral sulcus*
   9. *longitudinal*
   fissure

hemispheres is exchanged via the commissures.

– Interconnections between cerebrum and parts situated lower in the central nervous system, to which in the first place the ascending and descending tracts belong, have been discussed under the spinal cord and the brain stem.

The ascending (sensory) tracts pause in the thalamus, then the majority terminate in the postcentral gyrus.

There are two types of descending (motor) tracts:
The *pyramidal tracts* emerge from the precentral gyrus. The neurones of these tracts have large, pyramid-shaped cell bodies and, apart from the usual dendrites, a remarkably long axon. The axons of the pyramidal tracts travel caudally in a bundle. Approximately 80 % of the fibres cross the median line (decussatio pyramidum) at the level of the medulla oblongata.

They terminate thereafter at the anterior horn motor cells of some spinal segments. The 20 % of fibres which initially did not cross, cross the median line at the level where the majority terminate.

The *extrapyramidal tracts* contain all those motor fibres which do not belong to the pyramidal tract. These extrapyramidal fibres do not travel as much in bundles as the pyramidal fibres. They meet on their way, in a caudal direction, a number of synapses, for instance in the red nucleus, the black nucleus and the reticular formation. A number of extrapyramidal fibres cross the median line in the brain stem, continue distally and terminate in the anterior motor horn of some spinal segment. Other extrapyramidal fibres do not cross until they have reached the spinal segment where they terminate.

Apart from these ascending and descending tracts there are also interconnections between cerebrum and cerebellum, between cerebrum

Figure 2.7.30

Frontal section through
the cerebrum

1. cerebral cortex
2. white substance
3. corpus callosum or
   great commissure
4. caudate nucleus
5. internal capsule
6. lentiform nucleus
7. third ventricle
8. hypothalamus
9. cortex of the basal
   nuclei, over which
   other lobes are
   folded (insula)
10. longitudinal fissure
11. lateral ventricles

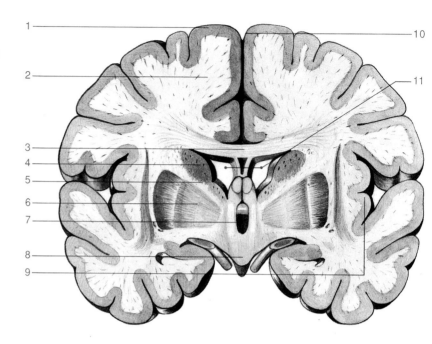

and brain stem, between the sensory centres in the head (senses of smell, taste, hearing, sight) and the cerebrum. These will be discussed later.

Cavities and nuclei

In a frontal section through the cerebrum a number of structures, apart from the grey substance of the cortex and the white substance of the medulla, are seen (Fig. 2.7.30).

In each of the hemispheres is a cavity, a *ventricle*. The first and the second ventricles are together called the *lateral ventricles*. Apposing the lateral wall of these ventricles is the *caudate nucleus* (Fig. 2.7.31).

There is another area with grey substance, the *lentiform nucleus*, between the cortical area of the insula and the thalamus.

A band of fibres, fanning widely from the cerebral peduncles to the cortex, is located between the lentiform nucleus and grey substances of the thalamus and the caudate nucleus.

This tract of fibres is called the internal capsule (Fig. 2.7.31). It contains the fibres of the descending pyramidal tracts and of the ascending sensory tracts.

8.2. Function

The cerebrum has a large number of functions and, as mentioned, special functions which man considers valuable. Not only does the cerebrum mediate in coordinating human motor functions with the somatic nervous system; functions such as consciousness, dreams, emotions, intelligence, artistic creativity and memory are attributed to the cerebrum as well.

The brain functions as a unit, just as the whole of the body functions as a unit under normal circumstances. Division into functions is sensible only in order to make it possible to discuss the whole.

It is known that some clearly distinguishable functions of the cortex occur in specific parts of the cortex, whereas other functions cannot easily be located.

A more or less limited part of the cortex with a specific function is called a *cortical area* (Fig. 2.7.32).

It is notable that the area anterior to the central sulcus has mainly a motor function, whereas the cortical area, posterior to the central sulcus, has mainly a sensory function.

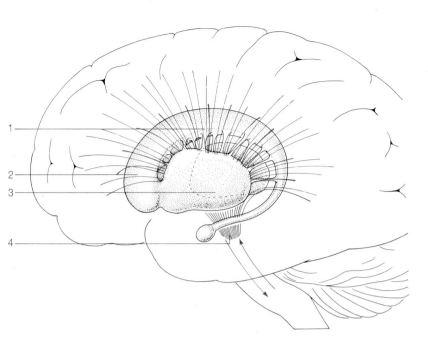

Figure 2.7.31
Cerebral nuclei and
internal capsule

1. *fibres of the internal capsule*
2. *head of the caudate nucleus*
3. *lentiform nucleus*
4. *cerebral peduncles (stalk)*

Motor function

The primary motor cortex is the cortical area of the precentral gyrus. In both hemispheres this convolution extends anteriorly to the central gyrus from the lateral sulcus almost to the corpus callosum. The pyramid-shaped cell bodies of the neurones which innervate the skeletal muscles are located in this area. They provide the animal motor function. The axons of these neurones leave the hemispheres via the internal capsules and the cerebral stalks. They form the *pyramidal tracts* mentioned before, which for the greater part cross the median line in the medulla oblongata. The right hemisphere supplies the muscles of the left half of the body.

The cells in a specific area on the motor cortex have a clearly defined relationship with specific muscle groups. When stimulated electrically by means of a micro-electrode, the same area always makes the same muscle or the same part of a muscle contract.

Each skeletal muscle has its own representative in the motor cortex. This is called the motor somatotropic representation.

When the representation is carefully recorded, via such stimulating experiments, it becomes evident that:
- there is a topographical pattern in the representation. The skeletal muscles in the feet have their neurones in the cortex near the corpus callosum and the muscles of legs, trunk, arms and head follow distally and laterally (Fig. 2.7.33a).
- the number of neurones available to a certain muscle is not directly proportionate to the size of the muscle, but to the precision with which the muscle can move (see Fig. 2.7.16). The more extensive the 'wiring' of a muscle is, the better this muscle can be controlled. Muscles with poor control have small motor units and consequently a relatively large number of descending fibres pertaining to them. For instance, there are many cell bodies in the precentral gyrus which pertain to them.

When the areas containing the neurones for the skeletal muscles in specific body parts are reproduced as if they were the body parts themselves, in other words if the neuronal area for the thumb were reproduced as a thumb, the

Figure 2.7.32
The cortical areas

1. *secondary motor cortex (premotor cortex)*
2. *primary motor cortex*
3. *sensory cortex*
4. *association area*
5. *secondary optic cortex*
6. *motor speech area (Broca's area)*
7. *auditory area*
8. *primary optic area*

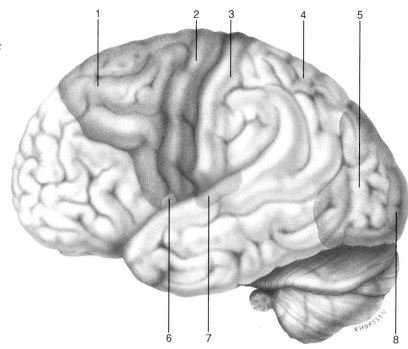

motor homunculus comes into being (Fig. 2.7.33a). Its largest part is taken up by the hand and the fingers, its smallest part by the trunk: of course, the hand muscles make more precise movements than the superficial muscles of the trunk.

The *premotor cortex* or secondary motor cortex is the cortical area of the frontal lobe, situated immediately anterior to the precentral gyrus (Fig. 2.7.32). In addition to relatively many cell bodies of the extrapyramidal tracts, neurones which are involved in the coordination of complicated movements are located in this area. Motion programmes are stored in the premotor cortex. These motor programmes can be executed when they are sent from the premotor cortex to the motor cortex.

A well-known part of the premotor cortex is the speech area (Broca's region) (Fig. 2.7.32). It is situated anterior to the central gyrus just above the lateral sulcus, near the homuncular part that pertains to it. The motor speech centre contains the programmes needed to control the muscles involved in speech.

Strangely enough the speech centre is located in one hemisphere only, notably in the dominant one, i.e. the left hemisphere in right-handed people.

A centre with the programmes for coordinated eye and head movements is located above the speech centre.

Sensory functions

The area posterior to the central sulcus has a primary sensory function. In relation to the division of the motor cortex there are *primary receptive areas* (the *somatosensory cortex*, the *auditory cortex* and the *visual cortex*), and also *secondary receptive areas*.

The *somatosensory cortex* comprises the cortical area of the postcentral gyrus, parallel to the primary motor cortex (Fig. 2.7.32). The neurones at which the ascending tracts of the sensory function terminate are located in this area. Somatic sensations, such as the senses of pressure, touch, pain, temperature and motion, are received here. The neurones involved in the sensation of taste are located in the cortical area

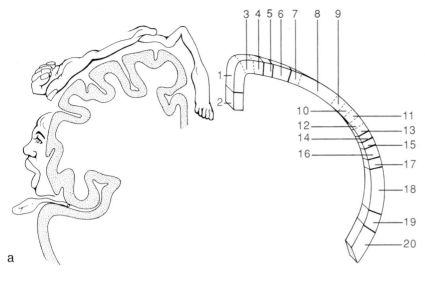

3 4 5 6 7 8 9

1
2
10 — 11
12 — 13
14 — 15
16
— 17
— 18
— 19
— 20

a

Figure 2.7.33
Topography

a. The motor
   representation

   1. ankle
   2. foot
   3. knee
   4. hip
   5. trunk
   6. shoulder/elbow
   7. wrist
   8. hand
   9. little finger
   10. third finger
   11. middle finger
   12. forefinger
   13. thumb
   14. neck
   15. eyebrow
   16. eye/eyelid
   17. face
   18. lips
   19. jaw
   20. tongue

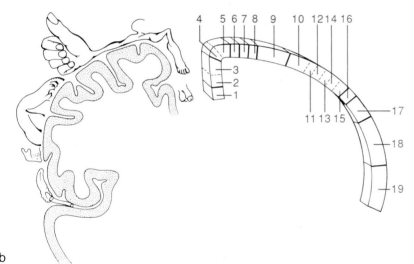

4 5 6 7 8 9   10 12 14 16

3
2
1
11 13 15
— 17
— 18
— 19

b

b. The sensory
   representation
   in the cortex

   1. sex organ
   2. toes
   3. foot
   4. lower leg
   5. upper leg
   6. trunk
   7. neck
   8. head
   9. arm
   10. palm of the hand
   11. little finger
   12. third finger
   13. middle finger
   14. forefinger
   15. thumb
   16. eyes
   17. face
   18. lips
   19. tongue/throat

above the insula (Fig. 2.7.30). A somatotopic representation can also be made of the sensory cortex.

Electric stimulation of a specific point on the postcentral gyrus always leads to a sensation in exactly the same area of the skin. The body areas have representatives of their own in the sensory cortex. The *sensory homunculus* is more or less similar to the motor homunculus, when the *sensory somatotropic representation* is mapped (Fig. 2.7.33b).

Body areas which are highly sensitive, such as the finger tips, the lips and the tongue, are extensively 'wired' and consequently have many neurones on the postcentral gyrus. Less sensitive body areas, such as the arms, legs and trunk, have relatively few neurones.

The *auditory cortex* is the cortical area of the gyrus immediately beneath the lateral sulcus, at the level of the sensory cortex (Fig. 2.7.32). The fibres which conduct the auditory stimuli from the auditory organ terminate here.

The *visual cortex* is the cortical area of the posterior part of the occipital lobe (Fig. 2.7.32). The fibres which carry the stimuli from the retina of the eyes terminate here. The stimuli from the left halves of both retinas arrive at the left hemisphere, the right retinal stimuli arrive at the right hemisphere (see Fig. 2.8.33). The central parts of both retinas, the yellow spots, are mainly connected with both hemispheres which results in *double-sided projection*.

Sensory stimuli are recognized in the primary sensory cortex. The meaning of the recognition, however, is realized by means of connections with the posteriorly located *secondary sensory cortex*, which is the location of the sensory memory. This enables the individual to relate the immediate sensation with past experience of that sensation.

In the primary auditory cortex, sensations of pitch of tone and volume of sounds are registered. In the primary visual cortex sensations of brightness and colours are received. In order to know the meaning of these auditory and visual stimuli, the related and near-located secondary auditory and visual cortical areas respectively must be addressed (Fig. 2.7.32). The secondary sensory cortical areas supply the *interpretation* of the received stimuli.

Association

A person watching a fire hears the crackling of the fire, sees the orange glow of the flames, smells the burning of the wood, feels the heat. All of these sensations are conducted to the cerebrum via separate 'wires'. In order to be able to understand that the separate sensations are interconnected, the cortical areas, which are situated between the areas so far discussed, are needed. These are the *association sensory centres*. It is self-evident that these areas are interconnected with all other sensory cortical areas.

There are interconnections between sensory and motor cortical areas as well. These enable, for instance, a person to read aloud. The text read (sensory) is interconnected with the speech areas (motor).

Interconnections between cortical areas inside a hemisphere are called *association tracts*, interconnections between hemispheres are called *commissures*.

Memory

The presence in the brain of experiences and knowledge is called *memory*. The memory is the human archive. This archive, however, is not located at a specific site in the brain, but spread over all of the cortex. Little is known of the way in which experiences are and have been stored. In some way or other, the cerebral tissue is capable of changing when information is offered. The changes can be either brief or lengthy, even life-long.

There are three distinguishable stages in the memory:

– the *ultra short-term memory*, the stage in which the senses and the afferent tracts contain the experience in the form of action potentials. This is a very short-lived *electric memory storage*. Based on the action potentials a picture appears, and disappears within a few seconds. Most of the information that reaches an individual does not go any further than the ultra short-term memory whose filtering function is very important. If this sieve were not there, the brain would be flooded. Only matters considered important should filter through.

– The *short-term memory (STM)*, the next stage of storing. This stage may last for about thirty minutes. Storage is thought to take place in the form of RNA-activity, released by the action potentials. It could be said that due to the transition of (electric) impulses to (chemical) molecules the experiences can whirl around in the brain for a while.

The transition to the short-term memory is the first filter. The short-term memory enables people to interrelate everyday activities, without the need to have them permanently stored;

– the *long-term memory (LTM)*, the last stage of storage. The transition from STM to LTM is a second filter. Changes, probably life-long, are made in the brain tissue. There are indications that protein synthesis takes place, via RNA activity. The proteins formed could be called memory-molecules. It is not known how such memory molecules are organized. Sometimes the term *engram* or *memory trace* is used. This, however, does not make the images of possible memory units any clearer.

The memory can be divided into a motor memory and a sensory memory. The motor memory contains the motor programmes for already known movements. These programmes can be executed by conducting them to the primary motor cortex.

The experienced sensations are stored in the sensory memory. We can recall the smell of cinnamon, we can imagine a specific movement, we can recognize the sound of breaking glass, etc.

Remembering - learning - training

'His name is on the tip of my tongue', 'Did I switch off the gas?', 'Where did I leave the car keys?' In terms of one's memory, it is well-known that storing is one thing and remembering another. Remembering can be described as the capability to select from the memory.

It is largely unknown how the reproduction of experiences is brought about. If storage takes place in the shape of memory molecules, the molecules should be recognized by the active brain, in some way or other.

Learning is the capability to feed and to fill the memory. In learning it is essential to make the transition from the short-term memory to the long-term memory. Little can be said about the functional-anatomical basis of learning. It is known, from psychology, that feeding the memory is more successful when a person is concentrating, when the information to be learned is meaningful, or when the new subject is inter-connected with something known or familiar. Learning, remembering and reproducing is easier if complex information is offered in

separate steps, and afterwards is put together to form a whole.

Time also plays a role in learning and mastering: the learning process is easier when the the start is slow and the pace gradually increases. This is evident in movements such as typing or playing the piano, or in reciting poems, etc.

The tyres of an unused bicycle soon become flat. Unused muscles become thin and weak. Unburdened bones lose calcium phosphates. Therefore, it may be better to wear something out than let it become rusty.

Nerve tissue reacts similarly. Learning and reproducing capabilities are dependent on *training*, exercising again and again that which is already known, in order to reach and maintain a certain level.

Probably the memory itself does not wear out, but the capability to efficiently select from the memory does, perhaps, with the passage of time.

Thinking and intelligence

The perceptions of *thinking* and *intelligence* are pre-eminently linked with the function of the human brain. It is hard to say what their meaning is, from a functional-anatomical point of view.

Thinking can be understood as the combining and recombining of different memory volumes. The result of thinking may be that a new memory volume is made, or that an old volume is changed.

Intelligence can be described as the whole of the cerebral capabilities of man. Consequently, in-telligence has a number of aspects. There is physical or motor intelligence, creative intelli-gence, logical intelligence, social intelligence, etc.

Thinking and intelligence are in no way related to the size of the brain. There are indications, however, that the number of connections, and therefore the number of synapses, may play a role.

Personality

Each individual is unique. Each individual has a character of his or her own. What is it? Where is it situated?

If one accepts the premise that personality is determined by the brain, then there are clear indications that the frontal lobe has a key role in personality formulation. Higher mental processes, such as the solution of problems, the formation of judgments and the creation of ideas, are attributed to the frontal lobe. Alternative views suggest that the personality is also determined by emotion, sex, appearance, age, socio-economic status, etc. This is the province of psychology.

Two different hemispheres

Many human organs appear in pairs or mirror images. Two arms, two legs, two lungs, two eyes, two ears, etc. The spinal cord has a mirrored symmetrical structure, and the cerebrum also appears symmetrical, with a left and a mirrored right hemisphere. Although in terms of structure the two hemispheres are similar, there are considerable differences in their function. One of the hemispheres is in fact *dominant*.

The dominant hemisphere may be typified as cognitive, analytic, determining, arithmetical, problem solving. The dominant hemisphere is sometimes designated the *logical brain*. This part 'thinks' in concepts.

The non-dominant hemisphere can be typified as the *artistic brain*. This hemisphere specializes in three-dimensional orientation, in the whole of images, in understanding and appreciation of music or other works of art. This part 'thinks' in images and emotion.

The speech centre of almost all right-handed people is located in the left hemisphere, the speech centre of less than 50% of the left-handed people is located in the right hemisphere. In about 15% of the left-handed the speech centre develops in both hemispheres. Left-handedness is certainly not always accompanied by a dominant right hemisphere; it is rather an indication that the specialization of each of the hemispheres is less obvious.

The left and right hemispheres seem to be identical, but they are opposite sides of the same coin. Together they provide the enormous potential with which man is gifted.

Emotionalism

The brain stem is said to control and coordinate the vital body functions; the cerebral cortex is especially responsible for intellectual functions. A coherent whole of cerebral elements involved in controlling the emotions of man is situated between the brain stem and the cortex. This is called the *limbic system* (Fig. 2.7.34). It is a complex structure which consists of elements of the cerebrum and the diencephalon. The limbic system extends like a dome over the end of the brain stem. The areas of the limbic system have intricate interconnections with each other and with both lower and higher parts of the brain.

The limbic system provides harmony, but it is also the seat of aggression, unrest, anger, fear, sexual behaviour, excitement, etc. All these functions together form emotionalism. Sensations of reward and punishment originate in the limbic system as well. The emotional feelings that an individual experiences and perhaps expresses, are the result of pluses and minuses in the limbic system.

The various emotional feelings are accompanied by various physical reactions or manifestations. There may be changes in the heart beat, blood pressure, respiration, size of the pupil, production of sweat, volume of blood reaching the skin, etc., which give rise to the expression 'the body is the mirror of the soul'.

Electroencephalogram

Cerebral activity can be examined by recording the changes in the average electrical activities of the millions of nerve cells. This is done by means of the EEG, the electroencephalogram. Electrodes are placed at standard places on the skull (Fig. 2.7.35). These electrodes are connected to an amplifier and the electrical activity is recorded as a trace on a piece of paper. When the test subject is resting physically and mentally, with the eyes closed, a rhythmic pattern of waves with a frequency of 8–13 Hz (1 Hz = one vibration/sec) is recorded. This is called the *alpha-* or *rest rhythm*.
The rhythm is interrupted when the test subject

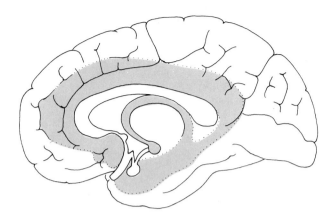

Figure 2.7.34
The structures of the limbic
system (in red)

opens his eyes: the visual cortex immediately becomes active and smaller amplitudes with a higher frequency (15–30 Hz) are recorded. This is the *beta-rhythm*.

The *theta-rhythm* (4–7 Hz) is found when a person falls asleep and the *delta-rhythm* (0.5–3 Hz) occurs during deep restful sleep.

Sleeping and dreaming

The reticular formation is responsible for the rest – activity cycle of man. Sleep shows various stages (Fig. 2.7.36).

There are two kinds of sleep: deep restful or synchronized sleep, and dreaming or desynchronized sleep.

In accordance with the differences in the EEG, *deep restful sleep* is divided into four stages. The stage of falling asleep precedes sleeping stage I (light sleep). The beta-rhythm is interrupted more and more by a theta-rhythm.

The person falling asleep may have all kinds of visual sensations such as scenery, abstract patterns, alternately light and dark. He may have the sensation of falling or getting a shock, he may have sudden muscular contractions that sometimes wake him up.

During light sleep an alpha-rhythm is seen which soon changes into a theta-rhythm, stages II and III. These are the stages of dreamless sleep in which the muscles are relaxed. Thereafter the sleeping person enters stage IV, or deep restful sleep, with the delta-rhythm. The

sleeping person cannot easily be woken up and is absentminded when woken. Somnambulism (sleepwalking), talking during sleep, and bedwetting occur during deep restful sleep. After having slept in stages III and IV for a while, the sleeping person finds himself, via stage II, in another kind of sleep, dreaming sleep.

Dream sleeping is characterized by loss of muscular tension, irregular heart rate and respiration, lively visual sensations, and involuntary muscular jerks and contractions in the extremities and the face (grimaces). Men can have an erection during dream sleeping. The EEG of dream sleeping shows waves of high frequency and low amplitude, similar to those during wakefulness. Moreover, the eyes make rapid movements underneath the closed eyelids. This stage is called the *REM-sleep* (rapid eye movements).

There are approximately five REM-sleeping periods per night. The period of dream sleeping gets longer and that of the deep restful sleep becomes shorter. There is often a short period of dream sleeping just before waking up. In many cases one wakes up because of an event or an image in the last dream.

It is found that the brain stays active during sleep. The exact functions of sleeping and dreaming are not clear. Sleeping and dreaming are, however, indispensable for man. It is

Figure 2.7.35
The electroencephalogram
(EEG)

a. standard position of the
   electrodes
b. EEG-amplifier and
   recorder
c. EEG with the eyes closed
   and with the eyes open

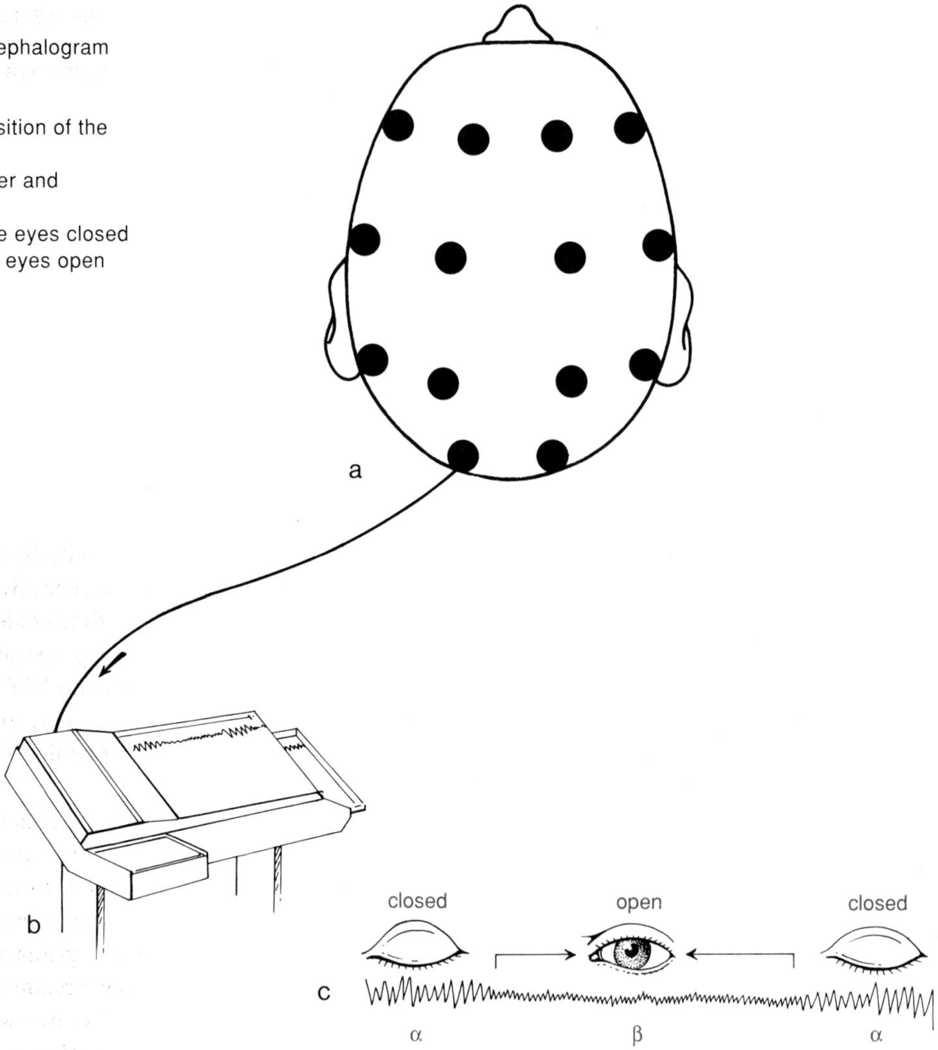

a

b

c

closed   open   closed

α   β   α

plausible that they are a kind of recovery for the organism. It is, in any case, a blessing that no pain is felt during sleep.

9. Vegetative system

The vegetative (autonomic, involuntary) nervous system regulates vegetative integration. This means that, generally independently of the will and the consciousness, the functions of the five vegetative systems (circulatory, alimentary, urinary, respiratory systems and skin) are regulated and made compatible. Vegetative integration is aimed at maintaining the homeostasis of the m.i. In addition, it regulates

certain functions which are involved in procreation, especially coitus, and it influences reticular formation and consequently the level of functioning of the brain.

Physiologically, the vegetative system can be divided into two opposing parts, the *sympathetic system* and the *parasympathetic system*.

The *sympathetic system* is active when the organism is (externally) active. The activity of the sympathetic system takes the body into a state of alertness and affects a large number of organs. Cardiac activity and respiration are stimulated, blood pressure rises, blood sugar level rises, adrenaline secretion of the adrenal medulla is

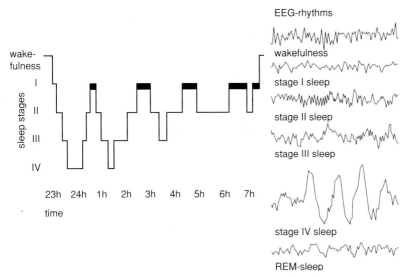

EEG-rhythms

wakefulness

stage I sleep

stage II sleep

stage III sleep

stage IV sleep

REM-sleep

Figure 2.7.36
The stages of sleep in relation to the EEG rhythms.
The black bars on the transition between the stages I and II indicate the REM-periods

activated. There is vasoconstriction of the alimentary tract, vasodilation of the skeletal muscles. The blood vessels of the skin may constrict or dilate. The pupils become wider, the production of sweat increases.

Excitement, anger and fear are accompanied by increased sympathetic activity.

The *parasympathetic system* is in direct opposition to the sympathetic system. It is active when the organism is (externally) passive. Its effects on the organs mentioned earlier are opposite to those of the sympathetic system.

One could say that the parasympathetic system is directed at making the organism recover and at replenishing energy supplies.

The two systems, the *sympathetic* and the *parasympathetic,* are complementary in their effects. Increased activity of the one is accompanied by decreased activity of the other. It should be noticed that the vegetative system comprises both efferent and afferent fibres. After all, the vegetative centres must be informed on going about their vegetative processes, and there are a large number of vegetative reflexes in which afferent fibres are of vital importance.

From the anatomical point of view, the vegetative nervous system consists of a central area (located inside a bony covering) and a peripheral area (Fig. 2.7.37).

The central area of the sympathetic system consists of nuclei in the hypothalamus and the brain stem, and of the cell bodies in the lateral horns of the spinal cord (C_7 up to L_2 inclusive), with which they are connected.
From the source cells in the spinal cord, the sympathetic fibres leave the vertebral canal together with the anterior roots. Outside the vertebral column the sympathetic fibres separate from the spinal nerve and travel in a small bundle to a ganglion next to the vertebral body, where they synapse. These ganglia are situated on either side of the vertebral column, from the neck to the sacral bone. They are interconnected by sympathetic fibres. On both sides of the vertebral column is a sort of beaded string, the sympathetic trunk. The *sympathetic trunk* extends both distally and caudally further than the sites where it is connected with the spinal cord. Above C_7 on either side there are two cervical ganglia which supply the neck, the head and the arms; below L_2 on either side there are four ganglia which innervate the pelvic organs and the legs. The cell bodies of sympathetic neurones, from which the fibres extend to all kinds

Figure 2.7.37
The vegetative nervous
system.
Red: sympathetic
Blue: parasympathetic

1. *spinal nerve*
2. *vagus nerves*
3. *lateral horn of the spinal
   cord*
4. *sympathetic trunk*
5. *autonomic ganglia in the
   abdomen*

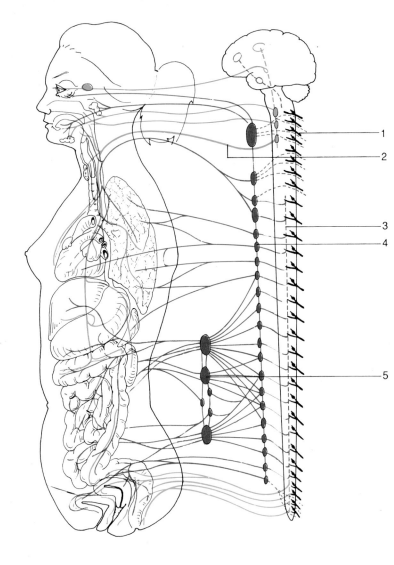

of organs in the body, are located in the ganglia. A number of sympathetic nerves (such as those which extend to the blood vessels of the skin and muscles) join the spinal nerves in their travels. Others, such as those that innervate the heart, lungs and digestive organs, take their own route. Apart from the ganglia in the sympathetic trunk, there are a number of ganglia very near to the organs to be innervated. Special attention should be paid to the innervation of the adrenal medulla. The adrenal medulla originated from developing nerve tissue. The glandular cells of the adrenal medulla can be seen as modified nerve cells, specialized in the production of the

adrenaline and noradrenaline hormones. The sympathetic fibres that innervate the adrenal medulla come straight from the spinal cord and terminate on these hormone-producing nerve cells. The sympathicus, accordingly, stimulates the production of hormones in the adrenal medulla. The hormones produced have, via the bloodstream, basically the same effect as the sympathetic system via nerve fibres.

The central part of the parasympathetic system consists of parasympathetic centres in the brain stem, especially in the medulla oblongata, and of parasympathetic origin cells in the lateral horns of the sacral spinal cord. The most important parasympathic nerve, the vagus nerve, emerges from the medulla oblongata. This cranial nerve is justly called a 'vagabond nerve': it supplies branches to the head, travels caudally through the neck, and innervates in a complicated course through the thoracic and abdominal areas in virtually all organs.

From the source cells in the lateral horns of the sacral spinal segments, parasympathetic nerves travel straight to the sex organs, the bladder and the lower parts of the large intestine.

In many cases, parasympathetic ganglia are located near, and parasympathetic plexus networks are situated on, the target organs, as in the sympathetic system.

10. Membranes

The central nervous system is enveloped by three meninges, the *pia mater*, or innermost meninx, the *arachnoid mater*, or middle layer, and the *dura mater*, or outer meninx. They are often referred to as the cerebral meninges, but as they also cover the spinal cord, this name is not accurate. The names are usually shortened to pia, arachnoid and dura (Figs. 2.7.38 and 2.7.39).

Pia

The pia mater consists of a very thin layer of connective tissue with many capillaries. The pia adheres closely to the brain and the spinal cord following the contours of these organs.

The pia mater supplies the nerve tissue with nutrients and oxygen.

Dura

The dura mater is a tough, thick and strong membrane consisting of connective tissue. It covers the inner surface of the skull and is scarcely distinguishable from the periosteum of the skull.

The dura mater has a fold between the two hemispheres, the *falx cerebri*, which forms a falciform medio-sagittal separation, extending to just above the corpus callosum and posteriorly to above the cerebellum (Fig. 2.7.40). Posteriorly, it connects with the *tentorium*

cerebelli (literally the tent of the cerebellum). This is a fold of the dura mater situated between the occipital lobe of the cerebrum and the upper surface of the cerebellum.

The tentorium cerebelli extends left and right, horizontally and anteriorly to the upper side of the petrous part of the temporal bone. There is a wide central opening where the brain stem is located.

The dural folds separate the cranial cavity into a number of compartments, as a result of which

Figure 2.7.38
Frontal section through the skull and the upper part of the cerebrum; the meninges

1. subdural space (for blood)
2. granulation of the arachnoid
3. skin
4. skull
5. dura mater, outer and inner layer
6. arachnoid
7. subarachnoidal space
8. pia mater
9. cortex (grey substance)
10. white substance of the brain
11. falciform process of brain

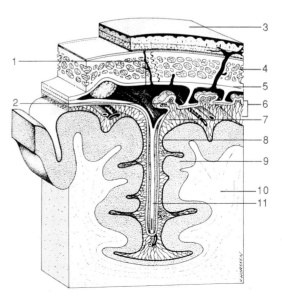

Figure 2.7.39
Dorsal view of a part of the
spinal cord;
the meninges enveloping
the spinal cord

1. *ligamentum flavum
   (yellow ligament)*
2. *peridural capillary
   network, mainly venous*
3. *dura mater*
4. *pia mater*
5. *cut edge of pia mater*
6. *spinal ganglion*
7. *posterior column*

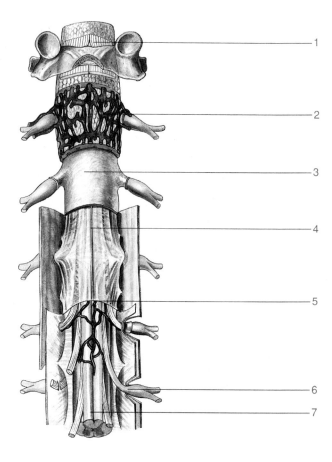

the mobility of the different parts of the brain is limited. Therefore, when the head is bumped or shocked, the brain tissue will not be easily damaged. The dura mater has a protective function. Between the two layers of the dura and at the site where the falx is attached to the tentorium, there are spaces which serve as blood vessels. These channels drain blood from the skull. They are called *venous sinuses of dura mater* or *cerebral sinuses* (Figs. 2.7.40 and 2.7.41).

Near the foramen magnum the dura mater is still continuous with the periosteum, but in the vertebral canal there is a space between the dura mater and the periosteum, the *epidural cavity* (Fig. 2.7.42).
The epidural cavity contains loose connective tissue, epidural fat, blood vessels and lymphatic vessels. At the points where spinal nerves leave the vertebral canal, the dura changes into the *epineurium*, the outer sheath of connective tissue covering a peripheral nerve. Caudally, the dura mater extends to approximately S_2 - S_3 level, where it ends in a closed cone. The dura mater continues further into the vertebral column than the spinal cord, but not to its very end (see Fig. 2.7.14). The area between the caudal end of the spinal cord and the end of the dura mater is called the *dural sac*. The dural sac envelopes the cauda equina.

Arachnoid
The arachnoid is closely related to the dura mater. It has numerous thin connective tissue connections, trabeculae, with the pia mater. Due to this structure the brain and the spinal cord are, as it were, suspended. The space between the connective tissue connections is called the *subarachnoid space*. This name correctly implies that only the part apposing

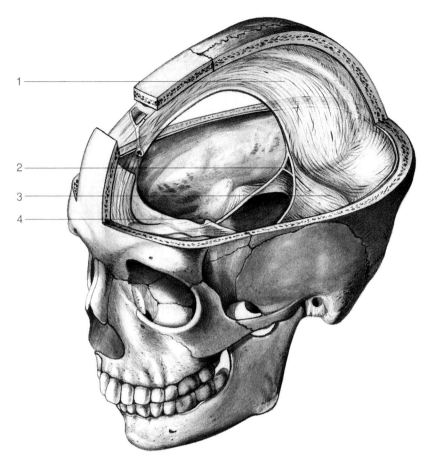

Figure 2.7.40
The dura mater in the open skull

1. *falx cerebri*
2. *cerebral sinus*
3. *tentorium cerebelli*
4. *opening for the brain stem*

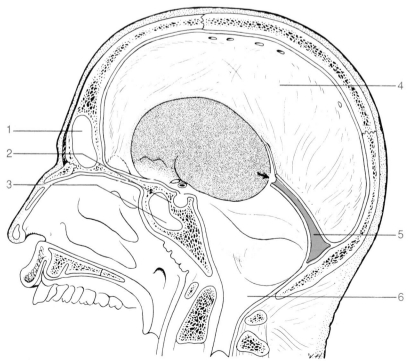

Figure 2.7.41
Near-sagittal section through the skull

1. *frontal sinus*
2. *sella turcica*
3. *sphenoidal sinus*
4. *falx cerebri*
5. *cerebral sinus, sinus of the dura mater*
6. *foramen magnum*

Figure 2.7.42
Meninges enveloping the
spinal cord

1. dura
2. pia
3. arachnoid
4. subarachnoidal space
5. spinal cord
6. epidural cavity
7. periosteum
8. spinal ganglion
9. vertebral body
10. epineurium
11. spinal nerve
12. perineurium

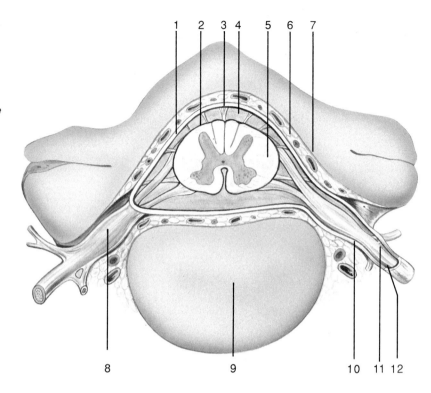

the dura mater, belongs to the arachnoid (Figs. 2.7.38 and 2.7.42).

There are a number of vessels in the subarachnoidal space. The *cerebral spinal fluid* circulates in it. The presence of fluid further protects the central nervous system against bumps and shocks.

The subarachnoidal space is relatively narrow in most places. But it is wide when there is a large distance between dura mater and nerve tissue, for instance, immediately below the cerebellum (cisterna magna) and in the dural sac (Fig. 2.7.43a).

The arachnoid gradually changes into the *perineurium* at the points where the spinal nerves leave the vertebral canal. At the point of exit the spinal nerves are somewhat protected against mechanical influences as they lie in a sheath filled with fluid (Fig. 2.7.42).

11. Cerebrospinal fluid

The central nervous system is totally surrounded by a circulating fluid, the *cerebrospinal fluid*. The *cerebral ventricular system*, a number of intercommunicating cavities in the brain, also contains fluid.

The inner fluid cavity (ventricular system) and the outer fluid cavity (subarachnoid space) are interlinked distally to the cerebellum (Fig. 2.7.43a).

Cerebral ventricular system
Developmentally, the central nervous system is a hollow organ. Its wall develops into the solid part of the central nervous system (see Fig. 3.2.17).

The internal fluid cavities are rudimentary parts of this development.

There are four ventricles (Fig. 2.7.44):
– The *first* and the *second ventricles* are located in each of the hemispheres. They are also called *lateral ventricles*. The lateral ventricles develop together with the hemispheres. They are irregularly shaped cavities with posterior horns. The anterior and superior parts are located in the frontal and the parietal lobes, respectively. The posterior horns are situated in the occipital

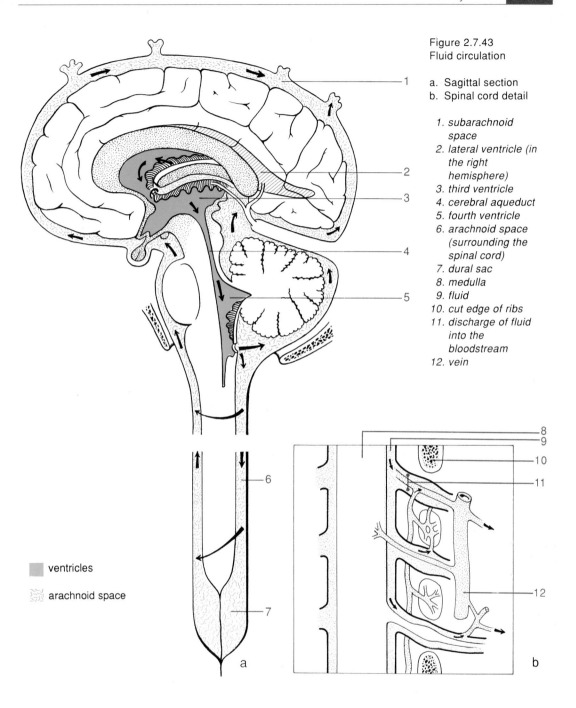

Figure 2.7.43
Fluid circulation

a. Sagittal section
b. Spinal cord detail

1. subarachnoid
   space
2. lateral ventricle (in
   the right
   hemisphere)
3. third ventricle
4. cerebral aqueduct
5. fourth ventricle
6. arachnoid space
   (surrounding the
   spinal cord)
7. dural sac
8. medulla
9. fluid
10. cut edge of ribs
11. discharge of fluid
    into the
    bloodstream
12. vein

ventricles

arachnoid space

lobes and the inferior parts in the temporal lobes.
- The *third ventricle* is a narrow, vertical cleft in the diencephalon. The first and second ventricles are each connected with the third ventricle via a small duct, the *interventricular foramen*.

- The *fourth ventricle* is a rhomboid-shaped space in the brain stem, inferior to the cerebrum. It is connected with the third ventricle via a narrow duct, the *cerebral aqueduct*. There are three openings in the lower part of the fourth ventricle median aperture (Magendie's) and two lateral apertures

Figure 2.7.44
The cerebral ventricles

a. Left dorsal view. The
   brain is transparent
b. Near-lateral view
c. View from above,
   projected into the brain

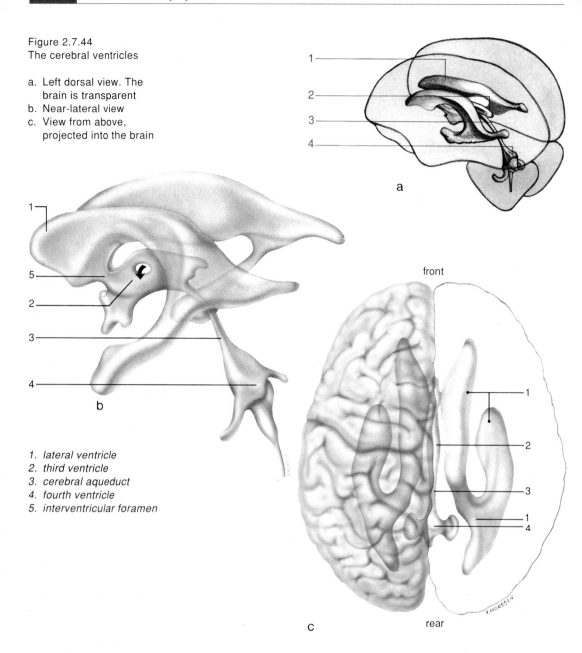

1. *lateral ventricle*
2. *third ventricle*
3. *cerebral aqueduct*
4. *fourth ventricle*
5. *interventricular foramen*

(Luschka's), which form the connections between the inner and outer fluid spaces.

Fluid production

Fluid is produced by the *choroid plexus*. This plexus is a highly vascularized membrane, which is located on the roof of the third ventricle, on parts of the floors of the lateral ventricles and on the posterior wall of the fourth ventricle (Fig. 2.7.43a). The endothelium of the capillaries is continuous, there are no intercellular pores. It forms the *blood/brain barrier*. The composition of the fluid is slightly different from that of blood plasma.

The fluid flows out of the lateral ventricles, via the interventricular foramen (foramen of Monro), into the third ventricle and thereafter into the fourth ventricle via the cerebral aqueduct.

Through the median and lateral apertures the fluid reaches the arachnoid space. The fluid circulates in this space until it is reabsorbed into the venous bloodstream. The site of reabsorption is made up of bulbous protrusions of the arachnoid extending into the cerebral sinus, the *granulations* (see Fig. 2.7.38). A minor part of the fluid is absorbed into the venous capillaries at the sites where the spinal nerves leave the spinal cord (Fig. 2.7.43b).

Shock absorption and nutrition
The main function of the fluid is to protect the central nervous system. The central nervous system is suspended in the fluid. On account of this, shocks and bumps are extremely well absorbed.
In addition, the fluid supplies the surrounding nerve tissue with nutrients, especially glucose.

12. Vascularization

The oxygen consumption of the brain, approximately 60 ml per minute, is high and generally constant. The oxygen supply must be meticulous, as the brain itself has no oxygen reserves and operates wholly aerobically. When the circulation is interrupted, unconsciousness occurs after five seconds. In normal conditions the brain is supplied with a fairly constant volume of blood, approx. 750 ml/min. The brain is, in fact, in terms of blood supply relatively unfavourably located. An individual is in an upright position for about 60% of his life, so during this time blood must be transported to the brain against the force of gravity. The average arterial blood pressure in the brain is only 75 mm Hg (10 kPa), in lying position it is 95 mm Hg (12.6 kPa). This apparently hampered supply is, however, compensated by a venous return transport which is improved due to the force of gravity.
A sufficient difference in arteriovenous pressure, needed for circulation, is in this way guaranteed.

The brain is supplied with blood by two routes: via the left and right internal carotid arteries, and via the basilar artery (Fig. 2.7.45).
The *internal carotid arteries* are branches of the common carotid artery, which enter the skull at an opening in its base. These arteries also supply the eyes with blood.
The *basilar artery* is formed from the junction of the two vertebral arteries. These branches of the subclavian artery ascend along the vertebral column, through transverse openings in the vertebrae, and enter the skull via the foramen magnum (see Fig. 2.1.34).
The internal carotid artery branches into the anterior cerebral artery and the *middle cerebral artery*. The basilar artery branches into two arteries, each of which arches posteriorly: the *posterior cerebral arteries*. Figure 2.7.46 shows the areas these three arteries supply.
The cerebellum is supplied by branches of the basilar artery.
Between the posterior cerebral arteries an anastomosis is situated, the anterior communicating artery. The posterior communicating arteries are anastomoses located at the left and the right between the middle and the posterior cerebral arteries (Fig. 2.7.45). In this way an arterial ring is formed around the sella turcica, the *cerebral arterial circle (circle of Willis)* or circulus arteriosus.
The arterial ring makes a collateral circulation possible. The anastomoses, however, are so thin that if one of the supplying arteries gradually becomes clogged, the blood supply to the brain can be maintained via a detour.

The cerebral arteries branch into small arteries and arterioles on the surface of the cortex (Fig. 2.7.47). They travel in the pia mater and follow the contours. Capillary networks enter the deep brain from the cortical surface. The venules and veins then drain the brain, discharge into the cerebral sinuses. At the level of the base of the skull the blood from the cerebral sinuses comes into the right and left *internal jugular veins*, from which it is carried to the superior vena cava (see Fig. 2.1.32).

There is a remarkable difference between the blood supply to the brain and that to most other organs. Most organs have a hilum, a gap where vessels enter or leave. The brain does not have a hilum. In most organs the ramifications of arteries and veins are situated between them. In the

Figure 2.7.45
Blood supply to the brain

a. View from underneath
b. The supplying arteries

1. *anterior cerebral artery*
2. *anterior communicating artery*
3. *internal carotid artery*
4. *middle cerebral artery*
5. *posterior communicating artery*
6. *posterior cerebral artery*
7. *basilar artery*
8. *vertebral artery*

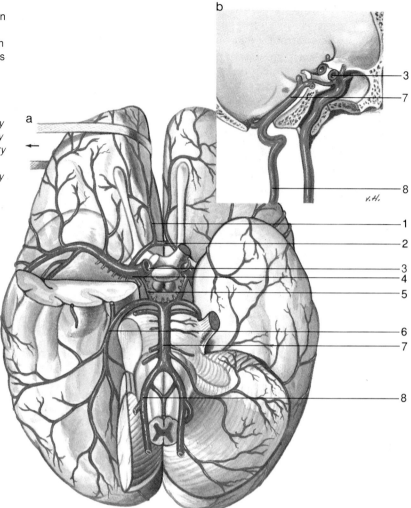

Figure 2.7.46
Supplying areas of the cerebral arteries

a. Lateral view of right hemisphere
b. Sagittal view of right hemisphere

1. *anterior vertebral artery*
2. *middle cerebral artery*
3. *posterior cerebral artery*
4. *corpus callosum*

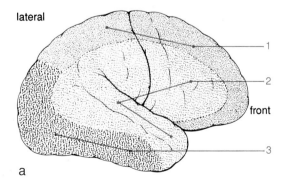

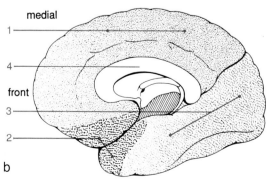

brain the blood supply comes from the surface and the venous drainage takes place via the cerebral sinuses.

One advantage is that the function of the neurones is not disturbed by pulsating blood vessels.

The structure of the capillaries is also different. The endothelium is continuous, it has no intercellular pores with the result that the exchange of substances between blood and m.i.takes place in a different way. This is the *blood/brain barrier*, a selectively operating barrier: fat-soluble substances, for instance, pass freely, whereas carbohydrates and proteins are dependent upon active transport for their passage. The blood/brain barrier can prevent certain harmful substances from reaching the brain tissue.

The meninges and the interior of the cranial bones are supplied with blood by a branch of the external carotid artery. It enters the skull through an opening in the base of the skull and, thereafter, branches between periosteum and dura mater. These branches are called the *meningeal vessels* (Fig. 2.7.48). The 'beds' of these arteries are readily visible on the inside of a prepared skull.

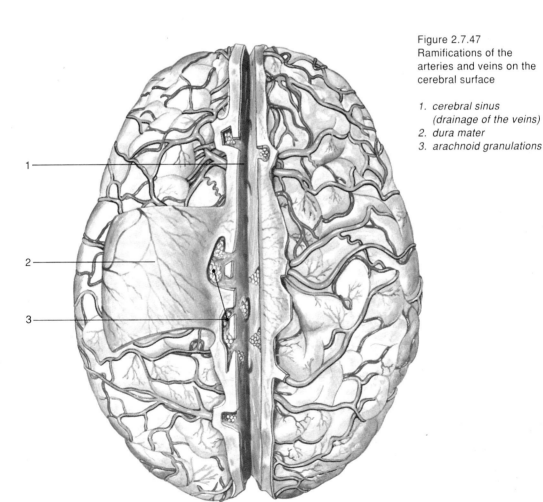

Figure 2.7.47
Ramifications of the arteries and veins on the cerebral surface

1. *cerebral sinus (drainage of the veins)*
2. *dura mater*
3. *arachnoid granulations*

Figure 2.7.48
Meningeal vessels

1. branches of the
   meningeal artery
   between dura and roof of
   the skull
2. external carotid artery
3. internal carotid artery

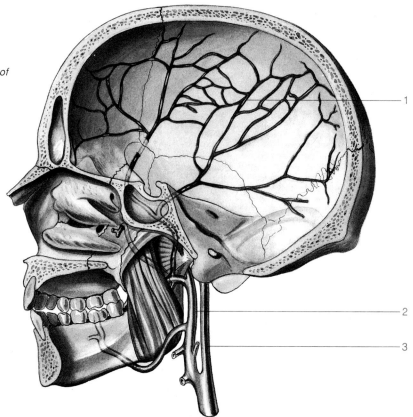

Sensory system

8

Introduction

Writing with one's eyes closed turns beautiful handwriting into a mess. Cycling with one's eyes closed is perilous and should not be attempted. Smelling a meal, hearing a piece of music, touching the skin of a loved one, these are all examples of the manner in which man obtains information about his environment, be it important information or not. Toothache, stomach ache, sore muscles, pain in the joints, hunger and thirst are examples of information about the body and coming from within the body itself.

All pieces of information offered to the nervous system are summarized in the term sensibility. The somatic nervous system is the system which is concerned with information and interpretation of this information into a form that is acceptable to the nervous system. The processing and interpretation of the information in and by the nervous system do not belong to the sensory or somatic nervous system, but are discussed here for the sake of unity.

After the introduction of the concept of sensor, the general mechanisms of the senses which are concerned with reception, transmission, perception, analysis, and recognition will be discussed in this chapter.

Then the five sensory organs will be discussed, and also other sensory mechanisms which exist within the body.

Learning outcomes

After studying this chapter you should have sufficient knowledge and understanding of:
- the overall composition of the sensory or somatic nervous system;
- the possible classification of the sensors and the general characteristics of the sensors;
- the structure and function of the five sensory organs: nose, tongue, skin, eyes and ears;
- the structure and function of the organs of balance, the muscle spindles, the tendon and articular sensors.

1. Sensor

When the structure of the cell membrane and the effects of hormones was discussed, the concept of *receptor* arose. A receptor is a molecule in the cell membrane which is specifically sensitive to one or more stimuli. When the receptor comes into contact with its specific stimulus-substance, the cell can be stimulated to some kind of activity. A receptor is the biochemical letterbox of the cell. Though receptors process information, they do not, strictly speaking, belong to the somatic nervous system. The starting point in the somatic nervous system is the *sensor*.

A sensor is a cell which specializes in receiving and converting stimuli for the benefit of the nervous system. Sensors are closely related to nerve cells. Often the terms *receptor cells* or *sensory cells* are used instead of sensor. In order to avoid confusion these terms are not used in this book.

1.1. Classification

The sensors which are available to man can be classified in two ways according to their location and according to the character of the stimuli.

Using the location and the origin of the stimulus as a starting point, the following groups of sensors can be identified:

– *exterosensors*. These sensors are located peripherally and form the interface between the outside world and the human body. The stimuli come from outside. The exterosensors are situated in the *sensory organs*. Sensory organs are highly specialized organs, the cores of which are formed by large numbers of organized sensors. All kinds of additional structures serve the sensors in order to make them function optimally. The five sensory organs (nose, tongue, skin, eyes and ears) belong to the *animal* sensibility. They play an important role in the interaction between man and his environment. Some exterosensors are so-called *telesensors*. The stimulus affects the sensor, as it were, from a great distance. The nose, the eyes and the ears are telesensors.

– *proprioceptors*. These sensors give information about the locomotor apparatus. They are located in the muscles, the tendons, the joints and the organ of balance. As the organ of balance is a highly specialized organ, of which the sensors are the central part, it is sometimes seen as a sensory organ. It is not, however, an exterosensor for, although the source of the stimulus (disturbance of equilibrium) is outside, the stimulus itself does not occur at the border of the external

world and the body. This is just as well, or else we could not have a 'sixth' sense.

– *enterosensors*. These sensors are located in the wall of hollow organs. They are situated in the internal part of the body but on the border with the m.e., for instance in the oral cavity, the intestinal tract, the lungs, the urinary tract, the blood vessels, the vagina. The stimuli originate from the cavities within the organs.

A different classification is possible, using the type of stimulation as a starting point:

– *chemosensors*. These sensors are specifically sensitive to chemical stimuli, such as scents, flavours, carbon dioxide or oxygen, acids. Well-known chemosensors are the nose and tongue. There are chemosensors in the aortic arc which react to the acidity of the blood.

– *mechanosensors*. These sensors are specifically sensitive to mechanical stimuli, such as pressure, vibration, movement of fluids, tension. Among the mechanosensors are the touch and pressure sensors in the skin, the muscle spindles, the tendon and joint sensors, the sensors in the auditory organs and the organs of equilibrium, the pressor sensors (blood pressure) and the stretch sensors in the lung tissue.

– *thermosensors*. These sensors are specifically sensitive to changes in temperature. There are heat and cold sensors in the skin. Central thermosensors, which control the temperature, are located in the hypothalamus.

– *electromagnetic sensors*. These sensors are specifically sensitive to damage or threat of damage. They are located in the skin, but also in the walls of hollow organs, in the peritoneum, and in the muscles and joints. Pain, however, can also be experienced when other types of sensors are excessively stimulated.

1.2. Characteristics

Sensors are cells which specialize in receiving stimuli. Generally speaking, stimuli are changes in the environment of the sensor. Sensors are connected to afferent nerve fibres. They have a number of general characteristics: uniform conversion of the stimulus, specific sensitivity, specific sensation, specific range, sensory adaptation, differential threshold and discrimination.

Uniform conversion

Whatever the stimulus concerned, it is always converted, by the sensor, into a form which is acceptable to the nervous system, i.e. electrical energy (Fig. 2.8.1). The sensors transform a certain type of energy (for instance, chemical, mechanical, electromagnetic, thermal) into a form which can be used by the nervous system. The electrical charge in the sensor, which is elicited by the stimulus, is called the *sensory or generator potential*. When a certain generator potential has been elicited, a number of action potentials per second (with a defined charge) will arise at the attachment of the sensor to the afferent nerve fibres and the sensor *depolarizes*. The action potentials are propagated along the fibre to the central nervous system, where they are interpreted.

When the sensor receives stronger stimuli, the charge of the generator potential will increase proportionately. Such an increase is called an *analogous increase*. In terms of radio transmission, the sensor operates according to the mid-wave principle, or amplitude modulation (AM). When the generator potential increases, the afferent nerve fibres conduct, per second, larger numbers of action potentials of the same specific charge. The charge is digital (Fig. 2.8.1). It is translated into a changed number of impulses per second, the frequency modulation (FM on the radio).

Specific sensitivity

The generator potential must have a large enough amplitude to be able to elicit action potentials in the afferent fibre. The minimum amount of stimulus energy needed to be effective is called the *stimulus threshold* (Fig. 2.8.2).
Although many sensors are sensitive to more than one type of energy, and therefore can be stimulated in several ways, each sensor evidently is optimally sensitive to a specific form of stimulus. Such a stimulus is called an adequate stimulus. An adequate stimulus for the eye is light, for the tongue, chemical sub-

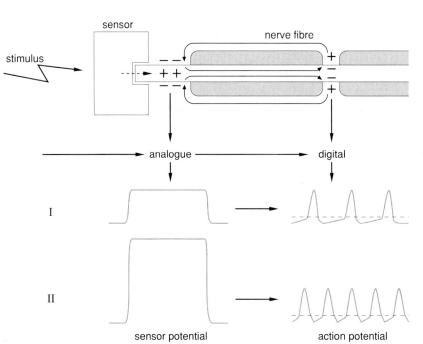

Figure 2.8.1
The uniform conversion of a stimulus. The sensor interprets the stimulus into an analogous generator potential. The nerve fibre conducts a specific frequency of action potentials (digitally)

I. pattern of signal by a weak stimulus
II. pattern of signal by a strong stimulus

Figure 2.8.2
The stimulus threshold.
When the stimulus is
sufficiently strong to cross
the stimulus threshold, the
afferent nerve fibre
depolarizes

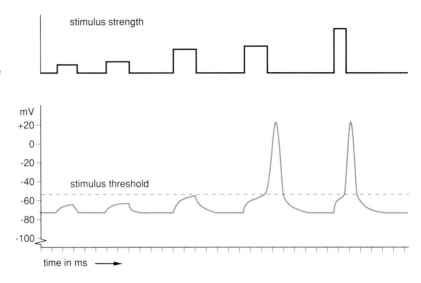

stances, for the auditory organ, sounds, etc. The adequate stimulus is the stimulus which depolarizes the sensor with a minimal amount of energy.

This means that the stimulus with the lowest stimulus threshold is designated an adequate stimulus. A relatively large amount of energy is needed to make the sensor depolarize by non-adequate stimuli. When a person closes his eyes and presses on the eyeball, all kinds of visual sensations may occur; a punch in the eye, also, is a non-adequate stimulus that can lead to the visual sensation of 'seeing stars'. The energy needed to make the sensor depolarize is many times larger, in these examples, than in the case of light entering the eye. When the electrodes of a 4.5 volt battery are held against the tongue membrane, the current is 'tasted'. The amount of energy is larger than when the taste sensors are stimulated in the normal manner.

Specific sensation

A sensor can be stimulated in many ways. All stimuli which generate action potentials to be propagated along the afferent nerve fibres, result in a perception specific to the sensor, the *specific sensation*. Whenever light reaches the retina, or the eyeball is pressed, or the retina is electrically stimulated, or the optical nerve is pressed, there will always be a sensation of light. This sensation is determined

by the sensor with its specific nerve fibres and cortical areas, and not by the character of the stimulation.

Specific range

When receiving adequate stimuli, the sensor has a specific *'band width'*. The electromagnetic sensors are sensitive only to light stimuli with a wave length between 400 and 800 nm (1 namometre = 10^{-9} metre). The mechanosensor of the ear is sensitive to acoustic vibrations with a frequency between 16 Hz and 20,000 Hz (1 Hertz = 1 vibration per second).

Heat sensors are specifically sensitive to temperatures between 35 and 45 degrees centigrade, cold sensors to temperatures below 38 degrees. It should be noted, however, that there are considerable differences between individuals as to the range of sensors. These differences can be related to age, sex, predisposition, etc.

Sensory adaptation

Sooner or later, when the stimulus remains constant, most sensors send less action potentials to the central nervous system. One could say that the stimulus threshold gradually becomes higher (Fig. 2.8.3). This characteristic is called negative sensory adaptation. This phenomenon can easily be demonstrated. When you push with the tip of your finger against a hair on the back of your hand or on your head, you 'sense'

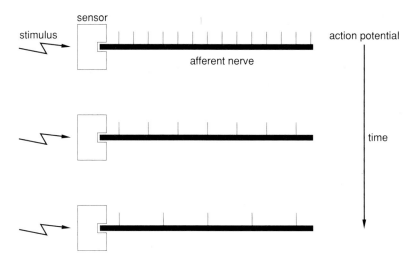

Figure 2.8.3
Negative sensory adaptation. After some time the frequency of the action potentials decreases, when the intensity of the stimulus is constant

this movement with the sensor that is located around the hair root. When you keep the hair in this position, you do not sense anything. When you lay a coin on the hand of a person with his eyes closed, the test subject initially feels the coin because the pressure sensors in the skin are stimulated. This sensation disappears after some time. When you enter a room filled with stale air, you can smell it immediately. After having been in the room for a while, you no longer smell the stale air.

In some sensors the stimulus threshold can become lower. This phenomenon is called *positive sensory adaptation*. The sensors become more sensitive to the stimuli. This is found in the rods, one of two types of electromagnetic sensors of the retina. When we enter a dark room, coming out of a bright hallway, initially we do not see anything. After a few moments the contours of objects in the room become visible and in a short while we see enough to be able to walk around the room without bumping against things.

The negative and positive sensory adaptations of the sensor itself (peripheral adaptation) are different from the negative and positive adaptations of the central nervous system (central adaptation).

The latter form of adaptation is related to the *attention* phenomenon. The ticking of a clock, the flowing of water in the pipes of the central heating, the 'humming' of a lamp are not normally perceived. Yet, impulses are continuously sent to the auditory cortex from the sensors in the auditory organ via the afferent nerve fibres. Negative central adaptation for these signals occurs in the auditory cortex. The opposite, positive central adaptation, also happens. When a person is concentrating, sounds can be heard which are not normally heard at all.

Difference threshold

When you put your hand in a bowl of water heated to 50°C and immediately thereafter in a bowl of water heated to 50.5°C, generally speaking you will not be able to tell the difference. There is too little difference in temperature between the stimuli. The same is true when an extra light is switched on in a well-lit room, or when a small amount of sugar is added to a cup of sweet tea. The least difference that can be detected between two stimuli, needed to pass the threshold again, is called the *difference threshold*. The different types of sensors have different types of difference thresholds.

Discrimination

When two blunt points at an interval of 1.5 mm are applied to the skin of the fingertips of a test subject, who has his eyes closed, the test subject will state that he feels two points. When the same is done on the skin of the back, for instance, or of the thigh, the test subject states, in nine cases out of ten, that he feels one point only. Furthermore, even at distances of a few centimetres apart on the skin of the back, no distinction can be made between one, two, or even more stimuli.

The ability to distinguish two stimuli as separate is called the *double-point discrimination*. The double-point discrimination of the skin of the fingertips is very high, while that of the skin of the back is very low.

The reason why there is a difference in double-point discrimination is due to the way in which the afferent nerve fibres are organized, in other words, to the difference in afferent 'wiring' (Fig. 2.8.4). In sensors which are connected in large numbers with one afferent fibre, via converging sensory branches, the ability to distinguish between several stimuli is small. The

Figure 2.8.4
Double-point discrimination

a. Each sensor has its own nerve fibre extending to the central nervous system; high double-point discrimination

b. More sensors are connected with one afferent nerve fibre; low double-point discrimination

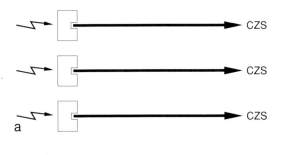

sensors   afferent nerves

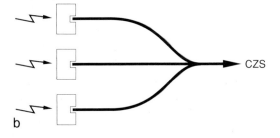

afferent fibre does not know where the action potential originates.

Such a sensory area has few 'wires', takes up little room in the peripheral nerve, in the corresponding ascending tract and the cortical area concerned. Double-point discrimination is high to extremely high for sensors which share an afferent fibre with only a few others, or which have an afferent fibre of their own. The 'wiring' from such a sensory area is very intensive and takes up more room in relative terms, in the peripheral nerves, the ascending tract and the cortical area concerned. Analogous to the *motor-unit* (i.e. one motoneurone together with the muscle fibres which it innervates; (see Fig. 2.7.16); the term *sensor-unit* (all sensors that are connected with one nerve fibre) can be used.

Small sensor-units have a high double-point discrimination, whereas large sensor-units have a low double-point discrimination.

Differences in discrimination not only occur in the sensors of the skin, but also, for instance, in the different light-sensitive sensors (rods and cones).

2. Five sensory organs

The sensory organs are *exterosensors*, in that the stimuli which affect the sensory organs come from the external world.

One could say that they are the instruments which enable man to perceive the quality of the outside world. The five senses are traditionally: *smell, taste, touch, sight* and *hearing*. They correspond with the five sensory organs: nose, tongue, skin, eye and ear. If we want to distinguish the sensory functions of these organs from other functions, we should talk of the smelling organ, taste sensors, skin organs, visual organ and auditory organ.

2.1. Nose

Olfactory sensors
The nose is the seat of smell. The *olfactory epithelium* lines the roof of the nasal cavity (Fig. 2.8.5). It consists of supporting cells with *olfactory cells* and a number of mucus cells (Fig. 2.8.6). The supporting cells are a continuation

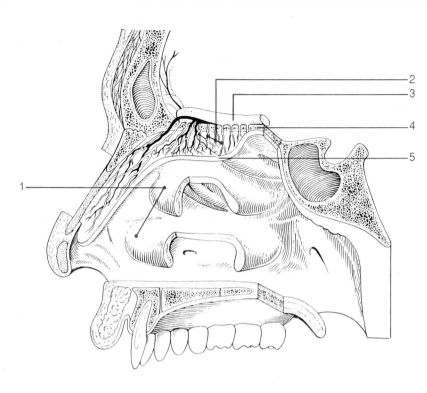

Figure 2.8.5
Sagittal section through
the nasal cavity; the
olfactory nerve

1. *nasal conchae*
2. *fila olfactoria*
3. *olfactory nerve (N. I)*
4. *ethmoid bone*
5. *olfactory epithelium*

Figure 2.8.6
Detail of the olfactory
epithelium

1. *afferent fibres of the*
   olfactory sensors
2. *cell body of the*
   olfactory sensor
3. *supporting cell*
4. *stimulus receiving*
   process of the
   olfactory sensor

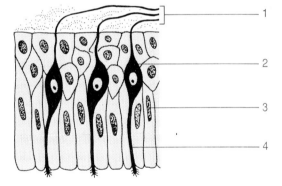

of the multi-layered epithelium of the rest of the lining of the nasal cavity. The mucus cells provide a thin superficial mucous layer.

The olfactory sensors are thin, elongated cells with very fine, hairlike processes (cilia) projecting into the mucous layer. On the proximal side the olfactory sensors change into thin, non-myelinated nerve fibres. A number of these fibres converge to small bundles, the *fila olfactoria*. They run through the openings of the ethmoid bone (see Fig. 2.9.12), and form the olfactory nerve (N.I) underneath the frontal lobe of the cerebrum. The olfactory nerve has connections with the brain stem, the limbic system, the hypothalamus and the cortex.

Smelling

The aromatic substances in the air which form scent accidentally end up in the nasal cavity (diffusion) or are actively inhaled during nasal respiration. There are thought to be six main types of smell: the smell of decay (rotten egg, sewage, faeces), smell of fire (smoke), the smell of ether-like substances (petrol, mothballs, white spirit), the smell of peppermint (vanilla, aniseed), the smell of flowers (roses, lavender) and the smell of fruit (apple, mango). Specific sensors are probably available for each of these main types of smell. The other sensations of smell are based on many combinations of the main types. The aromatic substances (chemical substances) are generally more easily soluble in the mucous layer of the epithelium. Once dissolved, they can reach the olfactory sensors and generate a generator potential. The olfactory sensors belong to the group called chemosensors. When the generator potential is high enough to cross the stimulus threshold, the sensors depolarize and action potentials are conducted along the nerve fibres of the olfactory nerve. The impulses are interpreted in the brain. This means that on the basis of prior experiences (acquired) and disposition (inherent) the meaning of the scent concerned can be determined. The scent may be experienced as pleasant, but it can, also, warn against bad food, harmful gases, insufficient hygiene (sweat, defecation) and so on. Involuntarily (reflexly), scent can lead to the production of saliva and secretion of gastric and intestinal juices. It is not certain whether certain scents affect the sexual behaviour of man. It is certain, however, that scent does give specific social information. Many parents can without

Figure 2.8.7
The tongue

a. View from above

   1. posterior palatine arch
   2. lymphatic tissue on the base of the tongue
   3. anterior palatine arch
   4. epiglottis
   5. tonsil
   6. circumvallate papilla

b. Categories of lingual papillae

   1. taste bud
   2. serous gland in connective tissue
   3. circumvallate papilla
   4. filiform papilla
   5. fungiform papilla
   6. mucous membrane
   7. blood vessel
   8. muscle tissue

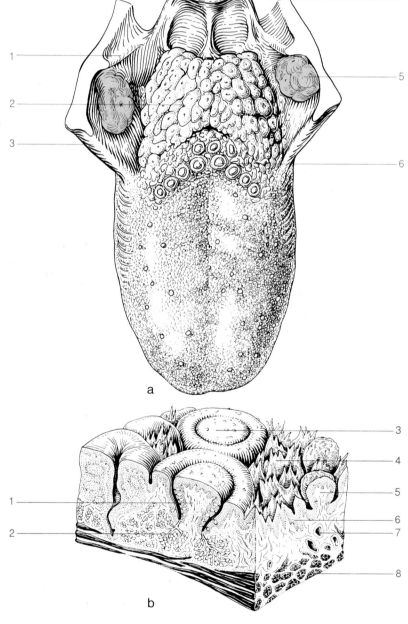

a

b

any difficulty identify by smell the day-old T-shirts of their offspring, from a pile of other children's T-shirts.

The effect of smell on taste is extremely significant. When you hold your nose while you are eating a delicious meal, more than half of the sensations of taste are lost.

2.2. Tongue

Gustatory sensors or taste sensors

The tongue is said to be the seat of taste. A large number of *taste sensors* are scattered over the multilayered pavement epithelium of the mucous membrane of the tongue. There are, however, also a number of taste sensors on the soft palate, the palatine arches, parts of the pharyngeal cavity and the epiglottis. Taste sensors are largely found in the *taste papillae* or *corpuscles*, also called *taste buds* (Fig. 2.8.7). There are three categories of taste papillae. *Circumvallate papillae* are located anterior to the base of the tongue. They are arranged in a V-shape and are visible to the naked eye. *Fungiform papillae* are situated mainly on the tip of the tongue. *Filiform papillae* are mainly located on the sides of the tongue and scattered over its surface. The taste sensors are elongated and located be-tween supporting cells (epithelium). They have, like the olfactory sensors, cilia that project into the mucous layer of the tongue (Fig. 2.8.8). There are, at the proximal side, contacts with thin, myelinated nerve fibres. The afferent fibres converge into two pathways to the brain stem; the anterior two-thirds of the tongue sends impulses via the facial nerve (N. VII), the posterior one-third does so via the glosso-pharyngeal nerve (N. IX).

From the brain stem the gustatory impulses are conducted to the thalamus, whence they end up in the gustatory centre of the somatosensory cortex (post-central gyrus).

Taste

Taste is more than the sum of the stimuli of the taste sensors. When you taste something, not only is the information supplied by the smell important, but also the senses of touch and temperature in the oral cavity (is the tasted substance warm or icy cold, hard, soft, tough, dry, and so on?).

Limiting tasting by means of the taste sensors on the tongue, there are four *qualities of taste*: sweet, salty, sour and bitter. Other qualities of taste are mixtures of these four or based on olfactory sensations. Sweet is experienced

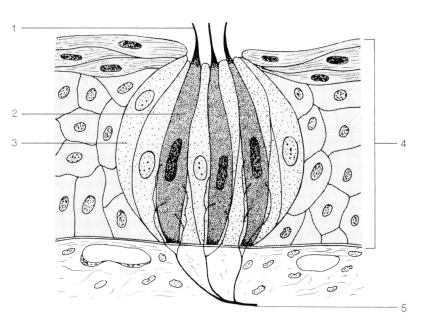

Figure 2.8.8
Detail of a taste bud

1. *stimulus receiving process of the taste sensor*
2. *cell bodies of the taste sensor*
3. *supporting cell*
4. *non-keratinizing multi-layered pavement epithelium of the tongue*
5. *afferent nerve fibre*

mainly via the tip of the tongue, bitter via the base of the tongue, sour by the edges of the tongue and salty by all of the surface of the tongue.

The taste buds, therefore, have their specialisms. Flavours (chemical substances) are dissolved in the mucous membrane of the tongue and stimulate the chemosensors so that a generator potential is elicited. When the stimulus threshold is crossed, the sensors depolarize and action potentials are conducted via the afferent nerves, the brain stem and the thalamus, to the gustatory centre in the post-central gyrus, where the impulses are interpreted. As with smell, prior experiences and disposition play a part in the interpretation of taste.

Taste reflexively stimulates the secretion of saliva, gastric and intestinal juices. It can be a pleasant sensation or, on the other hand, a warning against decayed or poisonous food.

2.3. Skin

2.3.1. Cutaneous sensibility sensors

The skin is the seat of *cutaneous sensibility*. *Cutaneous* or *dermal sensibility sensors* are located mainly in the dermis, just underneath the epidermis. A smaller number are found in the subcutaneous connective tissue.

The sensibility sensors are different as to structure and type of sensitivity. On this basis they are subdivided into:
- Krause's (bulboid) corpuscles
- Vater-Pacini (lamellated) corpuscles,
- Meissner (touch) corpuscles,
- Golgi-Mazzoni (lamellated) corpuscles,
- Merkel's (tactile) corpuscles,
- follicle sensors and terminal nerve corpuscles.

This classification has not only immortalized a number of anatomists, but has also led to a separate, specific function being attributed to each type of sensor. The latter is only partly true. The cutaneous sensibility sensors are connected to afferent nerve fibres. These can – depending on the body area where they orginate – be part of all kinds of afferent nerves or of nerves with an afferent part. The afferent fibres arrive at the spinal cord and are transmitted to ascending tracts. All sensory tracts arrive at the thalamus. This large centre in the diencephalon is an extensive connector station for sensory signals (and consequently for sensibility sen-

Figure 2.8.9
Types of cutaneous
sensibility sensors

1. *Krause's corpuscle*
2. *terminal nerve corpuscle*
3. *Ruffini's corpuscle*
4. *Vater-Pacini's corpuscle*
5. *Merkel's corpuscle*
6. *Meissner corpuscle*
7. *follicle sensors*

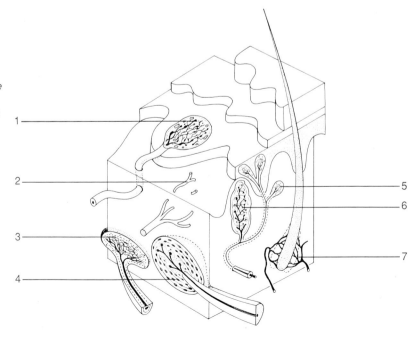

sations, as well). From the thalamus the sensibility signals are transmitted to the cortex, to the neurones of the post-central gyrus (see Fig. 2.7.29).

The post-central gyrus shows a somatrotopic representation, which means that each area of the skin has its own representation on the somato-sensory cortex (see Fig. 2.7.33b). It is remarkable that some areas of the skin take up a relatively large part of the post-central gyrus (the fingertips, the lips, the tongue, for instance), whereas other areas take up a relatively small part (the skin of the trunk, the arms and the legs, for instance). This is not only related to the number of sensors per cm^2 of skin, but also to the differences in double-point discrimination in the different areas of the skin. Areas with a high double-point discrimination have an extensive afferent 'wiring', many ascending fibres and consequently many neurones on the sensory cortex.

2.3.2. Touch

Touch is a collective term. The sense of touch comprises the possibility to feel warmth and cold, to feel the hardness of an object, to find out the shape of an object by touching, to feel pain as a warning in case of imminent or actual damage.
All of these *qualities of touch* (temperature,

pressure, feeling and pain) are sensed by dermal sensors. The dermal sensor is stimulated by an adaequate stimulus, and when the stimulus threshold is crossed, the sensor depolarizes and action potentials travel to the central nervous system via the afferent fibre. They travel along ascending tracts and are transmitted in the thalamus, ending up in the post-central gyrus, where they are interpreted. The significance of the signals is determined on the basis of both inherent and acquired knowledge. Determining the type of quality of touch involved in a specific case is only partly dependent on the type of corpuscle which is stimulated. The follicle sensors register movements of the hairs, Meissner's corpuscles are thought to register touch, Vater-Pacini's corpuscles pressure, and so on. There are, however, areas in the skin where such clearly defined sensors are lacking and where all qualities of touch can be perceived. An example of such an area is the auricle, where only terminal nerve corpuscles are located. As yet there is no explanation for the phenomenon that not only pain, but also touch, and changes in temperature or pressure can be perceived in the auricle.
Distinctions can be made not only in the qualities of touch, but also in sensibility, such as:
– sense of temperature;
– tactile and pressure sense;
– sense of pain.

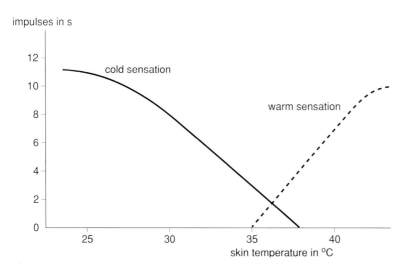

Figure 2.8.10
Depolarization pattern of two types of thermosensors

impulses in s

skin temperature in °C

Sense of temperature

The sense of temperature is provided for by *thermosensors*. These sensors react to changes in temperature. They are divided into *sensors of cold* and *sensors of warmth* on the basis of range and patterns of activity (Fig. 2.8.10). The sensors of cold are active below a skin temperature of 38°C. When the temperature falls, the frequency with which the sensors of cold depolarize increases. The colder it gets, the more active they become. The sensors of warmth are active when the skin temperature is over 35°C. When the temperature rises, the frequency of depolarization increases.

Krause's corpuscles (Fig. 2.8.9) are considered sensors of cold. They are globular organs which consist of a tangle of fine nerve branches, embedded in supporting cells and surrounded by a capsule. They are scattered over the skin, just underneath the epidermis. They are not only located in the skin, but also, in large numbers, in the cornea, the glans of penis and clitoris, and the mucous membranes of the walls of oral and pharyngeal cavities. These epithelial areas are extra sensitive to cold stimuli.

Ruffini's (spray) corpuscles (Fig. 2.8.9) are considered sensors of warmth. They are built similarly to Krause's corpuscles, but are differently shaped. They are flattened, and are scattered in the skin, in the depth of the dermis, very near to the borderline with the subcutaneous connective tissue.

They are not only found in the skin, but also in the iris, the dura mater, the mucous membranes of nasal, oral and pharyngeal cavities, and in the oesophagus. They are absent in the glans of penis and clitoris, so temperatures exceeding 38°C are no longer perceived in those regions.

The terms *sensors of warmth* and of *cold* suggest a greater degree of specialism than is actually the case. When we put our hands in a bowl of water of 20°C and thereafter in a bowl of water of 30°C, the frequency of depolarization of the Krause's corpuscles will decrease (Ruffini's corpuscles are not yet active at this temperature). On the basis of a smaller supply of impulses, the central nervous system will conclude: 'the water in the second bowl is warmer'. Krause's corpuscles, therefore, are able to sense a rise in temperature. Conversely, the same is true for sensors of warmth.

It is remarkable that both types of sensors are active in the area where the temperature is more or less equal to the body temperature. The advantage is that changes in temperature in this area, which are so important for the functioning of the body, are monitored by as many thermosensors as possible.

Senses of touch and pressure

Mechanosensors provide the senses of touch and pressure. These sensors react to mechanical stimuli.

The stimulus for the tactile sensors is by being touched (passive) or by touching (active).

The stimulus for the pressure sensors is caused by a change in pressure on the tissue of the skin due to the application of weight on it (passive) or to the exertion of pressure on something (active).

Four types of tactile sensors can be distinguished:
– *terminal nerve corpuscles;*
– *follicle sensors;*
– *Merkel's corpuscles;*
– *Meissner's corpuscles* (Fig. 2.8.9).

The terminal nerve corpuscles arise between the epithelial cells of the epidermis. Parts of them are sensitive to touching or being touched.

A follicle sensor consists of a mesh of nerve fibres around the follicle. Touching the hair ever so slightly is registered by the follicle sensor. The hairs on the back of the hand and on the fingers are especially sensitive.

Merkel's corpuscles are also called tactile discs. They consist of flat, oval cells, interconnected by a flat and broad part of a nerve branch. They are located in the area where the epidermis and dermis border, but also in the tissue of the follicles. They perceive touch on the skin and movements of the hair, respectively.

Meissner's corpuscles are the largest tactile sensors. They are oval-shaped corpuscles, composed of layers of flat cells surrounded by a thin capsule. Profusely coiled nerve branches are located between them and are linked with them. They leave the corpuscle at its base. Every cutaneous displacement is transmitted to the flat cells and leads to stimulation of the nerve branches.

Meissner's corpuscles are located in the taste-buds of the dermis (see Fig. 2.5.1). They are found in hairless skin, especially in the palm of the hand and the sole of the foot. The largest numbers are found at the distal ends of the fingertips, especially of the thumbs and the index fingers.

Vater-Pacini corpuscles are considered pressure sensors. They are particularly sensitive to changes in pressure. They consist of a large number of concentric lamellae surrounded by a capsule. They could be compared with the layers of an onion. The lamellae consist of flat connective tissue cells with fibres. There is fluid between the lamellae. The afferent nerve fibre arises in the centre of the corpuscle. Compared to other dermal sensors the pressure sensors are large: 4 mm long and 2 mm wide. They are clearly visible with the naked eye.

When there is a change in pressure on the skin, the position of the lamellae will be changed. Because of these changing positions a general potential is elicited. When this potential is large enough to cross the stimulus threshold, the pressure sensor depolarizes.

In this way the pressure sensors are not only able to perceive the weight of an object, but also to feel vibrations. Whether water is flowing in a pipe or not can be established by laying one's hand on the pipe. Flowing water makes the pipe vibrate, still water does not.

Vater-Pacini's corpuscles are deep in the dermis, very near to the borderline with the subcutaneous connective tissue. They are therefore not stimulated when a person is touched. The relatively largest number of tactile and pressure sensors per cm^2 is found in the connective tissue of the surface of the tongue, in the lips and in the skin of the fingertips and the toes, but there are also large numbers in the palms of the hands, the soles of the feet, the areolae, the glans

of penis and clitoris. There are fewer sensors per mm^2 in the skin of the back, the upper arms and legs, for instance. This difference in distribution and the difference in double-point discrimination are also seen in the sensory homunculus (see Fig. 2.7.33b). The difference in sensitivity is easily discernible. When the points of two pins are carefully placed on someone's tongue or lip (the test subject has her eyes closed), the distance between the points need only be 1 mm for two points to be felt. By comparison the smallest equivalent distance on the fingertips is 1.5 mm. At the back of the hand the distance should be at least 3 cm or more for two points to be felt; at certain places on the skin of the back more than 6 cm space is required; while on the outside of the thigh an 8 cm space is necessary.

Sense of pain

The *pain sensors* provide the sense of pain. These sensors react to damage or imminent damage of the tissue in which they are embedded. The damage can be done by mechanical violence, by fire or by the effects of some chemical substance.

Specific terminal nerve corpuscles are considered pain sensors. Large numbers of these corpuscles are found in the lower layers of the epidermis and the upper layers of the dermis (Fig. 2.8.9). Pain is felt when the nerve corpuscles themselves are damaged, as is the case in serious wounds. Usually, however, the sensation of pain is a result of stimulation of the nerve corpuscles by substances which are released into the tissue as a result of the damage. One of these substances is the hormone histamine. Excessive stimulation of other types of sensors is also experienced as pain by the central nervous system.

2.4. Eyes

2.4.1. Structure

The eyes are the seat of vision. Of the five sensory organs, the eye probably appeals most to the imagination. The eye literally gives mankind a view on the world, with its shapes, its colours, and its movements, both at close quarters and from a distance.

Figure 2.8.11
Layers of the wall of the
eyeball

1. vitreous body
2. ciliary body with fibres
   of lens capsule
3. iris
4. pupil with lens
   behind it
5. cornea
6. scleral venous sinus
   (canal of Schlemm)
7. superior oblique
   muscle of eyeball
8. sclera (white of the
   eyeball)
9. choroid
10. retina
11. optic nerve

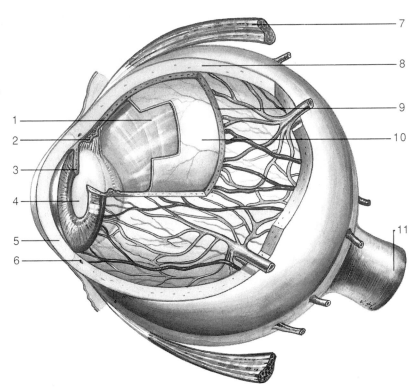

Figure 2.8.12
Transverse section through
the eyeball

1. conjunctiva
2. ciliary body
3. blind spot
4. optic nerve
5. cornea
6. anterior chamber
7. iris
8. posterior chamber with
   fibres of the lens
   capsule
9. lens
10. vitreous body
11. retina
12. choroid
13. sclera (white of the
    eyeball)
14. retinal macula
15. optic axis

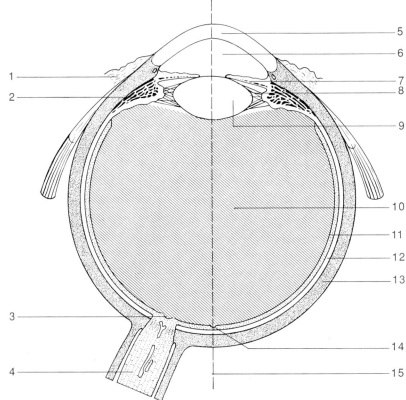

Vision is almost indispensable for cooperation between the somatic nervous system and the motor system. Adjustment to motor function is done mainly on the basis of visual information and feedback. Just think of games such as badminton, volleyball, tennis, or sports such as motor racing and cycle racing.

The eye is a complicated organ, the core of which consists of the photosensors. One could say that the other parts of the eye and the corresponding support organs are made available to *photosensors* in order that they may function properly.

Eyeball

The frontal section of the eyeball is approximately 24 mm. The optic axis, passing through the centres of the cornea and lens, is a little longer (25 mm).

The wall of the eyeball consists of three layers: the sclera, the choroid and the retina (Figs. 2.8.11 and 2.8.12).

The *sclera* is a thick and tough connective tissue capsule, built mainly of collagen fibres. It is white-coloured. The minute blood vessels stand out beautifully against the background of the white of the eyeball. The globular shape is maintained by the pressure of the fluid in the eye and by the tough sclera.

Posteriorly the sclera is continuous with the sheath of the optic nerve and thereafter, in the cranial cavity, with the dura mater. Anteriorly the sclera is continuous with the crystalline *cornea*. The line of separation between the sclera and the cornea is called the *limbus*. The cornea is more convex than the sclera, as a consequence of which that part of the eyeball protrudes slightly. The cornea is not vascularized.

Six eye muscles attach to the sclera; this tells something about the toughness of the sclera.

Interiorly the *choroid* covers the sclera. It is a thin vascular membrane, whose many blood vessels supply the eyes. The choroid posteriorly surrounds the optic nerve, fused with the sclera. In the cranial cavity it is continuous with the pia mater. At the level of the limbus the choroid changes into the *iris*, a flat circle located anteriorly in the eyeball. The pigmentation of the iris determines the colour of the eyes. When the iris is rich in pigment, the eyes are dark brown, when the iris has little pigment, the eyes are powder-blue.

The round orifice in the iris is the *pupil*. Immediately posterior to the site where the outer edge of the iris is continuous with the choroid, the choroid becomes a little thicker. At this place a smooth muscle, which functions as a sphincter, is located the *ciliary muscle*. The round-shaped thickened ring in which the ciliary muscle is situated, is called the *ciliary body* or *corpus ciliare*. From here a large number of minute tendinous fibres radiate, behind the iris, to which the lens is attached.

The *retina* lies deeper than the choroid and is the third layer of the wall of the eyeball. It consists of two layers, the pigment layer and the nervous layer of the retina, often called the retina proper (Fig. 2.8.13).

Figure 2.8.13
The layers of the retina

1. *afferent neurones*
2. *connector cells*
3. *photo sensors (rods and cones)*
4. *nervous layer of retina*
5. *pigment layer*
6. *choroid*

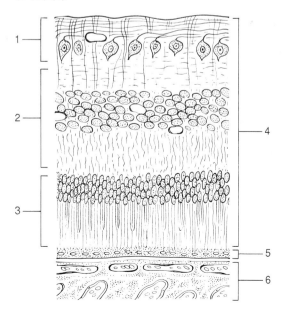

The *pigment layer* is very thin. It covers all of the interior of the choroid and also the back of the lens.

The nervous layer of the retina is yet deeper. It is much thicker than the pigment layer and continues anteriorly, almost to the site where the choroid shows the thickened edge of the ciliary body.

The nervous layer consists of a layer of *photo-sensors*, a layer of *connector cells* and a layer of *afferent neurones* (Fig. 2.8.13). The cell bodies and the axons of the neurones are separated from the interior of the eyeball by a single-layered basal membrane.

The photosensors are in apposition to the pigment layer. The axons of the afferent neurones run at the surface of the retina to the site where they leave the eyeball as the optic nerve.

At this site, called *optic papilla*, there are no photosensors. This is called the *blind spot*, as in this part of the eye creation of images is not possible. The blind spot can easily be demonstrated (Fig. 2.8.14).

The two blind spots of the eyes are not located at identical places, therefore there is no 'hole' in the field of vision when one looks with both eyes, as different parts of the object are projected on the blind spots. When we look with one eye only, we are not conscious of the 'hole' in our field of vision. This is due to the interpreting activity of the optic area in the cortex.

An artery enters the eyeball via the papilla (Fig. 2.8.15). This artery ramifies in the layer of afferent fibres into capillary networks. Blood is discharged into veins that also leave the eye via the papilla.

Before the light of the projected images can reach the photosensors, it must pass the blood vessels and pierce the two layers of the retina which are insensitive to light. This has a disadvantageous influence on the sharpness of the images. At the place opposite the pupil, at the end of the optic axis, the structure of the retina is somewhat different, as a result of which this disadvantage is counterbalanced. This place is called the *macula lutea* or *retinal macula*, or yellow spot (Figs. 2.8.12 and 2.8.15). At this place the retina is very thin, as the connector neurones separate and both the afferent fibres and the blood vessels of the retina arch around this area. With this part of the retina, vision is least 'blurred'.

Figure 2.8.14
Demonstrating the blind spot

Keep this page upright at a distance of approximately 30 cm in front of the right eye and look at the black cross. Close the left eye. The black dot is no longer visible, as it is projected on the part of the retina where no photosensors are located. Repeat this experiment for the left eye

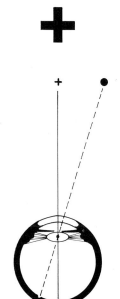

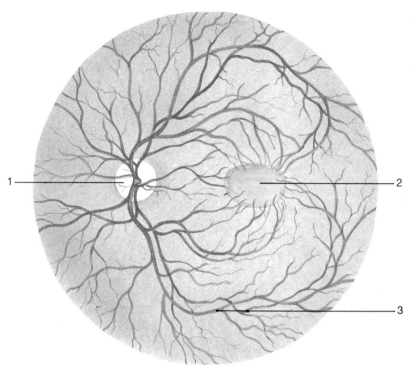

Figure 2.8.15
The fundus of the eyeball

1. *blind spot*
2. *yellow spot*
3. *blood vessels*

This structure is favourable for the function of the eye.

When we examine an object, our eyes are focused in such a way that the part which has our attention is projected exactly to the yellow spot.

With an ophthalmoscope (a magnifying glass with a lamp) the fundus of the eye can be examined through the pupil (Fig. 2.8.15). The blind spot is detectable as a whitish spot in the nasal half of the fundus. The artery ramifies into several branches (lighter, thinner), whereas the veinlets (darker, thicker) unite into a vein.

The yellow spot, yellowish as the name indicates, is located in the centre. No blood vessels run through this area. The blood is supplied from the edges.

The interior of the eye can be divided into the lens, the vitreous body and the chambers (Fig. 2.8.12).

The lens is suspended behind the pupil by a large number of minute tendinous fibrils, which radiate from the ciliary body, and form the *suspensory ligament*. The lens is transparent and has no blood vessels. It is a flexible structure which consists of elongated epithelial cells, the *lens fibres*, surrounded by the lens capsule. All through life the lens fibres are renewed.

The lens is a double convex disc. The anterior surface is less convex than the posterior surface. The structure of the lens is such, that it itself 'wants' to be as convex as possible. The refraction of light is greatest when the lens is most convex. When the eye is at rest, it is adjusted to longsightedness, and distant objects are seen sharply. This means that the lens is as flat as possible and the refraction of the light at a minimum. When the ciliary muscle relaxes, it makes the lens flatten. The diameter of the ring that forms the ciliary body is then at its largest. The tough suspensory ligament is tightened and flattens the lens.

The chamber behind the lens is filled with a colourless, transparent, gelatinous substance, the *corpus vitreum* or *vitreous body*.

Figure 2.8.16
Direction of flow of
intra-ocular fluid

1. *pupil*
2. *anterior chamber*
3. *posterior chamber*
4. *scleral venous sinus
   (canal of Schlemm)*
5. *ciliary body*

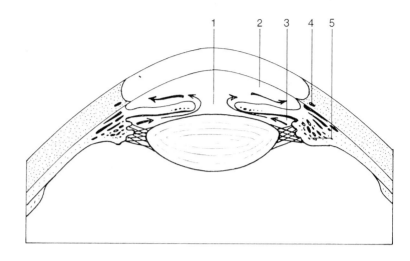

The *chambers* are filled with a clear fluid, the *aqueous humor* or *intra-ocular fluid*. The *anterior chamber (camera anterior bulbi)* is located between the cornea and the iris. The *posterior chamber (camera posterior bulbi)* is the ring-shaped space between the iris and the vitreous body around the lens. The chambers communicate through the pupil.

The aqueous humor is produced by a specialized capillary network in the ciliary body and the iris, in the posterior chamber. This function of the choroid is, therefore, similar to that of the choroid plexus in the cerebral ventricles. There is a continuous production of aqueous fluid. The amount of fluid stays at the same level as the fluid is drained via the scleral venous sinus (canal of Schlemm) (Fig. 2.8.12). This is a circular vein in the wall of the limbus that discharges the intra-ocular fluid into the bloodstream.

There is a continuous stream of aqueous humor from the posterior chamber, passing between the lens and iris, through the pupil to the anterior chamber (Fig. 2.8.16). The intra-ocular fluid is important for the supply of the lens and the cornea, as these structures have no blood vessels.

Eye socket and eyelids
The eyes are located in the *eye sockets* or *orbits* of the facial bones. The eye socket is a funnel-shaped cavity consisting of several pieces of bone (Fig. 2.8.17). A number of openings for structures entering and leaving are located at the back of the orbit, one of which is the optic nerve, which leaves the eye posteriorly/inferiorly at the medial side. The eyeball is embedded in adipose tissue which fills the eye socket. In this way the vulnerable eyeball is protected both by a strong bony cover and a shock-absorbing cushion of fat which keeps the eyeball in its place.

The eyelids are thin folds of skin which constitute the limits of the eye fissure. This skin

Figure 2.8.17
The structure of the
bony orbit

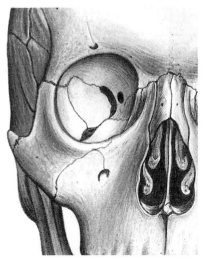

has virtually no subcutaneous fat and the folds are so large that they can cover the eyeballs; when covered, the eyes are 'closed'. The *medial angle* is located at the nasal side of the eye fissure, the *lateral angle* at the temporal side. Closing of the eyes is done by the eye sphincters (see Fig. 2.9.34). The upper eyelid can be raised (opened) by a muscle which travels superiorly in the orbit and attaches to the rim of the eyelid. The lower eyelid sinks mainly by its own weight (open).

At the inner surface of the eyelids, the skin continues from the rim as a mucous membrane. It runs in the direction of the edge of the orbit and there turns in the opposite direction. At the level of the limbus the mucosa attaches to the eyeball and becomes continuous with the front surface of the cornea (see Fig. 2.8.12). The mucous membrane consists of thin, transparent epithelium, and is called the *(tunica) conjunctiva*. As the conjunctiva is attached to the free edges of the eyelids, there is a space behind each eyelid, the *conjunctival sac* (Fig. 2.8.18). The conjunctiva completely separates the space in the orbit from the external world, but thanks to the conjunctival sacs the mobility of the eyes is not affected.

The *eyelashes* are implanted into the edges of the eyelids. These hairs keep growing until they have reached a length which is specific to each individual. Then the growth stops and only the lost hairs are renewed.

A large number of sebaceous glands are located between the implantations of the eyelashes. The sebum not only keeps the edges of the eyelids supple, but is also water resistant, as a result of which the lacrimal fluid is guided along the correct channels.

The eyelids are closed, for instance during sleep, to keep light from entering the eye.

When the eyelashes, the conjunctiva or the cornea are stimulated by a shortage of fluid, a dust particle or something similar, or when too bright a light suddenly enters the eyes, the eyelids are reflexly closed. This is caused by the *blink reflex* or *eyelid closure reflex*.

The eyebrows are located on the arch-like, bony edge over the orbits. The hairs of the eyebrows are thick and implanted in such a way that they are directed laterally. They prevent sweat from running from the forehead straight into the eyes.

Lacrimal apparatus

The *lacrimal gland* is located above the lateral angle of the eyeball (Fig. 2.8.19). This exocrine gland continuously produces *lacrimal fluid* or *tear water*, which mainly consists of water, with a small volume of NaCl and a bactericidal enzyme. The lacrimal fluid is discharged into the upper conjunctival sac via a large number of ducts, which pierce the conjunctiva. The tear water keeps the frontal surface of the eyeball moist. When the films of fluid on the conjunctiva and cornea become too thin, the blink reflex immediately occurs, as a result of which a new layer is evenly spread over the surface. At the same time the frontal surface is wiped clean. The layer of fluid keeps the tissue from becoming dehydrated and therefore blurring vision.

The tear water is transported via the two *canaliculi lacrimales* or *lacrimal ducts*, located in the medial angle of the eye, in a small protrusion at the edge of the eyelids where there are no eyelashes. When the lower eyelid is drawn slightly downwards, the beginning of the lacrimal duct is clearly visible. The lacrimal fluid is helped in the right direction by the water-resistant edges of the eyelids and the special positions of the openings of the lacrimal ducts in the medial angles. The lacrimal ducts extend in the nasal direction. They converge and discharge into the *lacrimal* or *tear sac*. Thereafter, the tear sacs discharge into the left and right *nasolacrimal* or *nasal ducts*. The nasal ducts run through openings in the bone and terminate beneath the lower nasal conchae in the lower meatus. The tear water evaporates here and contributes to moistening the inhaled air.

The production of tear water by the lacrimal glands is controlled by the vegetative nervous system. Stimulation of the conjunctiva by dust, dehydration, bright light or draught reflexly leads to an increased production of tear water. When the production is greater than the amount that the lacrimal ducts can hold, the lower conjunctival sacs overflow and the tear water flows

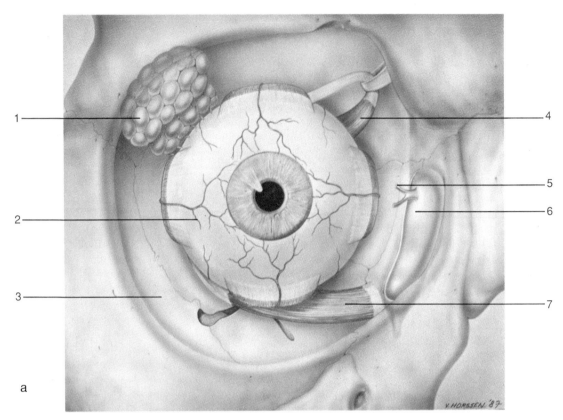

a

Figure 2.8.18
The position of the eye in
the orbit

a. Frontal view

1. *lacrimal gland*
2. *capillary*
3. *bony part of the orbit*
4. *superior oblique muscle of eyeball*
5. *lacrimal duct*
6. *nasolacrimal or nasal duct*
7. *inferior oblique muscle of eyeball*

b. Lateral view

1. *levator muscle of upper eyelid*
2. *superior rectus muscle of eyeball*
3. *adipose tissue*
4. *optic nerve*
5. *inferior oblique muscle of eyeball*
6. *eyelashes of upper eyelid*
7. *conjunctival sac*
8. *sphincter in lower eyelid*

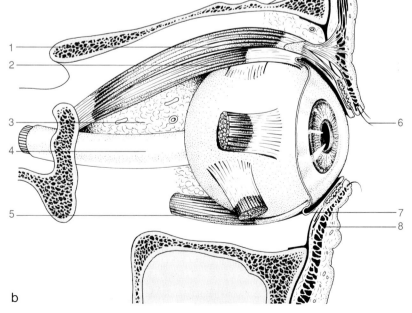

b

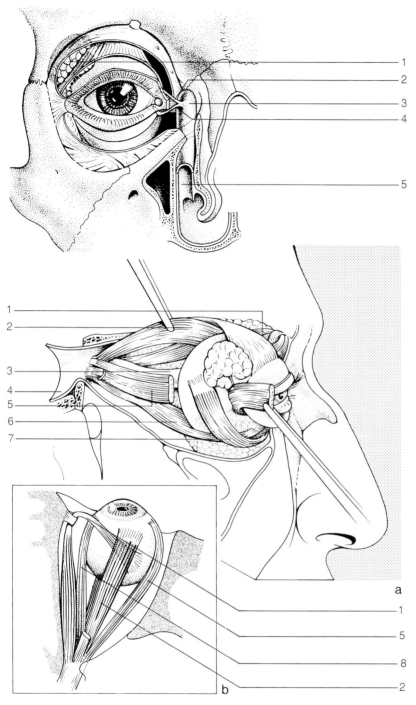

Figure 2.8.19
The lacrimal apparatus

1. *lacrimal gland*
2. *lacrimal duct of upper eyelid*
3. *tear sac*
4. *lacrimal duct of lower eyelid*
5. *nasolacrimal duct, discharging into lower nasal concha*

Figure 2.8.20
The eye muscles

a. Lateral view
b. View from above

1. *superior oblique muscle of the eyeball*
2. *superior rectus (straight) muscle of the eyeball*
3. *optic nerve*
4. *dura mater*
5. *lateral rectus muscle of the eyeball*
6. *inferior rectus muscle of the eyeball*
7. *inferior oblique muscle of the eyeball*
8. *medial rectus muscle of the eyeball*

over the edges of the lower eyelids on to the cheeks.

This happens not only when the conjunctiva is excessively stimulated (dust particles, scent of onions), but also when the production of tears is generated by emotion (weeping).

When the lacrimal fluid discharged via the nasal duct does not fully evaporate, this produces sniffing.

Eye muscles and eye muscle reflexes
The eyeball has considerable movement in all directions. The length of the optic nerve and the

position of the connective tissue allow it to do so. The movements are effected by six eye muscles, four straight and two oblique (Fig. 2.8.20).

The four *rectus (straight) eye muscles* all originate from a tendinous ring around the optic nerve at the site where the nerve enters the skull through the posterior wall of the eye socket. By means of a flat tendon, each of the four eye muscles is inserted on to a side of the eyeball, near to the margin of the cornea.

The *oblique eye muscles* are more medially inserted on to the eyeball. The tendon of the superior oblique eye muscle passes through a kind of pulley. The inferior oblique muscle is the smallest of all six.

The eyes are innervated by three cerebral nerves: the oculomotor nerve (N.III), the trochlear nerve (N. IV) and abducens nerve (N. VI). The eye muscles have remarkably small motor-units and are therefore meticulously controllable. Each eye not only works together precisely, so that the eyes can be focused over a range of distances within the visual field but also the coordination between the muscles of both eyes is perfect. In most cases both eyes move simultaneously in the same direction (Fig. 2.8.21), even when one or both eyes are closed. This is called *conjugated movement* of the eyes.

The eye muscles are the effectors of a number of reflexes of the eyes: the accommodation reflex, the ocular fixation reflex and nystagmus.

The *accommodation reflex* is the automatic change in the shape of the lens so that the image of the object within our visual field, which attracts our attention, is automatically focused on the part of the retina directly behind the pupil. In terms of quality, this yellow spot is the best part of the retina.

The *ocular fixation reflex* is a reflex fixation of the eyeball induced by the movement of an object. It does not matter to the eye muscles whether the object itself moves or whether the observer moves.

Nystagmus is a repetitive sequence of the two former reflexes, a (relatively slower) fixation reflex followed by a super-fast accommodation reflex, as seen, for instance in a person sitting in a train and looking out of the window. The eyes are able to follow a specific object for a short while, then jump back and accommodate again for the next object.

Nystagmus can also be seen in a person dancing the waltz. The slow phase (fixation reflex) of nystagmus makes the eyes move against rotatory direction and thus slows the rotation of the surroundings. During the fast phase (accommodation reflex) the eyes move with rotatory direction. A type of nystagmus also occurs in a person reading a book or a newspaper. The slow phase is the reading phase, the fast phase is the jump from the end of the line just read to the beginning of the next one. During the fast phase the realization of the images is suppressed. It is evident that this makes sense.

2.4.2. Vision

Vision is the most heightened form of tele-sensibility. Even stars which are many light years away from the earth can be observed with the naked eye.

Vision is effected by mediation of light-sensitive sensors: photosensors or electromagnetic sensors.

The light which falls on an object is reflected in a specific pattern of electromagnetic vibrations. Upon entering our field of vision, the reflected light, the *image*, is turned upside down by the light-conducting part of the eye, and is sharply and strongly reduced then projected on the *light-processing* part of the eye, the *retina*. The electromagnetic sensors of the retina are stimulated to depolarize by electromagnetic vibrations. Along the connector cells and the afferent neurones of the retina the action potentials are conducted to the central nervous system via the optic nerve. The impulses are interpreted in the brain.

The origin of vision will now be discussed step by step.

a. *Creation of an image*

Light

When white (sun) light passes through a prism, one can see a spectrum, a display of colours (Fig. 2.8.22).

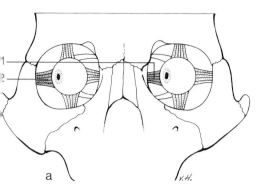

a

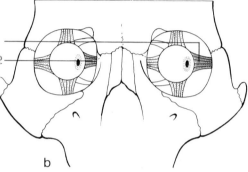

b

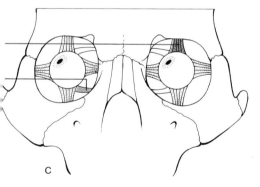

c

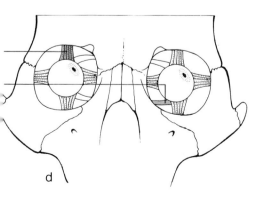

d

Figure 2.8.21
The coordination of the eye muscles of the left and right eyes in the conjugated movements of the eyes

a. looking to the right
b. looking to the left
c. looking upwards to the right
d. looking upwards to the left
e. looking downwards to the right
f. looking downwards to the left

1. *medial straight eye muscle*
2. *lateral straight eye muscle*
3. *superior straight eye muscle*
4. *inferior straight eye muscle*
5. *superior oblique eye muscle*
6. *inferior oblique eye muscle*

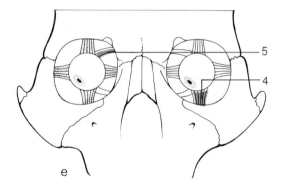

e

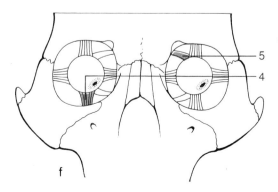

f

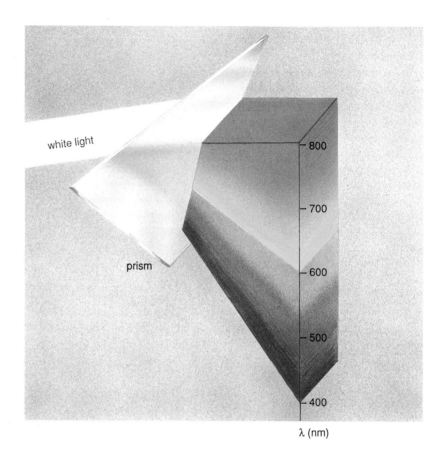

white light

prism

800

700

600

500

400

λ (nm)

Figure 2.8.22
The composition
of sunlight;
colour spectrum

All colours of the rainbow are represented in a sequence which is related to their specific wavelengths. Violet has a wavelength of approximately 400 nm, red of approximately 800 nm. White light has electromagnetic vibrations with shorter and longer wavelengths as well. Ultraviolet and infra-red are respective examples of these vibrations. Photosensors are not sensitive to these wavelengths and consequently the eyes cannot see these colours.

White light is a mixture of the colours of the rainbow. There are three primary colours within the range of colours of the spectrum, notably *red, green* and *violet*. All other colours can be composed of these three primary colours .

An object becomes visible when rays of light, coming from the object, enter the eye. This can happen when the object itself radiates as, for instance, the sun, the stars, a lamp, or when objects reflect the rays of light that fall on them. Such objects are visible without a source of light. A red lamp gives off light because rays of light with wavelengths of approximately 800 nm are radiated and reach the photosensors. A red object is seen as 'red' as the object only reflects the rays with a wavelength of 800 nm. Apparently, all other wavelengths present in the light source are absorbed by the object. A black object absorbs all wavelengths, there is no reflection, the photosensors are not stimulated. A white object does the opposite; all wavelengths are reflected, the colours remain mixed to white.

Range of vision and field of vision

In order to be able to perceive images they must be taken into the *range of vision* or *visual field*. The range of vision is the area that both eyes, by moving together, can cover. Objects can be taken into the range of vision, for instance, by turning the head or the trunk. It is self-evident that the range of vision is larger than the area which a non-moving eye can cover. The latter area is called the *field of vision* or *visual field* (Fig. 2.8.23). Only objects that are within that fixed area are seen.

The size of the visual field is related to the size of the part of the retina which can be reached by light rays.

When looking straight ahead, the horizontal visual field covered by both eyes is virtually 180°. The vertical visual field is confined by the upper and lower brims of the orbits. When the eyes are in a different position, the visual field is different.

b. *Conduction of light*

Before the light rays that enter the eye can reach the retina, they have to pass the cornea, the intra-ocular fluid, the lens and the vitreous body (see Fig. 2.8.12). These structures are called the *refractive media* of the eye, as they refract the light rays. The refraction is such that the images projected on the retina are *upside down* and *greatly reduced* (Fig. 2.8.24). Together, the re-fracting media of the eye form a biconvex (con-vex-convex) lens. As this lens turns the visual image upside down, it is called a *reversal lens*. The visual image is strongly reduced due to the radius of the curvature of the reversal lens, notably from that of the frontal surface of the cornea. Consider the sheer size of a large land-scape which fits on your retina when you are enjoying the scenery.

Accommodation, convergence and stereo-scopic vision

Normal functioning eyes allow us to see sharply, not only distant objects but close-up objects as well. This is possible only when the refraction of the light can be adjusted (Fig. 2.8.25). A change in refraction can be realized by adjusting the radius of the curvature of the

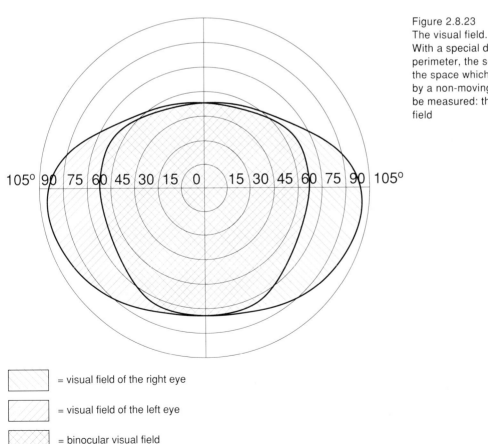

Figure 2.8.23
The visual field.
With a special device, a perimeter, the sector of the space which is seen by a non-moving eye can be measured: the visual field

105° 90 75 60 45 30 15 0 15 30 45 60 75 90 105°

= visual field of the right eye

= visual field of the left eye

= binocular visual field

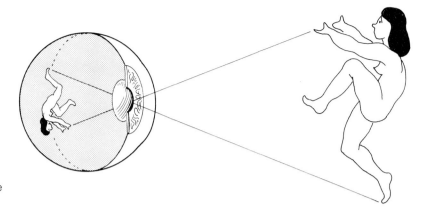

Figure 2.8.24
The reversal and the
reduction of the image by
the refracting media of the
eye

refracting media. The *lens* is the only element in the reversal lens with a variable radius of curvature.

When looking into the distance, the entering light rays run almost parallel to the optic axis. The eye is at rest, the ciliary body is relaxed, the lens fibres are pulled tight and keep the lens as flat as possible. The lens is therefore, also in a state of tension.

Suppose an object comes nearer. At a certain distance from the eye the light rays of the object will enter the eye at such a large angle in relation to the optic axis that the image projected onto the retina threatens to lose focus and become blurred. To prevent this, the radius of the curvature of the lens is increased. This is achieved by the ciliary muscle which reflexly contracts, as a result of which the diameter of the ring, formed by the ciliary body, gets a little smaller. This enables the flexible lens to become a little more convex. The lens fibres, by which the lens is suspended, remain tense. As a consequence of the larger radius of curvature of the lens, the refraction of the incoming light increases and the visual image remains in focus.

As the object gets closer, the activity of the ciliary muscle will enable the lens to become proportionally more convex. The opposite, of course, is also true. The adjustment of the radius of the curvature of the lens to the distance of the objects, is called *accommodation*. It is a reflex. The ciliary muscle, also called the accommodation muscle, is innervated by parasympathetic nerve fibres.

When we take the text of this page nearer to our eyes, we reach a point at which the text threatens

Figure 2.8.25
Accommodation

a. The lens has its largest
   convexity: maximum
   accommodation
b. The lens is flattened: no
   accommodation

1. *near point*
2. *far point*

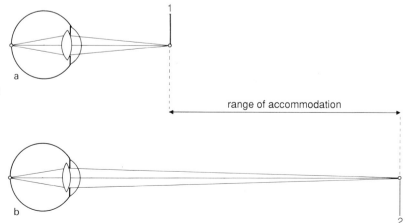

to become blurred. This is the point at which the lens has reached its natural maximum convexity. Further activity of the ciliary muscle would only relax the lens fibres.

The point at which the refraction of light is at its maximum and an object is projected on the retina just in focus, is the *near point*. The site of the near point is highly dependent on the flexibility of the lens. In babies the flexibility is maximal and the near point is at approximately 1 cm. As people grow older, the natural maximal convexity gradually decreases.

The opposite of the near point is the *far point*. This is the point at which adjustment of the radius of curvature is only minimal in order to get a sharp visual image. In normal eyes the far point is at approximately 6 metres.

The distance between the near point and the far point is called the *range of accommodation*.

The refractive strength of a lens is expressed in *diopter* (D) units. A convex lens is designated in positive diopters. The joint refraction strength of the eye at rest is approximately 60 D, of which 40 D are provided for by the cornea and the intraocular fluid, 18 D by the flattened lens and 2 D by the vitreous body. By accommodation of the lens another 14 D can be achieved.

When looking in the distance and staring, both eyes gaze straight ahead. The visual axes are parallel. When gradually gazing nearer, both eyes rotate simultaneously towards the midline; this is known as *convergence* (Fig. 2.8.26).

The visual axes cross at the point where the object being looked at is located. The convergence is brought about by contractions of the medial straight eye muscles. When objects go away from the eye, the opposite of convergence occurs. The visual axes become more parallel by contraction of the lateral straight eye muscles. So convergence occurs by reflex and takes place when objects at a distance of 100 m or less are focused upon. Analogous to range of accommodation one could speak of *range of convergence*. Normally, accommodation is accompanied by convergence.

Looking with two eyes is called *stereoscopic vision* or *binocular vision*. When an object is located in the area of the convergence range, the

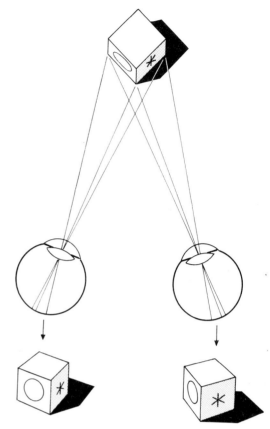

Figure 2.8.26
Convergence and stereoscopic vision. The visual axes cross (convergence). Either eye sees the object from a different angle

visual images on the left and the right retinas are not completely similar (Fig. 2.8.26). The object is looked at from different angles because of the distance between the eyes (about 6.5 cm). This can easily be demonstrated by holding an object, for instance a cube as in Figure 2.8.26, at a distance of about 30 cm from the nose. When alternately the left and the right eyes are closed, the right and the left sides of the object are seen, respectively. By interpretation of the different visual images on the retinas the dimension depth of field is added to the vision. Pouring a cup of coffee, with one eye closed, may suddenly pose a problem.

Focal depth is not only realized by binocular vision; there are other ways as well.

Size of pupil and depth of field

For a person to be able to see, the stimulus threshold of the photosensors must be crossed. An important factor is the amount of light admitted to the retina. One could say that the retina must neither be under-exposed nor over-exposed. When over-exposed, one gets dazzled. The light that falls on the retina can enter only through the opening in the iris, the *pupil*.

Light cannot pass through the iris itself. The adjustment to the amount of light occurs by proportionate dilatation or constriction of the pupil, caused by the light which enters the eye. This is a vegetative reflex: the *pupillary reflex*. In bright sunshine the pupil is as narrow as possible; when it is virtually dark the pupil will be as wide as possible.

Smooth muscle tissue in the iris controls the size of the pupil (Fig. 2.8.27). The *dilator muscle of the pupil* dilates the pupil, the *constrictor muscle* narrows the pupil. These muscles are antagonists.

The light refracting qualities of the edges of the lens are different from those in the centre. The consequence of this is that two objects which are within the accommodation range with a certain distance between them cannot be projected on the retina as a sharp image when the pupil is dilated (Fig. 2.8.28). By means of accommodation, either the one object or the other can be projected sharply. When the pupil is narrow, both objects can create a sharp visual image. In the latter case the edges of the lens are not used. The distance between two objects which are both projected sharply within the accommodation range is called the *depth of field*.

The depth of field grows when the pupil narrows, in other words when there is more light entering the eye.

c. *Processing of light*

After the light has been conducted, an upside-down, sharp and highly reduced image is projected on to the retina. The retina converts the image into impulses which can be conducted to the central nervous system.

The retina consists of two layers or sheets (see Fig. 2.8.13), *the stratum pigmentosum* or *pigment layer* and the *stratum nervosum* or *optic layer*.

The pigment layer consists of epithelial cells which, via processes, connect with the light-sensitive parts of the photosensors. The epithelial cells contain many dark pigment granules, which absorb all light that falls on

Figure 2.8.27
The pupil reflex

a. When the amount of light entering falls, the dilator muscle dilates the pupil
b. When the amount of light entering grows, the contractor muscle contracts, as a result of which the pupil narrows

1. *entering light*
2. *dilator muscle of pupil*
3. *pupil*
4. *iris*
5. *constrictor muscle of pupil*

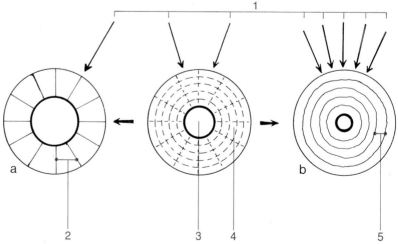

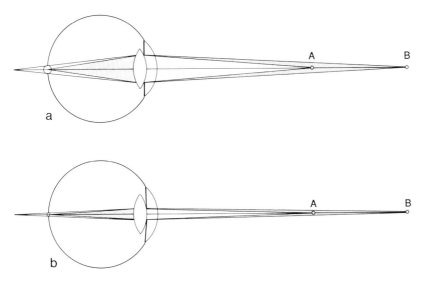

Figure 2.8.28
Depth of field

a. When the pupil is
   wide and the lens
   has a certain
   strength, point B is
   sharply projected on
   the retina (grey
   cones), whereas
   point A is projected
   as a small spot
b. When the pupil is
   narrow and the lens
   is similar in strength
   to case A, point A is
   projected as a much
   smaller spot, so that
   A and B are seen
   virtually equally as
   sharp. In this case
   the distance from
   A–B is the depth of
   field for this eye

them. This keeps the inside of the eyeball itself
from reflecting light rays, as this would be disadvantageous to the images projected on
the retina. Moreover, the pigment granules help
in keeping the photosensors from becoming
overexposed, when maximum narrowing of the
pupil is ineffective. In such a case the
pigment granulae move to the processes of the
pigment cells and thus apply a light-absorbing
layer around the light-sensitive parts of the
photosensors (Fig. 2.8.29).

The optic layer consists of a layer of *photosensors*, a layer of *connector cells* and a layer of
afferent neurones (Fig. 2.8.13). When the
photosensors are stimulated by an adequate
amount of light, they depolarize and action

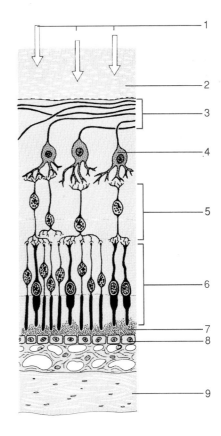

Figure 2.8.29
Cross-section through the retina

1. *entering light rays*
2. *vitreous body*
3. *neurites going to the optic nerve*
4. *afferent neurones*
5. *connector cells*
6. *layer with rods and cones*
7. *pigment granules*
8. *pigment layer*
9. *sclera*

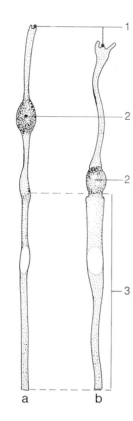

potentials go to the central nervous system via the connector cells and the afferent neurones.

Cones and rods

There are two categories of photosensors, named after their shapes, *cones* and *rods* (Fig. 2.8.30). They contain visual pigments, which are capable of absorbing light and converting the light stimulus into a generator potential.

The visual pigment *rhodopsin* or *visual purple* is produced in the rods (Fig. 2.8.31). The light stimulus decomposes the rhodopsin. One of the substances released induces a general potential. Thereafter more rhodopsin is synthesized, especially in darkness. Vitamin A plays an important part in this process. The rhodopsin reacts to light of all wavelengths of visible light and, therefore, is not used for the discrimination of colours. For this reason the cones are also called *light/dark sensors*. They perceive differences in *intensity of light*.

There are three types of cones, each with different visual pigments (Fig. 2.8.31). Each type reacts to light of a specific wavelength. The cones are *colour-sensitive sensors*.

On the basis of their specialization they are designated as *red/yellow* sensitive, *green* sensitive and *blue/violet* sensitive.

Figure 2.8.30
Structure of the
photosensors

a. Rod
b. Cone

1. *transmissive part*
2. *body*
3. *receptive part*

Figure 2.8.31
Types of cones

1. *blue-violet sensitive*
2. *green sensitive*
3. *red-yellow sensitive*

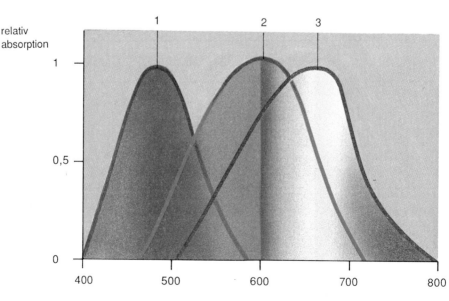

By means of these three types of cones, each of which is sensitive to one of the primary colours, the retina is capable of seeing all colours. Similar to the way in which all kinds of colours can be mixed from the three elementary colours, the central nervous system is able to distinguish about 160 colours on the basis of combinations of signals received from the cones. When the three types receive equally strong stimuli, we see the 'colour' white. When the cones are not stimulated at all, we see the 'colour' black. Dependent on the ratio at which the three types of cones are stimulated, we see all other colours.

The two types of photosensors are not equal in number and they are not evenly distributed over the retina. There are approximately 120 million rods and only 5 million cones. In the central part of the retina, the yellow spot, there are only cones (approximately 150,000 per mm^2). Towards the edges, the density rapidly becomes smaller (to 5,000 per mm^2). At the edges of the retina there are virtually no cones at all.
The distribution of the rods is virtually the reverse of the situation with the cones. Their density is great at the edges of the retina (approximately 30,000 per mm^2). In the direction of the central part of the retina, the density initially grows (to approximately 160,000 per mm^2) and then decreases quickly. There are no rods in the macula lutea (yellow spot).

The stimulus threshold of the rods is lower than that of the cones. Consequently, colours are not seen when the light intensity is relatively low. For that reason, everybody becomes colour blind during twilight.

The rods have a large *dark adaptation*. When one has been in a dark room for a while, it is noticeable that, after some time, one sees progressively more. This dark adaptation is caused by two factors: in the first place the rods build ever more rhodopsin and the light sensitivity grows. In the second place, the pigment granules are withdrawn as deep as possible into the pigment layer, which causes the rods to be maximally exposed to the small amount of light. One needs about 15 minutes for a reasonable

dark adaptation, but thereafter the sensitivity can grow considerably. Maximum light sensitivity is not reached until one has spent about ten hours in darkness. Dark adaptation is a form of *positive adaptation.*

Light adaptation is the opposite, but it holds good for the cones. When one comes out of a dark room and enters into bright sunshine, one is temporarily dazzled. Not until a considerable amount of visual pigment has been used and an absorbent layer around the photosensors has been provided by the pigment granules does one

Figure 2.8.32
The difference in discrimination of rods and cones (schematic)

1. *entering light*
2. *afferent neurones*
3. *connector cells*
4. *rods and cones*
5. *pigment layer*
6. *macula lutea*
7. *blind spot*
8. *optic nerve*

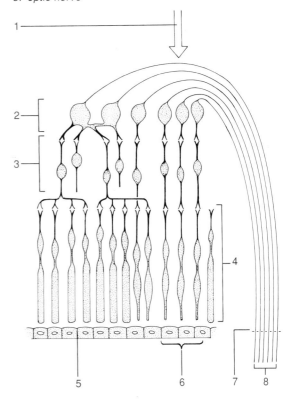

begin to see again. Sunglasses are used to help the pigment layer. Light adaptation is a form of *negative adaptation*.

The discrimination of the rods is considerably less than that of the cones: several rods are coupled to one connector cell and several connector cells are coupled to one afferent nerve fibre (Fig. 2.8.32). Consequently, the rods are less discriminative than the cones, because the nerve fibre 'does not know' from what spot on the retina the stimulus comes.

It can be seen from the differences between rods

and cones that vision functions least effectively during twilight. The comparatively large number of road accidents during that period may be related to this phenomenon. The centre of one's attention is always projected on the macula lutea, where only cones are located, the stimulus threshold of which has gradually become too high for the amount of light available. An important source of information, colour, disappears due to the cones being eliminated. The rods are not yet sufficiently dark-adapted and vision by means of the rods is less sharp because of their location underneath the capil-

Figure 2.8.33
Optic tracts

1. *optic nerve*
2. *optic chiasma*
3. *optic tract*
4. *optic radiation*
5. *eyeball*
6. *skull*
7. *lateral geniculate body*
8. *cerebrum*
9. *cortical visual area*

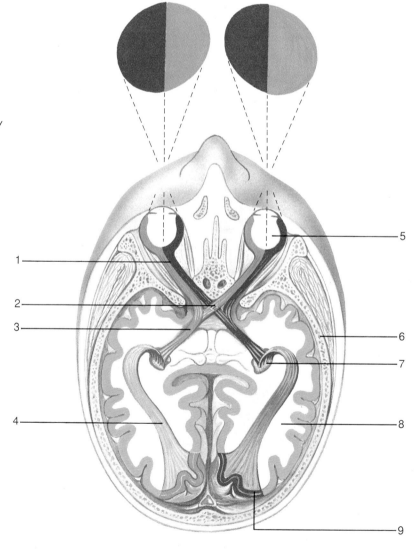

laries and nerve fibres, and their reduced discrimination.

Optic tracts
The signals from the photosensors are transmitted to the afferent nerve fibres via the connector cells. The axons of these fibres usually take the shortest pathway, going around the macula lutea (yellow spot), to the papilla. They leave the eye here and join the optic nerve. During their journey the fibres from the different parts of the retina maintain their specific positions (Fig. 2.8.33).

The optic nerve enters the skull through an opening at the back of the orbit. Just behind the pituitary stalk the two optic nerves converge, partly forming a crossing, the *optic chiasma* (see Fig. 2.6.4). The fibres from the nasal part of each retina cross in the chiasma. The fibres from the temporal half of each retina pass the chiasma and remain uncrossed.

After the optical chiasma there is a left and a right *optic tract*. These tracts terminate in the left and right *lateral geniculate bodies*, which are parts of the thalamus. The thalamus is therefore also a control centre for the visual sensory system. The last optic nerve fibres leave the geniculate bodies in line to go to the cortex. In the *optic radiation* they diverge to the visual cortex in the posterior part of the occipital lobe.

The result of the partial crossing of fibres in the optic chiasma is that all light stimuli coming from the left visual fields of both eyes (and therefore from the right retinal halves) arrive at the right cerebral half. Conversely, all light stimuli from the right visual field (and therefore from the left retinal halves) arrive at the left cerebral half.

The points at both retinas on which the same part of the visual fields are projected, the *identical points*, have a common terminal on the visual cortex. It is remarkable that the fibres from the yellow spot take a divergent course. Some of them cross, others do not. Moreover this small retinal area takes up the largest part of the visual cortex.

d. *Visual perception*

In the visual cortex the entering impulses are interpreted. The brain constructs recognizable images out of the numerous action potentials which come from the retinas. This function is called *visual perception*. In this process the reduced, reversed, two-dimensional retinal images are enlarged, made to stand upright and composed into three-dimensional images by the activities of the brain.

The way in which the images are understood is related to inherent and learned features. Seeing a fearsome dog generates an inherent impulse to

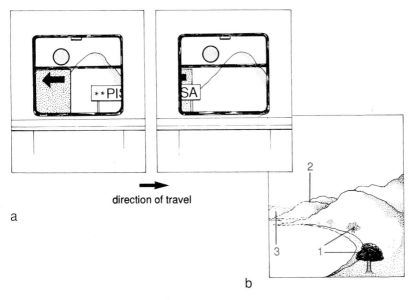

direction of travel

a

b

Figure 2.8.34
Depth perception

a. The observer, for instance a train traveller, is moving. Nearby objects move faster through the field of view than far away objects

b. The observer stands still. Measuring the distance is done by: differences in sizes (1), crossing contours (2), brightness (3)

flee or to resist. When seeing a meal, the reaction is highly dependent on former experiences; one person's mouth will water when a favourite dish is seen. The same dish may induce an urge to vomit in another person.

Depth perception

Depth perception at a shorter distance is effected by *binocular vision* or *stereoscopic vision* (Fig. 2.8.26).

At longer distances and when using only one eye, the brain uses a number of other phenomena for the perception of depth, such as relative motion, relative size, crossing contours, the location of shadows and the differences in brightness and colour intensity (Fig. 2.8.34).

Visual illusions

There are many examples which show that visual perception is not in accordance with reality (Fig. 2.8.35). We are deceived. This visual deception can take on many forms: one may see differences in contrast that are not there, colours, shapes or motions that are not there, and sizes and distances which differ from reality.

Visual illusions demonstrate that perception is often not factual. Illusions show the vulnerability of the otherwise extremely efficient visual perception.

2.5. Ears

The ears are considered the seat of the *sense of hearing* or *auditory sense*.

Sense of hearing is the capability to perceive auditory stimuli. From a functional anatomical point of view, although different to common usage, the ears are more than just the auricles which decorate the sides of the head. The ear is

b

Figure 2.8.35
Examples of visual
deceptions

a. Impossible object
b. Absent motion.
   Keep the figure at
   a distance of about
   30 cm from
   the eyes

a

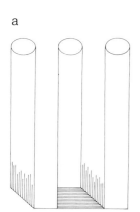
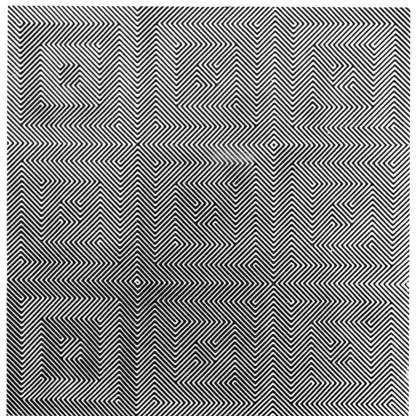

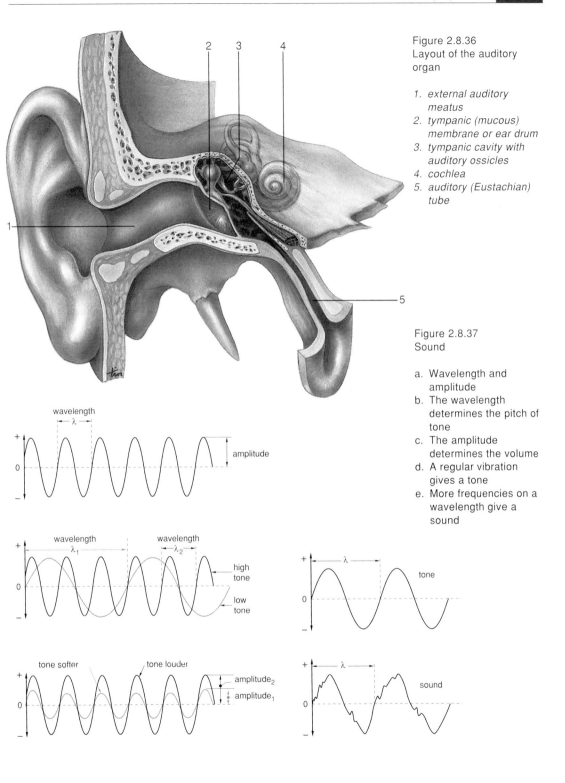

Figure 2.8.36
Layout of the auditory
organ

1. *external auditory
   meatus*
2. *tympanic (mucous)
   membrane or ear drum*
3. *tympanic cavity with
   auditory ossicles*
4. *cochlea*
5. *auditory (Eustachian)
   tube*

Figure 2.8.37
Sound

a. Wavelength and
   amplitude
b. The wavelength
   determines the pitch of
   tone
c. The amplitude
   determines the volume
d. A regular vibration
   gives a tone
e. More frequencies on a
   wavelength give a
   sound

an intricate sensory organ, the core of which is formed by *mechanosensors*. In the auditory organ, sound is converted in such way that these mechanosensors are stimulated. When the stimulus threshold has been crossed, action potentials are conducted along the nerve fibres of the auditory nerve. The arriving impulses are interpreted in the brain.

Three parts can be distinguished in the auditory organ: the *stimulus receiving* part, the *stimulus conducting* part and the *stimulus processing* part.

Sound

Sound is a pattern of vibrations in the air. These vibrations are usually composed of a large number of *tones*. When a tone is represented in a diagram a regular wave is seen (Fig. 2.8.37a). A whole wave (top plus bottom) is equal to one vibration. The number of vibrations per second, the *frequency*, determines the *pitch* (Fig. 2.8.37b). The more vibrations per second, the higher the pitch. The frequency (and therefore the pitch) is expressed in Hertz (Hz): 1 Hz is one vibration per second, 100 Hz is 100 vibrations per second.

The *volume* is determined by the size of the wave, the amplitude. The larger the amplitude, the greater the volume. Volume is expressed in decibels (dB). When several tones with different frequencies and/or amplitudes are put together, it is called a *sound* (Fig. 2.8.37e).

The auditory organ can perceive sound with a frequency of 20 Hz up to approximately 16,000 Hz. Females can often hear tones up to 20,000 Hz.

The largest sensitivity is in the frequency area between 2,000 to 4,000 Hz. In this area the stimulus threshold is lowest.

2.5.1. Structure

External ear

The *external ear* is the stimulus-receiving part of the auditory organ. It consists of the *auricle* and the *external auditory meatus* (Fig. 2.8.36). The *auricle* consists mainly of elastic cartilage, covered with skin. There are a number of folds in the cartilage which give the characteristic shape to the auricle. Somewhere in the centre of the auricle there is a depression which leads to the auditory meatus. The earlobe consists only of connective tissue, covered with skin, and contains relatively many dermal sensors.

Thanks to the shape and the position of the auricles, information is received about the direction of the sound. The auricles also serve as a kind of funnel to channel the sound into the auditory meatus. Sometimes the task of the auricle is assisted by placing a flat hand behind it. Some people can move their auricles up and down a little, due to the fact that they, unlike most people, can control the muscles which run from the skull to the auricle.

The *external auditory meatus* is about 2.5 cm long. The first part is a continuation of the cartilage of the auricle. Thereafter the canal continues in the petrous part of the temporal bone of the skull and comes to an end at the tympanic mucous membrane. Because of the shape of the auditory meatus the sound is amplified about three times. Specialized perspiratory glands in the skin of the auditory canal produce *cerumen* or *earwax*. This fatty substance keeps the skin and the tympanic membrane supple and water resistant. The skin of the first portion of the auditory meatus has hairs. This, together with the earwax, blocks dust particles.

The *tympanic membrane* or *eardrum* has a diameter of approximately 1 cm and is the partition between the external and middle ear. It is set obliquely and its upperside is somewhat laterally located. The sound vibrations make the very thin tympanic membrane vibrate.

Middle ear

The *middle ear* is the stimulus-conducting portion of the auditory organ. It is a narrow, rather irregular cavity in the petrous part of the temporal bone, called the *tympanic cavity*. The tympanic cavity is connected with the naso-pharyngeal cavity via the 4 cm long *auditory (Eustachian) tube*. The insides of the auditory tube and the tympanic cavity are covered with a mucous membrane, ciliary epithelium, similar to that of the pharyngeal cavity. Surrounding the site where the tube terminates in the nasopharynx, lymphatic tissue, part of the tonsillar (Waldeyer's) ring, is located (see Fig. 2.1.51). Normally, the auditory tube is 'squeezed' because of the tension in the tissue of the tube. When an increase in pressure occurs in the pharyngeal cavity, for instance during swallowing, the tube is temporarily pushed open, so that air can flow to or from the tympanic cavity. The connection is important for hearing. In order to prevent the tympanic membrane becoming tense, and consequently vibrating less easily, the pressure in the tympanic cavity must always be equal to the outside pressure.

During take-off in an airplane, the outer pressure falls. The tympanic membrane will, as it were, bulge outwardly. The sweets from the hostess are meant to stimulate swallowing, and consequently open the auditory tube. During landing the opposite occurs.

There are two small openings, each of them covered with a membrane, in the wall of the middle ear opposite the tympanic membrane.
The upper opening is called the *round window* or *fenestra cochleae*, the lower one is the *oval window* or *fenestra vestibuli*.
In the air-filled tympanic cavity lies the *ossicular chain*, which provides the connection between the ear drum and the oval window. It consists of three *auditory ossicles*, namely the *malleus*, the *incus* and the *stapes* (Fig. 2.8.38). The thin part of the malleus, the handle, is attached to the tympanic membrane, its head beside the broad part of the incus.

The incus in turn attaches by means of a thin process to the top of the stapes. The oval base of the stapes fits exactly to, and is continuous with, the oval window.

The vibrations of the tympanic membrane are transmitted to the oval window by the ossicular chain. Because of the joints between the bones they act collectively like a lever, doubling the vibrations. Due to this lever function and to the difference in size between the ear drum and the oval window, the vibrations are amplified approximately 20 times.

Each sound vibration, which makes the ear drum vibrate, is amplified and transmitted to the oval window. Very harsh sounds can be subdued by the activities of two muscles in the middle ear

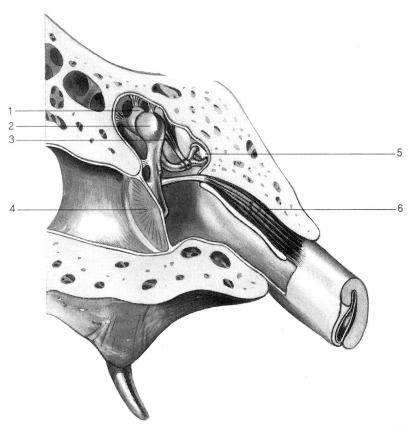

Figure 2.8.38
The tympanic cavity with the auditory ossicles

1. *incus or anvil*
2. *malleus or hammer*
3. *stapedius muscle*
4. *tympanic membrane*
5. *stapes or stirrup bone*
6. *tensor tympani muscle*

Figure 2.8.39
Osseous or bony labyrinth

1. semicircular canals
2. oval window
3. round window
4. vestibulocochlear nerve
5. cochlear nerve
6. vestibular nerve
7. cochlea

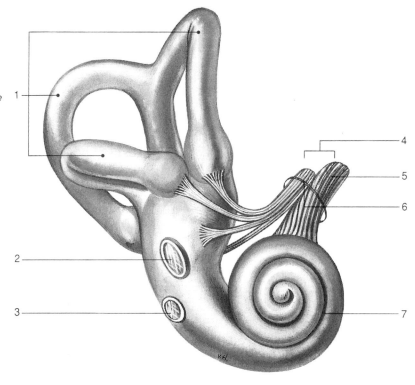

(Fig. 2.8.38). The first muscle runs to the handle of the malleus. When the sound becomes stronger, the vibrations of the tympanic membrane become greater and the tensor tympani muscle will contract as a result of the muscle spindle reflex. This tenses the tympanic membrane, so that it cannot vibrate so much. The stapedius muscle is inserted on to the top of the stapes and can similarly keep the movements of the stapes under control.

Inner ear

The *inner ear* contains the stimulus-processing portion of the auditory organ. It is part of a complex of cavities, passages and canals, lodged in the petrous part of the temporal bone, the *osseous* or *bony labyrinth* (Figs. 2.8.36 and 2.8.39).

The labyrinth (Fig. 2.8.39) contains:

– a centrally located space, immediately behind the oval window, the *vestibulum*;
– three *semicircular canals*, which terminate posteriorly at the vestibulum;

– the *cochlea*, a spiral-shaped structure which looks like a snail shell.

The cochlea is part of the auditory organ. The other portions of the bony labyrinth are parts of the equilibratory sensory system.
The bony labyrinth is covered with periosteum. Within the bony labyrinth is located the membranous labyrinth, which has an even more intricate shape (Fig. 2.8.40). It is a closed membrane, all parts of which are interconnected. It is filled with a clear, lymphatic fluid, the *endolymph*. In the space between the membranous labyrinth and the periosteum of the bony labyrinth *perilymph* is located. The composition of this fluid is a little different from that of the endolymph.

The cochlea has $2\frac{1}{2}$ spirals. The length of the cochlear canal is approximately 4 cm. The canal is divided into three spaces, comparable to three parallel, winding staircases: the *vestibular canal of cochlea* or *scala vestibuli*, the *scala*

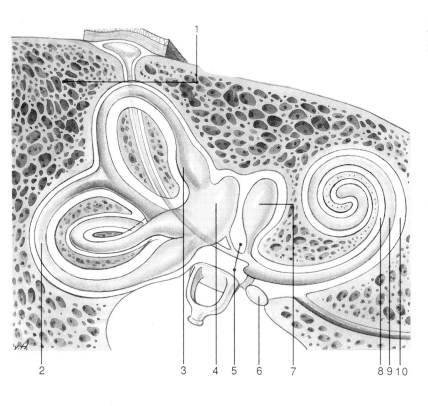

Figure 2.8.40
The membranous
labyrinth

1. petrous part of
   temporal bone
2. semicircular canal
3. ampulla
4. utricle
5. vestibulum
6. round window
7. saccule
8. scala vestibuli
9. scala media
10. scala tympani

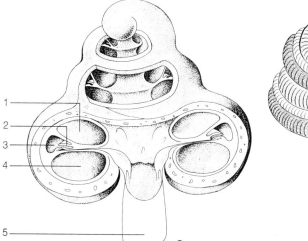

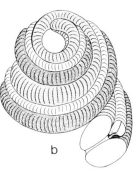

Figure 2.8.41
The cochlea

a. Open

1. scala vestibuli
2. organ of Corti or
   spiral organ
3. scala media
4. scala tympani
5. cochlear nerve

b. Impression of the
   meanderings of the
   three canals

media and the *tympanic canal of cochlea* or *scala tympani* (Fig. 2.8.41).

The scala media is the smallest space. It is formed by the part of the membranous labyrinth which bulges into the cochlea and is therefore filled with endolymph. The scala media is situated against the outer curve of the cochlea and does not extend into its apex.

The scala vestibuli begins in the vestibulum at the level of the oval window and communicates with the scala tympani at the apex of the cochlea. Both canals are filled with perilymph. The scala tympani terminates at the round window. The opening connecting the scala vestibuli to the scala tympani is called the *helicotrema*.

Figure 2.8.42
The spiral organ (organ of
Corti)

a. Location in relation to
the three canals.
(shaded: perilymph,
white: endolymph)

b. Detail

1. scala vestibuli
2. nerve cells of the
acoustic nerve
3. vestibular membrane of
cochlear duct (Reissner's
membrane)
4. scala media
5. spiral organ (organ of
Corti)
6. scala tympani
7. tectorial membrane
8. auditory sensors (hair
cells)
9. basilar membrane

c. Afferent nerve/origin

1. vestibulocochlear
nerve or auditory
nerve
2. cochlear nerve
3. vestibular nerve
4. afferent fibres

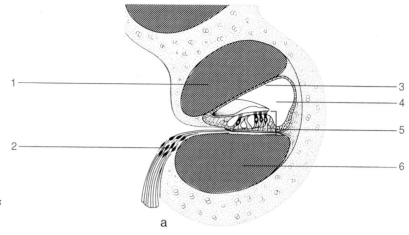

a

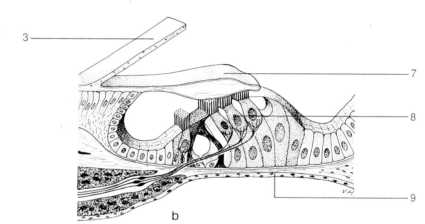

b

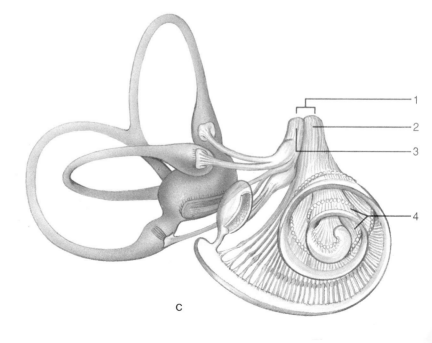

c

The combined diameter of the scala vestibuli and the scala media is about as large as that of the scala tympani.

Reissner's membrane or the *vestibular membrane of the cochlear duct* is the partition between the scala media and the scala vestibuli (Fig. 2.8.42). The basilar membrane is thicker and forms the separation between the scala media and the scala tympani in the outer curve of the cochlea. On the whole length of the $2\frac{1}{2}$ spirals of the basilar membrane the *organ of Corti* or *spiral organ* is located. This organ can be considered the centre of the auditory organ. At this site the acoustic vibrations are converted into action potentials which thereafter travel to the brain. The main elements of the spiral organ are the *auditory sensors* and the *tectorial membrane*. There is a layer of supporting cells on the basilar membrane. The auditory sensors are located on the supporting cells. They have at their upper sides short protrusions, cilia, which look like hairs, and which are sometimes called *hair cells*. The tectorial membrane rests on the 'hairs'. Nerve fibres go from the auditory sensors in the direction of the central axis of the cochlea (Fig. 2.8.42c). The afferent fibres join here to form the cochlear nerve.

Acoustic nerve tracts

The cochlear nerve joins with the vestibular nerve, or nerve of equilibrium, to form the eighth cranial nerve (N.VIII), the vestibulo-cochlear, or acoustic nerve (anatomical name) or N. stato-acusticus (physiological name). The auditory nuclei are located in the brain stem. From these nuclei auditory fibres enter the thalamus, usually after a number of synapses, and continue to the primary auditory or acoustic cortical area of the cerebrum (see Fig. 2.7.32).

2.5.2. Hearing

Hearing is a form of tele-sensibility. The sound vibrations which reach the ear are channelled to the tympanic membrane via the auricle and the auditory meatus. The tympanic membrane receives the vibrations and the oval window is, via the auditory ossicular chain, made to vibrate. As the sound vibrations take this path, they are considerably amplified for three reasons: the shape of the meatus (approximately 3 times), the lever function of the auditory ossicular chain, and the differences in the surface areas of the tympanic membrane and the oval window (approximately 20 times).

The fluid spiral in the scala vestibuli is made to vibrate by the vibration of the oval window. As fluid can hardly be compressed, each vibration of the oval window is mirrored in the round window, as the perilymph and endolymph are entirely surrounded by a bony cover. A bulging of the oval window away from the tympanic cavity is accompanied by a bulging of the round window in the direction of the tympanic cavity

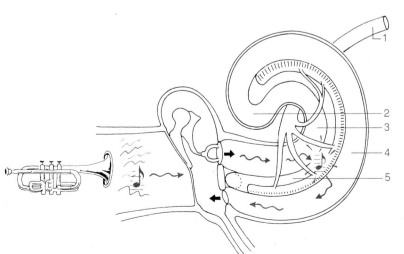

Figure 2.8.43
High tones cover shorter distances in the scala vestibuli than low tones

1. cochlear nerve
2. helicotrema
3. scala vestibuli
4. scala tympani
5. scala media

Figure 2.8.44
The basilar membrane and
its corresponding
frequencies

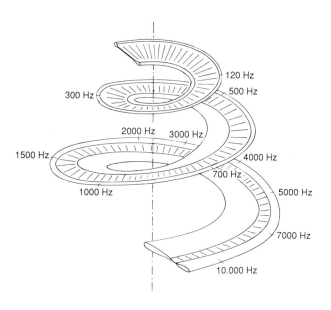

(Fig. 2.8.43). The scala media is also made to move by the motion in the scala vestibuli. The resulting movement of the basilar membrane leads to a displacement of the hairs of the auditory sensors in relation to the tectorial membrane (Fig. 2.8.42). This mechanical stimulation – auditory sensors are mechanosensors – leads to depolarization, whereafter action potentials are propagated along the afferent nerve fibres to the auditory nerve. Arriving impulses are then interpreted by the brain.

Depending on the pitch of the tone (the frequency), the vibration in the scala vestibuli will travel shorter or longer distances away from the oval window in the direction of the apex of the fluid spiral, before it descends to the round window. This is related to the structure of the basilar membrane.

High tones (high frequency) make the basilar membrane at the beginning of the cochlea move, the lowest frequencies make the basilar membrane at the apex of the cochlea move (Fig. 2.8.44). In this context, the basilar membrane could be compared to a harp in the shape of a winding staircase; the strings at the bottom are short, thin and tense (high tones), whereas the strings at the apex are long, thick and less tense (low tones).

A specific pitch corresponds with a specific part of the spiral organ. In the internal ear the sounds are broken down into the tones of which they are composed. The pitch and the volume are perceived in the primary acoustic or auditory cortical area. To interpret these impulses into meaningful sounds, noises, etc. one has to refer to the secondary acoustic cortex.

Vibrations in the skull also lead to acoustic perceptions. This *bone* or *cranial conduction* causes other persons to hear the sound of our voice differently from the way we ourselves hear it. This phenomenon is noticed when we hear our recorded voice.

3. Propriosensors

Propriosensors provide information about the locomotor apparatus. These sensors are located in the muscles, the tendons, the joints and in the sense organs of the vestibular organ. All of them are *mechanosensors*. They belong to the area of animal *sensibility*, as they provide information about the interaction between the individual and his environment.

3.1. Organs of balance

The organ of balance is often considered a sensory organ. Strictly speaking this is not true. The organ of balance, of which the centre is formed by mechanosensors, is a highly specialized organ, like the other five sensory organs which

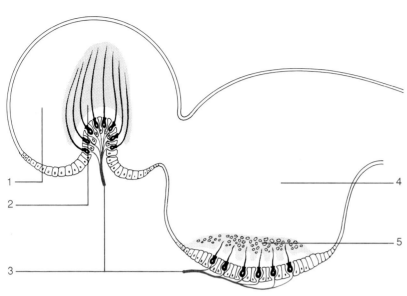

Figure 2.8.45
The ampulla with the
cupula and the utriculus
with the statoconium

1. ampulla
2. cupula
3. afferent fibres
4. utriculus
5. statoconium with
   statoliths

have been previously discussed. Mechanosensors give information on position and change of position (speed and direction) of the head, and therefore indirectly of the body, in space.

From an anatomical point of view the organ of balance forms a unit with the internal ear, namely the *labyrinth* (see Fig. 2.8.40). The elements which are important for the equilibrium are the vestibulum and the three semicircular canals. In this part of the osseous labyrinth, there is also a membranous labyrinth, as in the cochlea. There are two membranous spaces in the vestibulum: the *utricle* and the *saccule*. There are membranous canals in the semicircular canals which attach to the utricle with a dilatation, an ampulla. As to the way the spaces are filled with perilymph and endolymph, the same holds true as in the internal ear.

Within the utricle, the saccule and the three ampullae, the static sensors are located. The afferent fibres of these sensors form the vestibular nerve, the *nerve of equilibrium*. This nerve joins with the acoustic nerve to form the vestibulocochlear nerve (N. VIII).

Utricle and saccule

On the floor of the utricle (horizontally) and in the sides of the saccule (vertically), equilibratory sensors are located. They are embedded in supporting tissue, the *macula* (Fig. 2.8.45).

The sensors have 'hairs' (cilia) which project into a gelatinous substance, the *statoconium*, full of crystalline particles composed of calcite and protein, the *statoliths*.

The particles are under the influence of gravity and exert pressure on the mechanosensors in the equilibratory organs. With the head in an oblique position, the pressure of the statoliths will be different from that encountered in an upright position. Thus, the sensibility pattern of the sensors provides adequate information on each possible position of the head in space.

Information is not only given about the position of the head, but also about the change in its position. When one sits in an accelerating car, the statoliths initially lag behind due to the force of inertia. The opposite occurs when the car suddenly brakes. The same occurs, but then in a vertical direction, when one is in a lift. These examples show that specific stimulation of the sensors enables us to perceive the direction and the speed of the head – and therefore the body – in motion.

Semicircular canals

There are three semicircular canals, each lying at a right angle to the other two (Fig. 2.8.46). The planes in which they are situated are not parallel to the body planes.

Figure 2.8.46
Location of the osseous
labyrinth in the lower part
of the skull

1. *foramen occipitale*
2. *petrous part of*
   temporal bone
3. *sella turcica*

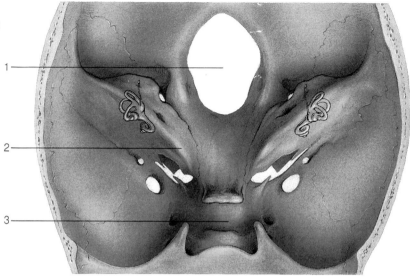

The end of each semicircular canal is dilated and forms an *ampulla* (see Fig. 2.8.40). There is a crest of sensory cells in the wall of the ampulla (Fig. 2.8.45). The afferent nerve fibres of these sensors join the fibres that come from the utricle and the saccule. Together they form the vestibular nerve. The sensors have hair-like processes projecting into a gelatinous substance, the *cupula*. The cupula functions like a kind of swing door in the passage. Depending on the direction the endolymph flows, the 'swing door' will move to the one or the other side. This movement makes the sensory cells depolarize.

For simplification, let us assume that the planes in which the semicircular canals are situated, do correspond with the three body planes. When one shakes one's head the endolymph in the 'transversal' canal lags behind the motion due to the force of inertia.

The fluid therefore moves in a direction which is opposite to that of the head. The cupula will bend with the flow and by doing so, cause the mechanosensors to depolarize. The fluid in the other canals does not move. When one nods, the endolymph in the sagittal canal will move, because of the force of inertia, and bend the cupula in that direction. When one bends one's head sideways there will be a flow of fluid in the frontal canal.

All movements of the head can be reduced to a combination of movements in the three planes in which the semicircular canals are situated. The movement concerned is 'reconstructed' from the specific supply of information that is carried to the brain via the ampullae.

When one rotates around one's longitudinal axis for some time, such as occurs during spinning, for instance, the endolymph in the transversal canal will gradually be carried along and develop the same speed as the wall. The cupula will enter a state of rest and provide no stimuli. When one suddenly stops, the force of inertia makes the fluid 'run on'. The cupula is bent, which generates for a while the sensation that one has been rotating in the other direction.

Everybody will have had the experience that excessive stimulation causes a sensation of dizziness. This can grow to such an extent that one becomes sick and even starts vomiting. Seasickness is a striking example of this.

Afferent equilibratory tracts

The majority of stimuli from the equilibratory organs travel to the nuclei in the brain stem. A small number travel to the cerebellum. The muscles receive impulses from these parts of the brain. In this way information from the

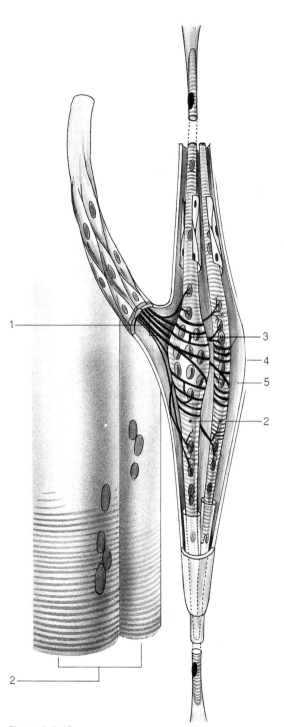

Figure 2.8.47
The muscle spindle

1. afferent fibres
2. striated muscular tissue
3. muscle spindle sensors
4. spindle-like capsule
5. lymph-like fluid

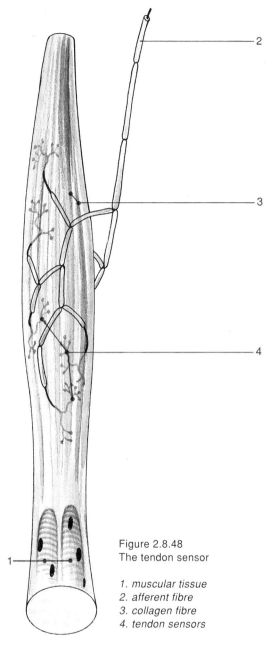

Figure 2.8.48
The tendon sensor

1. muscular tissue
2. afferent fibre
3. collagen fibre
4. tendon sensors

equilibratory organs can play an important part in regulation of posture and movements, notably in certain reflexes – for instance the crossed extensor reflex.

3.2. Muscle spindles

Muscle spindles are mechanosensors which occur in most striated muscles. They consist of a number of modified muscle fibres surrounded by a spindle-like capsule with lymph-like fluid (Fig. 2.8.47). Afferent nerve fibres are wound around the muscle spindles. When the muscle is stretched, action potentials travel to the central nervous system via the afferent fibres. The muscle spindles give information on the (passive) change in length of the muscle. A sudden stretch of a skeletal muscle often gives rise to the *muscle spindle reflex* (see Fig. 2.7.24a). The muscle spindle reflex is an important posture-regulating reflex.

3.3. Tendon sensors

Afferent nerve fibres in the tendon fibres inform the central nervous system of the tension in the tendons (Fig. 2.8.48). This is important in order to be able to estimate (consciously or unconsciously) what forces are exerted on the body. Moreover, they are capable of slowing down muscular activity when strong stimuli are received. This may help prevent damage to muscular tissue.

3.4. Articular sensors

Sensors, which are located between the fibres of the articular capsules and the ligaments, inform the central nervous system, via their afferent fibres, on the position and change in position of the joints. This information makes it possible, for instance, to faultlessly take the tip of the finger to the tip of the nose, with one's eyes closed.

4. Enterosensors

Enterosensors are sensors which are located in the walls of hollow organs. They are found in the oral cavity, the intestinal tract, the lungs, the urinary passages, the blood vessels, the liver, etc.

The enterosensors belong to the vegetative sensibility. The most important enterosensors have been discussed earlier. They are: stretch sensors in the walls of the bronchiae, pressor sensors in the aortic wall, blood acidity sensors in the respiratory centre, etc.

Motor system

9

Introduction

Movement is essential for man. Through the human capacity to move, the individual is able to establish active contact with the outside world. Here we are not solely concerned with 'daily' movements, such as walking and running, sitting down, bringing food towards the mouth, chewing and swallowing, scratching yourself when you itch, but also with special types of movements, such as ballet-dancing, playing tennis, driving a car, the proceedings during a heart operation or manipulating a defective satellite in space. Movement completes the communication and integration between man and the world. Movement is life. The heart moves in order to pump the blood around, the thorax and the diaphragm move to make breathing possible, the bladder moves to discharge urine, etc.

Making movements is at the same time a conditioning for life and a source of joy in life.

A human being moves incessantly. Generally there is a balance between activity and rest. Often this is regulated without any effort. Even when asleep man takes up a different lying position on average every two hours. Judging by the shape and the movements of his mouth you can often tell how someone is feeling, what his relation to the world is; a self-assured posture, a threatening move, a submissive attitude.

We regard both the animal motor system and the vegetative motor system as being part of the motor system. The vegetative system has been described with the corresponding body systems.

In this chapter the animal part of the motor system will be discussed, e.g. the skeleton, the joints and the muscular system. These subjects are often summarized under the heading of *locomotor apparatus*.

When discussing the locomotor apparatus we make use of subdivisions: *head, trunk* and *extremities* (arms and legs).

Learning outcomes

After studying this chapter you should have sufficient knowledge and understanding of:

– the general importance of the motor system;
– the structure and functions of the skeleton and the connections between bones;
– the structure and coherence of the skeleton of head, trunk and limbs;
– the general structure and function of muscles and tendons;
– the position and functioning of the muscular system of head, trunk and limbs.

Figure 2.9.1
The skeleton

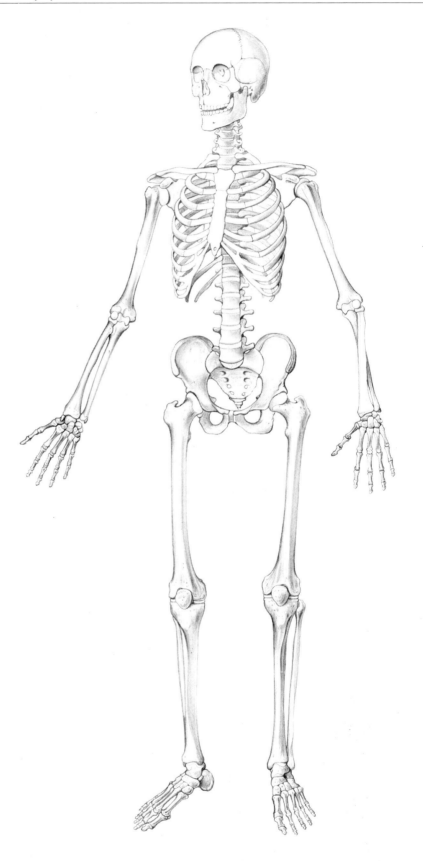

1. The skeleton

The *skeleton* supports the body from within (Fig. 2.9.1). It is an internal framework to which all the other elements are attached. The parts of the skeleton, the *bones* (ossa), are connected together in a more or less flexible way by means of *joints*, articulations. The muscles which bridge these joints make it possible for parts of the skeleton to move in relation to each other and thus the body as a whole develops motion. The skeleton offers a *counterforce* to *gravity* and enables *a wide range of movements*. In addition to this it has a number of further functions. The skeleton *protects vital* and *vulnerable organs*, such as the brain and spinal cord, the heart and lungs, the liver and kidneys.

Certain parts of the skeleton are producers of tissue; the red bone marrow, for example, is an important producer of blood cells.

Finally, the skeleton, is an important reservoir of minerals, mainly calcium.

Bones in types and sizes

The skeleton consists of more than 200 bones. On the basis of their shape they are divided into three types (Fig. 2.9.2):
- *long elongated bones*: these are slender bones. Their middle part is called a *shaft*. The ends are wider and form the 'head', the *ball* or the *socket* for a joint. The outer layer of the shaft consists of a very compact type of bony tissue. Within this shaft is a *marrow cavity* filled with *yellow bone marrow* (fatty tissue). In the widened ends is a porous, spongy, meshlike, trabecular network of cancellous bone (small sheets, stalks, plates and poles) and between these is the red bone marrow which produces blood. To the *longer* elongated bones belong the upper arm bone (i.e. humerus), forearm bones (i.e. ulna), radius, thigh bone (i.e. femur), shin bone (tibia) and calf bone (fibula).

The *shorter* elongated bones are located in the middle hand (carpal bones) and in the

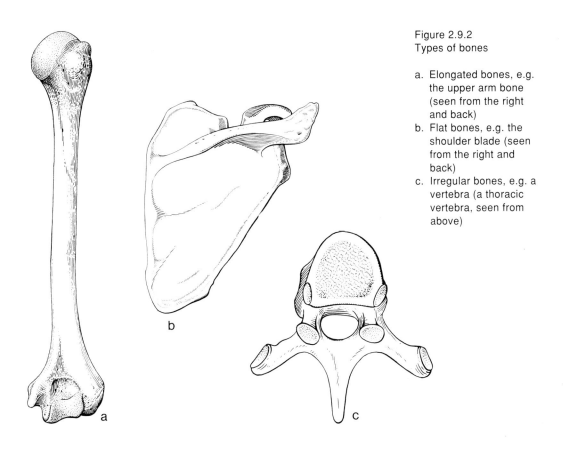

Figure 2.9.2
Types of bones

a. Elongated bones, e.g. the upper arm bone (seen from the right and back)
b. Flat bones, e.g. the shoulder blade (seen from the right and back)
c. Irregular bones, e.g. a vertebra (a thoracic vertebra, seen from above)

b

a

c

fingers (finger phalanges); and in the bones of the middle foot (tarsal bones) and the bones in the toes (toe phalanges);
- *flat bones*: the compact outer layers here are lying closely together. Between these compact layers we find the red bone marrow in a network of strips of bony material.
  To this type belong skull bones, ribs (costa), breastbone (sternum), shoulder blades (scapulae) and hip bones (ossa coxae);
- *irregular bones*: these belong neither to the first group, nor to the second group. Their compact outer layer is usually rather thin. As in the flat bones and the end parts of elongated bones, there is a network of little bony stalks with red bone marrow between them. To these irregular bones belong vertebrae, carpal bones (carpalia) and tarsal bones (tarsalia).

The division mentioned above has its limitations: there is another group, namely the collar bones (clavicles, claviculae), tongue bone (hyoid, os hyoideum), the components of the sets of teeth, and the sesame bones (pieces of tendon which have been calcified/ossified) such as the kneecaps (patellae). It is logical to count this other group in with the irregular type of bones.

General construction of bone

The structure of bone tissue received some attention when we discussed the tissues in Module 1, Chapter 4.
The 'in between substance' (m.i.) of two thirds of bone tissue consists of organic substances known as calcium phosphates. These give the bones their hardness. One third of the bony tissue consists of organic substances; bone cells and collagen fibres. The collagen gives the bones a certain elasticity. Without any collagen a bone would soon break. (We can illustrate this by comparing the supple living branch of a tree with a brittle dead one.) The name collagen ('glue substance') indicates that this substance when boiled yields glue.

People tend to think that bony tissue is a lifeless material. The bones are normally the only human remains to be found in archaeological sites. People tend to think that the living part of the body has decayed and that which was already dead has remained. Nothing is further from the truth: living bony tissue has a very intensive metabolism, is profusely supplied with blood and is in a constant cycle of demolition and composition. Because of this high level of metabolism a complete 'repair' of a broken bone (a rather serious injury) takes a

Figure 2.9.3

a. General construction of bone/bony tissue
b. Detail

1. *osteon*
2. *bone cells*
3. *lamellae*
4. *central canal (Haversian tube)*
5. *periosteum*
6. *blood vessel*
7. *substantia compacta*
8. *perforating (Volkmann) canal*
9. *spongy bone tissue (substantia spongiosa)*

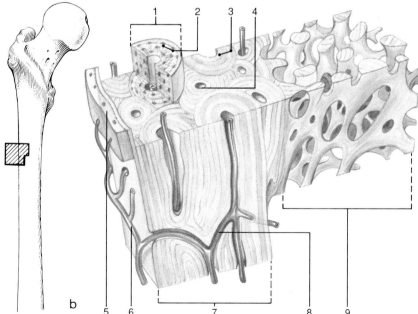

relatively short time (six weeks on average) compared to the 'repair' of connective tissue in capsules and tendons which have a far less generous supply of blood.

In principle, bones – independent of their form – all have the same structure (Fig. 2.9.3): a very solid firm outer layer (*compact bone tissue*) and within this a porous lattice of bone (*spongy bone tissue*). Amongst the trabeculae we find red bone marrow. Fortunately the bones are not solid bony tissue, or else the skeleton would be far too heavy.

The outer layer of the *compact bone tissue* consists of a number of *lamellae* of bony tissue which are closely linked together. These lamellae follow the entire outline of the bone. The direction of the fibres in successive layers of lamellae is always opposite. This gives an enormous firmness and strength.
This relatively thin crust or layer is called the *cortical bone*.
The next layer of the compact bone tissue (one step further inward) consists of long columns (strips, bridges) of an 'in between substance' known as *osteons*. An osteon is a kind of 'bone tube', consisting of a *central canal (Haversian tube)* surrounded by 8–15 concentric layers of lamellae. A comparison can be made with the annual rings of a tree. Just as in the corticalis the direction of the collagen fibres in successive lamellae of the osteons always lies in opposition. Between the lamellae are the bone cells, *osteocytes*. Between the osteocytes are a large number of linking channels as fine as hair. The tissue liquid within these very thin channels enables transport of substances between the bone cells and the capillaries.
In the central canals there are blood and lymph vessels. These are in communication with vessels which pierce into the compact layer from the outside via the more horizontally positioned *perforating canals (canals of Volkmann)*.

The *spongy bone* consists of a network of bone columns (trabeculae) and little cavities filled with red bone marrow. The bone columns, like the osteons, are constructed of lamellae, but there are no central canals. The bone columns have been arranged in such a way that the pressing forces which the bone has to endure most frequently can be absorbed (Fig. 2.9.4). This is especially noticeable in the upper end of the femur (thigh bone).

Constant breakdown of bony tissue takes place as a result of the activities of bone-consuming cells: the *osteoclasts* (bone dissolvers). These osteoclasts make tube-shaped holes of the size of an osteon. Into this hole there grows a blood vessel. Then the bone-producing cells, the

Figure 2.9.4
Arrangement of the bone columns in relation to the prevailing mechanical forces; illustrated on the lower extremity

1. hip bone (pelvis)
2. ball of the thigh bone
3. lowermost end of the thigh bone
4. uppermost end of the shin bone
5. lowermost end of the shin bone
6. skeleton of the foot

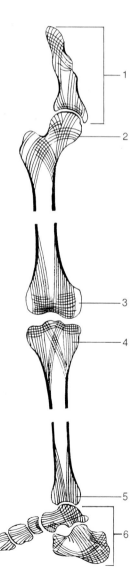

Figure 2.9.5
Longitudinal cross-section
of the proximal end of the
thigh bone

1. cartilage
2. spongy bone
3. compact bone
4. epiphysial plates
5. medullary cavity with
   marrow

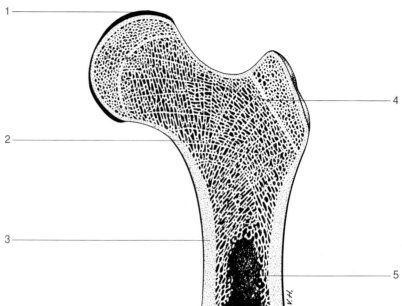

osteoblasts, start depositing calcium against the wall of the tube. Slowly the tube becomes filled up with concentric lamellae and some osteoblasts get caught between these lamellae. When they become immobile they are known as *osteocytes* (bone cells).

The middle part of a bone (see Fig. 2.9.2), is called the *diaphysis;* the ends of the bone are the *epiphysis.* These terms are used mainly in relation to elongated bones. So we can state: 'The diaphysis of an elongated bone contains a marrow cavity filled with yellow bone marrow'. In the epiphysis of a mature bone we often notice a characteristic solid disc of bony tissue: *the growth plate (epiphysial plate).* This is evidence of bone growth. On a cross-section of a bone (Fig. 2.9.5), or on an X-ray we see this as a line – the *epiphysial line.*

The outer surface of bone is covered by a bone membrane known as the *periosteum.* This is a tough membrane of tight connective tissue. The collagen fibres form a network which has been firmly anchored into the bone via ramifications into the corticalis. The tissues of the bone are supplied with blood from blood vessels in the periosteum which enter into the bone via the perforating (Volkmann) canals.

The periosteum plays an important role in the growth of the bone. Moreover, the tendons of muscles attach to the bone via the periosteum.

2. Connections of bones

Bones can be connected to one another in various ways. Every type of *bone connection* has its specific characteristics. We can divide them into three types: connections of connective tissue, cartilaginous connections and synovial connections.

In the case of a connection by means of *connective tissue* (junctura fibrosa) the skeletal parts involved are linked by firm networks of elastic fibres, mainly collagen fibres. An example of this type is the very firm, but at the same time springy, anchoring of teeth and molars in the jaws (see Fig. 2.2.14). Anyone who has ever had a tooth out can attest to the firmness of this anchoring. The sheet of tissue which largely fills up the space between radius and ulna is also a connective tissue connection. This *interosseous membrane* mainly serves as a muscle attachment.
We refer to *sutures* when the connective tissue is practically absent and the parts of the skeleton have more or less grown together as, for

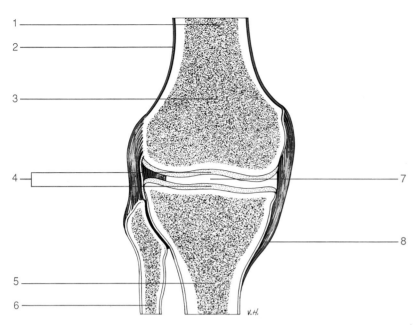

Figure 2.9.6
General structure of a
joint; illustrated on the
knee-joint

1. *thigh bone*
2. *periosteum*
3. *a bony tissue (spongy bone)*
4. *intra-articular cartilage*
5. *shin bone (tibia)*
6. *splint bone (fibula)*
7. *interarticular space, interarticular lubrication (synovia)*
8. *joint capsule with ligaments*

example, in the case of the bones of the skull (see Fig. 2.9.11).

In a *cartilaginous junction* there is a zone of cartilage between the bones. The length and depth of this cartilaginous zone determines the degree of flexibility. This can easily be illustrated in the (hyalin) cartilaginous linkages between ribs and breastbone (see Fig. 2.9.37a). If pressure is exerted onto the topmost ribs then we notice little movement; the lower ribs, however, yield rather more.

A further example of a fibrous cartilaginous linkage is the symphysis, e.g. the symphysis pubis (see Fig. 2.9.20). The symphysis unites the two pubic bones of the pelvis with one another.

The *synovial junction* is the linking of bones which in everyday English are called *joints* (articulatio). A joint, an articulation, is a connection between bones – usually quite flexible – in which the two adjoining surfaces have been adapted in shape to match each other, more or less perfectly.

Structure of a joint

Depending on the shape, one bone end is called the *ball* and the other the *socket*. Ball and socket are both lined with a layer of hyalin cartilage which has been anchored firmly in the bone. The layer of cartilage is on average 0.2–0.5 mm thick. The back part of the kneecap, however, can have a layer of cartilage up to 6 mm thick (Fig. 2.9.6).

Hyalin cartilage is extremely tough, and therefore it functions well as a shock absorber. The importance of this is easily understood if one calculates the enormous forces that develop in the joints of the legs, for example during normal walking.

Hyalin cartilage also has an extremely smooth surface. Because of this and also because of the presence of a lubricating joint-oil (synovial fluid) in the interarticular space between the joint surfaces, it is possible that in most joints movements can take place with a minimum of friction.

There are no blood vessels in cartilage. In relation to the forces it has to endure, this is an advantage, for blood vessels would constantly be damaged. Nutrition of the cartilage layer takes place by means of diffusion. This is a disadvantage for the maintenance or recovery of the cartilage when damaged.

From the edges of the joint cartilage and linking up with the periosteum of one bone runs an *articulatory capsule* to the other bone. This is attached in the same way to the periosteum. Such an articulatory capsule ensures that the

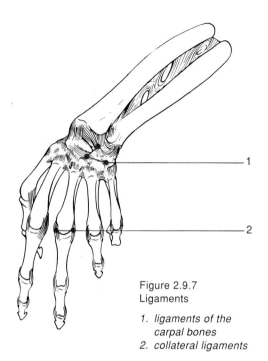

Figure 2.9.7
Ligaments

1. *ligaments of the carpal bones*
2. *collateral ligaments*

joint cavity is airtight. Thus ball and socket are kept pressed together; a vacuum would arise if the joint surfaces were pulled away from each other. The air pressure constantly pushes them back on one another. For this reason, the term *joint space* is much more accurate than joint cavity.

The articulation capsule consists of two layers: the *synovial membrane* and the *fibrous membrane*.

The *synovial membrane* is the inner layer. It is a thin membrane which contains elastic fibres, nerves and many blood vessels. The synovial membrane produces *synovia* (joint-oil) which is brought into the joint interstice. Synovia is a fluid with unique, lubricating qualities. Moreover, the synovia feeds the cartilage, as stated above.

The *fibrous membrane* is the outer layer of the joint capsule. It is a tough membrane (tight connective tissue) consisting of networks of collagen fibres which give the capsule its firmness.

In many joints at certain places in the fibrous membrane there are many extra bundles of collagen fibres lying in the same direction, known as *ligaments*. Such bundles (Fig. 2.9.7) give – locally – much firmness. They also influence the

options for movement in a joint. A sideward movement of the fingertips, for example, is not possible since to the left and to the right of the joint involved there are strong ligaments in the fibrous membrane.

The fibrous membrane is poorly supplied with blood vessels. That is why it takes a relatively long time to heal after damage is done (e.g. spraining) to the joint capsule or the ligaments embedded therein (at least six weeks). In the joint capsules and the ligaments there are sensors which provide information about the position and changes in the position of the joints, known as joint *sensors*.

Dynamic direction indications

In Module 1, Chapter 5 (topography) the *static direction indications* have been discussed. These make it possible to describe exactly the *position* of every organ and every structure in the body. To indicate the *changes in the position* of parts of the body (limbs, trunk, head) a number of extra notions are necessary. These *dynamic direction indicators* describe the *movements*, and constitute a subsection of topography.

The starting points in describing movements from the anatomic posture are the established planes (Fig. 2.9.8a).

A movement always takes place round an axis in the plane and at right angles to that axis. A group of three everyday examples will illustrate this. The movement of the front wheel of a bicycle occurs round a transversal axis in a sagittal (median) plane. A record disc turns round a longitudinal (vertical) axis in a transverse (horizontal) plane. The hands of a church clock turn round a sagittal axis in a frontal plane.

The same three axes are used in describing the basic movements of parts of the human body (Fig. 2.9.8). The axis involved is established every time reference is made to movement in a particular joint.

  – • *flexion* = bending movement
    • *extension* = stretching movement
      These movements take place round a transverse axis and in a sagittal plane. The notions of flexion and extension are used

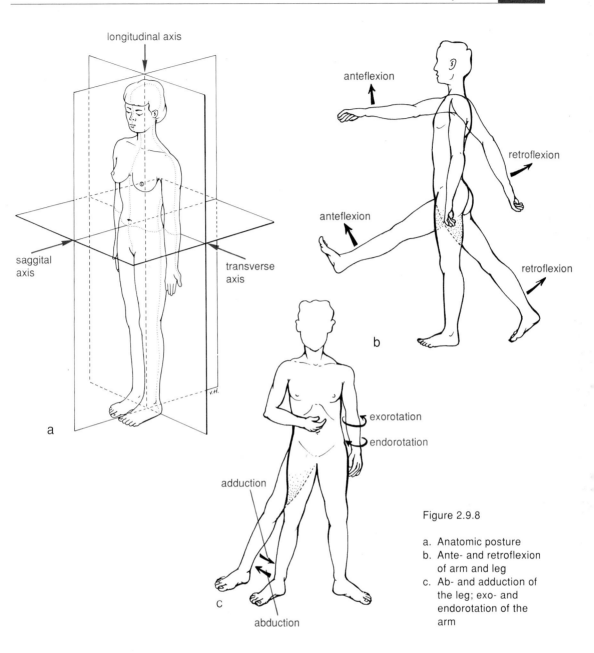

Figure 2.9.8

a. Anatomic posture
b. Ante- and retroflexion
   of arm and leg
c. Ab- and adduction of
   the leg; exo- and
   endorotation of the
   arm

especially for the movements in the elbow joint, the fingers, the knees and the toes. For the movement of the entire arm or leg to the front or back, or the trunk or the head, we use the terms: *anteflexion* and *retroflexion.*

A number of specific terms are in use for the movement of the entire hand in respect to the lower arm and the movement of the foot as an entity in respect to the lower leg. *Dorsal flexion* is to bend in the direction of the back of the hand and also of the instep of the foot. *Palmar flexion* is to bend in the direction of the palm of the hand. *Plantar flexion* is to bend in the direction of the sole of the foot (see Figs. 2.9.24a and 2.9.28a).

– • *abduction* = movement away from the middle line
• *adduction* = movement towards the middle line

These movements occur round a sagittal axis and thus in a frontal plane.

The sideward raising of an arm or a leg is abduction, and bringing it back to the body is adduction.

From the bones in the lower arm, the hand movements which take place in a frontal plane are called ulnar abduction and radial abduction (see Fig. 2.9.24b).

Sideways bending of head or trunk also takes place round a sagittal axis. However, for these movements the term *lateral flexion* (sideways bending movement) is used.

— • *exorotation* = rotation to the outside, turning to outside
• *endorotation* = rotation to the inside, turning inwardly
These movements occur round a longitud-inal axis in a transversal plane.
Charley Chaplin obtained his characteristic foot posture by exorotation of the upper legs in the hip joints.
The sideways turning of head or trunk is called lateral rotation. Turning back to the 'initial position' is called: medial rotation.

A very special type of movement is possible between the two bones in the lower arm, starting with a 90° bend in the elbow joint and the palm of the hand facing up. The turning down of the palm of the hand is called *pronation*. The opposite move is called *supination* (see Fig. 2.9.23d). These names are also used for the movements of the foot with regards to the lower leg. Pronation is rotation in which the medial foot-edge goes downwards; the opposite move (medial foot-edge coming up, coming higher) is called supination.

Finally the thumb has an important extra movement. The thumb can be placed opposite the other fingers of the same hand. This is called to *oppose*. The opposite movement is to *repose*.

Almost every movement, however complex it may be, can be traced back to movements of particular bones round a maximum of three axes. To illustrate this we will analyze a complex movement: someone grasps, with his right hand passing behind his back, the left upper arm

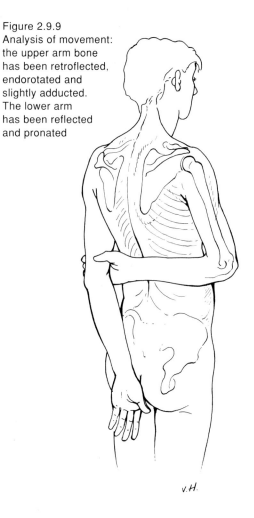

Figure 2.9.9
Analysis of movement: the upper arm bone has been retroflected, endorotated and slightly adducted. The lower arm has been reflected and pronated

v.H.

as in Figure 2.9.9. The upper arm bone is then retroflected, endorotated and slightly adducted. The lower arm is reflected and pronated. In the wrist joint dorsal flexion has taken place. All the fingers have been flected; moreover the thumb has been opposed.

Range of movements in joints

In a prepared skeleton there are more movements possible than in a living 'flesh and blood' human being. In a mere skeleton, for example, it is possible to move the hip joint through 360°. The possibilities of 'real life' movement in a joint are more limited. They depend not only on the shape of the joint surfaces involved but also on the placing of ligaments, on the presence and effects of muscles and tendons, and on the presence of other structures which might act as obstructions.

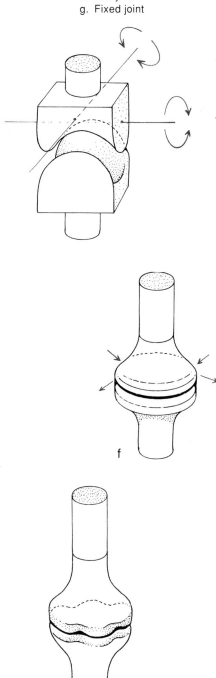

Figure 2.9.10
Division of the joints

a. Ball and socket joint
b. Ellipsoid joint
c. Hinge joint
d. Roll joint
e. Saddle joint
f. Flat joint
g. Fixed joint

a

b

c

d

e

f

g

In addition to this in some joints we find a fibrous cartilage structure that influences the scope of possible movements. Such a *joint disc* we find, for example, in the jaw joint (see Fig. 2.2.10b), and in the joint of the chest bone and clavicle (collar bone). In the knee there are two sickle-shaped cartilage structures, the menisci (see Fig. 2.9.27).

Joint discs have a number of functions. They act as shock absorbers (e.g. in the vertebral discs); they make those joint surfaces fit which did not originally fit at all, without the adaptation of cartilage (e.g. breast bone-collarbone); they increase the stability in a joint (e.g. in the knee); or they increase the options for movement (e.g. in the jaw).

In the region of certain joints we find *bursae*. A bursa is a flat little sac filled with synovia (a thick fluid) which is often in open contact with the joint-interstice. Bursae are found in those places where friction is likely to develop, for example between bones and weak parts (e.g. at the shoulder and knee joint).

With reference only to the shape of both the joint-surfaces and without consideration of the influence of muscles, tendons, ligaments and weak parts, the joints can be divided on the basis of their range of movement as follows (Fig. 2.9.10):

- *ball and socket joint:* it is possible to move in all directions, in other words round all three joint axes. The shoulder joint and the hip joint are ball and socket joints;
- *ellipsoid joint:* it is possible to move round the sagittal and the transverse axis. Rotation (i.e. here moving around the 3rd longitudinal axis) is impossible. The wrist joint as a whole (see Fig. 2.9.24) functions as an ellipsoid joint;
- *hinge joint:* it is only possible to move around the transverse axis. The elbow joint and the joints between the finger phalanges are hinge joints. Characteristic of a hinge joint are the ligaments lying on both sides in the joint-capsule, known as *collateral ligaments* (see Fig. 2.9.7);
- *roll joint:* it is only possible to move round the longitudinal axis. In the 'lower arm' the radius rotates in relation to the ulna (see Fig. 2.9.23d). The first vertebra, the atlas, rotates in relation to the second vertebra, the 'turner' (see Fig. 2.9.17c);
- *saddle joint:* as in the case of the ellipsoid joint it is only possible to move round two axes that are at right angles to each other. Rotation is impossible. The joint at the base of the thumb between the carpal and the metacarpal bone is a saddle joint;
- *flat joint:* it is only possible to make slight shifting movements in various directions. Between a certain number of the metacarpal bones and also in the tarsus is a more or less clear case of 'flat joints';
- *fixed joint:* both joint surfaces do fit well together, but only in one position. Moreover there are numerous firm ligaments, 'tethering' it on all sides. There is little range of movement; at the most there is a little bit of give. The best-known fixed joint is the one between sacrum (sacral bone) and ilium (iliac bone), the sacro-iliac joint (see Fig. 2.9.21).

From a scientific point of view the division of the joints described above is outdated, but it is still being used.

3. Bones and bone connections of the head

The shape of the head is mainly determined by the shape of the skull (cranium).

In two places at the bony – mouldable – parts of the skull, flexible cartilage is found: the cartilage of the shell of the ear (auricle, concha) and the cartilage of the nose (in the nostrils, and most of the internasal septum (i.e. partition of the nose)). These structures, though firm in shape, are flexible.

The skull serves to protect the brain, almost all the senses and the organ of balance. The face muscles are attached on to the skull. Moreover, the skull constitutes the entrance to the digestive and respiratory tract.

These various functions are reflected in the structure of the skull.

Two parts in the skull may be distinguished (Fig. 2.9.11): the brain skull and the facial skull.

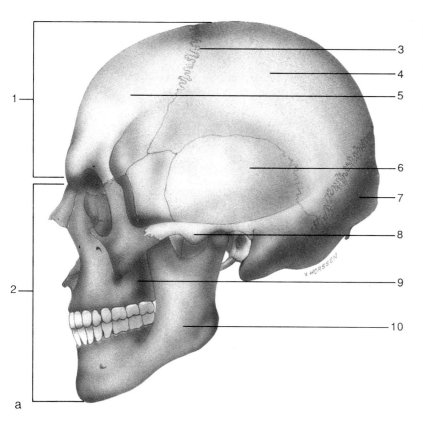

a

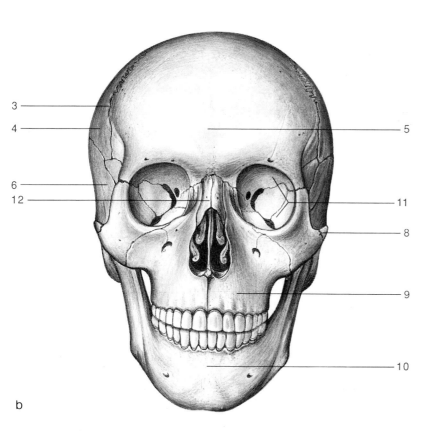

b

Figure 2.9.11
The skull

a. Side view/lateral view; brain skull and facial skull
b. Front view of the skull

1. brain skull (cranium)
2. facial skull
3. suture
4. parietal bone (os parietale)
5. frontal (coronal) bone (os frontale)
6. temple (temporal) bone (os temporale)
7. occipital bone (os occipitale)
8. cheek bone and zygomatic arch
9. upper jaw (maxilla)
10. lower jaw (mandible)
11. eye-socket, orbit
12. nose bone (os nasale)

Figure 2.9.12
The base of the skull;
seen from above

1. ethmoid bone
2. front cranial fossa
3. sphenoid bone
4. Turkish saddle (sella turcica)
5. middle cranial fossa
6. petrous bone
7. occipital opening
   (foramen magnum)
8. back cranial fossa

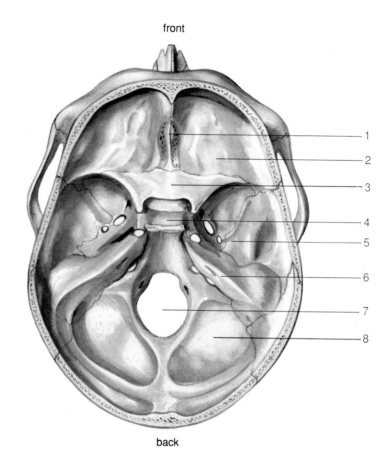

front

back

Brain skull

The brain skull encloses a space in which are placed the brain and the medulla: *the cranial cavity*. The brain can be divided into the large brain (cerebrum), the lesser brain (cerebellum) and the midbrain (several organs). The bony encasement offers excellent protection for the vulnerable nerve tissue. The cranium consists of a number of bent bone plates, closely and firmly united together by means of *sutures* (suturae). These sutures follow a winding course with a strong serration hooking firmly into one another, and are the remnants of the rudiments and growth period of the skull.

That part of the cranium which we can touch (e.g. by means of palpation or percussion) is the *skull roof*. That part on which the brain is resting is called the *base of the skull* (Fig. 2.9.12). The base of the skull has very many small openings for nerves and blood vessels and one big hole, the occipital hole (foramen magnum) for the spinal cord channel.

The cranium consists of the following bones (Figs. 2.9.11 and 2.9.12): the *frontal* bone (os frontale), *occipital* bone (os occipitale), the *parietal* bone (os parietale), the *temporal* bone (os temporale), the *ethmoid* bone and the *sphenoid* bone.

In the frontal bone is the frontal sinus (sinus frontalis). This paranasal cavity is connected to the nasal cavity just like the other sinuses and extends in the frontal bone to above the eye-sockets.

At the two temple bones can be distinguished the *acoustic duct* (auditory canal) with immediately ventral to it the joint socket for the lower jaw. Below the acoustic duct and a little bit to the back a protuberance known as the *mastoid* (processus mastoideus) can be felt.

The mastoid contains spaces filled with air (mastoid cells) that are connected with the ear (the tympanic cavity of the hearing organ). The *petrous bone* (pars petrosa, os petrosum) is a part of the temporal bone that protrudes to the inner side. In the petrous bone there is the bony *labyrinth* (see Fig. 2.8.46). This is a hollowing out of the petrous bone which fully deserves the name of 'labyrinth'. In a part of this labyrinth is found the inner ear; in another part the organ of balance.

On the outside of the occipital bone to the left and to the right of the foramen magnum are the *occipital condiles*. The joint surfaces on these occipital condiles together with the joint (i.e. articulatory) surfaces on the atlas (i.e. the first vertebra) form joints in which most of the movements which permit nodding 'yes' take place.

The ethmoid has a short upright crest in the middle of its roof. To the left and right of this are a large number of rather small apertures, through which the communicating fibres run between the smell epithelium and the nerve of smell.

This part of the ethmoid situated in the base of the skull is called the *cribriform* (i.e. sievelike) plate.

The lateral walls of the bony nose cavities consist of the three nose *conchae*, one above the other (Fig. 2.9.13). Beneath each nose concha is a passageway: the *nasal passages*. The lowest of the three is the widest one; the topmost one is the narrowest. A part of the intranasal septum *(septum nasi)* also belongs to the ethmoid bone. Within the ethmoid are found the ethmoidal sinuses (air-filled spaces). These paranasal sinuses have an extremely intricate shape.

The sphenoid has a transverse hollowing out in the middle, a fossa: the *sella turcica*. This contains the pituitary, a hormone gland with a large number of functions. In the sphenoid, also, we find paranasal sinuses connected to the nose cavity known as sphenoidal sinuses.

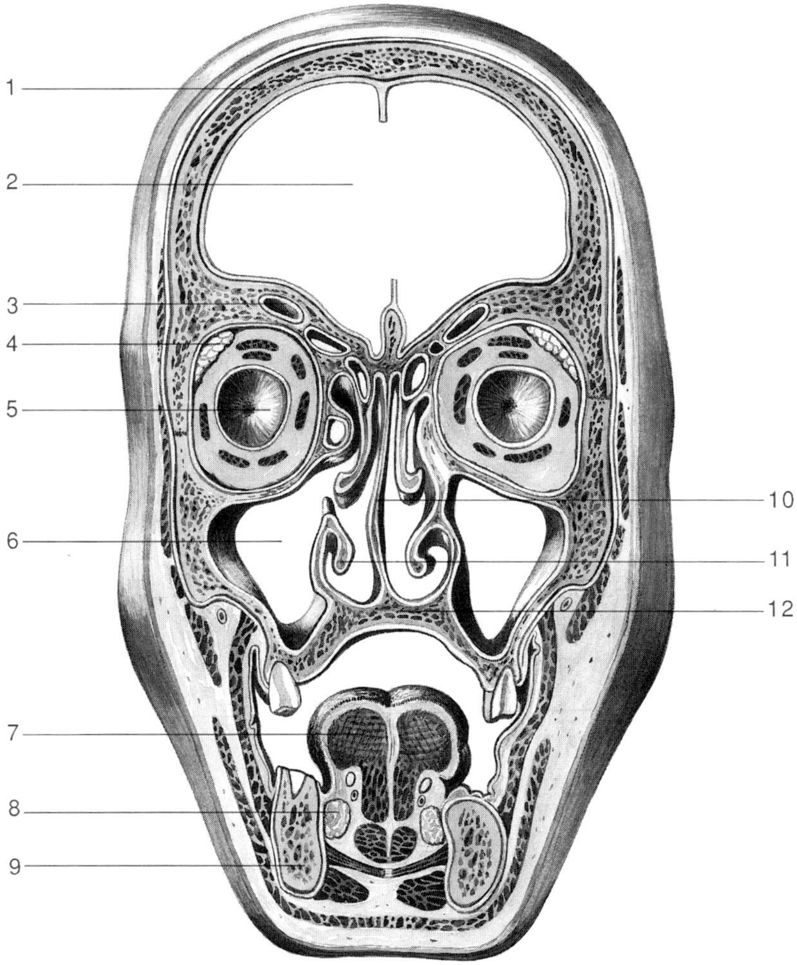

Figure 2.9.13
Frontal cross-section
through the head

1. frontal bone
2. cranial cavity
3. extension of the
   frontal sinus (sinus
   frontalis) in the roof of
   the eye-socket (orbita)
4. lacromal gland
5. eye surrounded by
   eye muscles
6. maxillary sinus
7. tongue
8. salivary gland
9. lower jaw
10. intranasal partition
    (septum nasi)
11. nasal concha
12. hard palate

Looking at the base of the skull from above, three hollowings can be distinguished one after another: *the cranial fossae* (Fig. 2.9.12).

The foremost cranial fossa is the one lying highest; its base supports the front lobes of the cerebrum.

The temporal lobes of the brain rest on the base of the middle cranial fossae. The backmost cranial fossa is on the lowest level and supports the occipital lobes of the cerebrum, the pons and the medulla oblongata.

Facial skull

Anatomists usually draw the borderline between brain skull and facial skull through the eye sockets. The facial skull consists of a number of bones which show far greater variety in structure than the bones of the cranium. The bones of the facial skull form the walls of the eye sockets, the nose cavity, the mouth cavity and the pharynx.

In the facial skull the following parts can be named (see Fig. 2.9.11): the *nose bone* (nasal bone), the *eye sockets* (orbits), the *left and right cheekbone* (zygomatic bone), the *upper jaw* (maxilla) and the *lower jaw* (mandible).

The nose bone consists of several pieces of bone; the bony nose partition (nasal septum) forms a part of the partition between the left and right nose cavity.

The eye sockets, too, consist of several bones. They have plenty of smaller openings for blood vessels and thin strains of nerves, and one bigger opening for the thick eye nerve, the optic nerve. On the medial side in the wall of the eye socket is the tearduct (lacrimal canal). It runs downwards from the corner of the eye and empties itself under the inferior nasal concha in the nose.

The cheekbone links up with a sideways protruding ridge of the temple bone and thus forms the *zygomatic arch*. This arch can easily be felt immediately above and in front of the jaw joint. The upper jaw does not consist only of the horseshoe-shaped bony ridge in which the upper teeth are anchored. It is a rather big piece of bone which connects with the frontal bone to the left and right of the root of the nose. The frontmost part of the roof of the mouth is the *hard palate*

(palatine bone); it is the bony, dome-shaped part of the skull under the nasal cavity (Fig. 2.9.13). In the bony tissue of the upper jaw, above the molars, is another airfilled space connected to the nasal passage: the *maxillary sinus* (upper jaw cavity). The connecting opening of these two paranasal cavities to the nasal cavities is found at the level of the middle nasal conchae (Fig. 2.9.13).

The lower jaw is the only movable bone of the skull. Directly in front of the left and the right outer auditory opening are found the joints involved. The construction of the jaw joints is closely related to the chewing function of the teeth (see Fig. 2.2.10). The part of the lower jaw into which the lower teeth are anchored has about the same horseshoe shape as the upper jaw. Thus usually the teeth of the upper jaw link rather well with the teeth of the lower jaw. This is known as occlusion.

The teeth and molars themselves form a part of the skull; their function in the digestive tract is described in Module 2, Chapter 2.

4. Bones and bone connections of the trunk

Strictly speaking the trunk skeleton consists of the *spinal column*, the *ribs* and the *breast bone* (see Fig. 2.9.1). The *pelvis* has an intermediate position. It constitutes both the bottom of the trunk and the upper part of the lower extremities.

4.1. Spinal column

The spinal column (Fig. 2.9.14a), consists of 33 or 34 *vertebrae* which are united together by *vertebral discs and joints.*

Between the cranial to caudal extremities are 7 *cervical vertebrae:* C_1–C_7; 12 *thoracic vertebrae:* T_1–T_{12}; 5 *lumbar vertebrae:* L_1–L_5; 5 *sacral vertebrae:* S_1–S_5 that have grown together to become the sacrum (os sacrum); and 4 or 5 *coccygeal vertebrae* which have grown together to become the tailbone or coccyx (os coccygis). The names refer to their position, but there are considerable differences in construction between the various types of vertebrae.

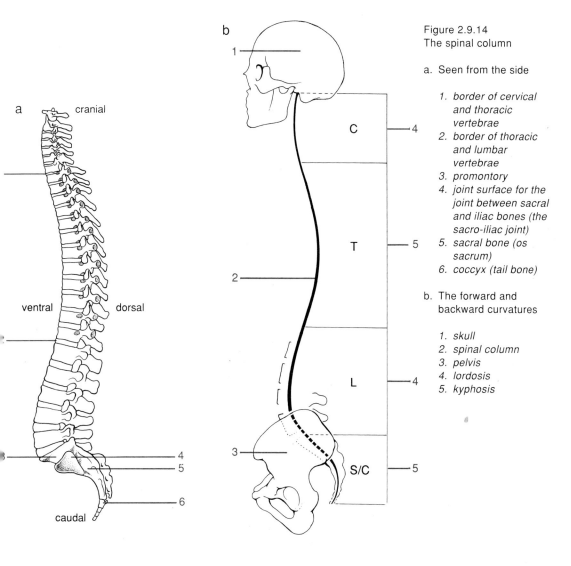

Figure 2.9.14
The spinal column

a. Seen from the side

1. border of cervical and thoracic vertebrae
2. border of thoracic and lumbar vertebrae
3. promontory
4. joint surface for the joint between sacral and iliac bones (the sacro-iliac joint)
5. sacral bone (os sacrum)
6. coccyx (tail bone)

b. The forward and backward curvatures

1. skull
2. spinal column
3. pelvis
4. lordosis
5. kyphosis

Curvatures

Seen from the side the vertebral column shows a number of forward and backward bends (Fig. 2.9.14b).

A *lordosis* is a forward curvature, a convex to ventral bulge. This is found in the region of the neck where it is called *cervical lordosis;* in the lumbar region it is called *lumbar lordosis.*

A *kyphosis* is a hollow, concave dorsal curvature. In the chest region is found the *thoracic kyphosis,* and in the region of the sacrum the *sacral* (or pelvic) *kyphosis.* The rather sharp kink in the transition of lumbar to sacral is called a *promontory.*

In addition to these S-shaped curves in the sagittal plane one or two slight sideways bends are often seen. Such a slight curve in the frontal plane is probably a consequence of small differences between the left and right part of the body. To maintain balance, the spinal column offers compensation known as functional *scoliosis.* A functional scoliosis disappears when the person concerned lies on his back. The functional scoliosis is reversible.

Vertebrae, similarities in structure

With a few exceptions we can distinguish in a vertebra a body (corpus) and a vertebral (neural) arch with a number of processes (Fig. 2.9.15). The *vertebral body* lies ventral. Its upper and lower surfaces are rather rough. The *vertebral discs* have grown together at these surfaces. The

Figure 2.9.15
General structure of a
vertebra, illustrated on a
thoracic vertebra

a. View from above
b. View from the side; two
   vertebrae with:

1. vertebral body
2. vertebral foramen
3. articulatory process
4. neural arch
5. transverse process
6. spinous process
7. vertebral disc
8. intervertebral foramen

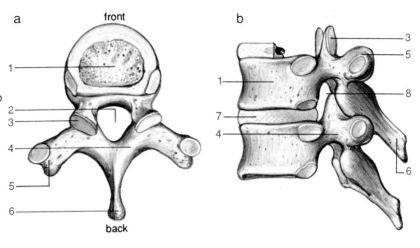

vertebral (neural) *arch* (arcus) is linked to the body on the dorsal side and encloses the *vertebral foramen* (foramen vertebrale). All the vertebral foramen on top of one another form the *vertebral canal* (canalis vertebralis) containing the spinal cord (or marrow), protected by the successive neural arches and the dorsal sides of the successive vertebral bodies (corpi).

Seen from the side the part of the neural arch which links with the corpus is rather narrow. This creates an opening between the successive arches known as the *intervertebral foramen*. Via these openings the spinal nerves are connected with the segments of the spinal cord.

The *processes*, numbering seven, are to be found at the neural arch. The *spinous process* (processus spinosus) protrudes backwards. As a result of the thoracic kyphosis they can easily be felt between the shoulder blades. The two *transversal processes* (processus transversus) are directed laterally. Of the four *joint processes* (processus articulares), two are directed cranially (upwards) and two caudally (downwards).

The joint processes of successive vertebrae together form the *intervertebral joints*, also called *facet joints* or facet articulations. They are synovial connections.

Figure 2.9.16
Types of vertebrae

a. Cervical (of the neck)
   vertebra
b. Thoracic (of the chest)
   vertebra
c. Lumbar (of the loin)
   vertebra

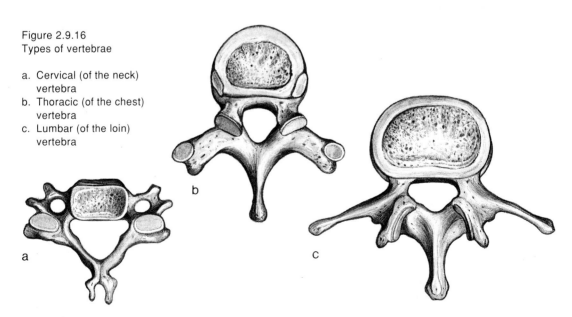

Vertebrae, differences in shape and function
The vertebrae, on top of each other, and each having a vertebral disc and two facet articulations in between, form the firm but yet flexible central 'spine' of the trunk. The support function of each vertebra increases from cervical to sacral (from neck to hips).

The difference in size of the vertebral bodies is a clear sign of this (Fig. 2.9.16). All the vertebral foramens together form the spinal, vertebral canal in which is the spinal marrow (so also called the marrow canal).

The area of the body yet to be 'served' by the spinal marrow decreases from cervical to caudal and accordingly the diameter of the spinal cord decreases. Along with this decrease is the decrease in size of the vertebral foramen from neck to loin (Fig. 2.9.16).

From cervical to sacral the positioning of the facet joints also changes. This has an influence on the range of movements possible at the various levels of the spine.

In the *cervical vertebrae* there is a hole in the transversal processes. Left and right through these successive openings (forming a canal) there runs the vertebral artery (see Fig. 2.1.34), which contributes to the blood supply of the brain.

The first (topmost) cervical vertebra, *the atlas*, shows a number of particular deviations from the 'model vertebra'. It has no spinous process. In the other cervical vertebrae the spinous process is relatively prominent and forked. The atlas is a ringshaped annular vertebra without a vertebral body (Fig. 2.9.17). On top of the ring, instead of the two joint-processes, are two articular facets (i.e. surfaces) jointing with the corresponding occipital condyles; and on the bottom of the ring are two articular surfaces for the facet joints with the second cervical vertebra, the *'pivot'* (the axis).

On the inner side of the anterior part of the atlas there is a kind of cutaway where you would have expected the vertebral body, the anterior ring. This is bridged by the *transverse ligament*. Anterior ring and transverse ligament form a round opening, lined with hyalin cartilage. This annular structure forms the articular surfaces of the atlas for the pivot of the axis. This pivot, with joint surfaces to the front and to the back, is attached onto the top of the rather small body of the second vertebra, the axis.

The range of movements possible in the joints between skull and atlas on the one hand, and the atlas and pivot on the other hand determine – to

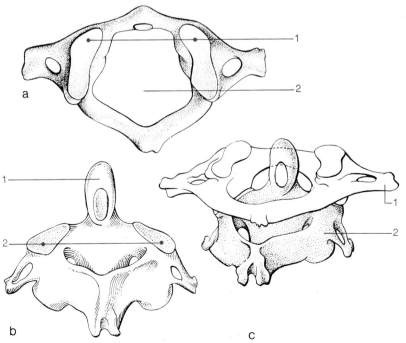

Figure 2.9.17
The first two cervical vertebrae

a. Atlas; seen from above

   1. *joint surfaces on top for the occipital condyles*
   2. *vertebral foramen*

b. Pivotal axis; seen from behind

   1. *tooth of the axis*
   2. *joint surfaces for the bottom plate facets of the atlas*

c. Atlas placed on to the axis

   1. *atlas*
   2. *axis, pivot*

a large extent – the flexibility of the head in relation to the trunk. In nodding 'yes' (ante and retroflexion of the head), movement takes place in the joints of occipital condyles and the matching joint surfaces on the atlas. Atlas and pivot do not move in relation to one another in nodding 'yes'. When shaking 'no' with the head (rotation of the head), the tooth of the pivot functions as the axis for the atlas; atlas and skull do not move in relation to one another.

Sideward movements of the head (lateral flexion) occur mainly in the joints of the lower cervical vertebrae.

Because of the position and the slope of the facet joints of the rest of the cervical area, a wide range of possible movement is found here: ante- and retroflexion, lateroflexion and a limited rotation.

In the case of the breast vertebrae (Fig. 2.9.16), it strikes us that the long spinous processes are pointing downwards. In addition to this, on the sides of the vertebral bodies and the frontplane of most transverse processes, are found the articular surfaces of the joints with the ribs.

The articulatory surfaces of the 'facet joints' in the thoracic area stand almost vertically upright in a frontal plane. This is why there is hardly any ante- and retroflexion possible in this area. On the contrary, lateroflexion and rotation are found here.

Lumbar vertebra. Here (Fig. 2.9.16), the processes are relatively large. The spinous process is flat and points practically straight backwards. The joint surfaces of the 'facet joints' are positioned in a predominantly sagittal plane. In the lumbar region rotation is practically impossible. But ante and retroflexion and slight lateroflexion is possible here.

The *sacral vertebrae* are fused together to become the *sacral bone* (os sacrum). It is a wedge-shaped bone with four pairs of characteristic apertures. Through these openings run spinal nerves. Spinous processes are hardly present. To the sides of the sacrum are the ear-shaped articular surfaces of the joint with the hip bones. These surfaces coated with hyalin cartilage are not smooth but have an irregular surface.

The *tailbone vertebrae* have grown together, fused to become the *tailbone* (coccyx – the os coccygis). The spinal canal ends in the tailbone.

Figure 2.9.18
The intervertebral disc

a. Superior aspect
b. Sagittal section
c. Structure

1. *anulus fibrosus*
2. *nucleus pulposus*
3. *intervertebral disc*

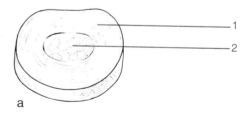
a

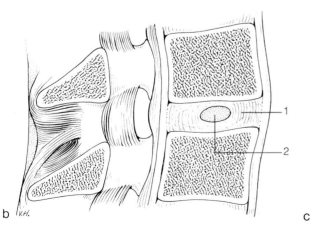
b

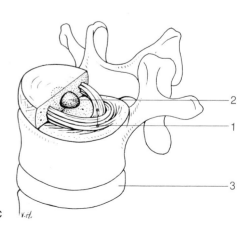
c

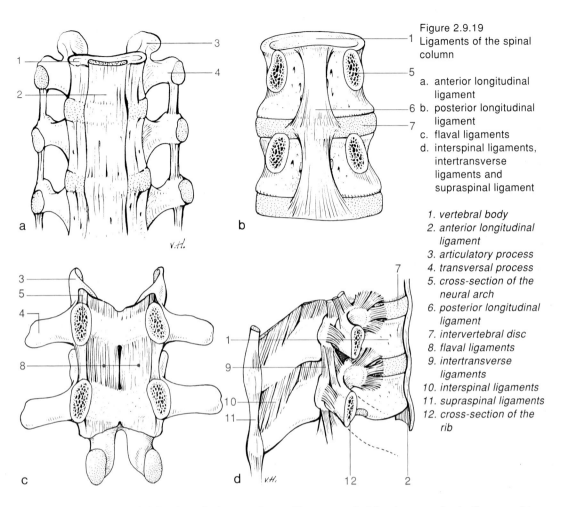

Figure 2.9.19
Ligaments of the spinal column

a. anterior longitudinal ligament
b. posterior longitudinal ligament
c. flaval ligaments
d. interspinal ligaments, intertransverse ligaments and supraspinal ligament

1. vertebral body
2. anterior longitudinal ligament
3. articulatory process
4. transversal process
5. cross-section of the neural arch
6. posterior longitudinal ligament
7. intervertebral disc
8. flaval ligaments
9. intertransverse ligaments
10. interspinal ligaments
11. supraspinal ligaments
12. cross-section of the rib

It has only slight flexibility in relation to the sacrum.

Vertebral discs and ligaments
Between successive vertebral bodies are found a vertebral disc (discus intervertebralis) and ligaments.

The *intervertebral disc* (Fig. 2.9.18) consists of a soft jelly-like kernel surrounded by a strong capsule constructed of fibrous cartilage, and within this kernel there are ring-shaped bundles of collagen fibres. The cartilage of the intervertebral disc has grown together on both ends with the bodies of the vertebrae.

The kernel, rich in water, is called the *nucleus pulposus*; the fibrous ring is called the *annulus fibrosus*.

The nucleus pulposus is a watercushion which is kept under pressure by the anulus fibrosus.

Because of this the vertebral disc combines two characteristics which are of essential importance for the spine: it can resist pressure and it can change its form.

Because it can stand pressure the vertebral disc can function as a shock absorber: the pressing forces which arise in the spinal column during standing, walking, jumping, etc are resisted.

The ability to change its shape contributes to a large extent to the flexibility of the total spine. It is not a rigid 'rod', but on the contrary a flexible central 'pole'.

The spine has several important ligaments (Fig. 2.9.19). Along the total length of the spine run the *longitudinal anterior ligament* and the *longitudinal posterior ligament*. They run respectively in front and behind the vertebral bodies. They are attached to the vertebral bodies and to the intervertebral discs. They inhibit retro- and anteflexion and they protect the intervertebral discs. The *flaval ligaments* stretch between parts of the neural arches. Furthermore ligaments are found stretching between the spinous processes (the *interspinal ligaments*), between the transverse processes (*intertransverse ligaments*), and there is a very long ligament over the tops of the spinous processes from C_1–L_5 (*supraspinal ligament*).

4.2. Thorax

The *chest cage* (thorax) is the semi-hard (bones and muscles) encasement of the chest cavity. It is a firm, but at the same time flexible, structure in which the relatively vulnerable organs, heart and lungs, find their well-protected place. It is by changing the volume of the thorax that respiration (breathing) occurs.

The thorax is formed by the thoracic part of the spine, the 12 pairs of ribs, the breastbone (sternum) and the zones of cartilage between this breastbone and the ribs (see Fig. 2.9.1).

The *ribs* have small synovial joints at the sides of the thoracic vertebrae.

There are 12 thoracic vertebrae and just as many pairs of ribs. *Ribs* (costae) are flat, narrow bones. The curvature of the ribs decreases from

T_1 down to T_{12} where the ribs have the smallest curvature.

In these 12 pairs of thoracic ribs, 7 pairs of *true ribs*, 3 pairs of *false ribs* and 2 pairs of *floating ribs* can be identified.

Sometimes the floating ribs are counted with the false ribs. Then there are 5 pairs of false ribs. The true ribs are provided with a cartilaginous connective strip of their own to the breastbone. The false ribs have no direct cartilaginous connection with the breastbone, but they are connected by cartilage strips to the next higher rib. The floating ribs have no connection to the breastbone at all, nor to the next higher rib. They are relatively very short. They 'float' individually.

Figure 2.9.20

a. The pelvis
b. The right hip bone

1. sacral bone (os sacrum)
2. obturator foramen
3. ischium (os ischii)
4. iliac crest
5. ilium (os ilium)
6. anterior superior iliac spine
7. promontory
8. borderline of the entrance to the pelvic cavity
9. symphysis pubis
10. pubic bone (os pubis)
11. ischial tuberosity (tuber ischiadicum)
12. the joint facet, socket for the ball of the thigh bone
13. acetabulum

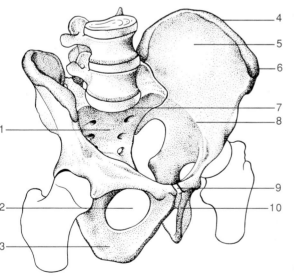

a

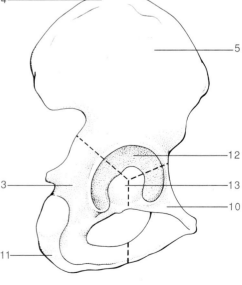

b

The *breastbone (sternum)* belongs, like the ribs, to the flat bones (red bone marrow, formation of blood cells). It can easily be felt, lying just under the skin.

Some parts of the breastbone are named (see Fig. 2.9.1), from top to bottom: the 'handle' (manubrium) on to which the clavicles and the first pair of ribs are attached, the body (corpus) to which the other true ribs (and indirectly the false ribs) are attached, and the sword-shaped appendix (xiphoid process) which can partly consist of cartilage.

4.3. Pelvis

The *pelvis* (pelvic girdle) is the solid bony structure at the bottom of the trunk which provides the framework for the articulation of the legs. It occupies an intermediate position. The pelvis (Fig. 2.9.20a) forms the bottom of the trunk and consists of four bones: 2 *hip bones* (flat bones) connected dorsally by the sacrum which fuses with the tailbone. Each hip bone or coxal bone (os coxae) originates from three bones which, without any clear seam (suture), have become fused together in an adult: *ilium* (os ilium),

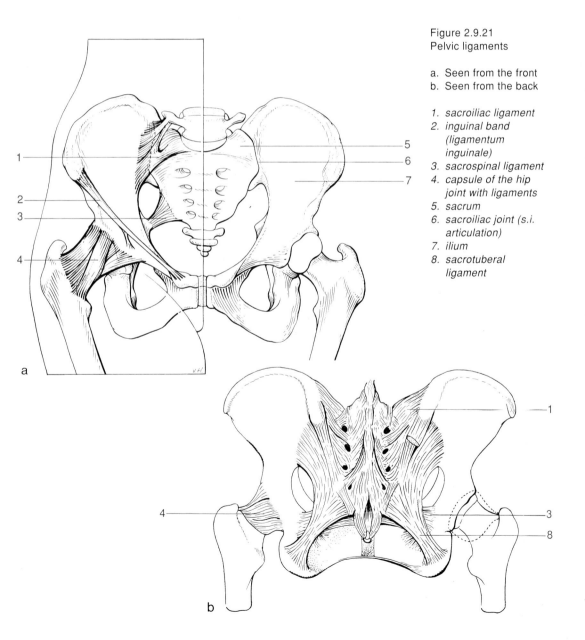

Figure 2.9.21
Pelvic ligaments

a. Seen from the front
b. Seen from the back

1. *sacroiliac ligament*
2. *inguinal band (ligamentum inguinale)*
3. *sacrospinal ligament*
4. *capsule of the hip joint with ligaments*
5. *sacrum*
6. *sacroiliac joint (s.i. articulation)*
7. *ilium*
8. *sacrotuberal ligament*

ischium (os ischii) and *pubis* (os pubis) (Fig. 2.9.20b). The upper rim of the ilium is relatively wide, the pelvic, *iliac crest* (crista iliaca). In the side of the body the front part of this crest can easily be felt. To the front the pelvic crest ends in a protuberance, the anterior superior iliac spine. To their medial side the iliums each have a joint surface for connection with the sacrum. In broad outline it has the shape of an ear. The ischium has a broadening to the back, *the ischial tuberosity.*

The pubic bone together with the ischium encloses an opening: the obturator foramen. This opening is almost entirely closed by a tendinous membrane which serves as a place of origin for muscles.

To the front both pubic bones do not quite link up directly to one another. Between the two ends is a cartilaginous zone (discus). This disc is enwrapped by and grown together with a large number of collagen fibres which connect the two pubic bones. This connection is called the *symphysis pubis*, shortened to symphysis.

The structure of the *sacral bone* (os sacrum) and the *coccyx* (os coccygis) has been discussed earlier. The connections between the sacrum and the two iliums are called *sacroiliac joints* (s.i. articulations). They are synovial connections, but there is relatively little flexibility: they belong to the rigid joints and have a very ligamentous capsule (Fig. 2.9.21).

The sacrum is jammed like a wedge between the two hip bones. The lowest part of the spine does not rest on the pelvic girdle, but is a constituent part of the pelvis. This enhances the stability of the complex.

Forces which suddenly occur in the pelvis, as when landing after a jump, can be effectively absorbed, in the sacroiliac joints and the symphysis. If the pelvis were a massive structure fractures would occur far more easily and far more often.

In the region of the pelvis are a number of ligaments which are not a part of any articulation capsule (Fig. 2.9.21). They increase the coherence of the pelvis and counterbalance the strong forces that the pelvis has to tolerate.

Large and small pelvis
Because of the shape of the pelvic bones there is an opening between the sacrum and the symphysis pubis (Fig. 2.9.20a). The part of the pelvis above this opening runs high and wide: *the large pelvis.* The part of the pelvis beneath this opening is smaller and is narrower: *the small pelvis.* The opening itself is called the *pelvic entrance.* This name has been chosen since the opening forms the borderline of the abdominal cavity and the transition to the pelvic cavity. The organs lying in the small pelvis, such as the bladder, rectum and uterus, are called *pelvic organs.* The opening that is formed on the bottom of the small pelvis by the lower edges of the pelvic bones, the sacrotuberal ligaments and the tailbone is called the *pelvic exit.* The size of the pelvic entrance and exit and other aspects of the pelvis play an important role in the delivery of a baby.

Sex differences
The shape of the pelvis generally offers an easy method of determining whether we are dealing with a male or a female skeleton (Fig. 2.9.22). A female pelvis is generally wider, more spacious and has flatter protruding iliums. It has a bigger pelvic exit. The pelvic entrance has an oval shape and beneath the symphysis the pubic bones form a blunt angle. In a male the iliums stand far more upright, the pelvic entrance is heart-shaped, and the angle of the bones beneath the symphysis is sharp. The volume of the pelvic cavity is smaller.

5. Bones and bone connections of the upper extremity

'He is a handyman.' 'She is handy with a pair of scissors.' 'A handbook.' 'The handbrake.' 'A handclasp.' 'A handout.' 'That comes in handy.'
Language indicates how essentially important the function of the hand is for a human being. 'To shake hands', refers to contact. 'Give me a hand', asks for help. A child explores and sometimes discovers the world by 'fumbling everything' with his 'paws' or fingers – groping precedes grasping. Hands can touch, caress, point,

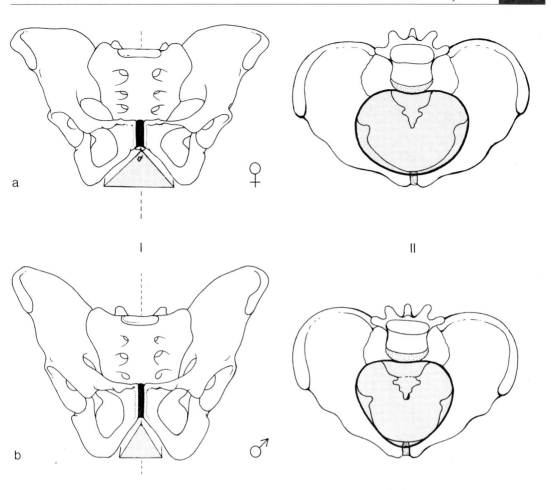

Figure 2.9.22
The differences between a female
and a male pelvis

a. Female pelvis seen (I) from the front and (II) from above
b. Male pelvis seen (I) from the front and (II) from above

carry, gesture, handle tools or an instrument, etc. The hand has an enormous freedom of movement not only because of its own construction, but also because the structure of the arm and the shoulder girdle grant the hand an extended scope for movement, even far behind the back.

Shoulder girdle

The *shoulder girdle* (see Figs. 2.9.1 and 2.9.23a), consists of the *breastbone (sternum)*, the *two collar bones (clavicles)* and the two *shoulder blades (scapulae).* Breastbone and collar bone have between them a joint in which there is a disc. When shrugging and rotating the shoulders one can feel the movements in that joint quite well. The clavicle in its turn has a

joint with the shoulder blade. In this joint there is less flexibility. The shoulder girdle is *'open'* and there is no joint between the shoulder blades and the trunk skeleton, or between themselves. This construction results in huge flexibility. The clavicles originate from cartilage which ossifies, calcifies. The shoulder blades belong to the flat bones and so contain red – blood producing – bone marrow.

Arm

The bones (see Figs. 2.9.1 and 2.9.23b) of the arm (brachium) in the upper arm are: *the upper arm bone (humerus)* and in the lower arm: *forearm (radius) and ulna.* They are long elongated bones. The ulna has a broad and rather flat proximal end known as the *olecranon.*

Figure 2.9.23
The upper extremity

a. Shoulder girdle

  1. *collar bone*
  2. *upper arm bone*
  3. *shoulder blade*
  4. *ribs*
  5. *breast bone*
  6. *costal cartilages*

b. The bones of the upper
   arm, lower arm, wrist
   and hand

  1. *radius*
  2. *upper arm bone*
  3. *elbow joint*
  4. *ulna*
  5. *carpal bones*
  6. *metacarpal bones*
  7. *phalanges (of the
     hand)*

c. Ligaments of wrist and
   hand

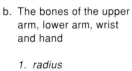

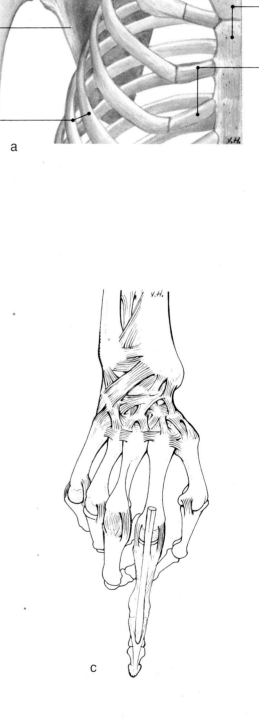

a

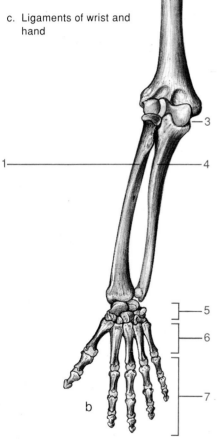

b

c

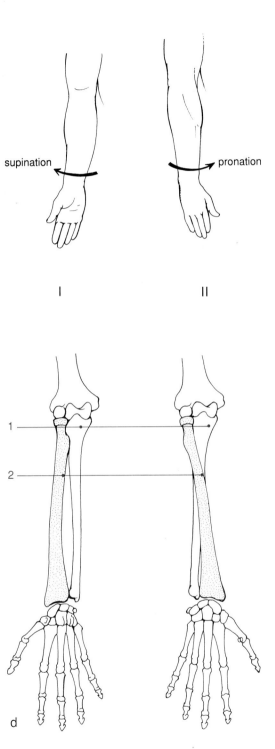

supination   pronation

I   II

d

Figure 2.9.23 (cont.)

d. Supination (I) and pronation (II); the
movement of the radius round the ulna

1. *ulna*
2. *radius*

The shoulder blade has a strikingly shallow and rather small articulation surface for the ball of the upper arm bone. The shoulder joint (gleno-humeral joint) is a ball and socket joint; in the capsule are strong ligaments.

In the elbow is the joint between the upper arm bone and the bones of the lower arm. It is a distinct hinge joint (Fig. 2.9.23b), with collateral ligaments. Bending and stretching are the possible movements.

Between the two bones of the lower arm in their middle part is an elongated membrane of collagen fibres (interosseous membrane). It is a connection consisting of a sheet of connective tissue used for muscle attachment. The dorsal and ventral lower arm muscles are separated by this membrane. If, with the elbow joint in an angle of 90°, one turns the palm of the hand down, in the lower arm the radius turns around the ulna and comes to lie on top of it. This movement is called *pronation*, and takes place in the two 'roll' joints at the ends of the two bones. The opposite movement is called *supination* (Fig. 2.9.23d).

In the wrist joint the arm becomes the hand.

Hand

The skeleton of the hand consists of a large number of bones: *8 carpal bones* (carpalia), *5 metacarpal bones* and *14 phalanges* (phalanx).

The carpal bones belong to the irregular bones; the metacarpal bones which border one another have synovial connections to each other, and these are strengthened by ligaments.

The wrist joint is a complicated joint. The socket is formed by the radius and the disc on the end of the ulna. The ball consists of the group of three proximal carpal bones. It is an ellipsoid joint with very many ligaments. The hand can achieve palmar and dorsal flexion, as well as radial and ulnar abduction (Fig. 2.9.24). The five metacarpal bones are connected to the distal row of carpal bones. They are – like the phalanges – short elongated bones. They can be found in the palm of the hand. Three phalanges form the skeleton of a finger, and the thumb has two phalanges.

The fingers can bend and stretch; the joints

Figure 2.9.24
Movements in the wrist
joint

a. Dorsal flexion and
   palmar flexion
b. Radial abduction and
   ulnar abduction

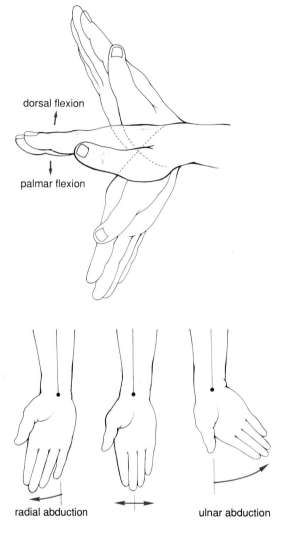

dorsal flexion

palmar flexion

radial abduction   ulnar abduction

Figure 2.9.25 ▶
Lower extremity

1. hip bone
2. hip joint
3. neck of the femur
4. greater trochanter
5. lesser trochanter
6. knee cap (patella)
7. splint bone (fibula)
8. thigh bone (femur)
9. knee joint
10. shin bone (tibia)
11. tarsal bones
12. metatarsal bones
13. phalanges (of the toe)

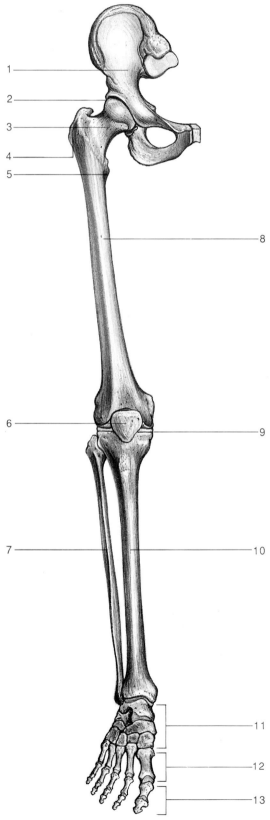

involved are hinge joints with collateral ligaments (Fig. 2.9.23c).

The distal ends of the metacarpal bones of the
four fingers are connected to one another (by
strong ligaments). The distal end of the metacarpal bone of the thumb is free. Combined with the
saddle articulation on the base of the thumb this
offers the thumb an important extra possible
movement. The thumb can be placed opposite to
each of the fingers of the same hand. This movement of the thumb is called, *to oppone*. The opposite movement is *to repone*. This accessory
additional motility in the thumb is essential for
the function of the hand. As a result of this we can
hold a pen or a violin bow, we can turn a page in
a book, take a firm grip on a handrail, etc.

6. Bones and bone connections of the lower extremity

Legs and feet are constructed for standing,
walking and jumping. Their construction is
aimed at support and stability, but at the same
time there is considerable freedom of movement. The heavy pressures that occur during
walking or jumping are effectively resisted.

Pelvic girdle, leg and foot are together called
the *lower extremity*. Compared to the upper extremity the difference in function and construction is striking: the bones are bulkier, the
joint can assume many positions and as a
result of the construction of the foot large
vertical loads can be carried. The foot has no
grasp function; the big toe cannot oppose.

Pelvic girdle
The *pelvic girdle* (see Fig. 2.9.20a) consists of
the pelvic bones and their connections. The
pelvic girdle is therefore *'closed'*. It is much
firmer than the shoulder girdle, but also far less
flexible. Each of the hip bones has a pronounced
hollow, a half cavity, the *acetabulum*. This is the
socket of the joint for the ball of the thigh bone
(see Fig. 2.9.20b).

Leg
The bones of the leg (Fig. 2.9.25) are: the *thigh
bone* (femur) which runs at a slant through the
upper leg; and the bones of the lower leg: *splint*

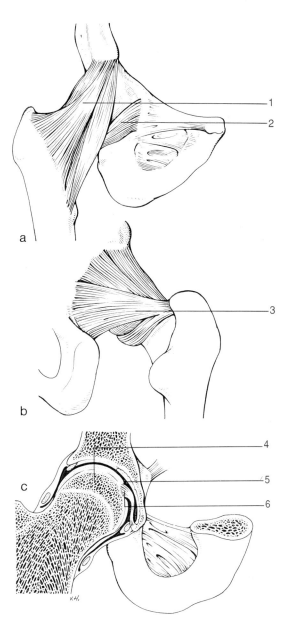

Figure 2.9.26
The ligaments of
the hip joint

a. Front view
b. Posterior aspect
c. Cross-section

1. *iliofemoral ligament*
2. *pubofemoral ligament*
3. *ischiofemoral ligament*
4. *head of the femur (thigh bone)*
5. *joint, articular interstice*
6. *ligament of head of femur*

bone (fibula) and *shin bone* (tibia). All three are long, elongated bones. In the knee is found at the front a sesamoid bone, the kneecap (patella).

The hip joint forms the connection between the pelvis and the leg. It is an almost perfect ball and socket joint. The socket is deep and offers much support to the round ball end of the femur. In the joint capsule there are four of the strongest ligaments in the entire body (Fig. 2.9.26). For example, the iliofemoral ligament possesses a tensile strength of about 350 kgm. In addition to that there is a ligament in the joint, the *ligament of head of femur*. The connection between thigh bone and pelvis is a very strong one.

The thigh bone is the longest elongated bone of all in the skeleton. Between the ball of the thigh bone and the actual shaft we find the *neck of the thigh bone* (collum femoris). In mature people this linkage forms, on average, an angle of 125° with the shaft. The pressures on this neck are enormous because of the weight of the body and of additional vertical forces when walking, jumping or falling. So the neck of the femur breaks more easily (and more often) than the shaft. An additional disadvantage is that the supply of blood – and consequently the rate of healing of a fracture – to this part of the bone is relatively poor.

The knob (tubercle) that can be felt through the skin on the top-end of the thigh to the side is called *the greater trochanter*. The *lesser trochanter* is slightly lower, to the medial side. These trochanters provide attachments for muscles.

The *knee joint* (Fig. 2.9.27) is the most complicated joint in the whole body. Three bones take part in this articulation or articulatory system

Figure 2.9.27
The arrangement of the knee joint

a. Seen from the front
b. Tibia-plateau with the menisci,
   seen from above

1. cruciate ligaments
2. lateral meniscus
3. collateral ligaments
4. shin bone (tibia)
5. splint bone (fibula)
6. thigh bone (femur)
7. medial meniscus
8. knee tendon
9. patella (kneecap)
10. tibia plateau

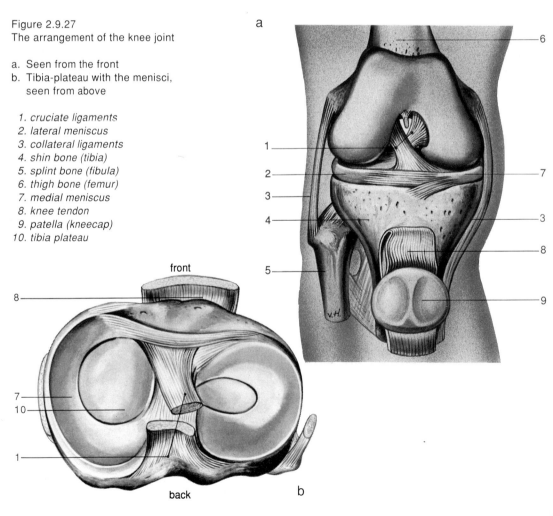

– femur, tibia and patella – but not the splint bone (fibula). It is called a hinge joint because bending and stretching are the movements which are most apparent. The joint has strong collateral ligaments. However, when bent, the shin bone (tibia) can also exo- and endorotate slightly. As a result of the shape of the joint surfaces of thigh bone and shin bone (tibia) the contact surface is relatively small. The distal (lower) end of the femur is rounded off, and the top or knee end of the tibia is flattened to become a kind of plane (*tibia plateau*). The joint looks like a door ajar. Two sickle-shaped cartilage structures (*menisci*) are fastened on to the tibia plateau filling up the gap, and thus enlarging the contact surface. As a result the pressures on the knee are spread out over a larger surface. The *inner meniscus* (medial meniscus) is half-moon shaped. It has developed with the medial collateral ligament and has relatively distant places of attachment. The *outer meniscus* (lateral meniscus) is almost round; its places of attachment lie close together and it has not developed with the lateral collateral ligament. As a result of these differences in construction the medial meniscus is far less flexible than the lateral one and consequently far more vulnerable to damage by rotation of the tibia in relation to the femur. In the middle of the joint are the two *cruciate ligaments* (Fig. 2.9.27).

The back of the kneecap is lined with cartilage. In knee flexion the joint surface of the femur end glides over this cartilage. In and near the knee joint we find numerous *bursae*. The biggest one is that above the kneecap (bursa supra-patellaris). It reduces the friction between the tendon attached to the patella and the tissues which lie deeper.

The *shin bone* (tibia) can be felt just beneath the shin along its entire length. The slender *splint bone* (fibula) lies on the outer, lateral side of the tibia. It does not take part in the knee joint and can hardly move in relation to the tibia. Between tibia and fibula stretches an interosseous membrana (see ulna and radius) that mainly functions – like the fibula itself – as a place to attach muscles.

The *inner ankle* (caudal end of the shin bone) and *outer ankle* (caudal end of the fibula) form a kind of fork in which the foot 'hinges'.

Foot

The foot skeleton consists of a large number of bones (see Fig. 2.9.25): 7 *tarsal bones* (tarsalia), 5 *metatarsal* bones (metatarsalia), 14 *phalanges*.

The tarsal bones belong to the irregular bones; the adjacent bones have synovial connections with strong ligaments where they have come to lie against one another.

Figure 2.9.28
Movements of the foot

a. Dorsiflexion and plantarflexion
b. Pronation and supination

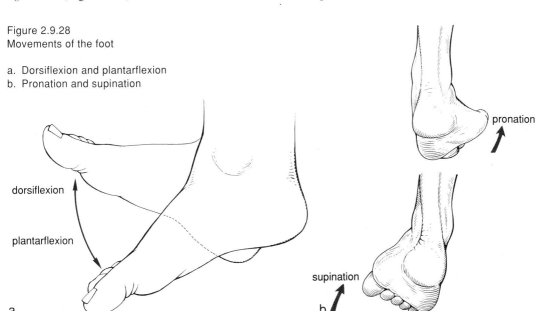

dorsiflexion

plantarflexion

pronation

supination

a

b

Figure 2.9.29
Print of a healthy right foot

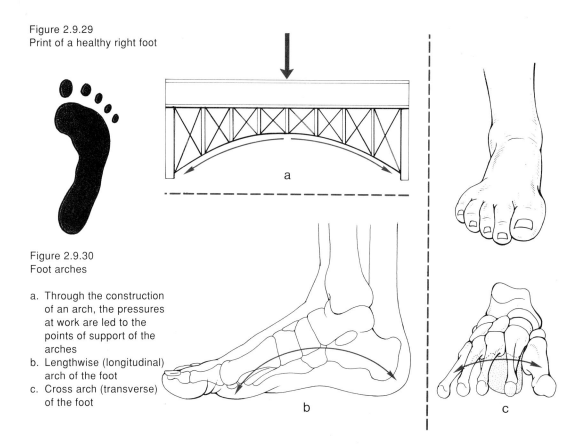

Figure 2.9.30
Foot arches

a. Through the construction
   of an arch, the pressures
   at work are led to the
   points of support of the
   arches
b. Lengthwise (longitudinal)
   arch of the foot
c. Cross arch (transverse)
   of the foot

The *ankle joint* is the joint between the bones of the lower leg and the foot (*talus*). In this hinge joint the foot can perform dorsiflexion and plantarflexion (Fig. 2.9.28a). There are very strong ligaments in the joint capsule. This joint is usually called the *upper hock*. The *lower hock* is the joint formed between the joint surfaces on the lower end of the talus on the one side and the joint surfaces of the *heel bone* (calcaneus) and of the os naviculare on the other side. Pronation and supination of the foot occur in the lower hock opposing sideways, and in slanting inclinations of the foot (Fig. 2.9.28b).

The 5 metatarsal bones are connected onto the distal row of tarsal bones. They are just like the toe phalanges – short elongated bones. They form, together with the tarsal bones, the 'closed' part of the foot. The big toe has two phalanges and the four other toes three each. The toes can bend and stretch.

When we study the print of a normal healthy foot (Fig. 2.2.29), we notice that the contact surface

is smaller than the surface of the sole of the foot. This is caused by the presence of the *arches of the foot*.

From architecture we know that through applying an arch construction it is possible to pass on the vertical pressures over the underlying structures (Fig. 2.9.30). In the foot we find two such arches: the lengthwise arch runs from the heel to the ball of the foot, the cross (transverse) arch runs from left to right.

7. Muscles and tendons

In every living tissue chemical reactions take place. In muscle tissue this occurs to achieve an external mechanical effect, *movement*.

There are three kinds of muscle tissue: smooth muscle tissue, striated muscle tissue and heart muscle tissue. (In Module 1, Chapter 4, a general introduction to these types of tissue was given.) The smooth muscle tissue is also called involuntary muscle tissue. It accomplishes the movements of vegetative integration.

When discussing the vegetative systems attention is always paid to the vegetative motor function. The heart muscle tissue has been described in greater detail in Module 2, Chapter 1. In this chapter the *striated muscle tissue* is discussed. This name derives from the fact that under a microscope this tissue shows a regular alternation of light and dark parts. Striated muscle tissue has a number of important characteristics. It can respond and contract quickly, and – in comparison to smooth muscle tissue – it is rapidly fatigued. Striated muscle tissue has been structured into *striated muscles*. These are mostly attached to parts of the skeleton, which is why it is often also called *skeletal muscle tissue* and *skeletal muscles*. On the basis of their innervation by and from the animal (voluntary) nervous system they are sometimes also called *voluntary muscle tissue* and so *voluntary muscles*. The human body numbers more than 600 skeletal muscles.

As has already been stated, they are the organs which bring about *movement*. As a result of their capacity to shorten and to lengthen, the bones to which they are attached can move in relation to one another or in relation to a point of support. Skeletal muscles also maintain the normal *bearing of the body* (posture); a certain level of muscle tension – even in a state of rest – ensures that the skeleton does not collapse like a house of cards. Striated muscles are *construction and shaping elements*. They give firmness to an otherwise weak body wall. The clearest case of this is the abdominal wall, in which abdominal muscles are the construction element.

When skeletal muscle tissue is active, it is a large *producer of heat*. In many cases this heat – being superfluous – must be removed (via perspiration, etc). The body temperature can be maintained through voluntary movements. In certain cases the skeletal muscles even start to contract uncontrollably. We may see little or no movement (trembling, chattering teeth), but there is *production of heat*.

General structure and function
Skeletal muscle has three distinct parts (Fig. 2.9.31): the *bulk of the muscle* (or the venter) and the two tendinous attachments. The bulk of the

Figure 2.9.31
The structure of a
striated muscle

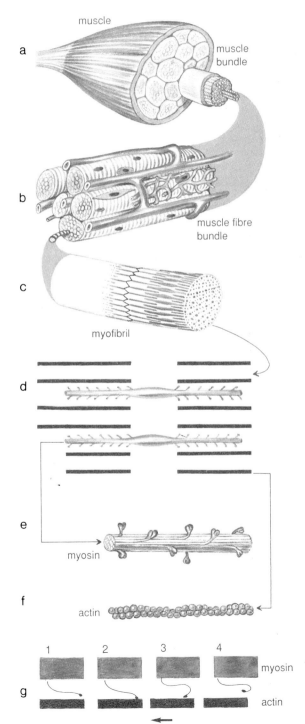

a muscle

muscle
bundle

b

muscle fibre
bundle

c

myofibril

d

e

myosin

f

actin

g   1   2   3   4

myosin

actin

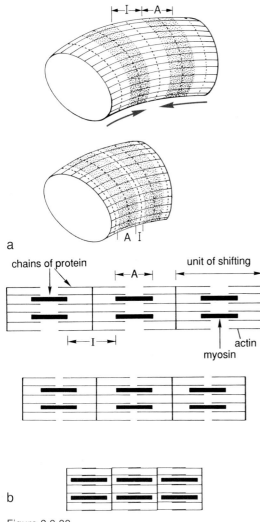

a

chains of protein

unit of shifting

actin
myosin

b

Figure 2.9.32
Contraction of a striated muscle

a. The change in the
   transverse stripes
b. The shift of actin and
   myosin

muscle is wrapped by connective tissue, the *muscle fascia*. This fuses into the tendons at the ends of the muscle. The muscle is attached to the bones via the tendons. As a rule the muscle is composed of a number of *muscle bundles*. Each muscle bundle in its turn consists of *bundles of muscle fibres* or *muscle cells*.

Muscle *fibre* relates to the length of the muscle cell. As a rule the muscle fibres are just as long as the length of the muscle to which they belong: they run from tendon to tendon.

Depending on the muscle involved their length

can vary from about 1 cm up to more than 15 cm. The muscle fibre is packed with longitudinal *myofibrils* (muscle fibrils). These are called the *contractile elements* of the muscle, since it is in the myofibrils that the shortening (the contraction) and lengthening of the muscles takes place. Under the microscope the myofibrils show, over their entire length, a regular alternation of light and dark parts. In adjacent myofibrils these stripes (striata) lie alongside one another so that bands of light parts and bands of dark parts run across the muscle tissue. The name *striated* muscle tissue is explained by this arrangement.

The pattern in the transverse stripes changes when the muscle cell changes its length (Fig. 2.9.32). This depends on protein fibres shifting in the myofibrils. The myofibril is composed of two types of protein fibres which are arranged in parallel lines: *actin* and *myosin*. The dark bands originate in those places where actin and myosin overlap. The *myosin fibres* are thicker than the actin fibres. On both sides they have protuberances, 'little feet' with which they can be anchored into the neighbouring actin strings (Fig. 2.9.31).

A voluntary movement depends on activity in a *motor end plate* (m.e.p.). Under the influence of nerve impulses the transmitter substance *acetylcholine* is set free out of little vesicles (see Fig. 2.7.6). This substance diffuses to the fibre membrane and depolarizes this membrane. This depolarization brings about a *shifting action* of myosin in relation to the actin. For this activity the presence of Ca^{2+} ions is required. The myosin fibres replace their 'feet' en masse in the actin fibres. The *tension* (tone) in the muscle increases. The parts overlapping each other become longer, so the dark parts widen and the lighter parts narrow. The muscle itself becomes shorter and thicker. The energy necessary for this '*shifting action*' is supplied by the 'energy packages' ATP. The large number of mitochondriae in the muscle fibres maintains adequate stocks of ATP, as far as possible.

The formation of linkages between myosin and actin is called *contraction* (pulling towards each other). When there are no bridges being formed or when these are being broken up, there is a state of *relaxation* (weakening).

Around the muscle fibres is an adapted endoplasmic reticulum. This system of tubes ensures – among other things – that calcium is supplied. Striated muscle fibres have many nuclei. These lie under pressure between the muscle fibres and the wall of the cell.

In the region of the muscle fibres are capillary networks of blood vessels. Muscles are well supplied with blood. This is understandable in the light of their high, extended level of metabolism. In the skeletal muscles are sensors which are specifically sensitive to alterations in the length of the muscles. They are the *muscle spindles* (see Fig. 2.8.47).

Enclosing the muscle fibres is a very thin layer of connective tissue, the *endomysium*, which is firmly attached to the fibres and which, at the ends of the muscle, transforms and becomes the tendon. A group of such fibres is enclosed in a thicker coating of connective tissue, the *perimysium*. The actual muscle length itself is wrapped in the connective tissue-like *muscle fascia*, also called the *epimysium*. Perimysium and epimysium merge at the end of the muscle to become the tendon.

The connective tissue in and round the muscle makes sure that the forces which are generated in the muscle are conveyed to the tendons (and thence to the bones). It contains a large amount of collagen fibres. Blood vessels, nerves and lymph vessels find their paths in and between the sheets of connective tissue in the fascia. The tendons themselves consist mainly of parallel collagen fibres running in the longitudinal direction of the muscle. At the end of the tendon the collagen fibres spread out – brush like, into the bone-membrane (the periosteum), even piercing through this membrane into the bone tissue itself, thus providing a really strong anchorage, generally even stronger than the actual bone. This is why people break a bone more often than they tear a tendon loose from the bone. The tendons are instrumental in the transfer of forces from the muscles to the bones.

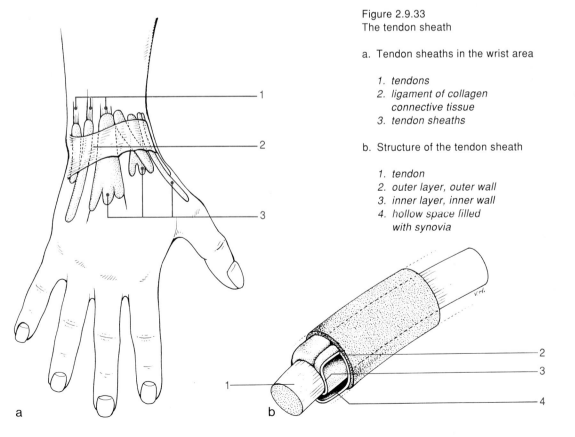

Figure 2.9.33
The tendon sheath

a. Tendon sheaths in the wrist area

1. tendons
2. ligament of collagen connective tissue
3. tendon sheaths

b. Structure of the tendon sheath

1. tendon
2. outer layer, outer wall
3. inner layer, inner wall
4. hollow space filled with synovia

a   b

In the tendons are found the sensors which provide information about the tensions in the tendon tissue, *tendon sensors* (see Fig. 2.8.48).

Synovial bursae and tendon sheaths

Synovial bursae are located between two muscles-lengths or, for example, between a tendon and a bony edge. Such bursae are simply extremely flat 'sacs' of synovial membrane filled with synovial fluid, synovia. Such an organ prevents too much friction occurring at certain places between tissues moving or rubbing across one another. An even more refined structure, preventing friction between two adjacent tendons or between a tendon and a bone, is the *tendon sheath* (Fig. 2.9.33). The tendon sheath is a 'tube' with a double wall around the tendon.

The inner wall has partly grown together with the tendon, the outer wall has partly grown together with the surrounding tissue. In the space between the inner and outer wall of the sheath there is synovial fluid. If the tendon now moves, the friction is reduced by this 'super-lubricant' synovia. For instance, in the wrist area the tendons of the lower arm muscles which move the hand and particularly the fingers run in tendon sheaths (Fig. 2.9.33). They reduce friction between the tendons by sliding over one another and between the tendons and the surrounding tissue.

Contractions of different types

Contraction can be thought of as the 'building of myosin/actin bridges' within the muscle fibrils. On the outside of the muscle a number of possible features related to contractions can be identified.

There are two forms of contraction: static and dynamic. In a *static contraction* the *length* of the muscle remains the same. This is *isometric* contraction. In fact, the thickness of the muscle increases minimally. The force which is being developed by the muscle depends on and is equal to the sum of a number of – sometimes opposing – factors, such as:
- the forces of muscles working in opposite directions;
- the force of gravity effecting the part of the body, or the limb to be moved (i.e. the weight of that part of the body, of that limb), sometimes with an additional extra load;
- the distance between the place where the muscle tendon is attached to the bone and the axis of the joint in which that bone is to move.

In a *dynamic (isotonic) contraction*, in contrast, the length of the muscle does change. This can result in two situations:
- if the sum total of the opposing forces is smaller than the force delivered by the muscle then the muscle can shorten: *the concentric contraction*;
- if the sum total of the opposing forces is greater than the force developed by the muscles then the length of the muscle is increased. That part of the body cannot be held up, nor that limb with an extra load, however hard we try: this is *the excentric contraction*.

An example. When a child (a gymnast) on the horizontal bar pulls herself up into a 'bent hang' her elbow flexors (the biceps) are concentrically active. During the 'bent hang' itself these flexors are in static contraction.

When this child lets herself sink down slowly, then we have a case of *excentric contraction* in the elbow flexors.

Synergists and antagonists

Muscles which have – when contracting – one and the same effect on a certain joint are called *synergists* (cooperators, collaborators).

The group of sural muscles consists of three muscle lengths which are called synergists, because when they contract concentrically they bring about *plantarflexion* in the ankle joint (as when you are going to stand on tiptoe).

Muscles which have an opposite effect on a joint, which move a bone in an opposite direction round the axis of a joint are called each others' *antagonists* (working 'anti' each other). For example the muscles to the front of the upper arm are the antagonists of the muscles at the back of the upper arm.

Muscle antagonism is the name of the following phenomenon; those muscles which have an opposite effect on a joint – when contracting – are (often) finely tuned together in their activities: when the one muscle is shortened, the other, opposing, muscle is lengthened at the same time and in the same degree, in an almost perfect balance of forces.

Fine and coarse movements

Muscles which can be controlled very precisely, such as the muscles of the hand and the fingers, need an extensive cable network (innervation). Muscles for a less refined type of movement, like those of the upper leg, can make do with a less intensive innervation. The muscle fibres in a muscle do not work in isolation. They are organized functionally in *motor units*. Such a motor unit consists of a *motor neurone* together with those muscle fibres which are in direct contact with that motor neurone (see Fig. 2.7.16). A refined type of movement (a finer motor system) is associated with small motor units. The ratio of the number of muscle fibres to the number of afferent neurones is small.

The less refined systems of movement go with large motor units: a large group of muscle fibres is innervated by one single efferent neurone.

Small motor units are characterized by small branching of the neurite, the axon. In large motor units, however, there are a large number of telodendrites per axon.

Power and speed

The force or power developed in a muscle depends largely on the number of muscle fibres being activated at the same time. This means that in a muscle which is working at half its capacity there are still quite a number of muscle fibres which have not yet been called into action. In other words, many motor units have not yet been activated. In a maximum muscular effort, however, almost every motor unit will be activated. At the same time we may conclude that muscles with many muscle-fibres – and thus with a large diameter – can produce more power than muscles with a smaller diameter.

In general we can state that a muscle working concentrically is able to yield less force than the same muscle working isometrically. This force in its turn is less than the force of this same muscle when it is working, contracting excentri-cally. In other words when jumping in the air, the knee-extensors are less strong than when absorbing the forces of the same jump from a height onto the ground.

The *velocity* with which a muscle can contract depends, among other things, on the type of muscle tissue in the muscle. *Slow* and *swift* types of muscle fibres can be identified.

Slow muscle fibres are more reddish in hue, swift muscle fibres are more whitish. This difference in colour is caused mainly by a relatively better capillarization of the red muscle fibres. These red muscle fibres work mainly aerobically and do not tire so quickly. In the case of the *white muscle fibres* it is the other way around.

In every muscle both these types of muscle fibres are present, but it is striking that those muscles which regulate posture have more red muscle fibres (e.g. the muscles of the back), whereas the 'white' muscles are motion muscles (e.g. the extensors of the elbows). Red muscles generally turn out to possess smaller motor units than white muscles. The velocity of contraction also depends on the force that must be developed. When the muscle is not under strain the velocity can be maximal. As the strain increases the possible velocity decreases. At a certain level of strain the muscle will only be capable of contracting statically; the velocity is zero.

Controlling the movements and function of the muscle

The cell bodies of the *pyramidal* and *extra-pyramidal* tracts are in the motor cortex. These consist of descending motoneurones which are end on to the motoneurones in the spinal ventral horn cells. Each motoneurone is a part of a motor unit. A motoneurone is excited to stimulate all the muscle fibres of the motor unit involved into contraction.

In controlling and coordinating movement the human body shows a finely tuned interplay, a sophisticated interaction of the *sensory* and *motor systems*.

We therefore refer to sensomotor activity. If you pick up a teacup from the table, the movements are initiated by commands from the motor cortex via the descending tracts to all types of arm

Figure 2.9.34
Muscles of head and neck

1. *masticatory muscles*
2. *temporalis*
3. *masseter*
4. *buccinator*
5. *sternocleidomastoid*
6. *mimic muscles*

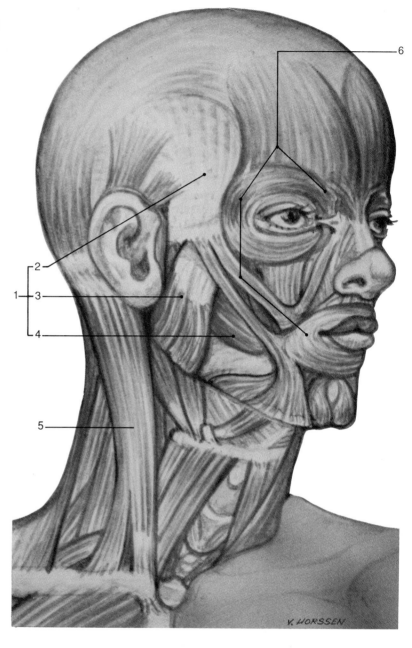

V. WORSSEN

muscles (motor). The small brain receives a 'copy' of this set of commands in order to take care of small adjustments. We see the effect of the movement, which can also lead to coarse adjustments. The moment we touch the cup (sensoric impulses to the brain), the rest of the programme for this movement can be started (motor): grasping it, picking it up, lifting it to the mouth, drinking, putting it back.

If someone suddenly bumps into the table and the cup threatens to topple over, an adjusted programme of movements (commands) is begun immediately. In the locomotive apparatus itself, in the muscles, the tendons and the joints there are little organs which are sensitive to changes (sensors) and can give information about the length of the muscle, the tension in the tendon, the angle in the joint, etc. These *propriosensors* play an important afferent part in movement coordination.

Listing in sequence the functions of a muscle is a hazardous affair. Most of the time the description of the effects of muscle contraction starts from the anatomical posture. In other postures and especially in the totality of the locomotive interplay the muscle function is highly dynamic. Contraction, relaxation, fixation and stabilization can follow each other extremely rapidly. Furthermore training by repetition is very influential. The control over muscles of a ballet dancer is quite different from the control over the muscles of, for example, a javelin-thrower. It is wise to keep these limitations in mind when describing the functions of muscles.

Electromyogram

The activity of muscles can be investigated by means of recording the changes in the electrical activities of the muscle cells; an electromyogram is made (EMG). At predetermined locations on the skin, detecting-electrodes are placed in order to record the activity of the muscles under that area of skin. By studying EMGs in relation to movements that have been observed, it is possible to gather more data about muscle function.

8. Muscle system

Of the approximately 600 striated muscles in the human body only a few will be discussed in this book because of limited space. Often muscles will be named as a group.

8.1. Muscles of head and neck

Head

In the region of the head there are four groups of muscles (Fig. 2.9.34): the *mimic muscles,* the *masticatory muscles, lingual muscles* and the *muscles of the eye.*
The *mimic muscles* are those muscles which bring about the facial expression; there are more than 30 of them.
They are attached to the skull and terminate, brushlike, in the skin. Almost all the mimic muscles are innervated by the facial nerve (N. VII).
The well-known facial muscles are: the muscles on the forehead (frowning), the muscles above

the eyebrow (looking surprised), the circular muscles around the eye (narrowing) and the ones around the lips, which round and spread the lips. Some people are able to move the nostrils by the muscles in their nose, others can move their ears slightly.
Mimics play an important part in the communication between individual human beings. It is often possible to tell how people feel by ob-

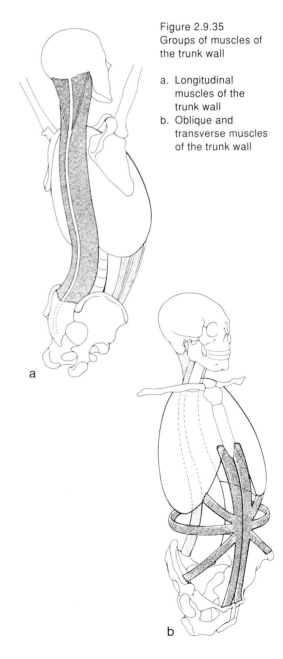

Figure 2.9.35
Groups of muscles of the trunk wall

a. Longitudinal muscles of the trunk wall
b. Oblique and transverse muscles of the trunk wall

a

b

serving the expressions on their face. The face can show if someone is happy, or sad, pondering, brooding, dreamy or tense, concentrating, angry, wronged, aggrieved, relieved, embarrassed, etc.

The *masticatory muscles* move the lower jaw in relation to the rest of the skull and thus in relation to the upper jaw. The names of the masticatory muscles are: *masseter, temporalis* and *buccinator* (Fig. 2.9.34).

The first two are extremely powerful muscles; you can easily feel them swell when you clamp your jaws together. The third, buccinator muscle, brings tension on to the cheek when chewing, exactly at that moment when the tongue pushes the food between the molars and the jaws are closed – so that no food can escape to the sides of the teeth. This muscle can also pull the corners of the mouth backwards, as occurs when weeping and laughing. To some extent the masticatory muscles also belong to the mimic muscles. They are innervated by the trigeminal nerve (N.V).

The *tongue* (lingua) consists of a number of cooperating lingual muscles (Fig. 2.2.16a), coated with the lingual mucus membrane. The special feature of the tongue is that one end of the muscles is 'free', the tip of the tongue. The tongue plays an important part in eating, chewing and swallowing. The lingual muscles are innervated by the hypoglossal nerve (N. XII).

To the *muscles of the eye* belong, besides the circular muscle round every eye, the six eye muscles which can direct the eye-ball very accurately (see Fig. 2.8.20).

Figure 2.9.36
Muscles of the back

a. Superficial muscles of the back

   1. trapezius
   2. deltoid
   3. latissimus dorsi
   4. superficial bundle of the erector spinae

b. Deep muscles of the back

   1. skull
   2. collar bone
   3. shoulder blade
   4. rib
   5. deep bundles of the erector spinae
   6. pelvis

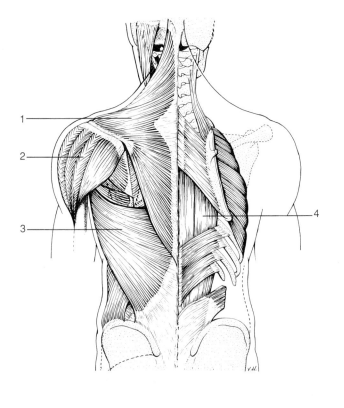
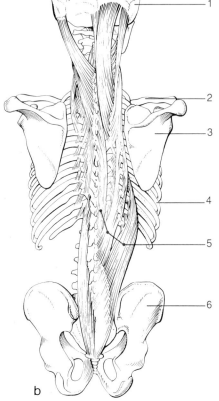

a   b

Neck

At the nape of the neck there are a number of powerful muscles, which are partly an extension of the muscles of the back which run on to the back of the head (the occipit). They are very important in keeping the head erect and in moving the head. The most superficial muscle is the most cranial part of the trapezius (the monk's hood muscle) (Fig. 2.9.36a).

To the front of the neck the mass of muscle consists only of slender, flat muscles. The *sternocleidomastoid* (Fig. 2.9.34), is the best known of these. The (extended) name refers to the places it originates from (sternum and collar bone) and the place it is attached to (the mastoid process – the tubercle right behind the ear). It is a muscle which can rotate the head. Further, it is an auxiliary respiratory muscle. In addition to this there are the muscles of the tongue bone (the hyoid bone) (see Fig. 2.4.7). These control the tongue bone during the processes of eating and swallowing.

8.2. The muscles of the wall of the trunk

A major part of the wall of the trunk is formed by skeletal muscle. In the thorax part of the trunk –

besides these muscles – a substantial part of the wall consists of skeletal bones; whereas there are hardly any bones in the wall of the abdominal part of the trunk – it consists mainly of muscles. On the basis of the direction in which these abdominal wall muscles are spun and contract, we can bring a number of them under three headings (Fig. 2.9.35):

– *longitudinal* muscles of the trunk wall: the fibres run vertically. On the dorsal side the deep back muscles and on the ventral side the abdominal muscles belong to this group;

– *oblique* muscles of the trunk wall: the fibres run obliquely. In the thorax region the intercostal muscles and in the abdominal region the oblique abdominal muscles belong to this group.

The groups of oblique muscles occur in pairs; with both the intercostal and the abdominal muscles there is a deep layer and a superficial layer. The direction in which their fibres run is left and right (sinistra and dextra) symmetrically as in a mirror reflection.

Figure 2.9.37
Intercostal muscles
a. A front view
b. The direction of the fibres of the outer intercostal muscles is practically at right angles with the direction of the fibres of the inner intercostal muscles

1. outer intercostal muscles
2. inner intercostal muscles
3. cartilagenous zones
4. rib
5. blood vessels
6. thoracic nerve

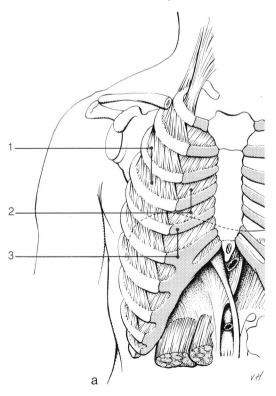

a

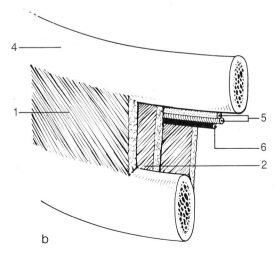

b

In the superficial layer the fibres run from cranial-lateral towards caudal-medial, as in the 'legs' of the letter V; in the deep layer they run exactly opposite in a 90° angle from cranial-medial to caudal-lateral as with the 'legs' of the letter A;

– *transverse* muscles of the trunk wall: the fibres run horizontally. These only occur in the abdominal area: the transverse abdominal muscles (m. transversus abdominis).

The muscles of the trunk wall are divided as follows: (deep) muscles of the back, muscles of the thorax wall, diaphragm, muscles of the abdominal wall and of the pelvic floor.

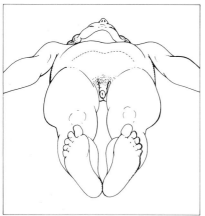

Muscles of the back

On the dorsal side there are a number of muscles which lie in layers on top of each other (Fig. 2.9.36). The most superficial ones (trapezius and latissimus dorsi) are not considered to belong to the muscles of the trunk wall. They are attached respectively to the shoulder girdle and partly to the arm, and for that reason they are regarded as muscles of the extremities. The erector spinae is a very strong muscle, one half to the left of the spinal column, and one half to the right, running from the pelvis to the occipit. The erector spinae is a collective name for all the shorter and longer muscles which – together – form this muscle. The erector spinae brings – as the name already indicates – the spinal column into an erect position (from a bent forward position) and bends the trunk backwards.

A number of short muscles of the back have their fibres in an oblique position, e.g. from the processus spinosus to the processus transversus of the next higher vertebra. They can effect rotation in the spine. The erector spinae is innervated by the dorsal branches of spinal nerves. The main antagonist of the erector spinae is the rectus abdominis.

Figure 2.9.38
The diaphragm seen from below

1. passage for the inferior vena cava
2. muscle fibres of the diaphragm
3. sword-like extension of the sternum: the xiphoid process
4. cartilaginous arch
5. central tendon plate (centre tendineum)
6. passage for the oesophagus
7. passage for the aorta
8. column of lumbar vertebrae
9. muscle of the back
10. part of the muscle iliopsoas

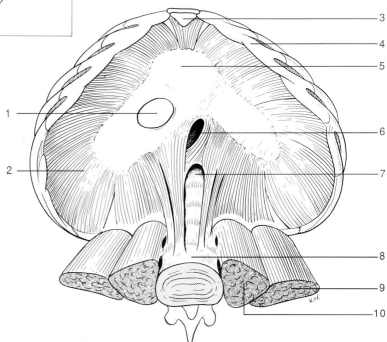

Muscles of the thorax wall

On the ventral side of the thorax we also find a number of muscles which are not considered to be muscles of the trunk wall. As they attach to the shoulder girdle or the arm they are muscles of the extremities (the pectoralis major and minor). The defacto muscles of the thorax wall are the *intercostal* muscles. These are stretched from the lower edge of a rib to the top end of the next lower rib (costa). Besides their bordering function they have a function in bringing about respiratory movements. In all there are 11 *pairs* of intercostal muscles. On the basis of their position and the direction of the fibres (Fig. 2.9.37), the *external intercostal muscles* (m. intercostalis externis) and the *internal intercostal muscles* (m. intercostalis interni) can be identified.

The external intercostal muscles can lift the ribs because of the direction of their fibres and their place of attachment, and so they can increase the volume of the thorax (breathing in).

The internal intercostal muscles can almost bring about an opposite movement: they can strengthen the lowering of the ribs (in addition to gravitational lowering) when breathing out, though the direction of the fibres differs no more than around 90°. The intercostal muscles are innervated by the intercostal nerve.

Diaphragm

The *diaphragm* marks the division between the thoracic and abdominal cavity (Fig. 2.9.38). It is a structure like a half-dome in shape, the high part of which consists of a tendinous plate, *the central tendon*. From the front it is clear that the

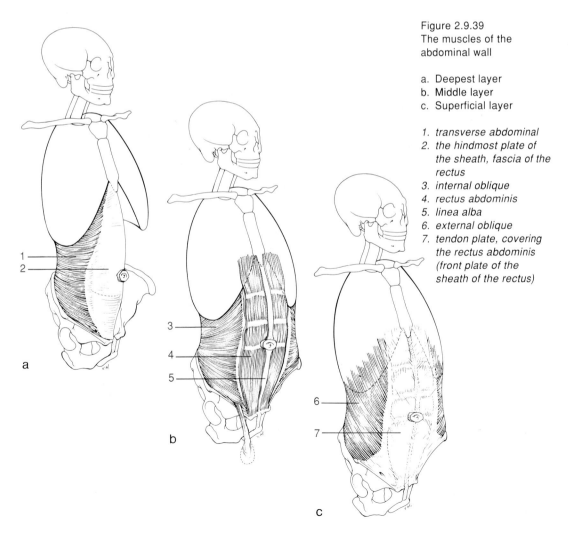

Figure 2.9.39
The muscles of the abdominal wall

a. Deepest layer
b. Middle layer
c. Superficial layer

1. transverse abdominal
2. the hindmost plate of the sheath, fascia of the rectus
3. internal oblique
4. rectus abdominis
5. linea alba
6. external oblique
7. tendon plate, covering the rectus abdominis (front plate of the sheath of the rectus)

Figure 2.9.40
The sheath of the
rectus and the
muscles of the front
wall of the abdomen

1. *linea alba*
2. *peritoneum*
3. *front sheath of
   rectus*
4. *left transvere
   abdominal*
5. *left internal oblique*
6. *left external oblique*
7. *subcutaneous fat*
8. *left rectus
   abdominis*

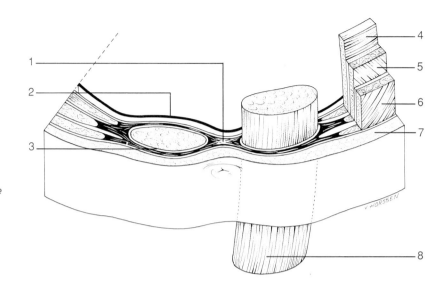

diaphragm, when resting, is higher to the right than to the left. In the middle, approximately where the heart lies, the diaphragm is somewhat pushed downwards (see Fig. 2.4.14a). Under the right dome of the diaphragm is the large right lobe of the liver. Under the left dome of the diaphragm is the stomach. From the tendon plate, a thin sheet of striated muscle tissue runs down to the lower edge of the thorax; the breast-bone, the rib arch, the floating ribs and a number of vertebral bodies.

The muscular part of the diaphragm is inner-vated by the phrenic nerve (see Fig. 2.7.21). When the diaphragm muscle contracts the half-dome is flattened and drawn down. When the muscle tissue relaxes the diaphragm returns to its original shape of a half-dome. These move-ments of the diaphragm belong to the most important of the respiratory mechanisms.

In the diaphragm there are a number of aper-tures. The inferior vena cava passes the dia-phragm via an opening in the central tendon. This is an advantage compared to the passages for the aorta and oesophagus, which both pass through openings in the more muscular part of the diaphragm. The advantage for the inferior vena cava is that during contraction of the dia-phragm muscles this vessel cannot be squeezed shut.

As mentioned earlier, blood pressure is zero or negative in this vessel, whereas the pressure in the aorta and oesophagus is high enough to resist

the force of the contracting muscle tissue. More-over, the direction of the muscle fibres round the passages (apertures) for aorta and oesophagus is not of a type to cause the effect of a circular muscle or sphincter.

Muscles of the wall of the abdomen

Between the lower edge of the thoracic cage and the upper edge of the pelvis there is a rather large area, more or less diamond-shaped, which is completely bridged by the muscles of the *ab-dominal wall*. They are innervated by the ventral branches of spinal nerves. The muscles of the abdominal wall have different shapes and directions at different depths (Fig. 2.9.39).

The *transverse abdominal muscle* (transversus abdominis) lies deepest and innermost. At the dorsal side this muscle starts as a tendinous plate around the erector spinae and further forward to the crest of the pelvis (iliac crest). At the front this muscle ends in the tendinous plate which is an extension of the fascia of the *rectus abdominis* (Fig. 2.9.40).

Lying on this deepest layer are the *oblique abdominal muscles* which consist of two layers: the *internal oblique abdominal muscle* (m. obliquus internus), and the *external oblique abdominal* muscle (m. obliquus externus). The inner one starts at the crest of the pelvis (iliac crest) and ends partly at the arch of the ribs and

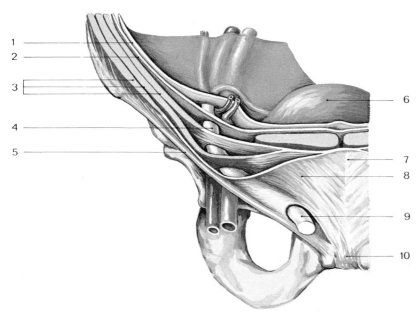

Figure 2.9.41
The inguinal canal

1. *peritoneum*
2. *layer of connective tissue between peritoneum and abdominal muscles*
3. *three layers of abdominal muscles; from left to right the external oblique, internal oblique and transversus abdominis*
4. *seminal duct passing the inner inguinal opening*
5. *inguinal ligament*
6. *dome of the bladder covered by the peritoneum*
7. *linea alba*
8. *the blade of the rectus sheath lying most superficially*
9. *the seminal duct lying in the outer, superficial inguinal ring*
10. *symphisis pubis*

partly with a tendinous plate into the sheath fascia of the rectus abdominus. The external oblique begins at the front of the 5th–12th rib (Fig. 2.9.39), and ends with a tendon plate which fuses into the sheath of the rectus.

The *'straight* abdominal muscle', *rectus abdominis,* consists of vertically running muscle. It is divided in two parts, left and right of the navel, and it begins at the sternum and the lower edge of the thorax, running all the way down to end at the pubic bone and the symphysis. This muscle is divided into 'blocks' by three or four transverse tendon bands. (The rectus is interrupted by three tendinous intersections.) This can easily be seen when a thin person contracts this muscle. Round the left and round the right part of the rectus abdominis there is a tendinous wrapping: the rectus sheath.

The rectus sheath is made, partly, out of tendon fibres radiating brushlike out of the transversal and oblique abdominal muscles, particularly their fascia. At the height of the median line, the left and right rectus sheaths come together as the *linea alba* (the white line). The *navel* (umbilicus) forms an interruption of the linea alba.

The firm connection of the linea alba with, for example, the surrounding tissues and between the constituent parts of this muscle complex, ensures that the left and right systems of abdominal muscles can operate as a unit. In the case of the transverse muscles this is easy to understand, as the left and right systems 'cross over' each other. In the case of the oblique muscles there is a similar type of 'cross-over'. The direction in which the fibres of the external oblique muscles run is in line with the direction of the inner oblique ones of the opposite side of the body. This may be imagined more easily when the letters V and A (mentioned earlier), representing respectively the directions of the external and internal oblique muscles, are laid on top of one another.

From the foremost point of the pelvic crest (i.e. anterior superior iliac spine) there is a strong ligament running to the pubic bone. It is the *inguinal ligament* (ligamentum inguinale). The inguinal ligament is the lowermost edge of a shared corporate tendon plate of the transverse and oblique abdominal muscles (Fig. 2.9.39). Above the attachment of the inguinal ligament

Figure 2.9.42
The pelvic floor (♀)

a. Seen from below

    1. *symphisis pubis*
    2. *levator ani*

b. Seen from above

    1. *sacrum*
    2. *muscles of the*
       pelvic floor
    3. *symphisis pubis*
    4. *rectum*
    5. *vagina*
    6. *urethra*

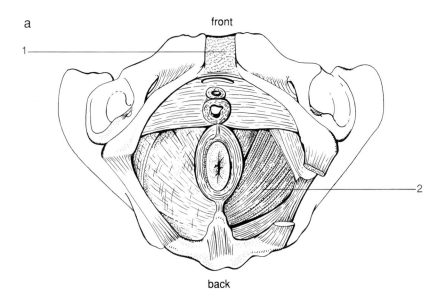

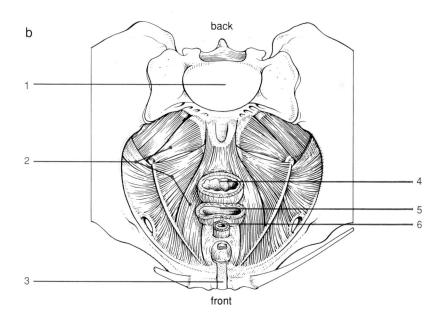

to the pubis, there is an opening in the tendon plate (Fig. 2.9.41). It is here that the *inguinal canal* (canalis inguinales) ends. This canal runs from the abdominal cavity and slants downwards, piercing several layers of the abdominal wall. At the inguinal canal, an *internal* and an *external inguinal* can be identified. In the case of a male the inguinal canal runs into the scrotum, and contains the spermatic cord or seminal duct terminals. In a female the ligamentum teres coming from the uterus runs through the inguinal canal, which ends up in the labium majus of the vagina.

The muscles of the abdominal wall have a number of functions. Since there are no skeletal parts in the abdominal wall, the digestive organs do not come under pressure when the trunk goes through all kinds of movements. The walls are elastic. The firmness of the muscular tissue – particularly in a contracted state – and the wrapping fasciae, generally offer sufficient protection to the abdominal organs.

When the abdominal muscles are forcefully contracted, even a powerful blow can be resisted.

The transverse and oblique abdominal muscles enclose the abdominal cavity and act as a muscular corset. In addition to this the transverse abdominal muscles can enable rotation of the trunk. The straight (rectus) abdominal muscles are very strong muscles which can bend the trunk forward. They are the antagonists of the deep back muscles. The volume of the abdominal cavity can be reduced by contracting the abdominal muscles. The effect of this reduction can be increased further by flattening the diaphragm when breathing in. Since the abdominal cavity is a closed space, the pressure rises there after this type of contraction; the rule $P \times V = c$(onstant) accounts for this. This complex is called the *abdominal press*. The enhanced pressure in the abdominal cavity increases the pressure on the abdominal organs, including those which have a connection with the outer world, in other words those abdominal organs of which the contents belong to the milieu extérieur. Thus the rectus abdominis plays a part in defecation, in urination, in giving birth and in vomiting. The abdominal press is also important when breathing out with force (sneezing, coughing, blowing one's nose, inflating a balloon, etc.).

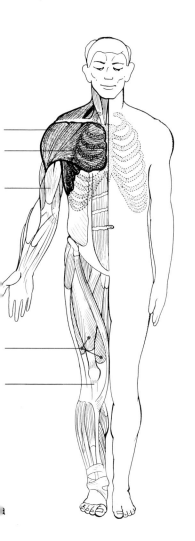
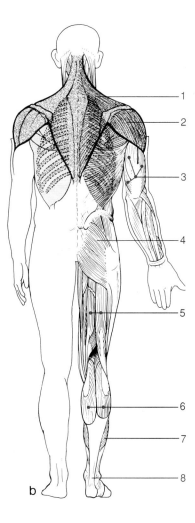

Figure 2.9.43
Extremity muscles

a. Seen from front

1. deltoid
2. pectoralis major
3. biceps brachii
4. quadriceps femoris
5. knee tendon

b. Seen from the back

1. trapezius
2. deltoid
3. triceps brachii
4. gluteus maximus
5. hamstrings
6. gastrocnemius
7. soleus
8. Achilles tendon

Muscles of the pelvic floor

The small pelvis is bordered by and closed caudally by the *pelvic floor* (diaphragma pelvis). It consists of a number of striated muscles, which mark the caudal end of the small pelvic cavity. These muscles are spun from the inner edges (crests) of the small pelvis in a bowl shape (Fig. 2.9.42). The most important representative of the muscles of the pelvic floor is the *levator ani*. The pelvic floor has a number of openings, one for the rectum with the anal sphincters and one for the urine duct (urethra). In a female there is an additional one for the vagina.

8.3. Muscles of the extremities

The muscles of the extremities are usually considered to include the muscles of the shoulder girdle, the arm and the hand and also the muscles of the pelvic girdle, the leg and the foot. The extremity muscles ensure that the freedom of movement offered by the skeletal bones involved can indeed be effected. In effecting these movements it is strikingly clear that – in line with the shape of the skeleton and of the joints – the muscles of the lower extremities are mainly intended for stability and locomotion whereas the muscles of the upper extremity assist the manifold functions of the hand, such as seizing and manipulating.

The upper extremity is for the most part innervated by branches of nerves, emerging from the cervico-brachial plexus, mainly the radial nerve, the ulnar nerve and the median nerve (see Fig. 2.7.20).
The lower extremity is for the most part innervated by nerve branches from the lumbo-sacral plexus. The most important nerves here are the ischial nerve and the femoral nerve (Fig. 2.7.22).

The upper extremity

The shoulder girdle can be moved in relation to the trunk by a number of smaller and larger muscles which are spun from the front of the thorax to the collar bone or from the back of the thorax to the shoulder blade (Fig. 2.9.43). Specific reference is made to the *pectoralis minor* (the small breast muscle) to the front and

the *trapezius* (monk's hood muscle) to the back. The trapezius received its English name after the hood that used to distinguish, for example, the Franciscan monks.

There are also muscles which run directly from the trunk to the upper arm, such as the *pectoralis major* (the large breast-muscle) and the *latissimus dorsi* (the broad back muscle). The pectoralis major covers the larger part of the ventral chest wall. The latissimus dorsi is the widest muscle of the body; it covers the larger part of the back.

The pectoralis major and minor are regarded as auxiliary respiratory muscles.

The deltoid (delta muscle) is a wide triangular muscle which runs from collar bone and shoulder blade to the upper arm. The deltoid muscle determines the outline of the shoulder.

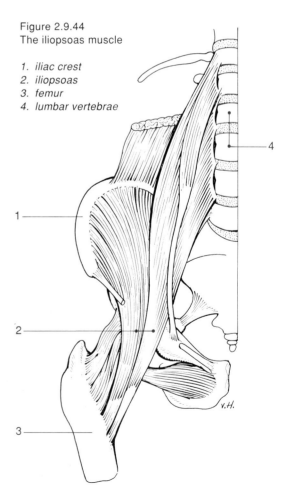

Figure 2.9.44
The iliopsoas muscle

1. iliac crest
2. iliopsoas
3. femur
4. lumbar vertebrae

To the front of the upper arm there lie those muscles which can bend the lower arm at the elbow joint in relation to the upper arm. They are called flexors (benders). The best known representative of this group is the *biceps brachii* (the two-headed upper arm muscle), in short the biceps. The extensors (stretchers) are at the back of the arm. They are the antagonists of the flexors.

The most important of this group is the three-headed upper arm muscle (the *triceps brachii*), in short the *triceps*.

To the front of the lower arm – in the anatomical position – are found those muscles which bring about *palmar flexion* of the hand in the wrist, and flexion in the hand and finger joints. On the back of the lower arm are the antagonists; these affect *dorsal flexion* in the wrist joint and the stretching of the hand and the fingers.

The muscles of the hand flexors and hand extensors are entirely located in the lower arm. They originate respectively from the ventral and dorsal sides of the ulna. They have remarkably long tendons; some of these tendons reach to the most distal finger-phalanx. This construction has the advantage that the hand is very slender but at the same time a lot of movement is possible. This is the factor which enables such a sophisticated functioning of the hand.

In the wrist area all tendons run beneath a band (retinaculum) which keeps the tendons in their place. In order to diminish friction in the wrist area, these tendons are embedded in tendon sheaths (see Fig. 2.9.33a).

In the hand itself there are numerous short muscles which allow subtle movements of the hands and fingers: spreading and closing the fingers as an entity, bending and stretching the hand and finger joints. For the opposing and reposing of the thumb, other muscles are available.

The lower extremity

One of the most important muscles of the pelvic girdle is the *gluteus maximus*. It is a strong muscle running from the pelvis to the upper leg bone. Its most important function is retroflexion (bending backwards) of the upper leg in the hip joint. It goes without saying that there are also muscles with an antagonistic effect running from trunk and pelvis to the upper leg bone (femur); the most important one of this group, running within the pelvis (pelvic cavity) is the *iliopsoas* (Fig. 2.9.44).

To the front of the femur are mainly those muscles which can stretch the lower leg in the knee joint. The most important of this group of extensors is the *quadriceps femoris* (the four-headed thigh bone muscle), in short the quadriceps. To the front of the knee the four 'heads' of this muscle come together into a common tendon that attaches to the front of the shin bone. The patella (kneecap) is a sesamoid bone in this tendon. The *knee tendon* is that part of the tendon that is found between the patella and the tibia.

To the back of the upper leg there are muscles which can stretch the lower leg in the knee joint. The most important knee-flexors are the 'hamstrings': the semimembranosus, the semitendinosus and the biceps femoris.

To the front and to the side of the lower leg are muscles which can effect *dorsal flexion* of the foot and stretching of the toes (in the direction of the knee.

To the back of the lower leg are the *plantar flexors;* the *sural muscles*. These are composed of the two-headed *gastrocnemius* and – lying beneath – the *soleus*. They have a collective tendon, which attaches to the calcaneus (heel bone). This is the well-known *Achilles tendon*. The muscles which move the foot lie in the lower leg area. Similar to the muscles that control the hand, they have very long tendons with and within tendon sheaths.

In the foot itself there is a large additional number of short muscles for the movement of feet and toes.

10

Reproductive system

Introduction

Every form of life is characterized by some form of *reproduction*. Reproduction is the process that ensures the preservation of the species. Remarkably enough *preservation of the species* is a more important characteristic of life than self-preservation. Self-preservation of individuals is useless in the long term if life is not passed on to progeny.

Human reproduction has a number of different components: searching for and finding a heterosexual partner, mating (copulating), pregnancy, giving birth and nurturing children.

Moreover the reproduction of the human being is embedded in the totality of *sexuality*. So by sexuality we mean all the aspects, physical and mental, of a human being's sexual relationships.

In view of the specific purpose of this book, this chapter will focus on the functional anatomy of human reproduction.

Human reproduction is *sexual reproduction* (i.e. procreation), that is to say that a new individual comes to exist out of the fusion of two extraordinary cells; a male reproductive cell and a female reproductive cell. It makes use of two sexes, not only where the production of the reproductive cells is concerned, but also for copulation.

The reproductive system of man consists of the reproductive organs of two individuals who differ sexually; a male and a female. Therefore this chapter discusses the female reproductive organs and the male reproductive organs, copulation and fertilization.

Learning outcomes

After studying this chapter you should have sufficient knowledge and understanding of:
- the importance of the reproductive system for the preservation of the species;
- the reduction division of cells;
- the sexual characteristics of man and woman;
- the structure and functions of the female reproductive organs: ovaries, uterine tubes, uterus, vagina and vulva;
- the structure and functions of the male reproductive organs: testicles, epididymis, passageways for semen, prostate gland and penis;
- the process of copulation and fertilization.

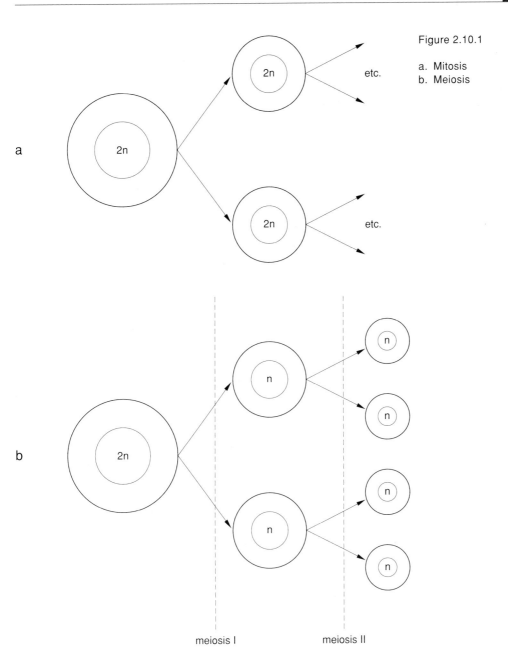

Figure 2.10.1

a. Mitosis
b. Meiosis

meiosis I   meiosis II

1. Meiosis

Sexual reproduction occurs when a male reproductive cell fuses with a female reproductive
cell. In Module 1, Chapter 3, it was pointed out
that all body cells contain 46 chromosomes, 23
pairs of chromosomes – except for the reproductive cells. In mitosis (normal cell division) the
original cell, the 'mother cell' splits up into two
exact copies of itself. As a result the two 'baby

cells' have 23 pairs of chromosomes (Fig.
2.10.1a). These are *diploid*, which means the
chromosomes occur in *pairs*. If the reproductive
cells (i.e. gametes) also contained 23 pairs of
chromosomes, then at the fusion of two gametes
the number of chromosomes would be doubled
(92). However, the children of successive gen-
erations have alway had the same number of
chromosomes as their parents. This is because
the chromosomes are divided into *two exact*

halves at the moment when the reproductive cells (i.e. gametes) are formed (gametogenesis) and then stored in this form in anticipation of the moment of fusion. This is a unique feature of this type of cell.

The 23 pairs of chromosomes of each forerunner of the reproductive cell are – in a number of steps – divided into two cells of an intermediary type. This process is called *meiosis* (reduction division, as distinct from the normal division, mitosis). In meiosis only one chromosome of each pair of chromosomes goes to one new cell (daughter cell) and the other twin-chromosome

from the pair goes into a second new daughter-cell (Fig. 2.10b).

Reproductive cells are *haploid*, which means the chromosomes are unpaired (or more loosely paired).

Meiosis I and II

In the various phases of meiosis there are two dividing processes: meiosis I and meiosis II (Fig. 2.10.2). It is important to note that the initial phases of meiosis I strongly resemble the initial phases of mitosis and that meiosis II can even be regarded as a normal cell division. The

Figure 2.10.2
Meiosis

1. centriole
2. chromatin-threads
3. nuclear membrane
4. centromere
5. pull threads

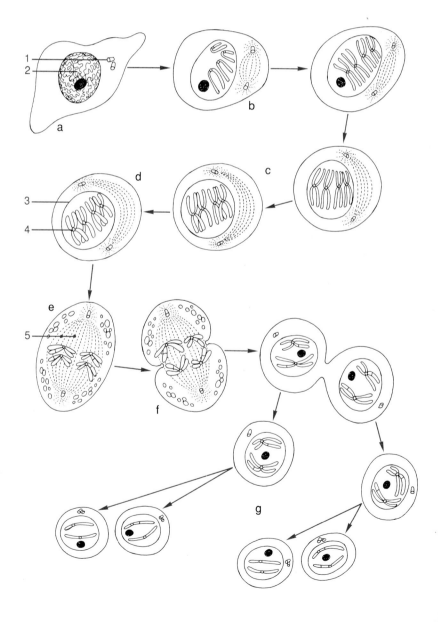

centrosome and the centrioles originating from it follow exactly the same scheme in meiosis II as in mitosis (see Fig. 1.3.10).

During the phase of growth of the forerunners of the reproductive cells (Fig. 2.10.2a), the DNA-strings of the chromatin network are copied. The differing, though doubled, chromatin-threads (filaments) start to make themselves far more compact by means of a threefold coiling; as a result of this the 46 chromosomes' coils become visible under a microscope (Fig. 2.10.2b).

After that the homologous chromosomes (twin-chromosomes) seek one another out (Fig. 2.10.2c), and in predestined places make contact with one another. Then the homologous chromosomes exchange matching parts of DNA at these places of contact, after which they separate from each other again (Fig. 2.10.2d). There develop completely new combinations of genes via this pairing of the chromosomes.

In the beginning of the next phase represented by 2b and by 2e, the nuclear membrane (2e) and the nucleolus (2b) disappear. The homologous chromosomes place themselves opposite each other on both sides of the equatorial plane (Fig. 2.10.2e). The 'pull threads' attach to the centromeres of the 23 single chromosomes on the 'home' side of the equator.

The factual reduction division now starts (Fig. 2.10.2f). The 'pull threads' shorten. Contrary to what happens in mitosis the centromeres do not split in two. The homologous chromosomes are separated. And at the same time the 'mother cell' starts to constrict at the equator.

In the last phase of meiosis I (see Fig. 2.10.2), we see the nuclear membrane and the nucleolus appear again in each of the two new 'baby cells', 'daughter cells'; in both daughter cells the number of chromosomes is *haploid* (23 single chromosomes). The hereditary material in every haploid cell, however, is also doubled. Every chromosome – you will remember – consists of two identical chromatids which are connected to each other by the centromere.

In meiosis II the division of the chromosomes themselves takes place in each of the two daughter cells (Fig. 2.10.2g). This division

(meiosis II) has exactly the same sequence as in mitosis; the only difference is to be found in the number of chromosomes involved in the division. In mitosis there are 46 chromosomes involved, whereas in meiosis II there are 23 chromosomes involved in each daughter cell. These granddaughter cells each contain a haploid number of *chromatids* (23 single ones) after meiosis II.

So out of a mother cell (Fig. 2.10.1b), with 23 pair of chromatids ($2n$; $n = 23$), four granddaughter cells develop each with 23 single chromatids (4 times n; $n = 23$) (Fig. 2.10.2g).

Meiosis ends with the uncoiling of the chromatids out of their hyperspiralized state to become the threads of the chromatin network again. This makes them so thin that they are again invisible under the microscope.

Sex differences

In the previous section the general scheme of meiosis was discussed. Meiosis takes place in the germ glands (gonads) of both the male, the *testes*, and the female, the *ovaries*.

Meiosis in the two different sexes shows remarkable differences (Fig. 2.10.3). In *spermatogenesis* (the formation of sperm cells, spermatozoa), the father cell (primary spermatocyte) transforms into four functional grandson cells (sperm cells, sperms, or spermatozoa).

In *oogenesis* (the formation of the ovum, plural: ova, the egg cells), the mother cell (the primary oocyte) transforms into one granddaughter cell (egg cell, egg, ovum or oocyte) and three luteal corpora or bodies.

The ovum contains almost all the cytoplasm; the luteal bodies are hardly anything more than a dumping-place for superfluous DNA.

The development of the germ cells in their most primitive state, from which the forerunners of the reproductive cells originate, starts in both sexes in one of the first weeks of the embryonic phase.

In the male the actual development of the real sperm cells, spermatozoa, out of the germ cells, spermatogonia, does not start until the beginning of puberty.

Starting with this phase in development the

Figure 2.10.3
Formation of germ cells;
sex differences

a. Spermatogenesis
b. Oogenesis

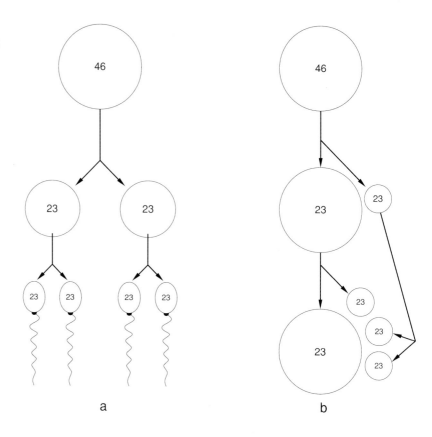

a

b

production of spermatozoa becomes a more and more continuous process that continues until later life. The development to a ripe spermatozoon takes about two months and on average there are about 200 million spermatozoa produced every 24 hours.

In the female the first phases of development take place at a time well before she is born. At the moment of birth all the primary oocytes are already present in the ovary; their number is estimated to be from half a million to two million. In the years after birth the vast majority of them perish so that at the beginning of puberty there are only some 40,000 left. Assuming that between the ages of 14 and 50 there is one egg becoming 'ripe' every four weeks, (36 years × 13 periods of 4 weeks in a year = 478) there are at maximum no more than some 500 eggs that fully develop to 'ripeness'. The rest also perish. The egg cell is one of the biggest human cells (diameter approx 0.035 mm). An egg cell can just be seen by the naked human eye. A sperm cell on the other hand is extremely small; about one tenth of the size of an egg cell.

2. Sexual characteristics

People can be distinguished according to sex on the basis of a number of characteristics. The differences in the gametogenesis, described above, could very well be regarded as sexual characteristics. However, the determination of sexual characteristics takes place on a macroscopic scale, and furthermore there is differentiation between *primary and secondary sexual characteristics.*

Primary sexual characteristics are signs of differentiation which are already present at birth. In the female they are: the two ovaries (ovarium, ovaria), the two uterine tubes (tubae), the uterus (uterus) and the vagina.
In the male they are: the testicles (testes), the epididymis, the spermatozoa tubes (ducti deferens), the seminal vesicles, prostate gland and the penis.
Secondary sexual characteristics typify sex differences which develop under the influence of gender hormones. These sexual hormones

are produced in quantity from puberty onwards. *Secondary sexual characteristics* of the female are: the further development of the ovaries, the uterus, the vagina and clitoris, the appearance of the first menstruation (menarche), the development of the breasts, the labia minor, the growth of hair in the armpits and pubic region, the widening of the pelvis, the increase of subcutaneous fat at certain places (upper arms, hips, upper legs) producing a rounder body-shape.

The secondary sexual characteristics of the male are: the further development of the testes and the penis, the growth of the paranasal sinuses and of the larynx (breaking of the voice), the development of hair growth in the armpits, the pubic area, on the face (beard and moustache) and in some cases on other parts of the body (chest, arms, legs and on the back), a bigger bone and muscle development.

3. Female genital organs

The female genital or sexual organs (the genitals) are divided into internal and external reproductive organs. The internal ones are in the small pelvis; they are the ovaries, the oviducts, the uterus and the vagina (Fig. 2.10.4).

The external organs are in the pubic area of the female; the labia majora and labia minora, the mons pubis, the clitoris and the vestibule. The external genital organs are referred to using the collective noun: the vulva (Fig. 2.10.5).

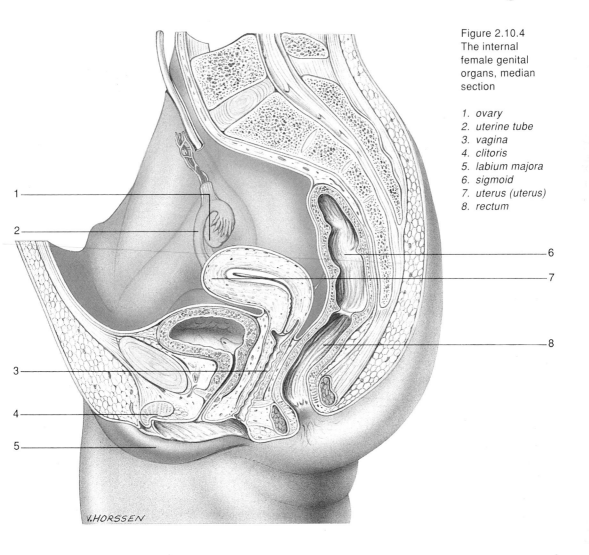

Figure 2.10.4
The internal
female genital
organs, median
section

1. ovary
2. uterine tube
3. vagina
4. clitoris
5. labium majora
6. sigmoid
7. uterus (uterus)
8. rectum

V.HORSSEN

Figure 2.10.5
The external female
genital organs

a. Normal

  1. *labium majora*
  2. *labium minora*

b. The labia have been
    spread for
    illustrative purposes

  1. *mons pubis*
  2. *clitoris*
  3. *urethra*
  4. *vestibulum
      (vestibule)*
  5. *anus*
  6. *perineum*
  7. *hymen*
  8. *vagina*

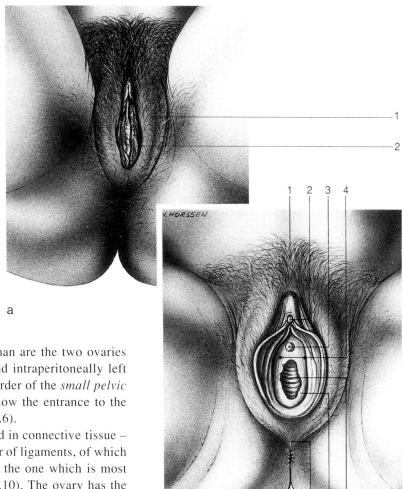

a

b

3.1. The ovaries

The gonads of the woman are the two ovaries
(ovaria). They are found intraperitoneally left
and right against the border of the *small pelvic
cavity*, immediately below the entrance to the
pelvic cavity (Fig. 2.10.6).
The ovaries – embedded in connective tissue –
are attached to a number of ligaments, of which
the ovarian ligament is the one which is most
apparent (see Fig. 2.10.10). The ovary has the
shape of a large almond; about 3–4 cm long,
1–2 cm wide and 1 cm thick.
The beginning of the uterine tube is bell-shaped
and arches over the upper part of the ovary.

Structure

The surface of the ovaries (Fig. 2.10.7) consists
of one-layered epithelium, covered by the visceral peritoneum. Beneath this lies the *cortex*,
consisting of dense connective tissue.
Towards the centre of the ovary the connective
tissue becomes more and more an open network
known as *medulla* (marrow). In the medulla are
found blood vessels, lymph vessels and nerve
fibres.

Development of egg cells (ova)

The development of egg cells (oogenesis) is an
extremely protracted process. The beginning of

this process takes place in the first few months
of fetal life. The primary germ cells migrate
from the wall of the yolk-sac towards the developing ovaria (see Fig. 3.2.22a).
Here they arrive at the end of the fourth week
of fetal life and develop into germ cells
(oogonia).
In the next months the approximately 1500
oogonia divide mitotically. In the fifth month
of fetal life the number of oogonia is at its
peak: about six million.
However, during the further stages of development more and more of these germ cells
(oogonia) perish. At birth only between half a

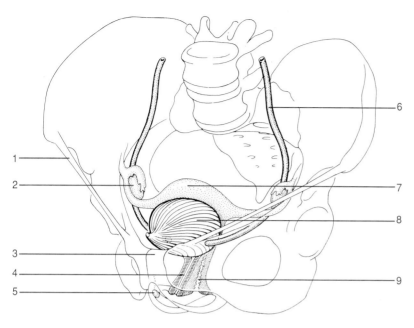

Figure 2.10.6
Position of the genital
organs in the pelvis of
the female

1. inguinal ligament
   (ovary)
2. ovary with uterine
   tube
3. symphysis
4. urethra
5. clitoris
6. ureter
7. uterus
8. bladder
9. vagina

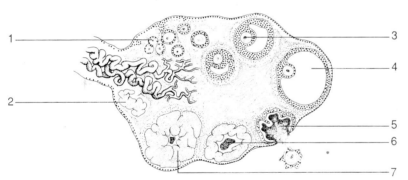

Figure 2.10.7
Structure of the ovary
with all the stages of
ripening of the follicle

1. primordial follicles
2. corpus albicans
3. the developing egg
   cell
4. Graafian follicle
5. rupture of the follicle,
   release of the ovum
6. corpus rubrum
7. corpus luteum

million and two million remain. By now these oogonia have grown in size and developed further. In the meantime they have begun the first stage of meiosis I. They are called *primary oocytes*. The primary oocytes are to be found in the cortex of the ovaries and are wrapped in one-layered epithelium (Fig. 2.10.7). Such a primary oocyte together with the one-layered epithelial tissue surrounding it is called a *primary follicle*. From birth until puberty there is a standstill in the development of all the primary oocytes. Meiosis I stopped in the very first phase. There are continually cells, and thus follicles, that die, with the result that at the start of puberty there are only about 40,000 left.

The enclosing epithelial cells, the *follicle cells,*

do develop in the meantime and gradually build up more layers of epithelial cells around the remaining primary oocytes.

The beginning of puberty can be defined as the beginning of a number of *ovarian cycles*. Each ovarian cycle lasts on average 28 days and consists of a series of processes which take place at about the same intervals and always in the same order (see Fig. 2.10.12).

The 'monthly' (every 28 days) bleeding (menstruation) marks the end of the cycle. This is why it is also sometimes called the *menstrual cycle*. In the first two weeks of the ovarian cycle, under the influence of the follicle-stimulating hormone (FSH) and the luteinizing hormone

(LH) there are a number (around 15) of primary follicles which ripen quickly. FSH and LH are therefore gonadotrophic hormones.

Only one of the 15 primary oocytes comes to complete ripeness, the others degenerate. During the process of ripening the primary oocytes start to grow in size, whereas the follicle cell walls at first become cuboidal and then start to divide. Between the follicle cell walls and the primary oocyte develops a translucent substance, *the zona pellucida*.

After some time, a fluid-filled hollow develops in the follicle (the follicular cavity).

The follicle fluid strongly resembles in its composition the fluid of the m.i. The primary oocyte is in an excentric position and is enclosed in follicle cells. Meanwhile the primary oocyte has completed meiosis I and is now called the *secondary oocyte* or *egg cell* (ovum). On the edge of the egg cell is the first *polar body*.

During ripening the follicle cells produce the hormone *oestrogen*, under the influence of the LH-hormone. This hormone oestrogen decreases production of FSH.

Towards the end of the period of ripening (when the ovarian cycle is about to start its third week) the production of FSH and LH increases strongly and reaches a peak. Consequently the production of oestrogen also reaches a peak. The ripened follicle arrives in a position just beneath the surface of the ovary. Due to the follicle-fluid the bulge can be seen with the naked eye (Fig. 2.10.7). It is known as the *Graafian follicle*.

The cloak of epithelium is so strong, that the increase of fluid within the follicle causes an increasing pressure within the follicle.

The Graafian follicle bursts open at the start of the 3rd week due to influences of LH. The follicle fluid with the egg cell are expelled. This is called *ovulation* (release of ovum) and can involve some pain. Because of this pain a number of women can exactly pinpoint the moment of ovulation.

The development of the egg cell is not yet completed as meiosis II still has to take place. This, however, only occurs just before fertilization.

Red, yellow and white body

After ovulation the follicle space in the ovary diminishes in size and contains some clotted blood. It is called the *corpus rubrum* (red body). Under the influence of LH the 'empty' follicle undergoes a number of changes (Figs. 2.10.7 and 2.10.12). The epithelial cells start to divide rapidly and within a few hours they fill up the follicle space together with the capillaries which grow towards the centre. The cells themselves swell up and after some time they form a yellowish, granular substance known as *lutein*. The small organ which has developed is called the *corpus luteum* (yellow body), after this substance. The corpus luteum is a hormonal gland. It continues the production of oestrogen by the follicle cells, but also secretes the hormone *progesterone*. Progesterone is the hormone which – after ovulation – prepares the mucous membrane of the uterus for the eventual embedment of a fertilized egg cell. Progesterone diminishes the production of LH-hormone. One effect is that for a while no new follicles ripen. If fertilization takes place, the corpus luteum will stay in existence for almost four months. If there is no fertilization, the yellow body degenerates after approximately 12 days, the blood vessels disappear and instead there is a growth of connective tissue. The yellow colour gives way to the white colour of scar tissue. This explains the name *corpus albicans* (white body).

3.2. Uterine tubes (Fallopian tubes)

Between the ovary and uterus runs a small tube about 10 cm long: the uterine tube (oviducts salpinx) (Figs. 2.10.8 and 2.10.9a). The *uterine tubes* are in an intraperitoneal position in the topmost part of the ligamentum latum (the broad uterus band) (Fig. 2.10.10). Each uterine tube has a funnel-shaped end and hangs over the ovary. In principle there is an open space between the cavity of the uterus (which makes this a part of the m.e.) and the abdominal cavity. The widened end of the uterine tube is called the *infundibulum*. Its edge has fringes (fimbriae) which give the impression of the uterine tube grasping the ovary (the fimbriae looking like fingers) (Figs. 2.10.9a and c).

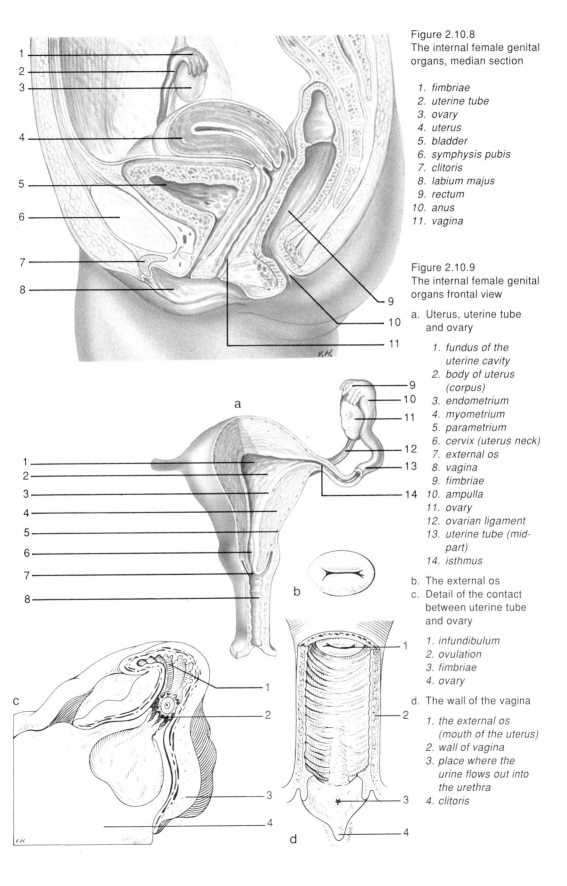

Figure 2.10.8
The internal female genital
organs, median section

1. fimbriae
2. uterine tube
3. ovary
4. uterus
5. bladder
6. symphysis pubis
7. clitoris
8. labium majus
9. rectum
10. anus
11. vagina

Figure 2.10.9
The internal female genital
organs frontal view

a. Uterus, uterine tube
and ovary

   1. fundus of the
     uterine cavity
   2. body of uterus
     (corpus)
   3. endometrium
   4. myometrium
   5. parametrium
   6. cervix (uterus neck)
   7. external os
   8. vagina
   9. fimbriae
  10. ampulla
  11. ovary
  12. ovarian ligament
  13. uterine tube (mid-
      part)
  14. isthmus

b. The external os
c. Detail of the contact
between uterine tube
and ovary

   1. infundibulum
   2. ovulation
   3. fimbriae
   4. ovary

d. The wall of the vagina

   1. the external os
     (mouth of the uterus)
   2. wall of vagina
   3. place where the
     urine flows out into
     the urethra
   4. clitoris

The rest of the tube is a muscular (smooth muscle tissue) pipe, which widens from the infundibulum to form the *ampulla*, and narrows near the entrance to the uterus to form the *isthmus* (Fig. 2.10.9a).

The inner sides of the uterine tube are coated with mucous membrane and have longitudinal folds. The mucous membrane consists of cilia cells and secretory cells. In the first part of the uterine tube the *cilia* cells prevail, and later on towards the uterus there is an increasing number of *secretory* cells. The menstrual cycle also influences the ratio between these two types of cells. During the first half of the menstrual cycle the ciliated cells are most prominent; after ovulation the secretory cells prevail.

There is fluid present in the narrow lumen of the oviducts. Some of this has been produced by the secretory cells in the mucous membrane, the rest has been drawn in from the abdominal cavity.

The cilia and peristalsis in the uterine tube create a flow of liquid towards the uterus.

A remarkable feature is that the fimbriae of the uterine tube touch and feel their way around all of the ovary after which the infundibulum places itself over and around the follicle, which is about to burst open. There is a slight pressure reduction in the uterine tube (and increased pressure in the follicle) due to the fluid flow in the tube. This has a drawing in effect on the egg as well as rupturing the follicle. If the infundibulum is not properly located over the Graafian follicle then there is a chance that the egg cell will enter the abdominal cavity (extra-uterine pregnancy).

The flow of the fluid and peristalsis are the factors which conduct the egg cell along the longitudinal folds of the uterine tube in the direction of the uterus. On its way the egg is fed by products of the secretory cells.

3.3. The uterus

The uterus (womb) is the organ where the embryo is carried during pregnancy. The mucous membrane of the uterus shows cyclical changes under the influence of hormones. The uterine wall adapts itself to the enlargement of the uterus and at birth it functions as a propelling force.

Shape, parts and position

The uterus is a muscular organ, lying in the middle of the small pelvis on the (empty) bladder and in front of the rectum (Fig. 2.10.8). A full bladder pushes itself in front of the uterus. The shape of the uterus (Fig. 2.10.9a) is more or less like a pear. Its total length is about 7–8 cm, its greatest width 5 cm. Both uterine tubes open into the sides of the thickest part.

The dome-shaped part of the uterus which lies above the uterine tube-openings is called the *fundus*. Beneath the fundus we find the *corpus* (the body). This constitutes by far the largest part of the uterus.

At the place where the thicker part becomes thinner, the corpus ends and the *cervix* (the uterus-neck) begins. The part of the uterus opening into the vagina is called the *portio* (door). The portio is about 1 cm long. The *hollow space* in the uterus (the cavum uteri) is confined to the corpus and fundus, and does not extend to the cervix.

The *cervical canal* forms the connection between the vagina and the uterine cavity.

The somewhat narrower passageway between uterine cavity and *cervical canal* is called the *internal os* (ostium internum) (Fig. 2.10.9a). The mouth of the cervix channel to the vagina is called the external os (ostium externum), or mouth of the uterus (Figs. 2.10.9b and d). Most of the time the cervical canal is blocked by a mucous plug in the uterine mouth, the *cervical plug*.

With the exception of the cervix the position of the uterus is intraperitoneal. On the sides of the uterus (Fig. 2.10.10) run the *broad ligaments*. There are double folds in the peritoneum. Between these two layers there are blood vessels and nerves to and from the uterus.

To the front the broad ligaments are bordered by *round ligaments* and to the back they are bordered by the *uterosacral ligament* which stretches to the rectum. From the region where the corpus changes to the cervix there stretches away a fan-like ligament to the side walls of

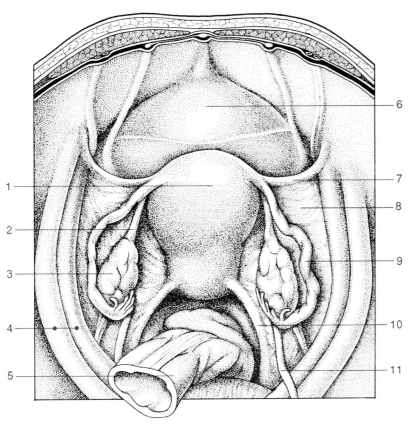

Figure 2.10.10
The uterus (womb) in the small pelvis, seen from above

1. uterus (corpus)/ fundus
2. ovarian ligament
3. ovary
4. artery and vein for the leg
5. rectum
6. bladder
7. round ligament (lig. teres)
8. wide ligament (lig. latum)
9. uterine tube
10. cardinal ligament
11. ureter

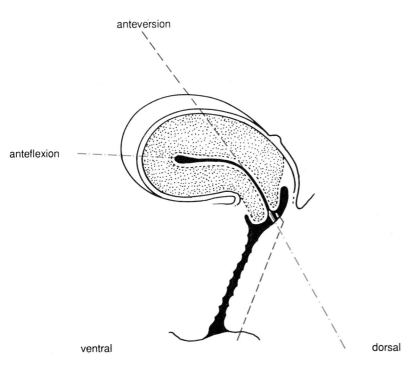

Figure 2.10.11
Anteflexion and anteversion of the uterus

the small pelvis, the cardinal ligament (Fig. 2.10.10). These ligaments keep the uterus in position. As the uterus curves over the bladder there are two peritoneal spaces before and behind the *broad ligaments* (see Fig. 2.2.40b):

- the *vesicouterine pouch*. This is the space between the bladder and the uterus;
- the *rectouterine pouch*. This space between rectum and uterus is the lowest part of the peritoneal cavity in the female. It is known as the pouch of Douglas.

The position of the uterine corpus and fundus, bent forward and over the bladder, is called *anteflexion*. The cervix being less bent forward, is in a position called *anteversion* (Fig. 2.10.11). The opening of the cervix points to the backmost wall of the vagina. When the bladder is full, anteflexion decreases; the anteversion of the cervix remains practically the same. A full rectum causes an increase of the anteversion position, in other words, the cervix is pushed into a more vertical position.

Structure of the uterine wall

This wall has three layers (Fig. 2.10.9a): an external serosal, middle muscular and an internal mucosal layer. Nearest to the uterine cavity is the *endometrium* (the mucous membrane of the uterus). Beneath this is the *myometrium* (muscular layer). This layer has grown firmly together with the outer third layer, the perimetrium. This *perimetrium* is part of the *visceral peritoneum*, which is also called the *serosa*. At the ridges to the sides of the uterus it becomes the broad ligament (Fig. 2.10.10). The myometrium consists of smooth muscle tissue and has an average thickness of 2 cm. Its bundles form external, middle and internal layers of variable distinctness.

The endometrium consists of epithelium and connective tissue with ciliated epithelium and mucous cells for its most superficial, most internal layer.

During the period of sexual maturity (from puberty till menopause, at around 50 years of age) the hormones of the corpus luteum have an influence on the structure of the endometrium, particularily in the body of the uterus. This is

why – when there is no pregnancy – a series of cyclical changes occurs, which spans some 28 days and is known as a *menstrual cycle*.

Menstrual cycle

The name menstrual cycle has been derived from one part of the cycle: the bleeding or menstruation (the period). However, given that – in essence – the events involved are the cyclical changing of the mucous membrane of the uterus, the name *endometrium* cycle would be more appropriate (Fig. 2.10.12). The cycle is divided into *phases*; always taking the first day of bleeding as the beginning of the cycle .

The first phase is the *menstruation phase* or *phase of bleeding* (1st–5th day) which occurs when there has been no fertilization. The production of progesterone has decreased sharply because the corpus luteum has degenerated. This leads to a spasm in the blood vessels of the mucous membrane, resulting in the endometrium dying and being shed and ejaculated. This is accompanied by loss of blood (50–150 ml) via the vagina.

At the end of the menstruation phase the 'wound' heals; there is hardly anything left of the endometrium but a thin epithelial layer with short tubular glands on the inside of the myometrium.

The second phase is the *proliferative phase* or *building-up phase* (5th–15th day). The thickness of the endometrium gradually increases due to an enormous number of cell divisions. At the same time the glandular tissue divides; the glandular tubes grow in accordance with the mucous membrane and the glandular epithelium spreads out over the surface of the endometrium.

At the end of the proliferative phase the epithelial layer of the endometrium might well have reached a thickness of half a centimetre. Between the numerous glandular tubes a large number of arterioles have developed. These have a remarkable winding pattern (spiral arteries). The blood flows from these spiral arteries to a capillary network which lies very near the surface of the mucous membrane.

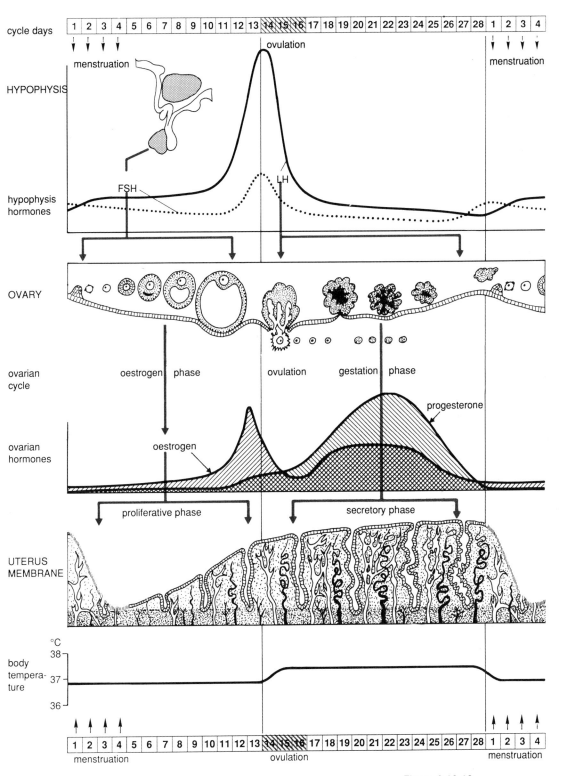

Figure 2.10.12
Hormonal and functional
changes during the
menstrual cycle

The proliferative phase is controlled by the hormone *oestrogen* from the follicle cells of the ovary. This is why this phase is also sometimes called the *oestrogen phase*. The proliferative phase ends after ovulation, when out of the 'empty pod' of the Graafian follicle the corpus luteum is developing.

The third phase is the *secretory phase* or excretion phase (15th–28th day). The corpus luteum continues the production of oestrogen, but it starts to produce a second hormone as well, *progesterone*. The increasing amount of this progesterone causes the glands to give off a mucous fluid. This secretion is amassed within the connective tissue of the endometrium, so that the mucous membrane swells up like a sponge. The capillarization increases even more and glycogen is accumulated in this innermost layer of the uterus wall.
The secretion phase is controlled by the *gestation* hormone progesterone.
The endometrium is now in an optimal condition for the embedding of a fertilised egg cell. When there is no fertilization of an egg cell, and therefore no 'nestling', then at about the 23rd day of the cycle the oestrogen and progesterone production will diminish. The lower level of the hormone progesterone causes vasoconstriction (vascular spasms) to occur in the spiral arteries. There is no supply of blood to the capillary network and therefore the endometrium perishes; after the 28th day the debris is ejaculated (bleeding), which marks the start of a new cycle.

The mucous membrane of the cervix (neck of the uterus) is far less subject to menstrual changes, and is not shed and therefore not ejaculated.

Body temperature during the cycle also shows a change (Fig. 2.10.12). The hormone progesterone influences the centre for temperature regulation in the medulla oblongata and in the hypothalamus. After ovulation the progesterone production increases rapidly, accompanied by an increase in the body temperature of around 0.5°C. At the end of this cycle progesterone production 'stops', and the body temperature returns to normal.

The menstrual cycle also has an influence on tissue in the female breasts. During the oestrogen phase the volume of glandular tissue in the breasts increases a little, and more blood goes to (and into) the breasts. With a number of women this leads to a somewhat painful and tense feeling in the breasts. In the secretory phase of the cycle these phenomena disappear.

3.4. Vagina

The vagina is a canal, about 8 cm long, which forms the connection between the uterus and the world outside. It is the female *mating* or *copulating* organ. At the end of pregnancy the vagina forms the last part of the 'parturition channel'. The vagina lies from high-back to low-front, approximately centred between and parallel to the urinary tubes and the mid-part of the rectum (see Fig. 2.10.8). At the upper end of the vagina is a part of the cervix (i.e. the uterus neck) which protrudes into the vagina. This part is called the *portio* (see Figs. 2.10.8 and 2.10.9).
The wall of the vagina consists of three layers (see Fig. 2.10.9d). The innermost superficial layer consists of multi-layered pavement epithelium (mucous membrane). This membrane has a large number of folds or ripples which run crosswise. In the lower part of the front wall there is also a fold running longitudinally. In this fold is the urinary tube (urethra). Below the mucous membrane there is a thin coating of smooth muscular tissue. Around this next layer (to inner or outer side), there is an elastic coat of connective tissue.
The wall of the vagina is extremely elastic. Under normal circumstances the front wall and the back wall touch each other, but during the delivery of a baby its diameter can easily widen to 15 cm.
The mucous membrane of the vagina is subject to changes because of certain hormonal changes, as is the mucous membrane of the uterus.
During the oestrogen phase there is also proliferation (growth) in the mucous membrane of the vagina. The epithelium grows, becomes more dense, and the cells start to accumulate glycogen. In the secretory phase many epithelial cells are shed, which results in a thinner and thinner

mucous membrane and more and more glycogen being released from those cells shed in the tissue of the vagina.

In the vagina there live the Döderlein bacilli; these use the glycogen for their metabolic processes. A waste product of this process is lactic acid which is released. This lactic acid produces a low pH in the vagina. Other micro-organisms are generally not capable of living in an acid environment. This is why the presence and functioning of the *Döderlein bacilli* protects the woman against infection to some extent. This barrier is the more important since in the female there is an open channel via the vagina, the cavity of the uterus and the oviducts from the external environment to the abdominal cavity. The Döderlein bacilli do, however, have a *symbiotic relationship* with humans similar to the coli-bacteria.

During menstruation approximately 90 per cent of the acid contents of the vagina are washed away and neutralized; in this phase the risk of infection is much greater.

3.5. Vulva

The word *'vulva'* is used to indicate the external female genital organs (Fig. 2.10.5). The constituent parts are the labia majora and minora pudendi, the mons pubis, the clitoris and the vaginal vestibule.

The *labia majora* are rather thick folds of skin with a growth of hair on the outside. They contain a considerable amount of subcutaneous fatty tissue and relatively many sweat and sebaceous glands.

The space between the two labia majora is called the *pudendal cleft*. Above the place where these two labia meet at the front is an elevation; the *mons pubis*, pubic mound or mound of venus (mons veneris). It is an area with a strong growth of coarse hairs, in front of the *symphysis pubis*. It is generally formed by a mass of subcutaneous, adipose (i.e. fatty) connective tissue. At their dorsal ends the labia majora merge into the tough cutaneous fold between vagina and anus. This tough fold of skin is called the *perineum*. The small *labia minora* are thin, hairless cutaneous folds, lying medially to the labia majora. There is no fatty tissue, but they do contain sebaceous glands. The left and right labia minora meet at their frontal ends. At their distal ends the connection is formed by a little band, a strip of skin. The space between the labia minora is called the *vestibulum*. When the legs are together, the labia majora – with the labia minora in between them – lie closed, pressed together. The vestibulum is not then visible to the eye.

The titles small and large for the labia can be rather confusing. With many females the small cutaneous folds of the labia are longer than the large ones are, so that they stick out of the pudendal cleft. The terms outer and inner labia are therefore more appropriate.

The vestibulum is the cavity into which the vagina and the urethra open through apertures in its back wall. In women who have had no sexual intercourse the vestibulum is partly closed by a thin, tough fold of skin tissue called the *hymen*. When sexual intercourse first takes place this membrane is usually torn and may bleed). If it is not torn then, it tears when the first baby is born. All that is left are a few frayed remains.

A number of mucous glands open into the vestibulum. They are the *vestibular glands* or Bartholin glands. With sexual stimulation these glands produce a quantity of mucus. This mucus functions as a lubricant during penetration of the penis into the vagina.

At the place in front of where the labia minora meet, the *clitoris* is located. It is a little cone-shaped organ, measuring only a couple of centimetres. It consists of two well-vascularized bodies, erectile structures or sacks (corpus clitoridis/corpora cavernosa), which originate at the front of the symphysis pubis (see Fig. 2.10.8). Then they bend sharply and run further downwards where they end in the glans of the clitoris (i.e. glans clitoridis). The front connection of the two labia minora forms a kind of hood or cowl over the clitoris. This is the *prepuce* of the clitoris. Under this prepuce it is possible that vaginal secretions or cell debris which can cause skin irritation, are deposited. The glans of the clitoris is densely packed with sensors. Stimulation of the clitoris can lead to sexual excitement and to *orgasm*.

Figure 2.10.13
The male genital organs

a.   Median section
b.   Front view of penis and scrotum

1.   *deferent duct*
2.   *seminal vesicle*
3.   *symphysis pubis*
4.   *the ejaculatory duct*
5.   *urethra*
6.   *penis*
7.   *epididymis*
8.   *the glans*
9.   *rectum*
10. *prostate*
11. *spermatic cord*
12. *testicle, testis*

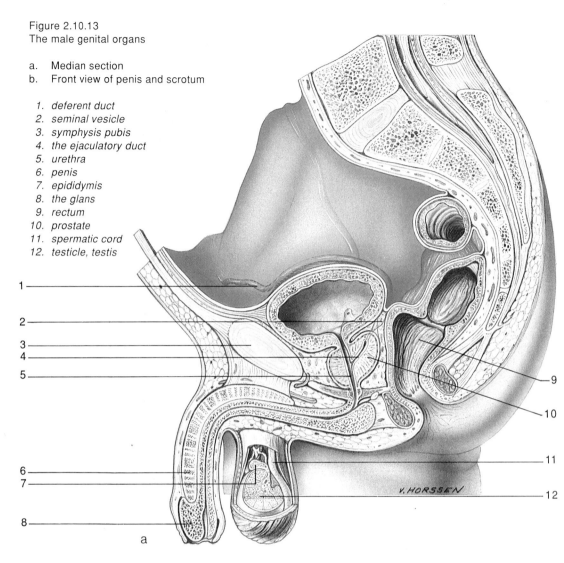

a

b

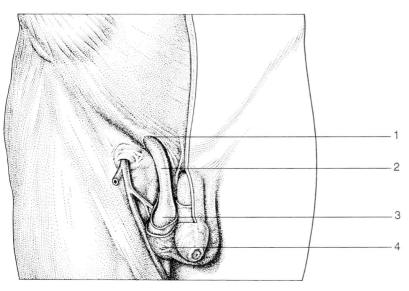

Figure 2.10.14
The position of the testis
in the scrotum; the route
of the spermatic cord

1. *external opening of*
   the inguinal canal,
   external ring
2. *spermatic cord*
3. *testicle*
4. *scrotum*

Figure 2.10.15
Structure of the testis

1. *urethra*
2. *spermatic cord*
3. *penis*
4. *epididymis*
5. *foreskin (prepuce)*
6. *seminiferous tubules*
7. *testicles*
8. *mediastinum*

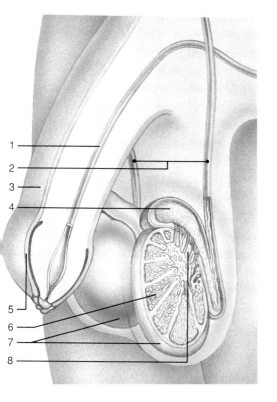

4. Male genital organs

The male genital organs (genitalia) are (Fig. 2.10.13): the testicles with the epididymises, the deferent duct, the seminal vesicles and the ejaculatory ducts, the bulbo-urethral glands, the prostate and the penis.

4.1. The testicles/testes

The gonads of the male are the two *testicles* (testes). Unlike the female gonads they do not lie within the abdominal cavity, but just outside this cavity, beneath the symphysis *pubis* in a cutaneous sack-like formation known as the *scrotum* (Fig. 2.10.14). The scrotum has a visible suture, a cutaneous raphe down the middle. The position of the testes outside the body is a result of the descent of the testes and is very important for their function. The temperature in the scrotum is approximately $2°C$ lower than in the abdominal cavity. A temperature as high as the abdominal temperature suppresses the production of male reproductive cells. Both testes hang in the scrotum from the spermatic cord (funiculus spermaticus) (Fig. 2.10.14). The left testicle is 1 cm lower than the right one.

Around the spermatic cord and in the wall of the scrotum there is muscular tissue (cremaster muscle). When the temperature in the scrotum increases the cremaster muscle relaxes: the testes descend further away from the heat of the body. When the temperature in the scrotum

Figure 2.10.16
The development of
germ cells

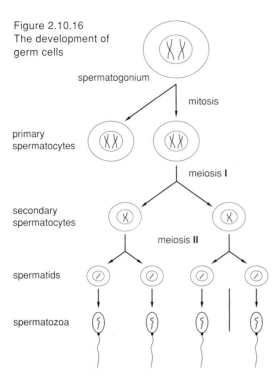

spermatogonium

mitosis

primary
spermatocytes

meiosis I

secondary
spermatocytes

meiosis II

spermatids

spermatozoa

falls, then the cremaster muscle contracts and
the testes are drawn closer to the body and
become warmer.

At the dorsal ridge of each testis is found the
place of entrance and exit of deferent ducts,
blood vessels, lymphatic vessels and nerves.
This passage is not called the hilum, but the
mediastinum (Fig. 2.10.15).

Structure

Each testis is egg-shaped and about 5 cm long.
The outer coat is tough connective tissue. From
the inner side of this connective tissue capsule
partitions of connective tissue are stretched towards the centre, the mediastinum. In this way
the content of the testis is divided in 200–300
compartments called lobules. Each lobule is a
wedge-shaped space in which there are some
coiled seminiferous tubules (Fig. 2.10.17). In
the very slight interstices between these coiled
tubules are the interstitial endocrino-cytes or
Leydig cells, also called interstitial cells. These

Figure 2.10.17
The seminiferous tubule

a. Cross-section through
    the testis
    1. deferent duct
    2. epididymis
    3. testis
    4. seminiferous tubule

b. Various stages of the
    development of sperm
    cells
    1. lumen
    2. the spermatocytes
        are set free in the
        lumen of the
        seminiferous tubule

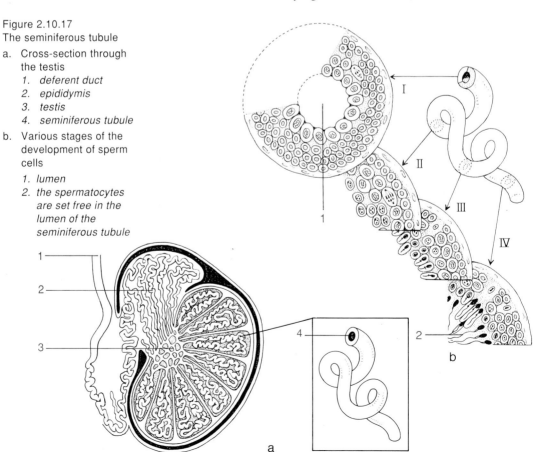

interstitial cells actually produce the male reproductive hormone testosterone under the influence of the LH-hormone. This is why the LH-hormone is referred to as ICSH (interstitial-cell-stimulating hormone) in the male. The hormone testosterone enhances the production of semen by stimulating both the production of semen cells and the function of the seminal vesicles and the prostate. In addition, it stimulates the development of the male primary and secondary genital characteristics. The seminiferous tubules open into the testis, a network of seminiferous tubules in the mediastinum. From the testis a number of seminiferous tubules run to become the head of the epididymis.

Development of the spermatozoa

The development of the spermatozoa (spermatogenesis) begins in the first months of fetal life – similar to oogenesis (the development of egg cells) in the female.
These spermatogonia migrate from the wall of the yolk sack towards the developing testes (see Fig. 3.2.22a). Here they arrive at the end of the fourth week of fetal life, and develop into germ cells (spermatogonia) in the wall of the seminiferous tubules. Besides these spermatogonia, the wall of the seminiferous tubules also consists of Sertoli-cells (nutrition and support cells). There are three stages in the production of sperm cells (Fig. 2.10.16). During these three stages the cells are shifted from the outer edge of the seminiferous tubule towards the centre, or lumen.

– Phase of increasing multiplication: after puberty, the spermatogonia start to divide themselves mitotically, under the influence of the hypophysis hormone FSH; the daughter cells are called primary spermatocytes. These seminiferous tubules are a gigantic production-line (Fig. 2.10.17), the tempo of which increases till about the 30th year. One of the daughter cells which develop through mitosis goes into the next phase, the other one divides again and so is the mother cell again.
– Meiotic phase: the primary spermatocyte is subject to meiosis I in which are formed two

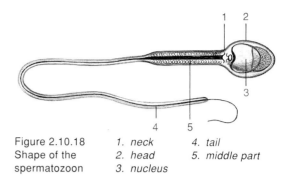

Figure 2.10.18
Shape of the
spermatozoon

1. neck
2. head
3. nucleus

4. tail
5. middle part

secondary spermatocytes. The number of chromosomes is 'n', which is only half of that in the primary spermatocytes, 2n; (2n : 2 = n). The secondary spermatocytes are haploid (having 23 chromosomes instead of the normal 46), however, the chromosomal material per chromosome is still doubled. This condition is ended by meiosis II which now takes place. In the following stage they become spermatids. Half of the spermatids contain an X-chromosome as the genital chromosome, the other half of the spermatids have a Y-chromosome.
During the meiotic phase the cells which are in the process of developing are lying closely against Sertoli cells. One such nutrition cell keeps contact with some 150 spermatocytes by means of a type of communication cable.
– Differentiation phase: the spermatids – now lying quite near the lumen of the seminiferous tubule – develop into the actual spermatozoa (sperm cells). They develop a long whiptail (flagella) which protrudes into the lumen, and then they break off contact with the Sertoli cells. The sperm cell becomes loose from the wall of the seminiferous tubule and finds itself floating free in the lumen.

The meiotic phase and the differentiation phase are controlled by the hormone testosterone from the interstitial cells.

Shape of a spermatozoon

A mature, ripened sperm cell has the following parts (Fig. 2.10.18): head, neck, middle part and tail (or principal part).

Seen from above, the *head* is eggshaped; seen from the side it is flat, piriform with a rather pointed tip. This tip-end contains enzymes to enable the penetration of the membrane of the egg cell. The rest of the head contains hardly anything but nuclear material; there is almost no cytoplasm.

The *neck* is a short and easily movable connection between head and middle part. The middle part is the motor of the spermatozoon. It contains a large number of mitochondria which provide the sperm cell with the energy for the movements of the tail.

Movement is brought about by the shifting of bundles of fibres (fibrils) which now contract on one side, and then on the other. These fibrils have originated from a centriole and can be compared to the fibrils in muscular tissue. This makes the tail move vehemently in an S-shape, which gives an enormous propelling force. If the spermatozoon had the size of a mature human being, the 100 metres freestyle would be swum in half of the world record time.

There are up to 200 million sperm-cells being produced every 24 hours, which is more than two thousand per second. An average of 10–20% of the sperms is misshapen; they are not fully developed, have more than one tail or seem to have broken their necks.

4.2. Epididymis

The *epididymis* consists of a system of little tubes on the posterior of the testis, enclosed by connective tissue (Fig. 2.10.17a). The spermatozoa which have been transported to the testis via the seminiferous tubules, then enter the coiled *efferent duct* (ductuli efferentes) of the epididymis. These efferent ducts have a thin epithelial wall with ciliated cells which produce a flow of fluid. These ten to twenty efferent ducts open into the body of the epididymis, which is 5 metres long and strongly coiled. This ductus epididymis runs along the posterior of the testis from high to low. The wall of the ductus epididymis consists of ciliated epithelium with a lining of smooth muscular tissue which gradually increases in thickness towards its lower end.

The epididymis is extensively vascularized.

The spermatozoa remain for longer or shorter periods in the epididymis depending on the sexual activity of the male. They are surrounded by a mantle of mucus-like protein which can protect them against an acid milieu. When the spermatozoa are not discharged (in the sexual act), they are broken down and reabsorbed after about two or three weeks.

4.3. Deferent ducts, seminal vesicles, ejaculatory ducts

Where the ductus epididymis bends back again at an acute angle (Fig. 2.10.15), the *deferent duct* (ductus deferens) begins. The course of the 50–60 cm long deferent duct shows that during their development before birth the testicles descended from the abdominal cavity into the scrotum.

The duct rises up out of the scrotum and lies there together with blood vessels, lymph vessels and nerves in the *spermatic cord* (see Fig. 2.10.14).

The spermatic cord (funiculus spermaticus) is about 20 cm long and ascends via the external inguinal ring to the edge of the pelvis (Fig. 2.10.19a). The spermatic cord terminates at the internal inguinal ring. The deferent duct bends over the crest of the pelvis downwards into the lesser pelvis. It then runs along the bladder and forms a widening, the *ampulla*.

The wall of the deferent ducts consists of a mucous membrane surrounded by a strong mantle of smooth muscular fibres. These ensure rapid transport of the spermatozoa by means of peristalsis.

To the side of the ampullae lie the *seminal vesicles* (vesiculae seminales). The term seminal vesicles indicates that in former days it was thought that sperm was temporarily stored here. A seminal vesicle is in fact a *gland*, about 5 cm long, with more or less the same structure as the ampulla. *Seminal fluid* is produced in the seminal vesicle. Seminal fluid is a slightly alkaline liquid (pH 7.2) which contains fructose, which is useful to the spermatozoa as a source of energy.

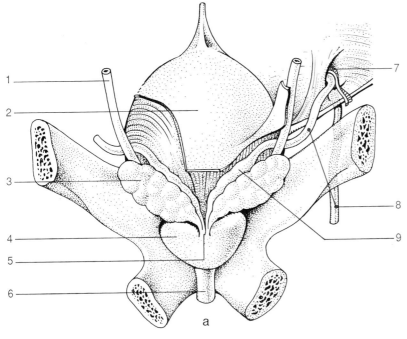

a

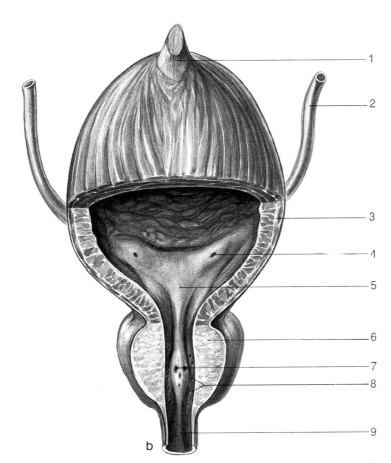

b

Figure 2.10.19

a. Bladder, seminal
   vesicles and prostate,
   seen from posterior

   1. ureter
   2. bladder
   3. seminal vesicle
   4. prostate
   5. ejaculatory duct
   6. urethra
   7. internal opening,
      orifice (entrance) of
      the inguinal canal
   8. deferent duct
      (ductus deferens)
   9. ampulla

b. Ejaculatory ducts
   opening into the
   urethra

   1. top of the bladder
   2. ureter
   3. section of the
      muscular wall
   4. ureter; opening out
      here
   5. triangle of the
      bladder (trigone)
   6. prostate
   7. ejaculatory ducts;
      opening out here
   8. discharge tubes of
      the prostate gland
   9. urethra

Figure 2.10.20
The penis

a.   Median section

1. bladder
2. deferent duct
3. prostate
4. corpus
    cavernosum
    penis
5. urethra
6. corpus
    spongiosum
    penis
7. epididymis
8. glans
9. testicle
10. ureter
11. seminal
    vesicle
12. rectum
13. bulbo-urethral
    (Cowper's)
    gland
14. anus

b.   Three dimensional
     representation of
     the relationship of
     the three erectile
     structures

1. large vein
2. urethra
3. corpus
    cavernosum
    penis
4. feeder artery
    of corpus
    cavernosum
    penis
5. corpus
    spongiosum
    penis

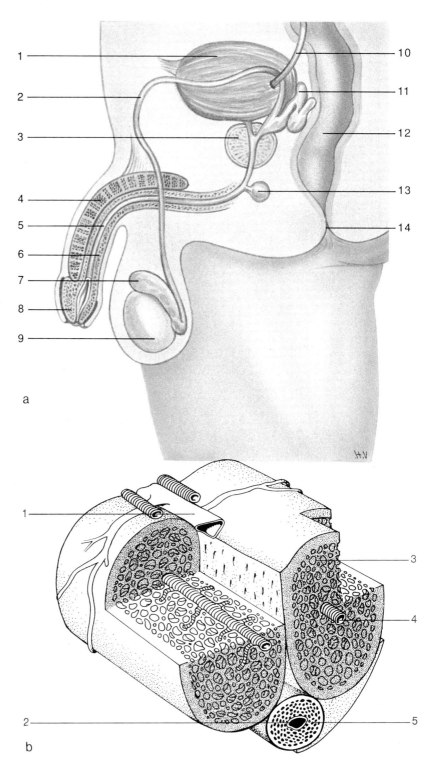

Just above the prostate the discharge tube of a seminal vesicle and of an ampulla meet (Fig. 2.10.19a). From this point the combined tube is called the *ejaculatory duct*. The left and right ejaculatory ducts run parallel, close together right through the prostate. They open next to each other in the *prostatic urethra*, which is that part of the urinary tube which – coming out of the bladder – runs through the prostate (see Figs. 2.10.13 and 2.10.19b).

The walls of the ejaculatory tubes also have a rather strong muscular mantle for peristaltic movements.

4.4. Prostate

The prostate resembles a chestnut in shape and size. It is situated between the symphysis and the rectum just beneath the bladder (see Fig. 2.10.13). The first part of the urethra, the prostatic urethra, runs through the prostate. The two ejaculatory ducts also pass through the prostate. The capsule of the prostate consists of dense, tough connective tissue, with extensive vascularization and many smooth muscle fibres. From the inner side of the capsule numerous fibromuscular septa radiate out towards the centre of the gland, dividing the prostate mass into some 40 lobes in which is found the glandular tissue. The wall of the glandular tubes consists of epithelium. The glands open into approximately 15 discharge tubes (2–3 lobes in one discharge tube), which flow into the prostatic urethra (Fig. 2.10.19b).

The glandular tissue of the prostate produces the slightly acid prostatic fluid which is conveyed to the urethra at a seminal emission.

4.5. Penis

The *penis* is the male copulatory organ. Expulsion of urine also takes place via the penis as the urethra runs through this organ.

The penis (Figs. 2.10.13 and 2.10.20) contains three *erectile bodies*. These are sponge-like structures with many collagen fibres, elastic fibres and smooth muscle fibres. Each of the erectile bodies is surrounded by a firm coat of connective tissue. The arteries run longitudinally through the erectile bodies. The blood transported to them finds itself in the meshes of the erectile bodies and is removed from them via veins (veiny networks) which are also around the erectile bodies.

The two upper erectile masses, named *corpora cavernosa penis*, are the largest. They begin as two separate organs but they immediately come together under the symphysis pubis. Between the two erectile organs occurs a fibrous septum, the *septum penis*.

The lowest erectile organ, the *corpus spongiosum penis*, begins immediately beneath the pelvic floor. There occur the *bulbo-urethral* (Cowper's) glands left and right of the erectile organ. These glands, the size of peas, produce a ropy mucus which enters the urethra through thin ducts (Fig. 2.10.20a). Further to the front the corpus spongiosum lies in the median groove on the urethral surface of the two combined corpora cavernosa. The top of the penis is formed by a thickening of the corpus spongiosum which fits as a kind of hood over the tip-ends of the corpora cavernosa. This thicker part is called the *glans penis*. The glans is coated with a type of epithelium which is not subject to keratosis and which contains very many sensors. The urethra leaves the lesser pelvis through the pelvic floor and after that traverses the corpus spongiosum. It opens eventually in the top of the glans penis.

The skin of the penis is remarkably thin, with no hair growth, and encloses the erectile organ rather loosely and amply. Penile skin is very easily movable, as the layer of connective tissue beneath it is very loosely meshed. The skin of the penis ends with an attachment in the groove behind the glans penis. There is so much skin tissue here, however, that there is a fold which under normal circumstances completely overlaps the glans penis (see Fig. 2.10.15). This is the foreskin or *prepuce* (preputium), and it can be slid back over the glans. As with the female, cell debris can easily get stuck beneath the prepuce, which can lead to skin-irritation.

The part of the penis beneath the symphysis is called the *root* or radix. This root is attached firmly by means of a ligament to the symphysis and surrounded by striated muscle tissue. The part that hangs freely is called the *corpus penis*.

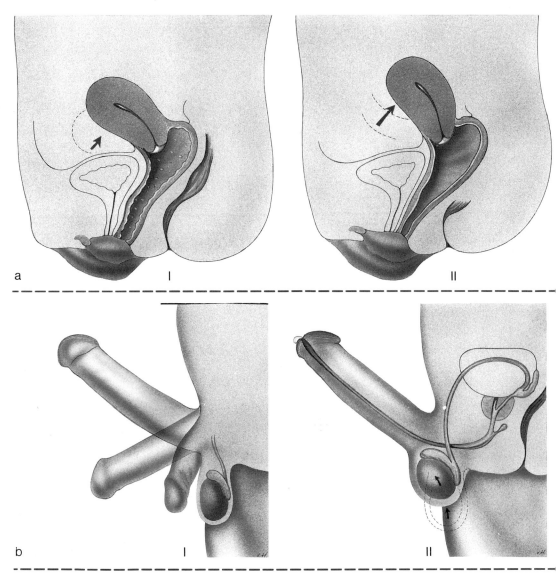

Figure 2.10.21
Sexual reaction and copulation

a. The sexual reaction phases in the female
b. The sexual reaction phases in the male

I   Phase of excitement
II   Plateau phase
III Orgasmic phase
IV   Relaxation phase

c. Copulation, median section

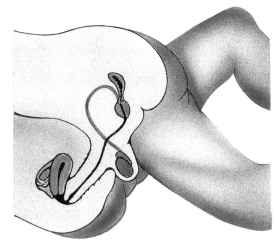

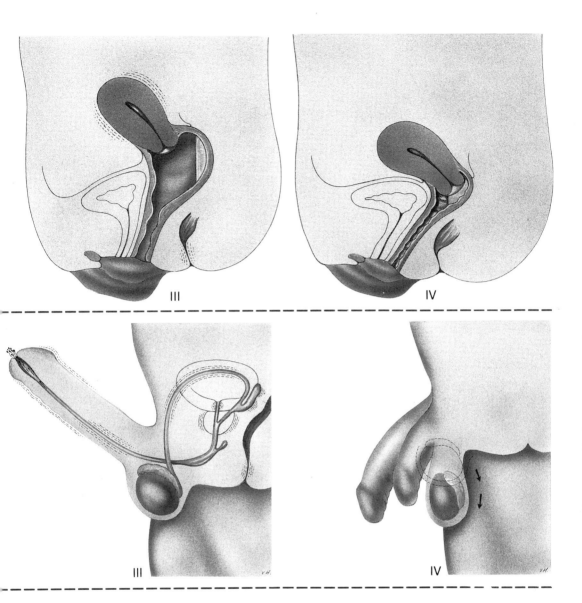

Under normal circumstances the penis is rather small (5–12 cm) and hangs down flaccidly. When sexually excited neuronal regulation ensures that the arteries deposit more blood into the sponge-like erectile organs, and at the same time the venous backflow of blood is diminished. The firm tunica of connective tissue are swollen. The penis becomes longer, thicker and harder. This results in an upright position, the erection, due to the structure of the erectile bodies and of the tunicas of connective tissue. The two corpora cavernosa are in fact the organs which bring about the erection; the corpus spongiosum remains more relaxed which prevents the urethra from being squeezed shut. Transport of sperm would then be severely hindered. The now stiff penis measures about 15–19 cm and can be moved by contracting the striated muscles around the root of the penis.

5. Copulation

Copulation (mating, coitus) or *sexual intercourse* is a complex activity between man and woman. It goes without saying that generally aspects other than functional anatomical ones are important in this activity. Nevertheless, this book restricts itself rather rigorously to the functional anatomical aspects of the process.

There are three stages of coitus: sexual excitement, copulation proper and relaxation (Fig. 2.10.21).

Sexual excitement

Sexual excitement can be brought up by a large number of factors, such as being alone with your partner, touching your partner, having erotic fantasies or reading an erotic story.

In *foreplay* sexual excitement will often be enhanced by means of kissing, caressing, etc.

The body possesses areas which are specifically sensitive sexually, areas which – when stimulated – enhance sexual excitement. They are called the *erogenous zones:* the lips, the neck, around the ear and the ear-lobes, the breasts (particularly the nipples), the insides of the thighs, the clitoris and the penis (particularly the glans). As sexual excitement rises, generally the sexually sensitive areas extend to include the buttocks, back, shoulders and abdomen.

A reddish glow can appear particularly in the female which often begins in the skin area of the breast, but can spread out over large areas of the body. The pulsation of the heart, blood pressure and respiratory frequency increase. The production of sweat can increase, the breasts swell, the nipples rise, the walls of the vagina fill up with blood. In the meantime erection of the penis and of the clitoris has usually taken place. This happens via a spinal cord reflex. The scrotum contracts and the testicles are lifted.

The labia majora and minora are engorged with blood, which results in a wider vaginal opening. The sphincter of the bladder contracts, so that during sexual intercourse no urine can enter the urethra.

The bulbo-urethral glands excrete a ropy mucus which makes the tip of the glans penis very smooth and slippery. The greater vestibular (Bartholin) glands also excrete fluid, which lubricates the entrance of the vagina. Over the total length of the vagina there is also an excretion of a lubricating mucus.

Copulation

Copulation begins with the introduction of the penis into the vagina. This is made easier by the lubricants excreted by the bulbo-urethral greater vestibular glands and the whole vagina wall.

The posture most commonly adopted is that: the partners face each other, the lower part of the male's body positioned between the spread legs of the female. The tip of the penis reaches to the end (internal end) of the vagina, and to the orifice of the uterus, the portio (Fig. 2.10.21c). The phase now beginning is called the *plateau phase*. The penis, especially the glans, increases in width even more; the upper part of the vagina also increases in width and length, whereas the lower part of the vagina around the penis, on the contrary, becomes narrower.

The condition of anteflexion of the uterus diminishes a little and the 'neck' of the uterus becomes somewhat wider.

The swollen clitoris is drawn back a little, which generally puts an end to its direct contact with the upper side of the penis. During the plateau phase one or both of the partners usually moves in such a way that the penis rhythmically moves up and down the vagina. All this tends to enhance the sexual excitement and the experience of enjoyment.

The *orgasm* or *orgasmic phase* now follows. The orgasm (coming) is an intense physical and psychic experiencing of delight. It can happen at the same time in man and woman, but this is certainly not a hard and fast rule. The male orgasm begins with a few powerful contractions of the muscles in the bottom wall of the pelvis. At the same time there are a few reflex reactions in the epididymis, the deferent ducts, the seminal vesicles, the ejaculatory ducts, the prostate and the urethra. Through contractions and peristalsis, the spermatozoa, the seminal fluid and the prostate fluid are deposited, enter the urethra and then with powerful waves are thrust deep into the vagina; this is *ejaculation* (the emission of sperm).

During female orgasm the vagina contracts rhythmically, the uterus 'climbs' to a more vertical position and displays rhythmical contractions, the clitoris is utterly sensitive to touch. In both partners orgasm is accompanied by muscle contractions over the entire body, and it is even possible that a temporary loss of consciousness occurs; the tension which has built up now fades away.

Relaxation

After the orgasm follows the relaxation phase; the erection of penis and clitoris disappear rather rapidly, the vagina relaxes and the uterus goes back to its original, normal degree of ante-flexion. The cervical channel, however, remains widened for about another half hour. The organs of both man and woman quickly return to a state of rest or relaxation. In the 'after play' it is possible to continue to enjoy the bodily and mental effects of the orgasm. After which, more often than not, many people will fall asleep.

Intensity, duration and frequency

There are enormous differences in intensity and duration of coitus. It can vary from time to time, and from partner to partner. There might occur an extremely intense orgasm or nothing at all. There can be one orgasm, but also two or more in succession, especially in the case of the female. Making love can take quite some time, but can also be over rather quickly. An orgasm can itself be short, or slightly longer. The frequency of sexual intercourse is to a very large extent dependent on the needs and wishes of the two partners. It can be that sexual intercourse takes place several times a week or only sporadically.

Sexuality is more than reproduction

Most of the time copulation is certainly not intended for explicitly reproductive purposes. The number of times that it actually takes place far exceeds the number of pregnancies. Fertilization can be avoided in more than one way. Sexual intercourse is mainly intended to experience a specific type of pleasure and a sense of well-being. These can also be achieved in non-heterosexual gratification.

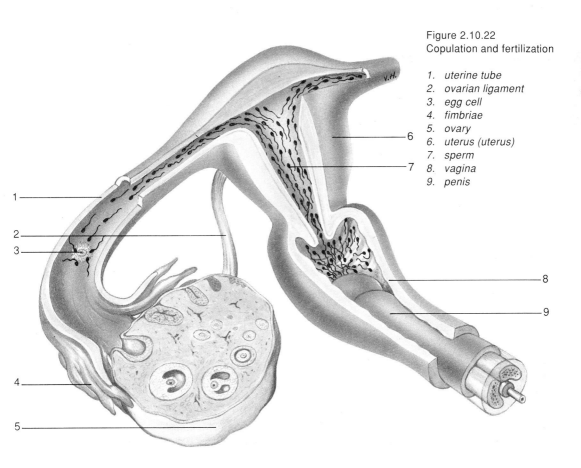

Figure 2.10.22
Copulation and fertilization

1. uterine tube
2. ovarian ligament
3. egg cell
4. fimbriae
5. ovary
6. uterus (uterus)
7. sperm
8. vagina
9. penis

Masturbation (self-gratification, onanism) is a common example of this. In masturbation an orgasm is produced without copulation. Girls and women masturbate through touching their genitals, especially the clitoris with their fingers. Boys and men masturbate by moving the skin of the penis rhythmically back and forth.

6. Fertilization

Fertilization (conception) is the key process in the creation of new life. In this process two unique half sets of genetic material are combined into a new unique complete full set. The full development of the new combinations of genetic codes in the hormones can start. Before this is achieved the sperm cell has to travel a long, long way.

Semen

The *semen* (sperm) which is set free in an ejaculation consists mainly of the products of seminal vesicles and of the prostate. The quantity of sperm cells which have come from the epididymis and the testes is relatively small (less than 0.5% of the total volume). In an ejaculation about 3 ml of sperm is set free, about one

thimbleful. There are found on average 60–120 million spermatozoa per millilitre sperm. The acidity (pH-value) of sperm is about 8.3. Because of this high pH-level the spermatozoa can temporarily stand the acid environment in the vagina. This acid milieu is a result of the metabolism-activity of the Döderlein bacilli. Moreover the sperm coagulates (thickens) temporarily in the vagina, which also makes it less vulnerable to the acidity. Yet 25% of the sperm cells die in the vagina.

On the way to the egg cell

The egg cell can be fertilized for about two days before it degenerates. Generally fertilization takes place high up in the uterine tube. Measured from the 'mouth' of the uterus, the sperm cell has to cover a distance of about 15 cm (Fig. 2.10.22). Converted to the world of the humans it would mean a road of about 4 miles. Sperm cells live a maximum of 4 to 6 days. Their route to the egg cell takes 12 to 24 hours. The living sperm cells set out to a less acid environment. They are helped on by certain excretions of the uterine cervix.

The sperm cells – it could be said – swim as a cooperating team; now and then it seems as if they are resting. They swim in the flow of fluid towards the end of the uterine tube and on into the body of the uterus.

They are helped here by contractions of the uterus and by movements of the ciliae in the uterine tube. The majority of the spermatozoa die during the relatively long voyage. In the end only a few thousand sperm cells reach the beginning of the uterine tube. About half of them will be in that uterine tube (of the two) in which an egg cell is waiting.

Fusion

High up in the uterine tube, in the ampulla, this swarm of sperm cells, the smallest human cells, meets the egg, the largest human cell. The egg (Fig. 2.10.23), is surrounded by the *zona pellucida,* with around that an 'aureole' (corona radiata) of follicle cells which serve as feeder cells. The sperm cells which reach the egg penetrate the corona radiata and run up against the zona pellucida. Remarkably enough the egg

Figure 2.10.23
Fusion of sperm cell and egg cell; meiosis II

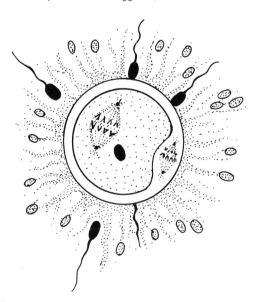

cell does not lie still. It turns round and round, always in the same direction: in the opposite direction to the spinning of the earth.

The enzymes in the tip of the sperm cell ensure that the zona pellucida is broken down locally. One sperm cell will be the first to pierce through this wall. As soon as this has occurred the zona pellucida immediately changes its structure: it becomes instantly impenetrable to other sperm cells.

At the moment of penetration the egg cell completes meiosis II; again a polar body is formed there. However, the polar body formed at meiosis I also divides. After meiosis II there is one egg cell and there are three polar bodies. The polar bodies are hardly anything more than a dumping place for superfluous genetic material. In due course they are disposed of by the egg.

The sperm cell loses its neck, its middle part and tail at the moment of penetration. The head of the sperm swells, the hormones are visible (under a microscope) because of the curling and coiling associated with meiosis II, as you will remember. The hormones of the egg cell also become visible, after which the two nuclei fuse with one another and the normal, diploid number of 46 chromosomes has been established. The egg cell has been fertilized; it is now referred to as the *zygote*. A new life has started.

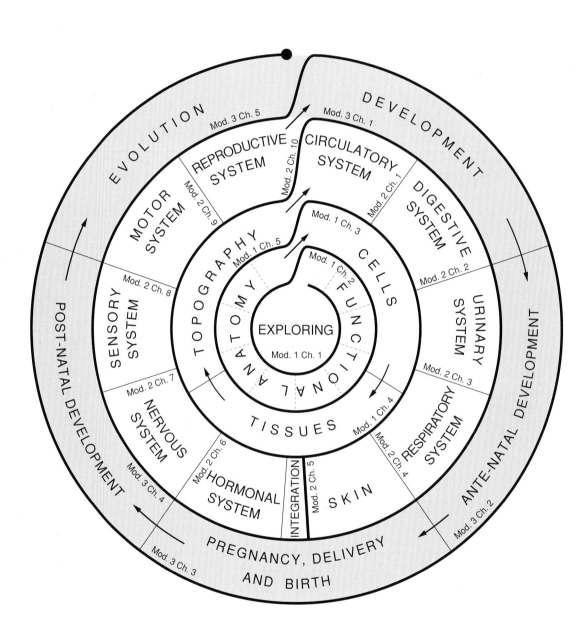

Module 3

THE LIFE CYCLE

Introduction to Module 3

Module 3 focuses on the human life cycle.

In Chapter 1 a number of general aspects of the phenomenon of development are discussed. The characteristics of development and the factors that influence development are examined.

In Chapter 2 development before birth is considered. In this period the fertilized egg cell develops into an embryo and then to a fetus. Childbirth marks the transition to the next phase in development.

In Chapter 3 pregnancy, delivery and childbirth are discussed.

In Chapter 4 developments after birth are described. After defining the phases of man's life, the development that each individual organ system passes through during life is outlined.

In Chapter 5, to conclude, the evolution of man is briefly considered.

Development

Introduction

Life is unfolding. During millions of years new beings have been created thousands and thousands of times. An egg cell and a sperm cell fuse: the process of development to maturity of a new human being has started. What is this development? Development is the intricate process of growth of a human being from the fertilized egg cell to the moment of death.
In this chapter the phenomenon of development will be discussed in general terms. Attention will be paid in addition to the characteristics of development, to the factors that control and influence development and to the methods which are being used to measure development.

Learning outcomes

After studying this chapter you should have sufficient knowledge and understanding of:
- the characteristics of the phenomenon of development;
- the internal and external influences on development;
- the methods used to measure development.

1. Characteristics

Development consists of three components; it has a number of phases and rates of change at different stages in the life cycle.

Growth, maturation and learning

We can distinguish a number of component parts in the process of development: growth, ripening and learning.

Growth is a *quantitative* term, an increase in measurements (length, width, depth and therefore volume).
At cellular level, growth not only implies an increase in size, but also an increase in numbers. The increase in numbers is a result of mitosis. The point where growth starts is the fertilized egg cell. This cell is the initiator of a considerable number of mitoses. In mitosis the parent cell transforms itself into two exact copies of

the parent. The 'daughter cell' and the 'mother cell' are as like as two peas in a pod.
If mitosis in this form went on for nine months, a mother would give birth to a great mass of copied zygotes instead of delivering such a unique baby with eyes, mouth, hair, arms and legs, a fast-beating little heart, etc.
During the process of development through which an individual goes it is evident that certain groups of cells gradually become different from other groups of cells. This *differentiation* is accompanied by *specialization*: the capacity to specialize in one single function. There are, for instance, cells that become distinct in having extremely long thin protuberances of the cell membrane, particularly suitable to transfer messages in the form of electric signals – nerve cells. Other groups of cells show neatly arranged protein chains within an elongated cell membrane.
These protein chains can shift in relation to one

another and in this way these cells specialize in shortening themselves – muscle cells.

In development, growth is synonymous with decay, demolition. The horny layer (stratum corneum) of the skin peels off and is replenished by deeper layers; red blood cells live for about 120 days and are then cleaned away while their 'successors' are already fully in 'active service'; bony tissue is constantly in the cycle of building up and breaking down. Breaking down cannot therefore be separated from growth.

In every human being development can be distinguished in roughly three phases. During the first phase of human development, total growth is greater than total decay; during the second phase growth is balanced against decay; and in the third phase total growth is not sufficient to replenish that which is broken down.

At the end of this process of development stands death.

Maturation is a *qualitative* term. It tells something about the efficiency with which cells, organs or the entire human being perform their tasks.

In most cases after growth there follows maturity. An apple first reaches a certain volume, and then it acquires taste through ripening. After growth, the blood cells become suitable for the transportation of oxygen by maturing.

Learning is absorbing knowledge and skills. In the development of man learning is of immense importance. Sometimes it seems as if everything we do has to be learned. We have *learned* language, how to dress, to write, to read, to do sums, to drive a car, to take care of our health, to swim, to play certain games, to feel responsibility for one another, etc.

Phases in life

The development of a human being follows an orderly sequence. Apart from greater and smaller inter-individual variations everybody goes through the same successive stages during their development. On the grounds of specific characteristics at a certain level of development, the following *phases in life* can be distinguished: the *newborn*, the *infant*, the *toddler*, the *schoolchild*, the *adolescent* and the *adult*. During adulthood there are again a number of life stages. Important elements in *growing old* are the *menopause, old age* itself and *dying*.

Tempo

The development of man does not occur at an even pace. There are alternatively periods of fast and slow growth. The best-known example of this is the spurt of growth during puberty. Moreover, enormous differences in development can exist between individuals with the same calendar-age because of a difference in time at which their period of accelerated growth starts. We speak of *precocious* physical development when it is ahead of the average development of people in the same age group. With *'late developers'* the opposite is the case; development takes place much later than the average.

Secondary developments

Research has shown that the average height of the Western European has increased during the last 2 or 3 centuries, by about 2 cm (= 0.8 in) every 10 years. This leads to beds formerly measuring a standard 1.80 m (5 ft 11) now being delivered by the manufacturer with a standard length of 2 metres (6 ft 7). In the Middle Ages houses were built with lower doorways than nowadays.

Between 1910 and 1980 in Western Europe the average life expectancy rose from 55 years to 73 years for a man; and from 57 to 80 years for a woman.

Development turns out to be a dynamic phenomenon; though the development of zygote to adult is more or less repeated in every human being, it is apparent that certain changes have taken place over the centuries. These types of changes are called *secondary developments*.

2. Influences

Man's development is influenced by *internal* and *external* factors. Internal factors are those factors that lie within man himself, such as hereditary qualities and hormonal factors.

External factors are factors which lie outside man, such as social, cultural and economic circumstances and environmental influences.

2.1. Internal factors

Hereditary factors

Growth and development seem to occur of their own accord. The fertilized egg divides again and again. Step by step the individual appears. What is the background of this astounding process of development? Who or what determines the course of this development?

The blueprint for the individual's developmental plan is laid down in the *genetic codes* of the zygote as in a computer program. One by one the instructions of this work scheme are located, read and implemented, thus permitting the next stage of development.

For the discussion of genetics, we must go back to the nucleus of the cell. Within the nucleus of the human cell 46 chromosomes can be counted. The chromosomes occur in pairs, so in a human there are 23 pairs of chromosomes. The 23rd pair consists of the *sex chromosomes*, or reproductive chromosomes (the heterosomes). The other 22 pairs of chromosomes are called *autosomes*. Each pair of autosomes forms an identical pair of chromosome twins. For these we use the term: homologous chromosome pairs. In a female the two sex chromosomes are also identical, namely XX. In a male it looks as if the two 'legs' of one of the X-chromosomes have been cut off. To distinguish them the sex chromosomes of the male are indicated as XY-chromosomes.

The chromosomes (Fig. 3.1.1) consist of *genes* which have been arranged in series. The total sum of genetic codes (instructions), the genome, of a human being consists of more than a million genes.

The specific place where such a gene can be found in the chromosome is called the *locus*. Homologous (twin) chromosomes are identical in composition in their genes. Each gene has a twin gene on the homologous chromosome. This 'twin gene' influences the same hereditary characteristic as the other half of the 'twin'. Such a pair of twin genes are called alleles.

Alleles occur in the homologous chromosomes in an exactly identical order in the exactly identical places.

Though alleles are identical pairs of genes, they are not always completely exact copies. An example will illustrate this; there is a certain gene which contains the code for the production of a certain protein in the wall of the erythrocyte membrane. There are one out of three possibilities: the code for protein A, the code for protein B, or the code for 'no protein' (actually no code, a zero code). In the example we have taken the code for protein A (Fig. 3.1.2).

The analogous alleles also have these three possibilities, A, B or zero. The two alleles are identical, so provided with possibility A (Fig. 3.1.2a), then the outcome of their fusion will be that the person has the blood group A. Such identical genes are called *homozygotes*.

When the genes are not exactly identical, then they are called *heterozygotes*. When in the gene in our example the second allele has the code for the production of protein B, then both types of protein, A and B, will be produced. The person has blood group AB.

Sometimes the characteristic that is eventually realized stands exactly midway between the characteristics which each of the alleles by themselves would have brought about. For example, this is the case when two people with a differing skin colour have a child. The skin colour of the offspring is a mix of those of the parents.

However, more often in the case of heterozygotous genes there is only one of the two alleles which comes to 'expression' (actually being realized). The other gene is suppressed. When one gene has the code for the quality of 'brown eyes' and the alleomorph to match the code for 'blue eyes', then the person will have brown eyes.

The gene that expresses itself at the expense of the other one is called *dominant*. The suppressed gene is called *recessive* (receding).

Dominant genes are commonly indicated by a capital letter (e.g. brown eyes: B), recessive genes with a small letter (blue eyes: b).

Figure 3.1.1
Chromosomes

a. Chromosome chart;
   firstly the chromosomes
   are coloured, dyed, then
   photographed and after
   that arranged in
   accordance with their
   shape and size.
b. Chromosome structure
c. Two examples of
   homologous
   chromosomes

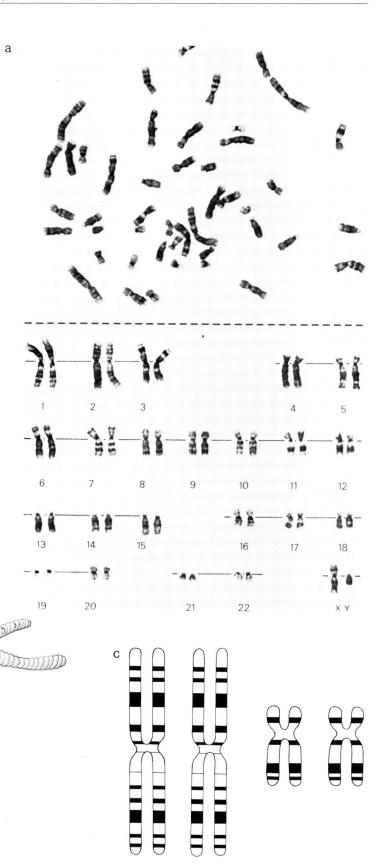

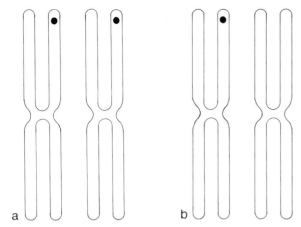

Figure 3.1.2
Genotype and phenotype
Assuming the black dot (●)
contains the genetic code for
the production of the protein
A in the erythrocyte
membrane.
The pair of chromosomes (a)
is homozygotous; the
genotype is AA and the
phenotype is (blood group) A.
The pair of chromosomes (b)
is heterozygotous; the
genotype is A0 (A and zero)
and the phenotype is (blood
group) A

By assessing the characteristics of people we can hardly ever tell the genetic composition of the individual person. Someone with brown eyes may or may not have alleles that both contain the genetic code for 'brown eyes'. This is why it is important to distinguish between the *genotype* and the *phenotype* (Fig. 3.1.2).

The genotype is the exact description of the genetic composition of an individual; the phenotype is the total sum of all the characteristics the person exhibits. Thus the phenotype is the result of addition and subtraction in the alleles.

Each human being originates from a unique combination of genes of which half stem from the father and half from the mother. The reproductive cells needed for this (egg cell and sperm cell) are produced by means of meiosis in the gonads. Through 'crossing-over' at the beginning of meiosis (see Fig. 2.10.2), there also come into being completely new combinations of genes.

Though the combination of genes in the fertilized egg cell is unique, it is possible to say something about the characteristics of the child on the basis of data about the genes of the father and mother.

This is achieved with the help of the laws of genetics, the law of segregation, the law of independent assortment, the law of dominance, etc. With the help of the genetic laws we can first of all explain, for example, why there are on average as many men as there are women.

As a result of meiosis all egg cells have an X-chromosome for the 23rd chromosome (Fig. 3.1.3a).

Of the semen cells 50% have an X-chromosome for their sex chromosome and 50% have a Y-chromosome (Fig. 3.1.3b).

Assuming that the chance of fertilizing an egg for both types of semen cell is equal, then the chance of achieving an XX-combination (a girl) equals the chance of an XY-combination (a boy), namely 50% (Fig. 3.1.3c).

At the same time the sex of the child is determined by the genotype of the semen cell, by the input of the father.

If the genotype for a certain characteristic of both parents is known, the chances of having a certain phenotype in their offspring can be calculated (Fig. 3.1.4).

When both parents have blood group AB (genotype AB) then the chance of blood group A (genotype AA) in their offspring is 25%; the chance of blood group AB is 50% and the chance of blood group B is 25% (genotype is BB).

When the father has blood group B (genotype B0, and the gene 0 is recessive) and the mother has blood group AB, then the chance of having a child with blood group AB is 25%; the chance of blood group B is 50% (on the basis of the genotypes BB and B0) and the chance of A (genotype A0) is 25%.

If both parents have blood with a positive rhesus factor (both with genotype Dd) then there is a 75% chance of the children having Rh⁺ blood

Figure 3.1.3
Heredity and sex
determination

a. Oogenesis; 100% of
   the egg cells have an
   X-chromosome
b. Spermatogenesis; 50%
   of the semen cells
   have an X-
   chromosome, 50% a
   Y-chromosome
c. Fertilization; the input
   of the male determines
   the sex of the fertilized
   egg cell

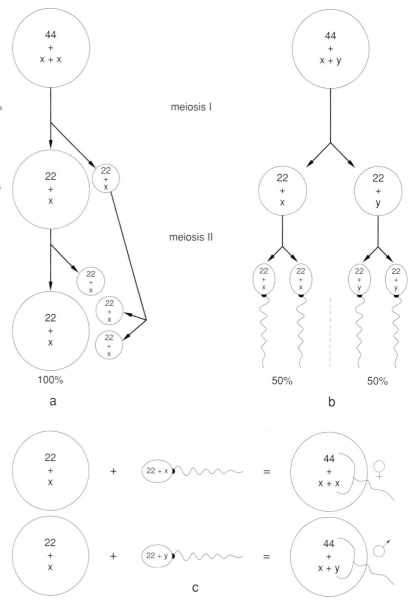

(phenotype). For all kinds of characteristics this type of scheme can be made.

A specific form of heredity is the type of heredity that is linked to *sexual reproduction*. The genes concerned are on the 23rd pair of chromosomes. They carry the codes for sexual characteristics, but by far the larger part of the genes on the sex chromosomes consists of genes which have got nothing to do with the differences in sex.

In sexual heredity a specific quality is shown by the genes that are situated on the 'lower legs' of the X-chromosome (Fig. 3.1.5). In a female all the sexually linked genes appear in pairs. In a male there are single genes, having to stand on their own. On the Y-chromosome (an X with the legs cut off, you will remember) the corresponding alleles are missing. This fact might account for some genetic defects, which are restricted to a specific sex, such as haemophilia which is restricted to males.

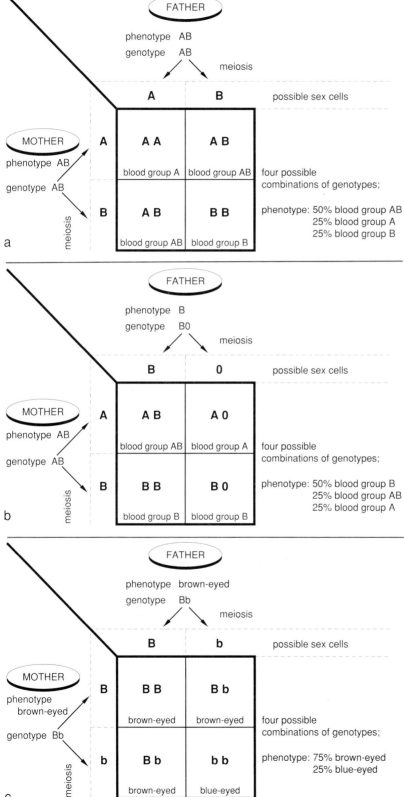

Figure 3.1.4
Heredity; examples

a. Blood groups; both
   parents have blood
   group AB, so the
   genotype AB
b. Blood groups; the
   parents respectively
   have blood group B
   (B0) and AB
c. Colour of the eyes
   (iris); both parents are
   brown-eyed

FATHER
phenotype AB
genotype AB
meiosis

A   B   possible sex cells

MOTHER
phenotype AB
genotype AB
meiosis

A | A A | A B
  | blood group A | blood group AB | four possible
  | | | combinations of genotypes;
B | A B | B B
  | blood group AB | blood group B | phenotype: 50% blood group AB
  | | | 25% blood group A
  | | | 25% blood group B

a

FATHER
phenotype B
genotype B0
meiosis

B   0   possible sex cells

MOTHER
phenotype AB
genotype AB
meiosis

A | A B | A 0
  | blood group AB | blood group A | four possible
  | | | combinations of genotypes;
B | B B | B 0
  | blood group B | blood group B | phenotype: 50% blood group B
  | | | 25% blood group AB
  | | | 25% blood group A

b

FATHER
phenotype brown-eyed
genotype Bb
meiosis

B   b   possible sex cells

MOTHER
phenotype
brown-eyed
genotype Bb
meiosis

B | B B | B b
  | brown-eyed | brown-eyed | four possible
  | | | combinations of genotypes;
b | B b | b b
  | brown-eyed | blue-eyed | phenotype: 75% brown-eyed
  | | | 25% blue-eyed

c

Figure 3.1.5
Heredity based on sex

a. The 23rd pair of
   chromosomes in a
   female; all the genes
   (for example ●) occur
   in pairs
b. The 23rd pair of
   chromosomes in a
   male; the genes (for
   example ●) on the
   lower part of the X-
   chromosome, do not
   occur in pairs

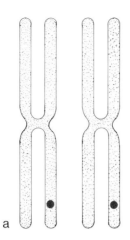

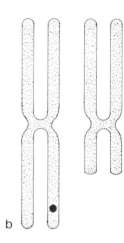

a   b

Hormonal factors

The production of certain hormones during certain phases of life is of enormous importance for
development. At different points in time hormones stimulate the growth and maturation of
the brain tissue, the development of primary and
secondary sexual characteristics and the growth
of skeleton and muscles. A specific hormone
production is characteristic for the transition to
puberty. During this change a person becomes
sexually mature. The *menopause* (i.e. change of
life) in women of about 50 years of age is
another example of a development directed by a
specific change of hormone production.

2.2. External factors

Development is also influenced by social, cultural and economic factors and by environmental influences.
Good food, good and inspiring working conditions, a harmonious relationship with the
people inhabiting the same house, a well-balanced ratio between exertion and relaxation,
clean air, clean unpolluted water, a sufficient
income: these are all small elements which
contribute to good development.

3. Measurement of development

Calendar age is a poor criterion for determining
the stage of a person's development. Two human
beings of the same calendar age (Fig. 3.1.6) can
show considerable differences in development.
This is why the term *developmental age* or

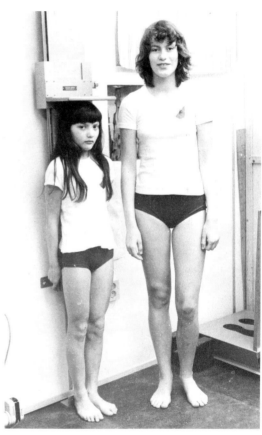

Figure 3.1.6
Calendar age and physiological
age.
Two girls of the same calendar
age (12.6 years).
The girl to the left is late in
maturing (physiological age 11.1
years), the girl to the right is early
in maturing (physiological age
13.5 years)

physiological age has been invented and adopted. It describes age in the *biological* sense of the word. Physiological age can be determined in the most accurate way on the basis of the stage of development in the skeleton: by *skeletal age*. Skeletal age is determined with the help of an X-ray photograph of the hand. The shape of the bones and the extent of ossification in the epiphysial discs are compared with average values (Fig. 3.1.7).

Specific information about development is obtained by regularly measuring the contours of the skull, by recording changes in body weight, by observing the development of the teeth (the *dental age*) or by comparing the development of the secondary sex characteristics (the *sexual age*) with average values.
An idea of intellectual development (the *mental age*) can be obtained by means of intelligence or mental tests.

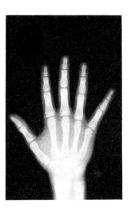 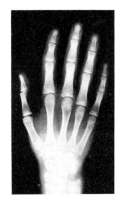

Figure 3.1.7
Skeletal age
X-ray photographs of a hand of each of the two girls in Figure 3.1.6. By comparing these X-rays with standardized indexes, the physiological age can be established

2

Ante-natal development

Introduction

This chapter will discuss how the zygote develops from the moment of fertilization till the moment of birth. In this period – generally nine months – two phases can be distinguished. The *embryonic phase*, which lasts for the first 8 weeks. During this time the placenta, the umbilical cord and all the systems develop. The embryonic phase ends when all the body systems have been established in rudimentary form. The *fetal phase*, which follows, and ends with birth, is mainly characterized by growth in size and quality of all the structures of the body systems.

Learning outcomes

After studying this chapter you should have sufficient knowledge and understanding of:
– the development from embryo to fetus;
– the development of the fetus to the moment of birth;
– the development of the individual body systems before birth.

1. The first three weeks

From zygote to blastocyst

About 30 hours after fertilization the first mitosis takes place. It is noticeable that the two 'daughter cells' are not as big as the 'mother cell' and that – just as in the next series of mitosis – the total quantity of cytoplasm does not increase significantly (Fig. 3.2.1). The zygote is called the *morula* when it is a mass of 16 cells. It is still enveloped by the zona pellucida. About four days after fertilization this morula leaves the uterine tube, enters the cavity of the uterus and develops into the *blastocyst* (Fig. 3.2.1). A hollow space has been prepared (the yolk sac) filled with fluid, surrounded by flat cells (trophoblast cells). There is a concentration of cells to one side within the hollow space: *the embryo bud*, or inner cell mass. After about 6 days this side of the hollow space, with the embryo bud, of the blastocyst comes into contact with the wall of the uterus, after

which the embedment or implantation takes place (Fig. 3.2.2). The blastocyst is completely absorbed into the mucous membrane (the endometrium) of the uterine wall.

Germinal discs

The embryo bud develops into a two-layered embryo disc and, in addition to the yolk sac which already exists, a second hollow space develops, the *amniotic cavity*, which is filled with fluid (Fig. 3.2.2). This hollow space is enveloped on the one hand by a layer, a disc of cells, called the *ectoderm*, and on the other hand by a layer, a disc of the amnion cells. The layer of cells bordering the yolk sac is called the *endoderm*.

In the meantime the embryo is in the second week of its development. Contact with the body of the mother becomes firmer and firmer. Little cavities (*lacunae*) have been formed in the layer of trophoblast cells, connecting with blood vessels from the uterus wall. This is the start of the formation of the placenta. The enclosed

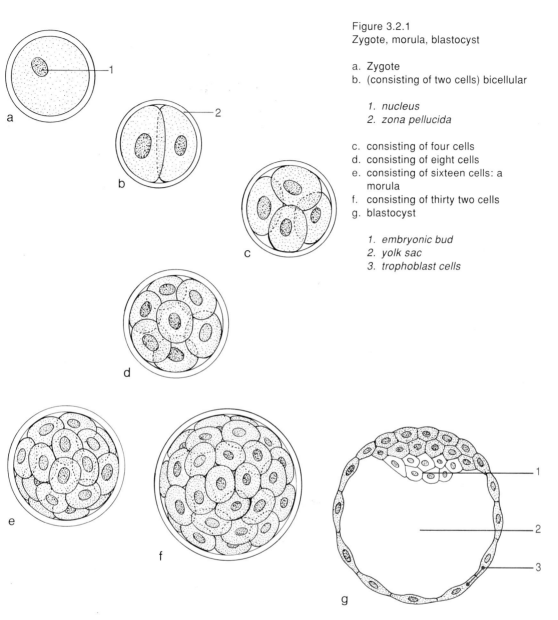

Figure 3.2.1
Zygote, morula, blastocyst

a. Zygote
b. (consisting of two cells) bicellular

    1. *nucleus*
    2. *zona pellucida*

c. consisting of four cells
d. consisting of eight cells
e. consisting of sixteen cells: a
   morula
f. consisting of thirty two cells
g. blastocyst

    1. *embryonic bud*
    2. *yolk sac*
    3. *trophoblast cells*

amniotic cavity, embryonal disc and yolk sac in their turn come to lie within a hollow space: the *chorionic* cavity. The *connecting stalk* secures the connection with the developing placenta. The umbilical cord will later develop from this connecting stalk .

In the third week of fetal development the third layer in the embryonic disc is formed: the *mesoderm*. This layer develops out of ectoderm cells which bulge out between endoderm and ectoderm.

Ectoderm, mesoderm and endoderm are called the *germ discs*.

Each of the germ discs gives rise to a specific group of body systems.

Out of the ectoderm arise the skin, the hair, the sensory system, the nerve system and parts of the hormonal system.

Out of the mesoderm develop the motor system, the urinary tract, the circulatory system, the reproductive system and parts of the hormonal system.

Out of the endoderm derive the metabolic tract, the respiratory tract and parts of the hormonal system.

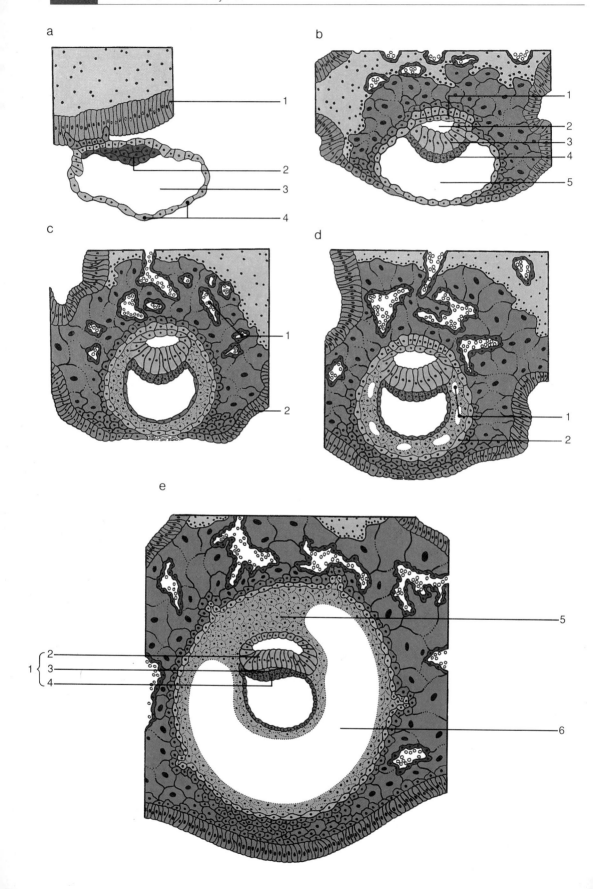

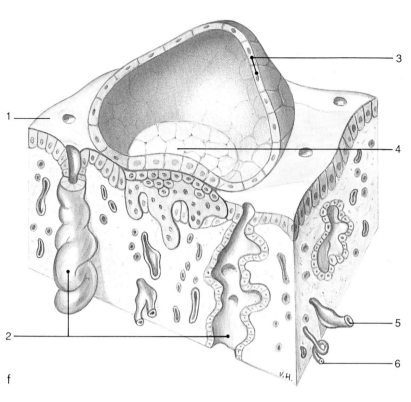

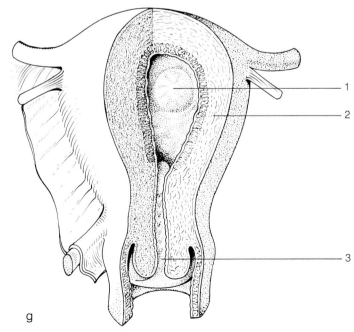

Figure 3.2.2
Nidation (or embedment)
and the development of
the germinal discs

a. Nidation (developing a
    nest)

    1. *mucous membrane
        of the uterus*
    2. *embryo bud*
    3. *yolk sac*
    4. *trophoblast cells*

b. The amnion cavity is
    formed

    1. *amnion cell*
    2. *amnion cavity*
    3. *ectoderm*
    4. *endoderm*
    5. *yolk sac*

c. Formation of lacunae

    1. *lacunae*
    2. *mucous membrane
        of the uterus*

d. Development of the
    chorionic cavity

    1. *development of the
        chorionic cavity*
    2. *trophoblast cells*

e. Three germ layers

    1. *germinal discs*
    2. *ectoderm*
    3. *mesoderm*
    4. *endoderm*
    5. *development of the
        connecting stalk*
    6. *chorionic cavity*

f. Spatial representation
    of the nidation

    1. *uterus epithelium*
    2. *uterus glands*
    3. *trophoblast cells*
    4. *embryonal bud*
    5. *vein of the mucous
        membrane*
    6. *spiral artery*

g. View of the uterus wall
    with nidated fruit

    1. *fetus*
    2. *myometrium*
    3. *uterus neck, cervix*

2. The fourth to eighth week

The first rotation and a beating heart

In the fourth week of fetal development the embryo – hanging from the connecting stalk – performs a rotation of 180° (Fig. 3.2.3). At the same time substantial changes take place in the germinal discs.

The embryo, which up till now was flat, starts to form the shape of a letter C. Moreover, sideways curvatures develop. The shape of the embryo is somewhat comparable to a mushroom; the cap a little bit oblong, the yolk sac as its stem.

By the end of its rotation, the yolk sac is integrated within the embryo. This constitutes the beginnings of the intestines.

By now the connecting stalk has developed into the umbilical cord, containing the newly-developed umbilical cord vein (v. umbilicalis) and umbilical cord arteries (arteriae umbilicales). These are connected to the first embryonic blood vessels (Fig. 3.2.4). Already in an early stage one of these blood vessels has a rhythmical contraction: the embryonic heart. This is observable on an ultrasound scan of a four weeks' pregnancy: a moving sign of life!

Amnions and segmentation

The membrane which encloses the amniotic cavity is called the *amnion;* the membrane around the chorionic cavity is the *chorion.*

During rotation of the embryo the chorionic cavity becomes more and more occupied by the expanding amnion cavity (Fig. 3.2.3). At the end of this process these two membranes (amnion and chorion) are lying against each other and grow together.

Amniotic fluid is produced inside the cavity formed by the amnion, and entirely surrounds the embryo. In this fluid the embryo moves about freely; an excellent protection against shock and pressure from outside. The amniotic fluid is refreshed completely every three hours by diffusional exchange with substances from the blood of the mother.

In the beginning of the fourth week a longitudinal groove in the middle of the back of the embryo is observable, as the ectoderm cells on either side of the 'groove' start to strongly divide. The folds formed in this way bend over the groove towards each other and grow together (Fig. 3.2.5). Instead of a groove there is now a tube with an opening on both sides, the *neural tube.* The spinal cord develops in the neural tube, both openings close and there is a strong swelling of part of the head: the brain is being formed.

Figure 3.2.3
The developing embryo
turns 180°

1. amniotic cavity
2. connective stalk
3. yolk sac
4. chorionic cavity
5. embryonal heart
6. chorion flakes
7. amnion
8. placenta
9. umbilical cord
10. primal intestines
11. chorion

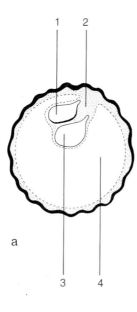
a

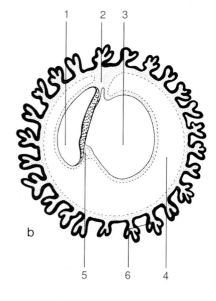
b

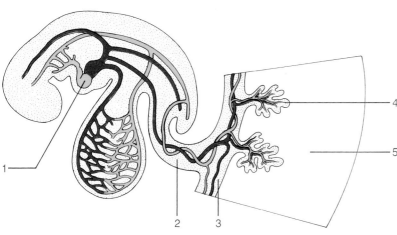

Figure 3.2.4
The vascularization of
the embryo (4th week)

1. heart
2. umbilical cord with
    blood vessels
3. chorion
4. chorion flake
5. placenta

From the 20th day onwards an indentation develops to the left and right of the spinal cord;
three segments (*somites*) per day over 18 days
(Figs. 3.2.5 and 3.2.6). In man the segmented
structure is mainly recognizable in the successive vertebrae, in the segments of the spinal
marrow and in the way the skin is provided with
nerves.
In every somite are distinguished a sclerotome,
a dermatome and a myotome. The vertebral
column segments derive from the *sclerotomes*.
The skin segments are related to the *dermatomes*. The muscles, also, are segmentally
arranged, in their rudimentary form and known
as *myotomes*.

Early fetal development
At the end of the fourth week there is still not
much of a human being to be recognized in the
embryo. Rather, it resembles a shrimp. A head
part has taken shape, in which rudimentary eyes
are visible. Moreover three of the six *gill arches*
are seen, promontories in the neck area, from
which the lower jaw and the larynx will be
derived, among other things. These gill archs,
the primitive mouth just above the heart, the
'little tail' and the only just apparent extremity
buds determine the rather peculiar appearance
of the embryo at the beginning of the fifth week.
Development is now tremendously fast in
relation to the development of a recognizable

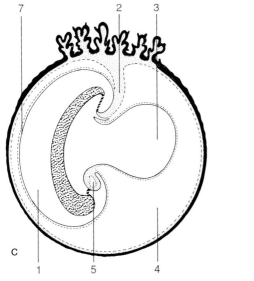

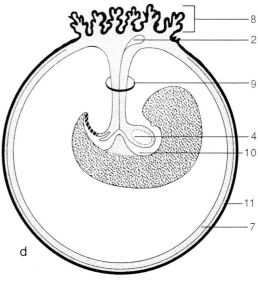

Figure 3.2.5
Neural tube and
segmentation

a. 18 days
b. 19 days
c. 20 days
d. 22 days
e. 23 days

1. *amnion*
2. *groove*
3. *somites*

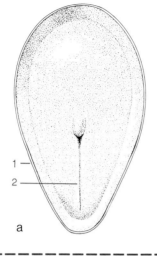

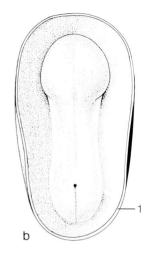

a

b

Figure 3.2.6
Development from embryo
to fetus

a. 4 weeks
b. 5 weeks
c. 6 weeks
d. 7 weeks
e. 8 weeks

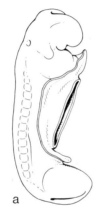

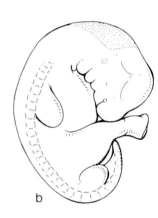

a

b

'mini human', the *fetus* (Fig. 3.2.6). The growth of the head is comparatively quick; the face is being further formed: the eyes (slightly to the side of the head), the nose and the mouth. The ears appear above the neck, which is vaguely taking shape. The embryonic little tail disappears. The development of the arms always slightly precedes the legs. The fingers appear before the toes.

3. The third to ninth month

The fetal period (Fig. 3.2.7) is characterized by rapid development of the rudimentary body systems.

In this period growth and maturation are far more important processes than differentiation and specialization (which were prevalent in the embryonic phase). The average increase in length and weight of the fetus is given in Figure 3.2.8.

The fetal measurement normally recorded is head-tail length; nowadays this can easily be estimated from an ultrasound scan. It is striking that the lead which the head took in growth during the embryonic phase is lost in the fetal phase – and during all the growth after birth – because of the relatively stronger growth of the rest of the body (Fig. 3.2.9).

At the beginning of the third month the length of the head is about half the overall head-tail length; at the beginning of the 5th month the length of the head is about one third of the total length. At birth the ratio of the length of the head

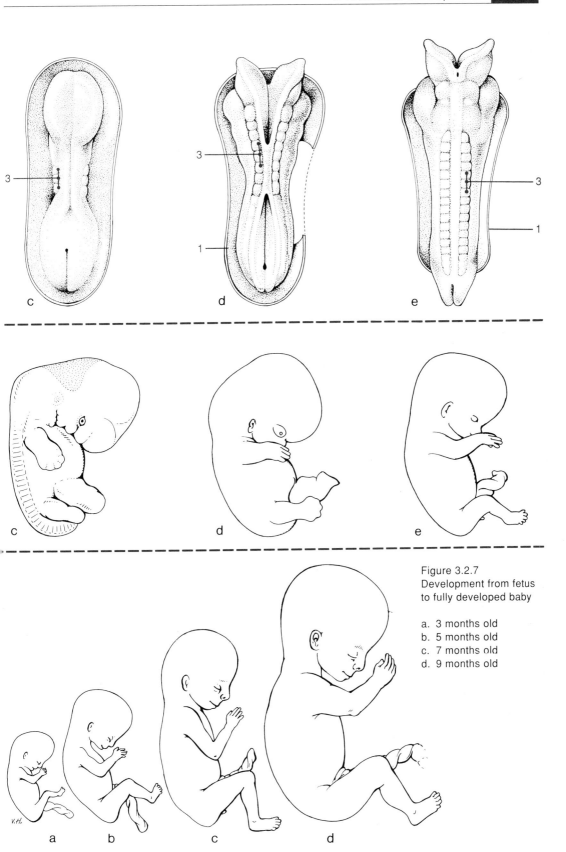

Figure 3.2.7
Development from fetus
to fully developed baby

a. 3 months old
b. 5 months old
c. 7 months old
d. 9 months old

Figure 3.2.8
Growth of the fetus

a. Increase in length
b. Increase in weight
c. Increase in the
   contours of the head

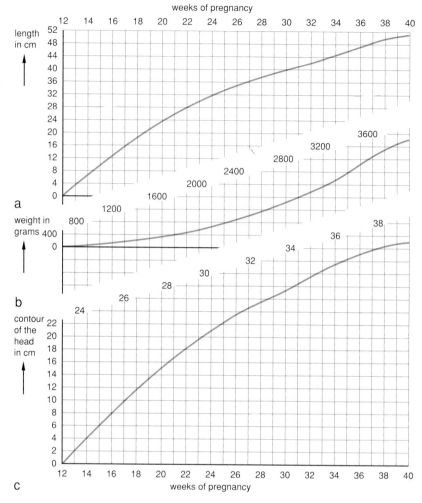

Figure 3.2.9
Proportional growth
As the standard the length
at the moment of birth has
been taken. In comparison
the trunk and limbs grow
considerably more than the
head

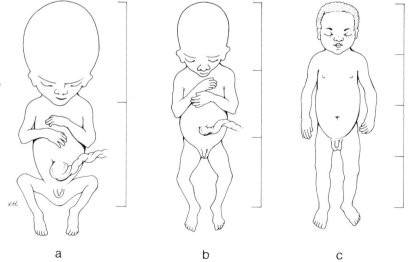

to the length of the body becomes 1:4. At the end of physical development, in the phase of adulthood, this ratio is 1:7.

The eyes of the fetus have come to rest more to the front. The eyelids have grown towards each other and cling to one another as if glued together. The eyelids do not open again until the 7th month. In the beginning a swelling in the umbilical cord is seen, near the body of the fetus, because of bulging intestines. In the 3rd month the swelling disappears when the intestines are incorporated into the abdominal cavity. In this period the sex of the fetus can be established, judging from the development of the external genital organs. During the 5th month the fetus takes a sip of the amniotic fluid every now and then, about 400 mls per day – approximately half of the total quantity of amniotic fluid. Urine is transferred into the amniotic fluid daily by means of the fetal kidneys. This urine consists to a large extent of water: waste materials, you will remember, are carried away by the placenta.

From the 3rd month onward the fetal fluffy hair

Figure 3.2.10
The line pattern of the fetal down, fluffy hair

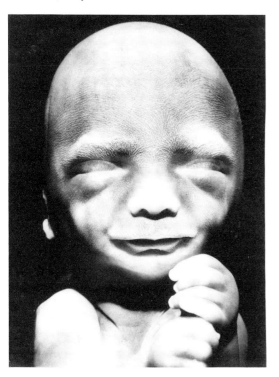

(*lanugo*) grows all over the fetal body in a typical line pattern (Fig. 3.2.10). A month later a coarser type of hair starts to grow on the head and the eyebrow-arches. Usually most of the fluffy hair (down) drops off before birth.

From the 6th month onwards the sebaceous glands excrete a white, fatty substance: *vernix caseosa*. This covers the entire skin. The remains of it are clearly visible after birth.

Internal signs of life

After about 18 weeks the movements which the fetus makes can be clearly felt by the mother. This is a sign that the developing nerves do indeed provide the growing muscles with commands. Legs and arms bump into the uterus wall. Until just before delivery the fetus can take up differing positions in the uterus as a result of its movements. Eventually this leads to a specific birth position.

Sometimes the fetus swallows the amniotic fluid the wrong way and gets hiccups. Then the mother can very clearly feel the rhythmical movements associated with hiccups.

The time has come

Generally life in the uterus lasts (from fertilization to birth) for 38 weeks; counting from the first day of the last menstruation, 40 weeks pass. The baby is said to be full term when this period is complete. The baby is said to be *premature* (literally: unripe) if birth takes place more than two weeks early. Generally a full term baby turns out to be slightly better prepared for the enormous changes which are associated with birth and the first few weeks of life outside the mother's uterus than a premature child. The child is called *serotine* (literally: late) if it is born more than a week later than the calculated date.

4. Systems

The fetus develops as an entity. Nevertheless it is true that in the successive stages of development different events become apparent.

In view of the purpose of this book it is appropriate now to briefly sketch a number of aspects

Figure 3.2.11
The development of
the heart

1. foramen ovale
2. septum cordis
3. endocardial duct
4. arterial pool
5. venous pool
6. ventricles
7. atria
8. atrium
9. ventricle
10. aorta
11. pulmonary trunk

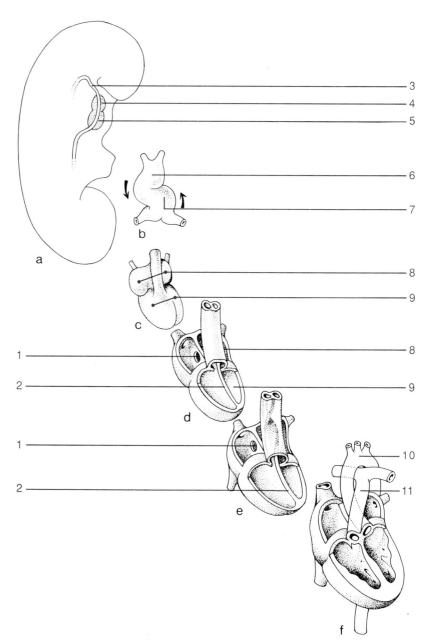

of the prenatal development of the 10 body systems of a human being.

(1) Circulatory system

The *heart* develops out of what is at first nothing more than a small pulsating blood vessel, the *endocardial tube* (Fig. 3.2.11a). The upper end of this embryonic heart is the arterial pool which will develop in the ventricle muscle; the lower end is the venous pool which will become the atrium muscle.

Up till now there are no separate compartments. The next phase is characterized by a winding of the endocardial tube (Figs. 3.2.11b and c): the atrium muscle runs behind the length of the ventricle muscle and settles more upwardly. At the same time the development of the annulus fibrosus and the septum cordis (Fig. 3.2.11d) is noticeable.

Also noticeable is the development of a spiral shaped ($180°$) continuation of the septum into the artery which arises from the – initially –

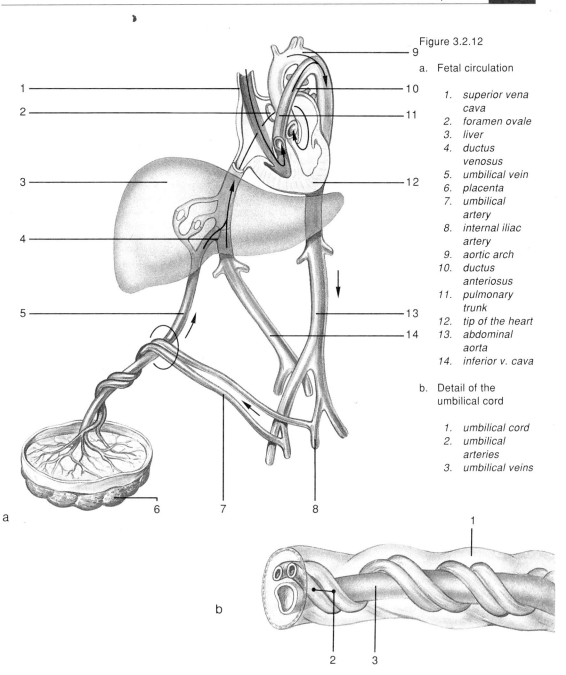

Figure 3.2.12

a. Fetal circulation

1. *superior vena cava*
2. *foramen ovale*
3. *liver*
4. *ductus venosus*
5. *umbilical vein*
6. *placenta*
7. *umbilical artery*
8. *internal iliac artery*
9. *aortic arch*
10. *ductus anteriosus*
11. *pulmonary trunk*
12. *tip of the heart*
13. *abdominal aorta*
14. *inferior v. cava*

b. Detail of the umbilical cord

1. *umbilical cord*
2. *umbilical arteries*
3. *umbilical veins*

shared ventricle cavity (Fig. 3.2.11e). Thus the beginnings of the aorta and the pulmonary trunk have started to develop. Later in the development these separate tubes (Fig. 3.2.11f) can be recognized.

The embryonic circulation (see Fig. 3.2.4) develops into fetal circulation (Fig. 3.2.12a), which differs markedly from the circulation after birth (see Fig. 3.3.14).

The fetal heart pumps blood towards and into the placenta through branches of each of the pelvic arteries. These umbilical arteries (arteriae umbilicales) run internally to the navel. They contain blood with carbon dioxide and waste products of fetal metabolism. In the placenta these arteries branch off into a vast capillary network of which the vessels lie very near the vessels of the capillary network in the uterine wall.

Here, oxygen and nutrients are exchanged for waste products and carbon dioxide particularly by diffusion. The capillary network and the venules (venulae) in the placenta converge into one umbilical vein (v. umbilicalis) which brings back the refreshed blood to the fetus.

The umbilical cord contains one vein and two arteries which lie spiralized around the vein (Fig. 3.2.12b). The vein is protected against blockage because of pressure by a vast package of firm epithelium. The umbilical cord is coated by amnion. After entering the body of the fetus the umbilical vein goes immediately in the direction of the liver. A small amount of blood takes care of the liver tissues. Most blood goes past the liver via a direct tube; the *ductus venosa*, and flows into the inferior vena cava. Here the 'refreshed' blood is mixed with 'polluted' venous blood. This mixed blood flows into the right half of the heart along with the venous blood out of the superior vena cava. The lungs of the fetus have not yet been unfolded and breathing is obviously impossible. The blood vessels in the lungs are barely penetrable to blood. In order to make a flow of blood possible there are two connections present between the right lung circulation and the left body circulation. From the right atrium the blood can flow into the left atrium, via an opening (the foramen ovale) in the atrium septum (Fig. 3.2.12a). Because of this blood pressure an epithelial valve is pushed open. This epithelial valve can close the foramen ovale on the side of the left atrium.

The second connection between lung and body circulation is the connecting tube between the pulmonary trunk and the aorta: the ductus arteriosus. The pulmonary arteries will only receive a small quantity of blood. Figure 3.2.12a shows that the direction of the blood flow in its differing compositions is relatively favourable. The blood which is richest in oxygen and nutrients flows from the inferior vena cava, via the foramen ovale, the left ventricle and the aorta to the heart muscle itself and to the brain. These are exactly the organs which need a good supply of oxygen for their function and development. It must also be pointed out that fetal blood is relatively rich in erythrocytes and the haemoglobin level is high (on an average: 13.2 mmol/litre). Thus the increased capacity to transport oxygen offers a compensation for the lower oxygen pressure in that blood. The blood which is less rich in oxygen and nutrients from the superior vena cava is pumped via the right ventricle, the pulmonary trunk and the ductus arteriosus into the aorta. There it is mixed with the blood already present in the aorta. After that it goes to the body of the fetus or to the two umbilical arteries in order to become 'refreshed' again in the capillary network of the placenta. Thus both halves of the heart pump blood into the body circulation. The blood pressure in both ventricles is equal; the contraction power and the thickness of the heart walls are also equal.

The thymus is very important for the development of the immune system of the human being. The rudiment of this organ is 'formed' in the 5th week in the throat area of the fetus. During its growth it sinks down and down to its final destination, on top of the heart (see Fig. 2.1.53). The excretion of the thymus prevents the embryonic lymphocytes producing antibodies against *endogenous antigens*. This is important, for without this auto-immunity life itself would be impossible: the developing tissue would immediately be broken down. What actually happens is that those thymus cells which would make possible the production of antibodies against endogenous proteins are destroyed.

(2) Digestive tract

The gastrointestinal canal develops out of a straight, open tube, which in a four week embryo runs from the head to the tail area. At first this *rudimentary intestinal canal* is connected to the yolk sac via a tube at about the mid-point (Fig. 3.2.13). In the primal intestine there is a fore intestine, a mid intestine and an end intestine. Out of the *fore intestine*, 'the foregut', develops the gullet (oesophagus), the stomach and the duodenum.

Out of the *mid intestine*, 'the midgut', are formed the small intestine and about half of the large intestine, to halfway along the section that runs transversal. Out of the *end intestine*, 'the hind gut', the remaining parts of the large intestine are developing.

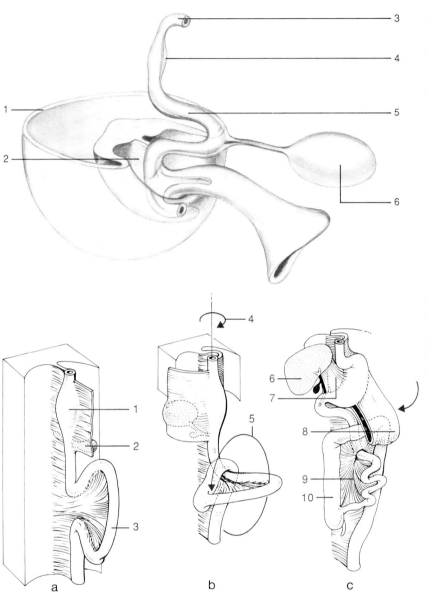

Figure 3.2.13
The development of
the gastrointestinal
canal

1. amnion
2. mesentery
3. gullet (oesophagus)
4. stomach
5. intestine
6. yolk sac

Figure 3.2.14
The development of
the peritoneum; twists
of the intestines

a. Starting point
b. Rotations
c. Further rotation and
   growth

1. stomach
2. developing liver
3. small intestine
   (a twisting of the
   bowels)
4. rotation of the
   stomach
5. rotation of the
   intestine(s)
6. liver
7. omentum minus
8. omentum majus
9. mesentery
10. ascending colon

The liver has a fast development from the 3rd week onwards out of a bulging fold of the fore intestine. In the 10th week the liver weight is about 10% of the total weight of the fetus.

The large size of the liver is related to its 'blood-forming function' of the liver, which in this stage works at full capacity. At birth the weight of the liver has decreased to 5% of the total.

The embryonic intestine is enwrapped by the peritoneum (Fig. 3.2.14). The double folded edge of the peritoneum which attaches the primal intestine rudiment to the backmost part of the abdominal wall is called the *dorsal mesen-*

tery. The part of the primal intestine rudiment which becomes the stomach is also still attached to the front wall of the abdomen with a double folded edge of the peritoneum. This *ventral mesentery* contains the rudiments of the liver and the gall bladder. Blood vessels and nerves can run to and from the intestines and other organs and tissues via the mesentery. It marks the hilus of the intraperitoneal organs.

By the 5th week there is a widening of the stomach portion of the intestine rudiment and a strong growth of the middle part, so that a loop arises in this area.

Seen from above, the primitive stomach makes a quarter turn round its longitudinal axis in a clockwise direction. After this turn the liver and the gall bladder come to lie to the right of the stomach. The loop of the intestines makes, as if it were the grip of a key, an anti-clockwise 3/4 turn. At the same time the intestines grow rapidly in length. The mesentery grows less fast so that the intestines become strongly convoluted, hampered as it were by the short mesentery folds which continue to hang the intestines on to the hindmost part of the abdominal wall. As a result of the processes described above the gastrointestinal canal and the peritoneum obtain their typical and intricate position and outline in the abdominal cavity.

(3) Urinary tract

In the development of the urinary tract a number of – partly overlapping – stages can be distinguished. In the 1st stage (2nd–4th week) there develop primitive kidney tubes (nephric tubules) per segment against the back wall of the embryonic body cavity, to the left and right of the centre. Immediately after that the successive kidney tubes grow together to become a common drainpipe (primary excretory duct) (Fig. 3.2.15). At the same time small 'branches' of the developing aorta form small bulges (glomeruli) in the wall of this drainpipe. A primitive glomerulus and the part of the drainpipe attaching to it together form an embryonic unit of excretion. The excretional units which

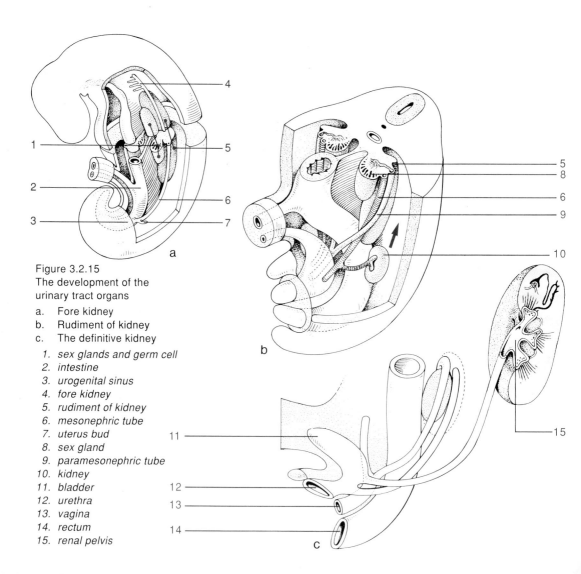

Figure 3.2.15
The development of the
urinary tract organs

a.   Fore kidney
b.   Rudiment of kidney
c.   The definitive kidney

1. sex glands and germ cell
2. intestine
3. urogenital sinus
4. fore kidney
5. rudiment of kidney
6. mesonephric tube
7. uterus bud
8. sex gland
9. paramesonephric tube
10. kidney
11. bladder
12. urethra
13. vagina
14. rectum
15. renal pelvis

are positioned highest become the *fore kidneys*. By the end of the 4th week the two fore kidneys (preliminary kidneys) have already disappeared.

During the second stage (3rd–9th week) the *primal kidneys* develop. A primal kidney derives from a second system of excretory units which develops at a place caudally of the fore kidney, and which finds a connection to the already existing discard tube (the primary excretory duct) (Fig. 3.2.15). This discard tube or *mesonephric (Wolffian) duct* flows out and widens at its lower end: the *urogenital sinus*. The end of the primal intestine is also connected to this widened hollow.

At the end of the second month the primal kidneys disappear; only the mesonephric (Wolffian) duct remains.

Meanwhile against the back of the abdominal cavity in the lumbar region, the *permanent kidneys* develop, the third stage in the developing process of the urinary tract.

At the same time the ureteric bud arises at the lower end of the mesonephric (Wolffian) duct.

This ureteric bud grows in the direction of the developing kidney tissue, establishes contact and starts to branch into the kidney tissue. In this way the renal pelvis and the renal calices are formed. At the distal side the ureter establishes contact with the developing bladder.

In the female embryo the mesonephric duct now also disappears. In the male embryo the mesonephric duct develops into the deferent duct. At the same time as the nephrons are developing within the kidney tissue, the associated blood vessels are developing, among them the glomeruli.

Finally the drainage tubes of the nephrons are connected to the renal calices; the embryonic development of the urinary tract organs is completed.

(4) Respiratory system

From the 4th week of embryonic development onward the respiratory system develops out of a bulge at the ventral side of the foregut (Fig. 3.2.16). The open connection between the developing air pipe and the developing oesoph-

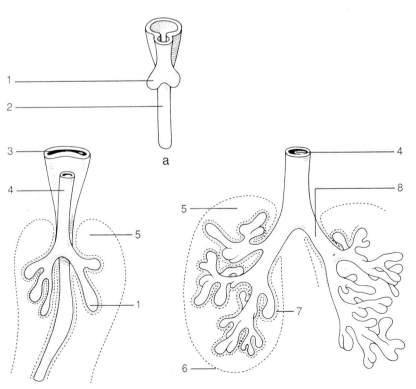

Figure 3.2.16
The development of the respiratory system

a. The development of the lung buds
b. and c. The growth of the lung buds; formation of the lung membranes

1. *lung bud*
2. *oesophagus*
3. *pharynx*
4. *trachea*
5. *pleura cavity*
6. *parietal pleura*
7. *visceral pleura*
8. *main bronchi*

agus is closed, separated by a septum. The air pipe grows and forms a bulge to the left and right, the *lung buds*. The right lung bud splits into three branches; the left lung bud into two branches. These branches will become the main bronchi.

During further development the main bronchi repeatedly branch off, as it were, at the expense of the primitive pleural cavity. The visceral pleura grows together with the lung tissue; the parietal pleura grows together with the inner coating of the chest cavity.

By the end of the 6th month the fine peripheral end tips of the smallest branches of the bronchi fold out and form the lung alveoli.

During the time that the lung tissue is developing, the associated blood vessels are formed, including the lung capillaries.

(5) The skin

The skin develops out of the ectoderm. During the first weeks, the embryonic skin consists of only one layer of epithelial cells. In the second month this epithelium starts to divide quickly and forms more layers. At the beginning of the 5th month the number of permanent epidermal layers is complete.

The sebaceous glands are relatively active before birth and cover the skin with a film of a whitish substance: the vernix caseosa. This grease protects the skin against maceration, i.e. the weakening and softening influences caused by the amniotic fluid.

(6) Hormonal system

The hormonal system originates from all of the three germ layers. The hypophysis, the epiphysis and the medualla of the suprarenal glands have developed out of the ectoderm. These producers of hormones have mainly been derived from differentiated, developing nerve tissues. The adrenal cortex is of mesodermal origin. The thyroid gland, the parathyroids and the pancreatic islets (Islets of Langerhans) are formed out of the endoderm.

(7) The nervous system

Something has already been said in section 2 of this chapter about the development of the brain

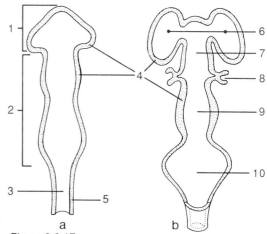

Figure 3.2.17
The development of the brain

a. 4 weeks
b. 6 weeks

1. origin of large brain(s), small brain(s) and interbrain
2. origin of the brain stem
3. the future central canal
4. developing nerve tissue
5. spinal cord, developing
6. lateral ventricles
7. future 3rd ventricle
8. optic cup
9. future aquaductus cerebri
10. future 4th ventricle

and spinal cord from the neural tube. The brain ventricles and the central canal relate to this period. In the fourth week the development of the cranial part of the neural tube into three primitive *cerebral vesicles* occurs (Fig. 3.2.17). The brain will develop out of these cerebral vesicles by an intricate process of bulgings, sharp bends and at the same time growth of the wall of the neural tube. In addition to this there are two bulgings on both sides of the foremost cerebral vesicle: the ocular vesicles, which will develop into eyes. Out of the brainstem and the spinal cord respectively will grow the cranial nerves and the spinal nerves. The spinal cord (Fig. 3.2.18), in the 3rd month of development, still has the same length as the vertebral canal. The 32 successive spinal segments are each enclosed at the same height by the 32 successive vertebral arches. After the 3rd month the vertebral column grows considerably faster than the spinal cord. This increases the distance between caudal end of the cord and the 'bottom' of the vertebral tube. The cord seems to have been

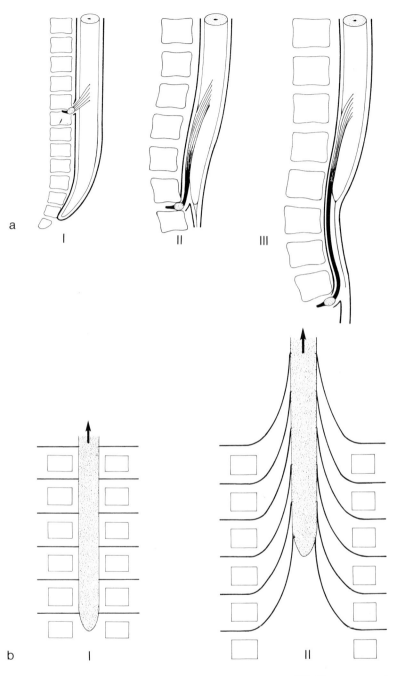

Figure 3.2.18
Ascensus

a. Side view: at first the
    spinal nerve leaves the
    vertebral tube almost
    horizontally (I).
    Because of the relative
    difference in speed of
    growth of the spinal
    column (fast) and of
    the spinal cord (slow)
    the spinal cord seems
    to rise up into the tube
    (II). When growth
    stops, there is only a
    'brush' of spinal nerves
    left in the *dura sac* (III),
    below the level of the
    cord.
b. Front view (schematic)

lifted up (ascensus). The 32 pairs of spinal
nerves which at first could leave the vertebral
tube horizontally, must now from cranial to
caudal tip descend even further before they
reach their 'own' intervertebral foramen in order to exit.

At birth the ascensus has advanced as far as
L_3/L_4; distally in the vertebral tube is only found
a 'pony-tail' of spinal nerves: the *cauda equina*.

(8) Sensory system

The sensory system, like the nervous system, is
of ectodermal origin. During the 2nd month a
rudiment of the five sense organs becomes
visible. Further discussion of their antenatal
development is beyond the scope of this book.

(9) Motor system

The fetus grows vigorously. The increase in

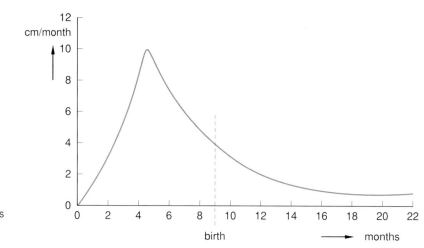

Figure 3.2.19
Curve of the speed of
growth before parturition
and during the first months
after it

Figure 3.2.20
The skull of the new-born

a. Seen from the side
b. Seen from above

1. *large fontanelle*
2. *small fontanelle*
3. *the side fontanelles*

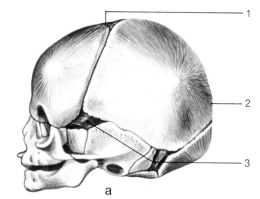

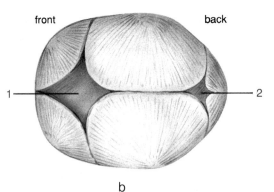

length per unit of time, the speed of growth, is considerably faster before birth than at any time after birth (Figs. 3.2.19 and 3.4.8).

In the 1st month after fertilization the beginnings of the rudiments of the skeleton are apparent. The cranial bones develop out of epithelial tissue. Nearly all the other bones derive from cartilaginous preliminary stages (hyaline cartilage). In characteristic places in the connective tissue of the cranial roof, in the course of the 6th week, the ossification-centres arise. Excentrically, from these ossification-centres, ossification of the epithelium occurs when calcium salts which have been carried by the blood are deposited. The edges of cartilage tissue between these plates of bone become smaller and smaller. At the end of the fetal stage ossification of the cranium has not yet been completed. So there are still cartilage-like cranial *sutures*, seams between two bony plates and fontanelles (between more than two bony plates) (Fig. 3.2.20). The best known fontanelles are: the *anterior fontanelle* (the large fontanelle) and the *posterior fontanelle* (the small fontanelle). In the newborn the large fontanelle can be felt as a soft diamond-shaped shallow, in the middle of the front part of the head. The small fontanelle is a similar smaller triangular shallowness in the middle of the back of the head. The suture which runs over the middle of the head from the front to the back of the neck is called the *sagittal suture*. The suture between the coronal and parietal bones is called the *coronal suture*.

The strong connective tissue of cranium sutures and fontanelles offers excellent protection to the brain tissue, but at the same time it grants to the

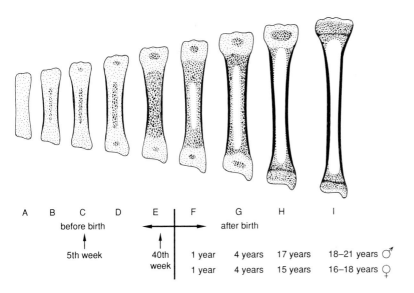

Figure 3.2.21
The development of a long
bone

A	B	C	D	E	F	G	H	I

before birth   ←———→   after birth

5th week   40th week   1 year   4 years   17 years   18–21 years ♂

1 year   4 years   15 years   16–18 years ♀

skull some degree of flexibility, which facilitates its passage through the relatively narrow birth channel.

The cartilaginous skeleton has a long period of development in which a general pattern is followed (Fig. 3.2.21). Stage one: in the centre of the bony pieces of the cartilages a central ossification centre arises and at the same time a central sleeve (*diaphysis*) of bony tissue is deposited by what later will be called the *periosteum*. The ossification round the ossification centre expands towards the ends (*epiphyses*); the periosteum continually deposits new layers of bony tissue against the sides. The thickness of the bone is kept under control, by bone-consuming cells (osteoclasts) which take away bony tissue on the inside of the bone. As a result of the activity of these osteoclasts a cavity develops, a hollow for marrow in the centre of the bone. The ends of the bone grow. In these ends – for each and every bone at a characteristic moment – appear ossification centres. In what later will become the largest bones this occurs before birth, in the smaller bones not until a few months or even 14 or 15 years after birth.

At birth the human skeleton still contains a relatively large amount of cartilage and connective tissue. This makes it much more flexible and elastic than the mature skeleton. Thus fractures will not occur so easily, but on the other hand the mechanical burden it can stand is much smaller.

The muscular system develops out of the *myotomes*, mentioned earlier. The formation of the nerves which innervate the muscles keep in pace with the development of the muscles.

(10) Reproductive system

The primary stage of the reproductive cells arises in the wall of the yolk sac (Fig. 3.2.22a), and migrates to the developing genital glands. They arrive here at the end of the 4th week of embryonic life and develop into germ cells. During the first six weeks, male embryos and female embryos look alike. The primitive gonads (genital glands) and the genital organs in their primary stages show no differences. Only in the 7th week do typical male or female genital characteristics begin to show.

The ovaries develop retroperitoneally in the loin area and medially of the rudimentary kidneys. As a result of faster growth of the neighbouring structures, the position of the ovaries becomes lower than that of the kidneys, viz. in the small pelvis and near the developing uterine tubes and uterus. The uterine tubes grow from the *paramesonephric (Müllerian) ducts*. The paramesonephric duct is a tube which is formed next to the mesonephric duct (Fig. 3.2.22b). At the lower end the left and right paramesonephric ducts grow together and so the uterus and the upper parts of the vagina also develop out of the paramesonephric tube.

Figure 3.2.22
The development of the
reproductive system

a. Migration of the primary
    reproductive cells

    1. head
    2. primary reproductive
        cells
    3. yolk sac
    4. tail
    5. reproductive gland
    6. primary kidney
    7. mesonephric duct
    8. intestine
    9. ureter bud

b. Undifferentiated
    reproductive system

    1. gonad
    2. paramesonephric
        duct
    3. mesonephric duct
    4. preliminary stage
        of the greater
        vestibular /
        bulbo-urethral
        gland
    5. primary kidney
    6. urogenital sinus

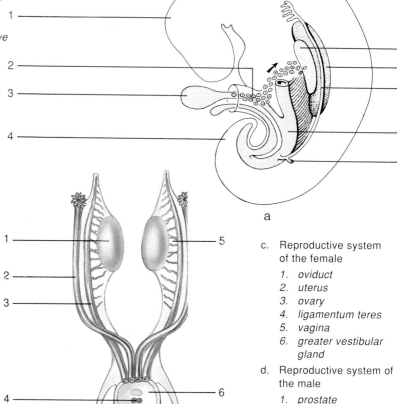

a

c. Reproductive system
    of the female

    1. oviduct
    2. uterus
    3. ovary
    4. ligamentum teres
    5. vagina
    6. greater vestibular
        gland

d. Reproductive system of
    the male

    1. prostate
    2. bulbo-urethral
        gland
    3. deferent duct
    4. testis

b

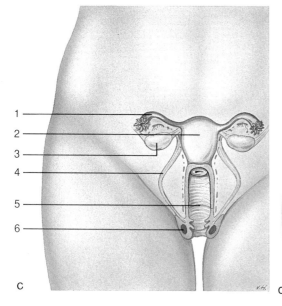

c

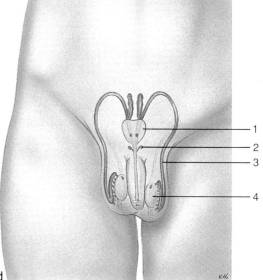

d

The origin of the testes lies in a similar place to that indicated for the ovaries. The descending of the testes during their development is a striking occurrence. From the 7th week onward till the beginning of the 14th week the testes descend down between the peritoneum and the abdominal wall (Fig. 3.2.23), towards the inguinal canal (canalis inguinalis). Here the testes remain from the beginning of the 3rd month until about the 7th month. Then the descent is completed: through the inguinal canal, across the rim of the os pubis into the scrotum. After that the entrance to the inguinal canal will be virtually closed by growths of tissue, leaving room only for the deferent duct (grown out of the mesonephric duct) and for some blood vessels which have grown in length with the descending testes.

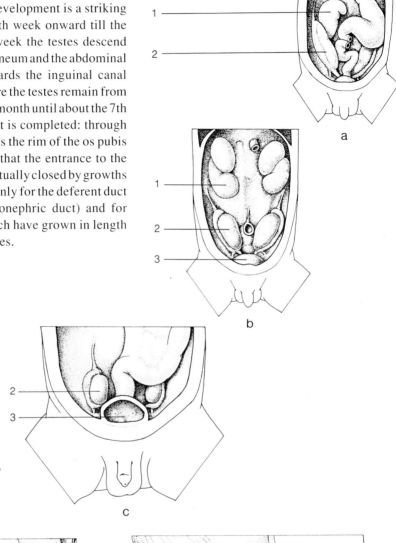

Figure 3.2.23
The descending of
the testes
a. the 3rd month
b. the 5th month
c. the 6th month
d. the position during
    the 8th month
e. the position of the
    testes in a mature
    male

1. kidney
2. testis
3. bladder
4. outward opening of the
    inguinal canal
5. spermatic cord
6. scrotum

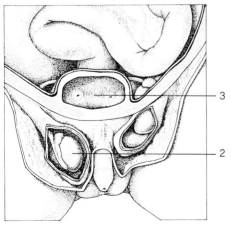

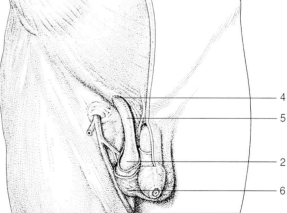

3

Pregnancy, delivery and birth

Introduction

With the fusion of an egg cell and a sperm cell to a zygote the basis of a new life has been established. Before this new life actually enters our world, there is a period called pregnancy with a duration of about 9 months. At the end of this period parturition takes places and the baby is born. This chapter will briefly describe the processes of pregnancy, delivery and birth.

Learning outcomes

After studying this chapter you should have sufficient knowledge and understanding of:
- the changes in the individual body systems during pregnancy;
- the process of delivery: cervical dilatation, expulsion and afterbirth;
- the process of birth: longitudinal turns, transition to life outside the mother;
- the experience of childbirth from the point of view of the woman.

1. Pregnancy

It is possible to refer to pregnancy some 6 days after fertilization when the blastocyst establishes contact with the uterine wall and starts to embed itself, in a process called *nidation* (Fig. 3.2.2). In many cases the woman will not yet have noticed that she is pregnant. The first clear sign indicating pregnancy is the continued absence of menstruation. This absence of menstruation is caused by the corpus luteum remaining intact, due to the effect of a very important pregnancy-hormone: the *human chorion gonadotrophin* (HCG). This hormone is produced by blastocyst cells which are in contact with the uterine wall. The HCG-hormone then takes over the function of the luteinizing hormone (LH), assisted in this task by the leuteotrophic hormone (LTH). The corpus luteum continues to produce oestrogen and progesterone. This not only keeps the endometrium in the secretion phase, but also prevents a new ovulation from taking place. In just over 3 months (13 weeks) the production of HCG stops in this organ, but by then the placenta – now fully developed – has already taken over the production of important pregnancy-protecting hormones.

1.1. Placenta

The development of the placenta begins in the second week of pregnancy. Hollow spaces (lacunae) are being formed. In the layer of trophoblast cells (see Fig. 3.2.2) is the embryo, and blood vessels in the uterine wall come into contact with these lacunae. The lacunae therefore fill with blood from the mother. Fingerlike projections, the *chorionvilli*, evolve from the trophoblast layer. With these the placenta becomes firmly anchored into the uterine wall. Blood vessels start to grow out from these villi. The maternal part of the placenta remains separated from the fetal part by means of cell membranes. Though the distance between the

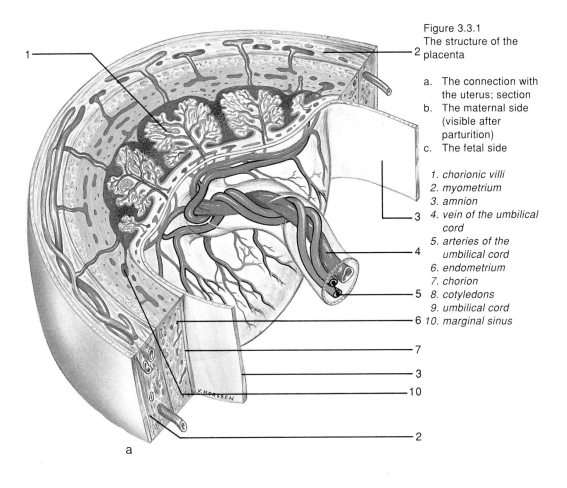

Figure 3.3.1
The structure of the placenta

a. The connection with the uterus; section
b. The maternal side (visible after parturition)
c. The fetal side

1. chorionic villi
2. myometrium
3. amnion
4. vein of the umbilical cord
5. arteries of the umbilical cord
6. endometrium
7. chorion
8. cotyledons
9. umbilical cord
10. marginal sinus

a

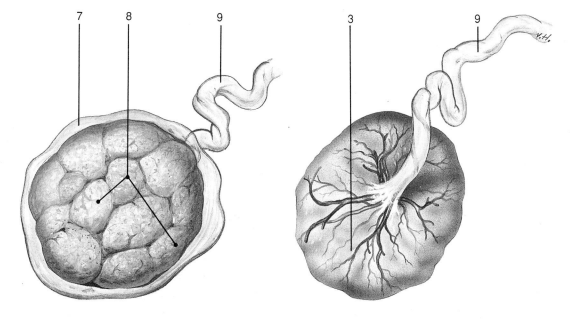

b   c

maternal blood in the placental blood and the fetal blood in the chorionvilli is very small, no mixing of blood takes place.

The chorionvilli provide a very extensive diffusion surface with the maternal blood.

The placenta is divided by separating walls (septa) into 15–20 flow-units, the *cotyledons*. Each of these cotyledons receives blood through a branch of one of the two umbilical cord arteries (arterii umbilicales). After passing through the dense capillary network (on the fetal side in the placenta) the blood is carried away again in veins which flow out into the vein of the umbilical cord (v. umbilicalis). This vein takes the blood, now rich in nutrients and oxygen and moderately well cleansed of waste products, to the fetus.

On the maternal side of the placenta a large number of spiral arteries supply blood, which is rich in nutrients and oxygen, to the diffusion-surface. Despatching the maternal blood out of the placenta occurs via large vessels, the *marginal sinus* (Fig. 3.3.1).

The placenta grows steadily in accordance with the growth of the fetus. By the end of the pregnancy the placenta weighs about 600 grams. At that moment it measures around 20 cm across and is about 2 cm thick. On the fetal side the placenta is covered with amnion. This is very, very smooth to the touch. When the placenta (or 'afterbirth') has been delivered it often turns out that the maternal side is hard, irregular and rough to the touch as a result of calcium deposits.

The average length of the umbilical cord is 60 cm and its diameter about 1.5 cm. Usually it is attached to the middle of the placenta. The umbilical cord gives the fetus a considerable freedom of movement within the uterine cavity.

In addition to its *transport function* the placenta is also important as a *producer of hormones*. The hormones produced are:
– *h*uman *c*horion *g*onadotrophin (HCG); this hormone is very much related to the LH-hormone; it maintains the corpus luteum;
– *h*uman *c*horion *s*omatomammotrophin

(HCS); the effects of this hormone are an increase of the glucose level in the blood and of the amount of free fatty acids, and a stimulation of the development of lacteal gland tissue;
– progesterone and oestrogen.

1.2. Systems

Pregnancy is a state which brings about enormous physical changes for the woman. In almost all of the 10 body systems adaptations take place which are related to the development of this new individual in the mother.

(1) Circulatory system

In the circulatory system numerous adaptations take place. The heart/minute volume (HMV) rises during the course of the pregnancy by 25–50%. This is not solely a result of an increase in the heart frequency (f_H) around 15 beats per minute, but also a result of an increase in the volume per heartbeat (V_{beat}).

The major part of the increase in the heart/minute volume benefits the bloodflow to the placenta, the kidneys and the skin. This last item means that many pregnant women notice that they feel less cold than before the pregnancy.

The *blood volume* also increases by about 20–30%. The plasma component increases more than the cell component. This is one of the reasons why generally there is a lower haemoglobin level during pregnancy.

There is an increase in the *tissue fluid* of around 15%. This is related to a decrease in the colloid-osmotic value (COV) as a result of which reabsorption has been diminished. Since in the last months of pregnancy the venous flow of blood back to the heart from the legs is hampered by the pressure of the pregnant uterus on the abdominal vessels, some pregnant women show a specific accumulation of serous fluid (oedema) in the lower limbs during this period.

Blood pressure changes little. Systolic pressure remains the same in principle; diastolic pressure often drops in the second trimester by 10–15 mmHg (1.3–2.0 kPA) and after that it rises again to the original level.

(2) Metabolism

The body metabolism is also subject to change. It goes without saying that the need for nutrients has increased. The *bodyweight* increases by an average of 10–15 kg.

Table 3.3.I gives an impression of the distribution of this extra weight at the end of pregnancy.

fetus	3500 g
placenta, umbilical cord and membranes	700 g
amniotic fluid	800 g
uterus	1000 g
blood volume	1200 g
breasts	400 g
fat in store	3000 g
tissue liquid	2000 g
Total	12,600 g

Table 3.3.I

Another *metabolism*-related feature is that the level of blood sugar is generally somewhat higher. This is connected to the production of the hormone HCS in the placenta.

As there are many minerals going to the fetus, there is a greater need for iron and calcium. Moreover, more vitamins are utilized. A normal diet can easily supply these substances.

Many pregnant women have *pregnancy eating-whims*, which means that at certain moments a strong preference arises for – often very special, unorthodox – foodstuffs. The opposite, a horror of certain dishes, also occurs.

About half of all pregnant women suffer from *nausea,* to different degrees, during the first three months, sometimes accompanied by *vomiting.* This is related to changes in the hormone levels.

As a result of increased secretion of the hormone progesterone, the intestinal peristalsis may be slowed down. This leads to a thickening of the faeces in the large intestine. A result of this can be *constipation.*

(3) Urinary system

During pregnancy considerable demands are made on the urinary system. Both the supply of blood and the filtration processes in the *nephrons* are intensified. Slightly more salt is retained, and the *threshold value* for glucose is often exceeded, producing a mild case of glucosuria. A higher level of amino acids in the urine is often also found. The amount of urine being produced remains the same.

There is an influence on the bladder function of the increased volume (and weight) of the pregnant uterus, especially during the last months of the pregnancy. The bladder comes under pressure. There is an urge to evacuate the bladder far more often, or difficulty in containing the urine (incontinence).

(4) Respiratory system

The respiratory system ensures that maternal blood is supplied with considerably more oxygen. *Oxygen consumption* during pregnancy rises by 20–30%. In order to achieve this the breath/minute–volume (BMV) is increased by 40% by increasing the volume per breath of air (V_b, volume per breath). The respiratory rate generally remains the same. Increasing the breath/minute-volume has little effect on the P_aO_2, but a much greater effect on the P_aCO_2. This is lowered by 25% from about 40 mmHg (5.3 kPa) to about 30 mmHg (4 kPa).

(5) Skin

The skin will show a number of phenomena indicating pregnancy. After six weeks of pregnancy the breasts start to develop further under the influence of pregnancy hormones. Glandular tissue grows, vascularization is intensified, fluid is accumulated. The breasts often feel unpleasantly tense, and often a strengthened vein-pattern on the surfaces of the breasts is seen. The areola and the nipple become darker in hue because of a stronger pigment production, and the tubercles in the areola become bigger. From the 20th week of pregnancy onward it is possible, from time to time, for small amounts of *foremilk (colostrum)* to be excreted from the nipples.

Because of continued stretching of the skin tissue in places and under the influence of hormones we can often see *pregnancy lines* (i.e. *striae*) appear on the breasts, but more commonly on the abdominal area (Fig. 3.3.2). They are little fissures in the connective tissue.

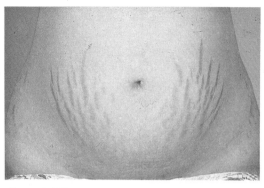

Figure 3.3.2
Pregnancy stripes (striae)

Fresh striae are blueish-red in colour; as time passes they fade.

Many pregnant women show an increased *skin pigmentation* in certain areas. Strengthened pigmentation in the face is known as *chloasma* (Fig. 3.3.3). Other zones affected are: the vulva, the line below the navel between the abdominal

Figure 3.3.3
Chloasma

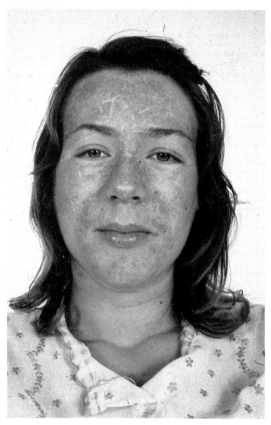

muscles and – as stated above – the nipples and the areolas.

A considerable amount of fatty tissue is deposited in the subcutaneous connective tissue. The sebaceous and perspiration glands are often far more active. This becomes apparent by – among other things – the thicker hair on the skull, and the body odour, which is more noticeable than usual.

(6) Hormonal system

It goes almost without saying that during pregnancy there are considerable changes in the hormonal system. The normal hormone cycle is disrupted. The most important pregnancy hormones are: HCG, HCS, progesterone and oestrogen.

The hypophysis produces more *melanocyte-stimulating hormone* (MSH), which affects the skin pigmentation. The hormone LTH is also produced. It is concerned with the maintenance of and secretion by the corpus luteum (LH = luteotrophic hormone) and it also stimulates milk production (LTH = lactotrophic hormone or prolactin). The thyroid gland produces more thyroxine, which considerably enhances metabolism in the cells.

The adrenal glands show an increased production of cortisol and the pancreatic islets secrete more insulin than normal.

(7) The nervous system

In the nervous system there are no changes which can be measured. However, the influence of pregnancy on the psychological condition of the woman is unmistakeable.

(8) Sensory system

The sensory system, too, seems to be unperturbed by a pregnancy. Though it is striking that often the senses of smell and taste produce new impressions. This frequently accompanies the pregnancy food whims discussed earlier.

(9) Motor system

The motor system of the woman is substantially affected by pregnancy. Her posture changes, which often gives rise to pains in the back. Connective tissue and cartilage are subject to softening. In delivering the baby this can be an advantage, especially in the pelvic connections. On the other hand it may also result in a degree of instability in the joints. In the case of the sacro-iliac articulations this may lead to pains radiating towards the lower abdomen and the legs. Many pregnant women suffer regularly from muscular spasms, especially in the sural muscles and usually at night.

(10) Reproductive system

The reproductive system, of course, is the one which is most directly involved with these changes. Usually the first sign of pregnancy is the absence of menstruation, *amenorrhoea*.

It is customary to count the period of pregnancy from the first day of the last menstruation. It might be more logical to take fertilization as the starting point, but this point in time is not used since it cannot be exactly pinpointed. The date at which it is presumed birth will take place – *the full term date* – is 40 weeks after the first day of the last menstruation. As a rule of thumb the following formula can be used:

full term date is:

      the first day of the last menstruation
      + 7 days
      + 9 months

If, for example, the first day of the last menstruation was 7th April 1992, then the anticipated full term date is 14th January 1993.

Already rather early in pregnancy, because of an increase in the blood supply, there is a change of colour in the labia minora, the wall of the vagina and the portio. The colour changes from the normal pale pink to a bluish red.

In the neck of the uterus a plug of mucus has developed, known as the *cervical plug*. This has a slight antibacterial effect.

The uterus gradually increases in size and weight. Normally the uterus is 7–8 cm long, about 5 cm wide and weighs around 50 grams. At the end of the pregnancy its length is more than 30 cm, the greatest width about 20 cm and it weighs around 1000 grams. During the first 16

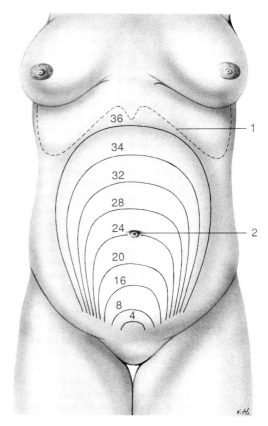

Figure 3.3.4
The position of the upper rim (fundus) of the uterus at different points in time (counted in weeks) during the pregnancy

1. *costal arch (thorax)*
2. *navel*

weeks of pregnancy the increase in volume of the uterus equals, and is caused by, the growth of the uterus itself. After that its size increases through the growth of the contents of the uterus (fetus, placenta, amniotic fluid).

The position of the uterus provides us with an indication of how far the pregnancy has progressed (Fig. 3.3.4).

After 12 weeks of pregnancy the upper rim (fundus) of the uterus can be felt quite distinctly above the symphysis.

In the 16th week the fundus has progressed to halfway between the symphysis and navel. In the 24th week the navel has been reached; in the 32nd week the halfway point between the navel and the lower costal arch. In the 36th week the

Figure 3.3.5
Sagittal cross-section and
side view of the pregnant
woman

a. 14 weeks pregnant
b. 23 weeks
c. 40 weeks

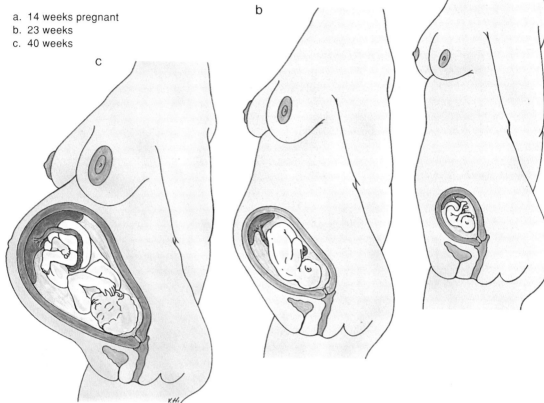

lower costal arch is reached. In the following weeks the *fundus uteri* could come to rest lower again as the head of the unborn child is already sinking into the small pelvis in a movement known as *descent*.

The growth of the fetus not only causes a steady abdominal growth of the mother, but also a steadily increasing pressure on the intestines of the mother (Fig. 3.3.5). Usually this does not cause too much trouble.

1.3. Multiple births

We speak of a *multiple pregnancy* when there is more than one fetus in the uterus. Twins occur in 1 out of 80 pregnancies, a triplet in 1 out of 7700 pregnancies.

Multiples can be uniovular (monozygotic), binovular (dizygotic) or multiovular (polyzygotic).

Uniovular (identical) *twins* result from the fertilization (resulting in the zygote) of one egg, which then splits up into two zygotes; and the two daughter cells start to develop, each separately. (This could be described as asexual reproduction or natural 'cloning'.) The two daughter zygotes are exact copies of the mother zygote. They are genetically identical and thus always of the same sex, the same blood group and have tissues of identical antigenic potencies. Uniovular twins also have the same protein specificity.

Binovular (dizygotic, fraternal) twins arise when a double ovulation has taken place and both egg cells (ova) are fertilized. This is a case of two genetically different zygotes. It goes without saying that these can have, for example, different sexes. The ratio between uniovular and binovular twins is about 1:2.

1.4. Position of the child

During the first months of the pregnancy the fetus has considerable freedom of movement.

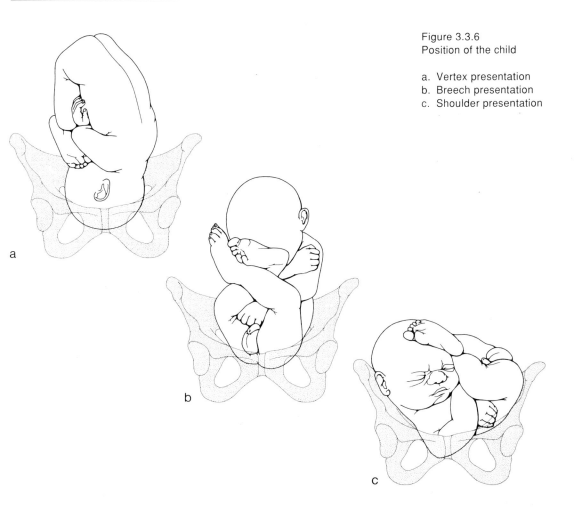

Figure 3.3.6
Position of the child

a. Vertex presentation
b. Breech presentation
c. Shoulder presentation

a

b

c

There is a comparatively large amount of amniotic fluid and the umbilical cord is long enough for turns and somersaults.

After 32 weeks the fetus has grown to such an extent that it almost completely fills the uterus. Then there is comparatively far less amniotic fluid. The fetus is allowed much less freedom of movement and is in a more or less fixed position. The description of the position of the child at the end of the pregnancy relates to the longitudinal axis of the uterus (Fig. 3.3.6).

Longitudinal and *transverse* positions can be distinguished.

Further description is based on that part of the baby which presents itself at the entrance to the pelvis.

The *vertex* presentation occurs most frequently (96%).

The *breech* presentation occurs in 3.5% of the cases.

Shoulder presentations (back, abdomen, side position) and oblique positions only occur in 0.5% of the cases.

In every type of position varieties occur, a fact which is important in obstetrics. The *occipital position* is the commonest. This position is promoted by the structure of the uterus and of the pelvic entrance in combination with the shape of the fetus. The oval shape of the head in occipital position fits very well in the pelvic entrance. When discussing delivery and parturition this type of position will be singled out as it is the most common.

1.5. Birth canal

The *birth canal* or *parturition canal* consists of the *bony pelvis* and some soft parts.

The bony pelvis (Fig. 3.3.7) has a *pelvic entrance* and a *pelvic exit*. The pelvic entrance,

Figure 3.3.7
The birth canal; bony pelvis

a. Seen from above: the entrance to the pelvis is a transverse oval
b. Seen from below: the exit of the pelvis is a longitudinal oval
c. Median cross-section: the axes of pelvic entrance and pelvic exit

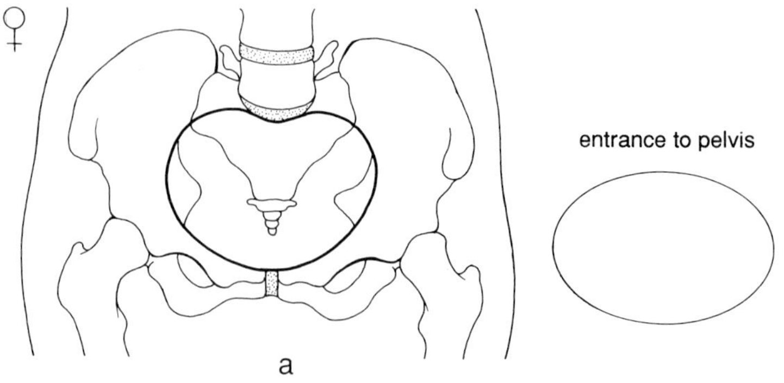

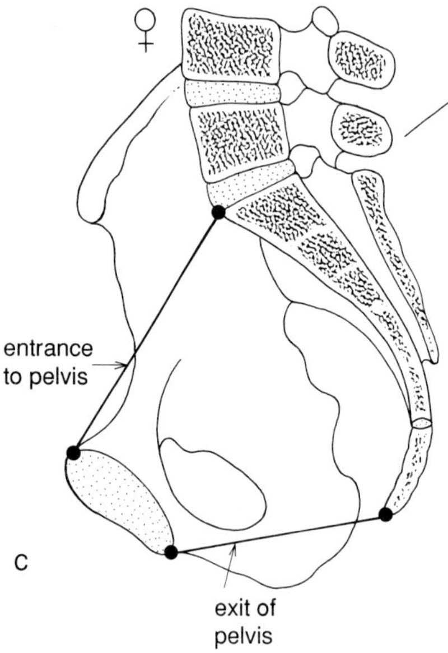

entrance to pelvis

a

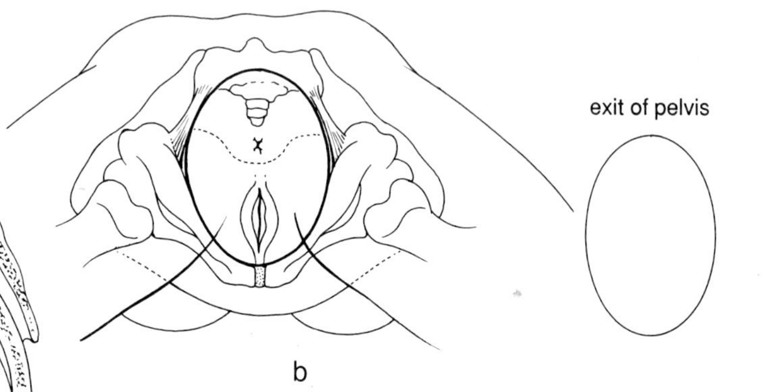

exit of pelvis

entrance to pelvis

c

exit of pelvis

b

fetus can clearly be felt externally before the descent. After descending the skull lies behind the symphysis and is barely palpable any more.

2. Parturition

Parturition or *giving birth* is the impressive event of a new human being being brought into this world. The physical changes accompanying birth are tremendous.

The mechanism initiating parturition is not yet fully understood. It is probable that hormones play a leading part. Throughout pregnancy there are contractions in the myometrium of the uterus, which are very weak and pass unnoticed. The uterus is, as it were, practising, exercising, training. From the 6th month onward the mother sometimes notices these exercise contractions and she speaks of 'practising labour'. The high concentration of progesterone probably limits the contractions.

which marks the transition of the large pelvis to the small pelvis, is an oval-shaped opening in a transverse plane. The pelvic exit (between the lower borders of the pelvic bones, the sacrotuberal ligament and the coccyx) is an oval opening in a longitudinal direction.

The directions of the two oval openings in the pelvis therefore differ by exactly $90°$ in relation to each other. This is important for the way in which it is easiest for the baby to pass through the birth canal.

During delivery, the soft parts (neck of the uterus or cervix, pelvic floor and vagina) are transformed into a tube which lines the bony parts.

Usually the descent into the small pelvis will have taken place in the 36th week. The skull of the fetus – oval in shape – presents itself in the transverse oval pelvic entrance. The skull of the

The start of the actual delivery is accompanied by a dramatic drop in the progesterone level. At the same time there is an increase in the neurohypophysis hormone oxytocin. Oxytocin

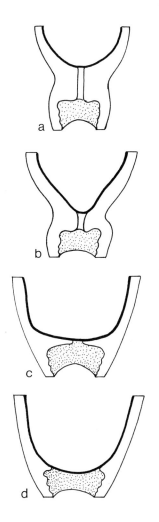

Figure 3.3.8
Dilatation

vehemently stimulates the uterine contractions. The woman's labour has started.

Three stages in parturition can be identified: *dilatation, expulsion* and the *afterbirth*.

Dilatation

The phase of dilatation is the phase in which the neck of the uterus, the pelvic floor and the vagina are transformed into a wide tube in line with the direction of the uterus (Fig. 3.3.8). Up to the beginning of dilatation the neck of the uterus and pelvic floor functioned as a protective door. Dilatation is caused by contractions of the uterus, the labour pains.

At first these *contraction*s are hardly perceivable. Gradually they become stronger and stronger. They follow one another after intervals of about 20–30 minutes. Then the frequency increases: after some time the intervals

are as short as 2–3 minutes. The time each contraction lasts becomes longer: they can continue for about a minute. Pushing during dilatation contractions is unnecessary, even disadvantageous. For pushing could cause a swelling of the tissues of the neck of the uterus, which makes dilatation more difficult.

The head of the unborn baby lies against the entrance of the neck of the uterus and gradually pushes open the passageway. Dilatation is stimulated by the amniotic fluid-pouch. This is a wedge-shaped bulging of the amniotic membranes in front of the head of the unborn baby. This pouch pushes open – softly but firmly – the neck of the uterus, the pelvic floor and the vagina. Step by step the dilatation progesses. A telling indication of this is the loss of the cervical plug. This mucus clot, together with some blood, is shed out of the mouth of the uterus.

After that the wall of the neck of the uterus becomes stretched out thinner and thinner, and so the diameter of the soft birth canal increases all the time. By the end of the dilatation, but often somewhat earlier, the amniotic *membranes break* and part of the amniotic fluid flows out. After this the contractions follow

Figure 3.3.9
The uterus in the expulsion-phase

1. oviduct
2. round ligament of uterus
3. inguinal ligament
4. symphysis

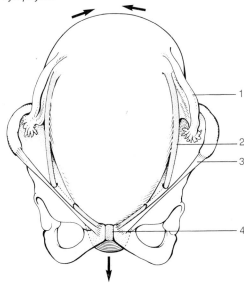

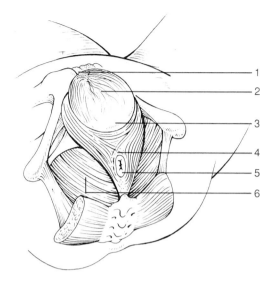

Figure 3.3.10
Parturition; the
'piercing through' of the
skull

1. symphysis
2. head of the baby
3. front
4. perineum
5. anus
6. levator ani muscle

each other even faster. *Complete dilatation* (CD) occurs when the outward opening of the birth canal has been widened to around 10 cm. The length of the dilatation phase differs considerably. In the case of women who are having their first baby dilatation takes 14–16 hours on average. At the second and following parturitions dilatation usually lasts considerably less, an average of 6–8 hours.

Expulsion
In this phase the baby is expelled out through the birth canal into the outer world.

When full dilatation has been reached the contractions follow one another even more quickly and become stronger. The woman giving birth often experiences a forceful *urge to push*. The expulsion is stimulated by pushing along with the rhythm of the contractions.
In this way the strong uterine contractions (Fig. 3.3.9) are combined with stronger pressure from the abdominal cavity on to the uterus.

Now the head of the baby quickly becomes visible; the skull appears. Then the skull pierces through the pelvic floor (Fig. 3.3.10). At this moment the tension on the perineum increases enormously. After the delivery of the head, the shoulders and the rest of the body follow.
The umbilical cord still connecting the baby to the mother is tied off and then cut through.
The period of time which is taken up by the phase of expulsion varies. With some women – especially when it is not their first delivery – it can be a matter of only a few minutes. With others it can last up to an hour.

Afterbirth
In the *afterbirth phase* the placenta is expelled. This phase averages 20 minutes. After the delivery of the child the contractions do not stop, as the calmness and relaxation of the mother would suggest. The uterus contracts and shrinks to a size of around 15 cm. The upper border of the uterus then lies at the level of the navel (Fig. 3.3.11). The placenta comes loose from the uterine wall and slides in the direction of the birth canal. The placenta, also called the *afterbirth*, is delivered as a result of pushing by the mother (contractions of the uterus) and mostly through assistance from the midwife. The bleeding of the mucous membrane on the uterine wall is stopped by vasoconstriction and continued uterine contractions.

3. Birth

Birth (Fig. 3.3.12) describes the occurrences from the 'point of view' of the baby who enters the world. What this baby has to endure can hardly be ascertained. It does not, however, require much imagination to see that it is biologically for the baby – just as it is for the delivering mother – a very dramatic, all-encompassing occurrence that literally turns everything upside down.

Longitudinal rotation
After descent the baby lies in front of the pelvic entrance. The skull with its oval shape has found the entrance with its transverse oval shape and the chin is pressed against the chest. Thus it is

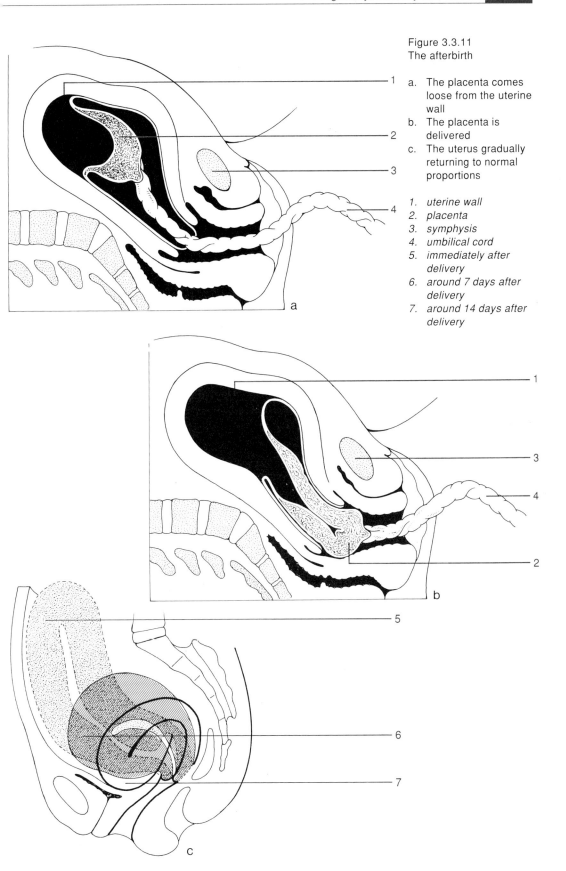

Figure 3.3.11
The afterbirth

a. The placenta comes
    loose from the uterine
    wall
b. The placenta is
    delivered
c. The uterus gradually
    returning to normal
    proportions

1. *uterine wall*
2. *placenta*
3. *symphysis*
4. *umbilical cord*
5. *immediately after
    delivery*
6. *around 7 days after
    delivery*
7. *around 14 days after
    delivery*

Figure 3.3.12
Parturition, seen from the
side. The drawings to the
right of a-h show the
position of the skull in the
birth canal at each stage

a. High position

    1.   *uterine wall*
    2.   *symphysis*
    3.   *urethra*
    4.   *vagina*
    5.   *cervix*
    6.   *rectum*
    7.   *promontorium*
    8.   *sacrum*

b. The descent starts: the
chin is laid on the chest
c. The skull lowers into the
transverse oval aperture
of the pelvic entrance
d. Dilatation is almost
complete
e. The internal axle-rotation
begins
f. The small fontanelle
turns over to the
symphysis
g. The brain skull is
delivered
h. The delivered skull
turns back to its original
position: external or
second axle-rotation

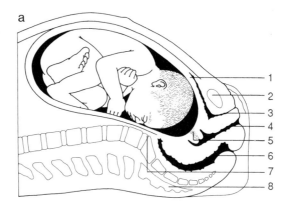
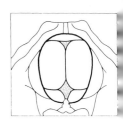
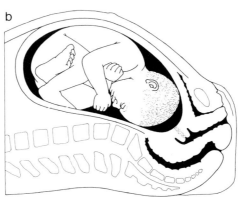
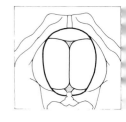
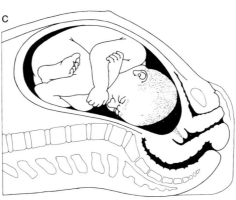
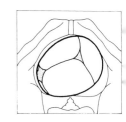
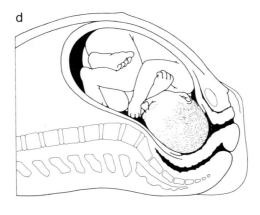
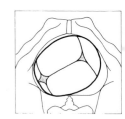

e
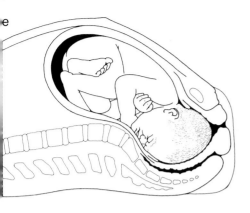
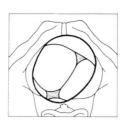

f
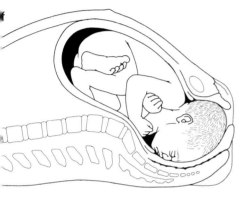
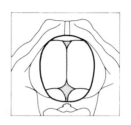

g
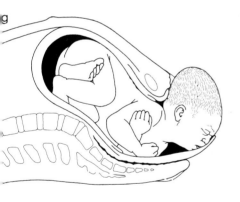

h
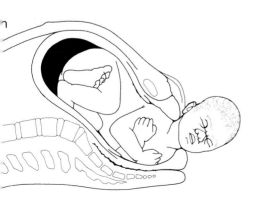
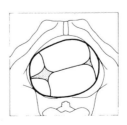

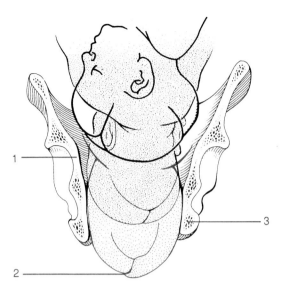

Figure 3.3.13
Moulage of the skull when
passing through the birth
canal

1. ischial tuberosity
2. moulage
3. transition: pelvic

ensured that the skull, with its smallest possible plane of cross-section, will pass through the birth canal. While dilatation is in progress the occipit passes further through the pelvic entrance and subsequently through the first part of the soft parturition canal. When the head arrives between the symphysis and the coccyx and the pelvic exit has been reached, the internal or first *axle rotation* takes place. The head makes a longitudinal rotation of 90° and in this position follows the lengthwise oval shape of the pelvic exit. The head is delivered with the face in the direction of the mother's back. While the head is being delivered the shoulders also come through the pelvic entrance transversally and subsequently make a rotation of 90° in order to pass longitudinally through the pelvic exit. At this moment the head, which has already been delivered, makes a further longitudinal rotation, returning to the position it occupied before the first or internal rotation. This is called the second or external rotation. The next step is that the shoulders are delivered one by one, and after that the rest of the body follows very quickly.

The baby's skull (Fig. 3.2.20) still has fontanelles and unfixed, unossified bone sutures.

This accounts for its remarkable flexibility (moulage) (Fig. 3.3.13), which considerably facilitates its passage through the parturition canal.

The first cry
Often when the rest of his body is still being delivered, the new person announces himself with his first cry. The incentive for this is shaped by the transition from intrauterine life to extra-uterine life. Outside the uterus it is colder, there is light, there are more and very different sounds, and there is less pressure on the thorax. All this leads to a strong stimulus to breathe, the lungs unfold and in breathing out the vocal chords can be made to vibrate for the first time.

Changes in the blood circulation
When the baby has been born the umbilical cord is tied off in two places and severed. This is absolutely painless since the umbilical cord has no pain sensors. The cutting of the umbilical cord marks the transition from fetal blood circulation to post-fetal blood circulation (Fig. 3.3.14).

By blocking the umbilical arteries (and *also* by vasoconstriction) the blood pressure in the aorta rises. At the same time the P_aCO_2 rises as carbon dioxide is transported to the placenta. This excites the respiratory centre and with the first cry the alveoli and the capillary networks involved are being unfolded. This causes a drop in the peripheral resistance in the lesser circulation, which is why there is a lower blood pressure in the right half of the heart. Now that the pressure to the left is higher than that to the right, there is a 'threat', a chance that blood will flow from the left atrium through the foramen ovale to the right atrium. This is prevented because now the connective tissue valve is pressed against the opening of the foramen ovale, so that it is closed. Within an hour of delivery the ductus arteriosus and the ductus venosus are closed by contraction of the vessel walls.

This valve grows tightly together with the atrial septum. Both the ducti and the two umbilical arteries and umbilical veins change into epithelial tissue.

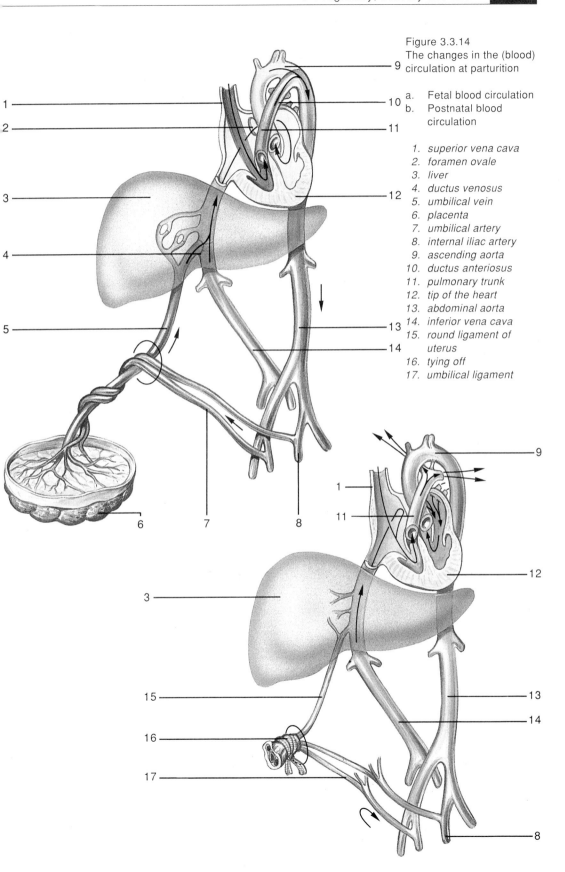

Figure 3.3.14
The changes in the (blood) circulation at parturition

a. Fetal blood circulation
b. Postnatal blood circulation

1. superior vena cava
2. foramen ovale
3. liver
4. ductus venosus
5. umbilical vein
6. placenta
7. umbilical artery
8. internal iliac artery
9. ascending aorta
10. ductus anteriosus
11. pulmonary trunk
12. tip of the heart
13. abdominal aorta
14. inferior vena cava
15. round ligament of
    uterus
16. tying off
17. umbilical ligament

Table 3.3.II
The Apgar score.
Immediately after birth
and 5 minutes later
there is a score of 0, 1
or 2 awarded to five
factors. The maximum
total score is 10

Factor	Points 0	1	2
heartbeat	absent	< 100/min	> 100/min
respiration	absent	infrequent, irregular	continued crying good/strong
muscle tone	weak	moderate	good
reflexes	absent	some motion	powerful reaction
colour	blue/pale	body pink, extremities blue	entirely pink

Apgar score

In order to obtain an idea of the condition of the newborn baby (neonate) the *Apgar score* is nowadays often recorded. This is done by noting a score of 0, 1 or 2 for some five factors, both immediately after delivery and about five minutes later, Table 3.3.II. Each time the scores are added up, the result is the Apgar score. A normal healthy newborn scores at least a 7 or 8 (points) immediately after delivery and scores 10 after 5 minutes.

4. Woman in childbirth

After parturition it will take quite some time before a complete recovery and convalescence of the woman in childbirth has been reached. This is the phase of the puerperium.
A couple of hours after giving birth the intestines have already returned to their normal position. The uterus takes about 12 days to return to practically its normal size (involution). The recuperation of the uterine mucous membrane and of the vagina takes longer. For the elasticity of the abdominal wall to be re-established is more a matter of weeks or months and depends on specific exercises carried out.

After giving birth the milk production in the breasts generally starts. This is caused by hormonal factors. After the placenta has been delivered the oestrogen and progesterone levels decrease. The negative feedback on the production of the hormone prolactin drops off. So the *prolactin* produced and present in the blood takes effect in starting the activity of the mammary glands. The glandular cells of the approximately 20 racimeform glands in the mammary tissue (see Fig. 2.5.5b), start to produce mother's milk. At first the amount is small, but normally it rapidly increases. The milk flows to the despatch tubes, which all open in the nipple. If the mother breastfeeds the baby, the suckling of the nipple causes a reflex stimulation causing prolactin production (positive feedback) and oxytocin production. The hormone oxytocin stimulates the erection of the nipple and the contraction of the smooth muscle tissue around the milk tubes which causes the milk 'to rush in'. The higher oxytocin level makes the uterus react in the form of postnatal contractions. This strengthens the involution of the uterus.

Post-natal development

4

Introduction

The life of a human being is one continuous line between birth and death. Every day of life has its ups and downs. We live our lives; each day using knowledge and skills which have been learned during the previous days. In the life-span clear periods, the phases of our life, can be identified. In each phase of life a human being reaches a characteristic level of development.

This chapter deals with development after birth. After naming the successive stages of life attention will be given to the development of all the individual organ systems during the course of life. The scope of this book makes it impossible to go into great detail. In conclusion, the book focuses on the phenomena of dying and death.

Learning outcomes

After studying this chapter you should have sufficient knowledge and understanding of:
- the different stages of life after birth;
- the development of the separate body systems after birth;
- the phenomena of dying and death.

1. Stages of life

In every human life it is possible to indicate a number of clearly marked stages of development. In some categorizations there are many stages, whereas in others the number is more limited.

A rather crude division is infant, child, mature person. The phase of infancy is characterized by almost total dependence; during the phase of childhood the way to independence can already be clearly seen. Maturity is characterized by what is called *self-determination*. In this crude division there is often a further subdivision distinguished nowadays with the following characteristic names: the newly born (neonate), the infant, the toddler, the pre-school child, the schoolchild, the adolescent and the mature person (i.e. adult).

The neonatal phase lasts about four weeks. This period is mainly characterized by adaptation to the enormous changes which have taken place during the transition from intrauterine life to extrauterine life, in other words by parturition. The infant phase lasts up to about the first birthday. Characteristically during this phase the basis is laid for establishing a feeling of trust in the world.

The toddler phase (c. 1–3 years) is dominated by the child's drive towards autonomy (or pursuit of autonomy). By wanting to do ever more things himself (or herself) independence grows and so dependence decreases.

In the pre-school phase (3–5 years), *initiative* is the central item. The child's main concerns are playing, investigating and enterprise.

The *schoolchild* (5–12 years) has a predominantly constructive attitude. He is at the same time competitive and has a sense of belonging to a group.

Adolescence is the period from about 12–20 years. This phase forms the *transition* from being a child to being an adult. Within adolescence is the phase called *puberty* (12–16 years); in this period enormous sexual development takes place. Puberty is also characterized by the search for a new identity and for intimacy.

With *maturity* a long period of physical and mental growth comes to an end. In maturity there is initially a period in which building up and deteriorating are more or less balanced. Then follows a period in which deterioration gradually becomes greater than building up. This *growing older* ends with *death*.

The development of the mental capacities continues steadily in the period of maturity. At first sight there seem to be no limits to the capacity to learn. Yet, for most people growing older often brings limitations in the capacity to learn.

2. Body systems

A human being develops as an entity. The development of the different body systems which can be distinguished, but not divided, occurs in a certain harmony. It is true, however, that in a specific phase of life the development of a certain body system is more noticeable than in a different phase. The division of the 'total entity' into distinguishable body systems in order to describe the development of the human being, has occurred in this book only in order to maintain a systematic approach to a complex subject.

(1) Circulatory system

At the moment of birth circulation experiences an enormous change. The system is extremely well prepared for its new and heavier task. The wide variation in the *heart frequency* of the neonate is remarkable. When the neonate is fast asleep the heart frequency may average 80 beats per minute. In a waking baby the heart beats some 120–150 times per minute. During any activity (being bathed, crying), a heart frequency of 180 is not unusual. In the case of the waking infant the heartbeat frequency decreases gradually over the months to about 100. During the toddler and pre-school child phase it decreases further to about 80–90 beats per minute. An average of 70–75 beats per minute when at rest is reached in adolescence. The systolic blood pressure of the neonate varies from 55–80 mmHg (7.3–10.6 kPa); the diastolic blood pressure varies less: 40–45 mmHg (at 5.3–6.0 kPa). Infant blood pressure values gradually reach 90/60 mmHg (12.0/8.0 kPa). In the toddler and pre-school child phase, blood pressure remains constant. In the schoolchild we see a gradual rise in blood pressure to an average of 110/65 mmHg (14.6/8.6 kPa). This rise is a result of the growth of the heart and thus of the volume per heartbeat. Because of a further growth of the heart during adolescence, blood pressure shows a further rise and eventually reaches the average 'mature' values of 120/80 mmHg (16.0/10.6 kPa). For people in advanced ages much higher values are normal, e.g. 170/90 mmHg (22.6/12.0 kPa).

These blood volumes during development are to a large extent related to a certain percentage of the body weight. In the newborn the volume of blood is almost 10% of the body weight. In the course of growth this percentage diminishes gradually to around 7.5% in maturity.

The composition of the constituent parts of blood is also subject to change. The development of the capacity of the blood to transport oxygen is remarkable. This capacity is at a considerably increased level before birth and also during a number of days after birth, partly because of an increased amount of erythrocytes, but mainly because of an enhanced level of haemoglobin. This increased capacity for oxygen transportation before birth is an important compensation for the relatively low oxygen supply via the placenta compared with the oxygen supply via the lungs. During the pre-school phase the composition of the blood stabilizes into the mature composition.

At parturition the red blood-producing bone-marrow is found in the marrow cavities of all bones. Between the toddler phase and the beginning of adolescence this marrow is gradually substituted in the long bones by yellow bone-marrow (fatty tissue).

For data about the role of the thymus in the body's immune systems, see the discussion of

standard value

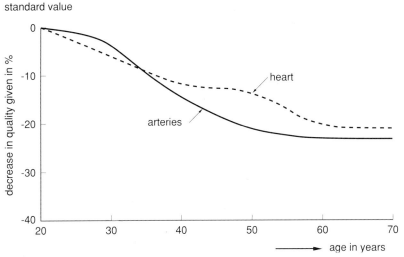

Figure 3.4.1
A rough representation of the decrease of the mechanical characteristics of the heart and the arteries. The reference point (0) has been chosen as the mechanical quality at 20 years of age

the thymus in Chapter 1 of Module 2. The development of the nasopharyngeal tonsils (adenoids) and the palatine tonsils is striking. They often continue to grow to a considerable size until well into late childhood, when they generally decrease in size again.

Degeneration of the circulatory system begins at an average age of 25. The strength of the wall of the heart and the volume per beat decrease gradually. The elasticity of the walls of the vessel also decreases (Fig. 3.4.1).

(2) Digestive system

Most neonates sleep during the first 10 or 12 hours after birth. They show no need for nutrition. During the next 48 hours the need for food gradually makes itself felt. At first the body loses on average about 10% of its weight, because of a loss of fluid and the need to draw on the stock of fat. Generally this loss of weight is quickly replenished by the end of the first week. Normally the ingestion of the necessary fluids and the required nutriment takes place solely by breast (or bottle) feeding. Breast milk contains all the substances which are necessary for the baby (Fig. 3.4.2). During the first days in which the mammary glands are active, the milk has a different composition than afterwards. This is called colostrum. Colostrum is relatively rich in proteins and it also contains many immune bodies which offer the newborn a certain protection against infection.

After around three days the production of milk (lactation) is well under way; then the milk contains mainly fat and sugar.

The enzyme *rennin* is produced in the neonate's stomach. This enzyme ensures that the milk proteins of the mother's milk are coagulated. Thus the proteins stay longer in the stomach and assist digestion.

Peristalsis of the gastrointestinal tract is still irregular and often painful (causing colic). The capacity of the stomach is only 50 ml at birth. During the first few weeks this capacity increases to 100–150 ml and further in the first year to 200–350 ml. In the toddler phase it

Figure 3.4.2
Composition of breastmilk
a. Colostrum
b. The normal mother's milk

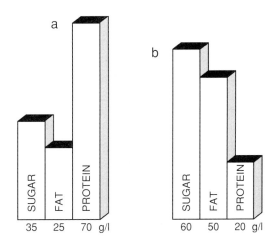

increases to 500 ml. After that the stomach volume increases slowly until puberty, when during the spurt in skeletal growth the stomach also increases considerably in size to its mature form. The relatively small size of the stomach and the greater need for food are the main factors which account for the more frequent feeding requirements of the neonate and infant.

The production of gastric acid has already reached its full strength after the fourth month and by the first birthday there are sufficient enzymes produced to be fully capable of the digestion of carbohydrates, fats and proteins. The changeover to solid food is gradually made during the first year. The length of the gastro-intestinal tract is about 3.5 metres in the neonate, which is more than enough for the processes of digesting food. The gastrointestinal tract grows steadily; in the toddler it already measures 4.5 metres and after the spurt of growth it reaches a length of just over 6 metres. Because of the increased length it is now also possible to digest a wider variety of foodstuffs.

The output of faeces by the neonate generally occurs for the first time even before he has been fed. The first evacuation of the bowels, called meconium, is dark green to pitch black in colour, paste-like and without smell. Meconium consists of thickened remains of amniotic fluid (lanugo, smegma, skin cells), digestive juices, intestinal mucus, intestinal cells and gall. As soon as the baby is fed there are gradual changes in the meconium; its composition, consistency, colour and smell. After one week of breastfeeding a yellow-gold, less solid and more watery type of faeces with a sour smell is seen.

The frequency of evacuation varies from 1 to 4 times per 24 hours.

When the transition to solid food has been made the faeces have the same characteristics as those of the mature person.

As a result of the increased stomach volume and of the slower intestinal peristalsis, the frequency of evacuation decreases, for the toddler, to the 'mature' number of 1–3 times per 24 hours. Generally in the course of the toddler phase a child becomes 'potty trained'. Becoming potty trained is the process of getting the external constriction muscle of both the anus and the bladder under voluntary control. These sphincters are striated and thus voluntary. As soon as the toddler learns how 'to use' these muscles, defecation and the voiding reflex can be suppressed. The child learns that in the case of 'an urge' he must warn others with respect to his needs. In the toddler phase the child becomes adroit enough to go to the toilet himself. Learning how to control the sphincters almost always takes place with the help of external motivation (small rewards, etc.). Till the end of the pre-school child phase it is possible, especially at night, that a spontaneous passing of urine takes place (wetting the bed). Generally this is no longer a problem with the schoolchild.

The place of the liver is remarkable in the development of the digestive tract. At birth the liver is, relatively, extremely large, weighing 5% of the total body weight. This decreases to 2.5% of the total weight in a mature person, because of the comparatively stronger development of the other organs.

Development of the teeth

At birth there are no elements of a set of teeth to be seen. Yet in the upper and lower jaw the incisors and canine teeth are already present in rudimentary form. Together with the molars which develop later these elements will form the set of milk teeth. This is a temporary set preceding the permanent set (Fig. 3.4.3a).

On average the teeth and molars of the milk teeth break through in a fixed order and at characteristic points in time (Fig. 3.4.3b). The same is true for the substitution of the milk teeth by the permanent teeth. Dental-age can be established by comparison with average values.

Milk teeth usually start to come through around the sixth month. Saliva production has been stronger since the third month, leading to frequent drooling. At the time of teething saliva production is enhanced even further. The eruption of teeth is often painful. Red cheeks are another indication of teething.

The elements of the first set of teeth are smaller and less in number than those of the permanent set. By the end of the toddler phase the first set of teeth is completed. Then 2 incisors, 1 canine

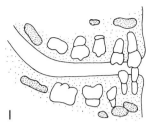

I

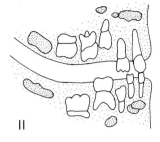

II

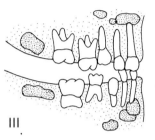

III

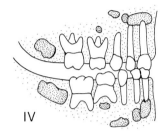

IV

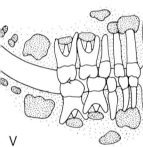

V

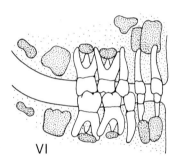

VI

a

Figure 3.4.3
Milk teeth

a. Eruption of the teeth
   I   9 months old
   II   12 months old
   III   18 months old
   IV   24 months old
   V   3 years old
   VI   4 years old

 milk

 permanent

b. The development of the teeth and molars up to around 11 years old

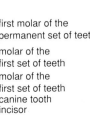

UPPER JAW

LOWER JAW

b

	ERUPTION	DROPPING OUT
incisor	7.5 months	7.5 years
incisor	9 months	8 years
canine tooth	18 months	11.5 years
molar of the first set of teeth	14 months	10.5 years
molar of the first set of teeth	24 months	10.5 years
first molar of the permanent set of teeth	6-7 years	
first molar of the permanent set of teeth	6-7 years	
molar of the first set of teeth	20 months	11 years
molar of the first set of teeth	12 months	10 years
canine tooth	16 months	9.5 years
incisor	7 months	7 years
incisor	6 months	6 years

Figure 3.4.4
The pattern of the first
set of teeth

right half of the set of teeth ◄─────► left half of the set of teeth

55	54	**53**	52	51	61	62	**63**	64	65	↑ upper jaw
85	84	**83**	82	81	71	72	**73**	74	75	↓ lower jaw

tooth and 2 molars in each half jaw are seen. The first set of teeth is also represented in a pattern (Fig. 3.4.4, compare Fig. 2.2.12).

Around the age of six the change of teeth begins. Part of the permanent elements have been present in the jaws for years in a rudimentary form. Between the ages of 12 and 16 the change is completed and by then two extra molars have appeared behind the premolars. The third molar, the wisdom-tooth, usually appears before the 26th year. In principle the name 'permanent' set of teeth is correct; providing we look after them, we can rely on them for biting and chewing until old age.

(3) Urinary system

The output of urine had already started within the uterus. By the end of pregnancy the fetus drank rather large amounts of amniotic fluid and the kidneys were already functioning quite well. The urine being produced by the neonate and infant is of low concentration. The reason for this is that the different crystalloid-osmotic values must still be adjusted in the nephrons (C.O.L/Cr.O.V./Cr.O.L.) At the end of the infant phase the urine has already become more concentrated; during the schoolchild phase the kidneys reach their functional maturity.

(4) Respiratory system

The process of respiration has reached its full-scale operational level within one minute after parturition, providing the blood with oxygen and removing carbon dioxide. Most of the alveoli will have unfolded within some two or three days. Breathing is very irregular during the first phase of life. The respiration frequency (f_R) when at rest varies from 40–70 times per minute. The volume of air per inspiration (V_I) also varies. Respiration is alternatively deep and superficial. In the infant the frequency of respiration has decreased to around 30 times per minute; the volume of air per breath remains irregular. During the toddler phase respiration starts to stabilize. Gradually the frequency of breathing decreases to about 25 times per minute and the volume of air per breath becomes regular. Due to the growth of the lungs their vital capacity (VC) increases considerably. At the age of 6 the vital capacity is on average around 1.3 litres; at the age of 12 the vital capacity has doubled to around 2.8 litres. At the same time the breathing frequency has decreased further to around 18–20 times per minute. At the beginning of adolescence the breathing frequency, the volume of air per breath and the vital capacity reach mature values.

After the age of 25 the vital capacity decreases in most people by 20–30 ml on average each year. The most important reasons for this process lie in the slow decrease of the thorax volume in more advanced age, the decline of the strength of the respiratory muscles and the diminishing elasticity of the lung tissue itself.

Learning how to control one's voice occurs in a rather short period of time, mostly before the 1st birthday. All the vowels and consonants can then be produced by the vocal chords – and in the widest combinations. Mastering the spoken language takes much longer.

The larynx gradually grows until puberty; then an acceleration in growth is seen in both sexes, but especially in boys, where the length of the vocal chords doubles within a year (from 8 to 16 mm). The result of this is known as 'his voice is breaking'.

(5) The skin

Generally the skin of the neonate still bears traces of intrauterine life (vernix and lanugo). Both substances disappear rather quickly. The skin of the newborn is soft and supple to the touch and is pink in colour. The darker colour of the nail matrixes is remarkable. Often a large number of little white dots known as milia are seen on the nose, the cheeks and the chin. These are the external openings of sebaceous glands, which are still a little swollen as a reaction to the sudden dry environment after birth. After some time they disappear.

When the umbilical cord is cut through it is usually shortened to a length of 4–5 cm. Within a few days this stump of the umbilical cord has dried up. During the second week the remnants drop off.

The changes in the ratios of body measurements during development prove to be very influential with respect to temperature regulation and the role of the skin. Figure 3.4.5 represents the body as a column, with a specific length (l), width (w) and depth (d). The surface of the column has the dimension L^2 (cm²). The volume has the dimension L^3 (cm³). When the column experiences , 'growth', the surface increases quadratically and the volume increases to the third power. The increase in volume is far greater than the surface increase.

An example may explain this.

Assume that column A (the body of a child) has the following measurements.

l = 90 cm
w = 15 cm
d = 7.5 cm

The total surface (the skin area) is:
$2 \times (l \times w) + 2 \times (l \times d) + 2 \times (w \times d) =$
$2 \times (90 \times 15) + 2 \times (90 \times 7.5) + 2 \times (15 \times 7.5) =$
$2700 + 1350 + 225 =$
$4275 \text{ cm}^2 =$
$\approx 0.43 \text{ m}^2$

The total volume (the contents) of column A is:
$l \times b \times d =$
$90 \times 15 \times 7.5 =$
$10{,}125 \text{ cm}^3 =$
$\approx 0.01 \text{ m}^3$

In order to be able to establish the influence of growth on these values we now take column B (body of the adult) in which all measurements are twice the size of column A, namely:

l = 180 cm
w = 30 cm
d = 15 cm.

w = width
d = depth
l = length

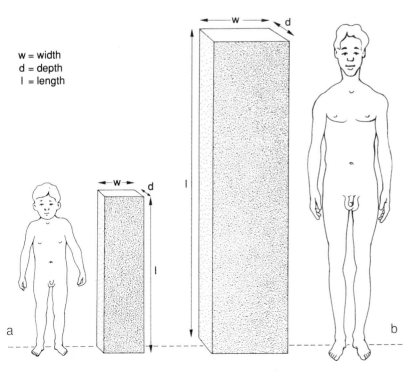

Figure 3.4.5
Comparison of the surface of the body and the volume of the body

a. In children
b. In adults

a

b

The total surface (the skin) of column B is:
$2 \times (180 \times 30) + 2 \times (180 \times 15) + 2 \times (30 \times 15) =$
$10,800 + 5400 + 900 =$
$17,100 \text{ cm}^2 =$
$\approx 1.7 \text{ m}^2$ (400% of A).

The volume (the contents) of column B is:
$180 \times 30 \times 15 =$
$81,000 \text{ cm}^3 =$
$\approx 0.08 \text{ m}^3$ (800% of A).

For column A the ratio of surface (m^2) and volume (m^3) is:
$0.43/0.01 =$
$43 \text{ m}^2/\text{m}^3$.

For column B this ratio is:
$1.7/0.08 =$
$21.25 \text{ m}^2/\text{m}^3$.

In other words: column A (the child) has more than twice the amount of surface (skin) per unit of volume (m^3) than column B (the adult).

For the neonate the value of this ratio is highest. During growth it gradually decreases.
Since regulation of temperature occurs through the skin there is a good chance of a drop in body temperature in our climate, especially in the case of neonates and infants, which is why precautions such as a warmed bed, sufficient covering and clothes are vitally important.
For the very same reason children cool down much quicker than grown-ups whilst swimming, because water is an excellent conductor of heat, and the relatively large surface is a relatively large area from which to conduct heat.

The elasticity (turgor) of the skin decreases with advancing age. The main reason for this is the decreased amount of elastin-fibres in the dermis, and also in advancing old age there is a decreasing percentage of intracellular fluid. When, for instance, you take the skin on the back of the hand of a 75 year old person in a fold, pull it up and then let go of it, it takes some time before the fold is completely smoothed out again. With children such a fold is instantly smooth again.

As far as the development of hair growth is concerned, the only phase worthy of attention is the hair growth during puberty. Under the influence of sex hormones in both sexes a strong growth of hair occurs in the genital area (Fig. 3.4.16), and in the armpits. In the case of the male stronger hair growth is also seen in the form of moustache and beard, combined with a more or less luxuriant growth of hair on the chest, arms, legs and even the back.
Getting old is generally accompanied by the hair becoming increasingly thin and grey or white. This last phenomenon is a result of the decreasing production of pigment and of little bubbles of air in the hair-shaft.

(6) Hormonal system and (7) nervous system

The regulatory systems of the newborn and the infant are generally well equipped for their tasks, though in the regulatory systems themselves stability has not yet been established. This is evident in the irregularities of, for example, the heartbeat, blood pressure, breathing, etc. During the first four months of life the baby is asleep most of the time; in the first few weeks up to 20 out of 24 hours. During these hours there are long periods of 'REM'-sleep. The periods of being awake are mainly used for drinking and changing. Feeling uncomfortable is without exception expressed by crying. This crying, however, occurs completely without tears. The tear glands (lacrimal glands) are only capable of producing sufficient tear fluid for crying after a couple of months. At first – just as it had been in the uterus – there is no trace of a diurnal rhythm. Probably this is being developed under the influence of sunlight. By the fourth month of their life most babies are sleeping through the night. In the daytime there are longer and longer periods of being awake. In the toddler phase, the daytime nap proves to be no longer necessary.
The ratio between waking and sleeping is about 50 : 50. During further development the need for sleep gradually decreases to around 6–8 hours on average.

Though the possibilities of the functioning of the nervous system increase from birth to advanced years, the number of nerve cells already starts to decrease halfway through the fetal period. This process of dying off continues steadily all through life. That this phenomenon is not an immediate disaster is apparent to all.

But there is also growth. The cerebral cortex increases in thickness after birth. The length of individual nerve fibres increases and the number of branchings and connections increases enormously. The increasing complexity of the connections is the basis for the enhanced capacity to function. At the age of 6 or 7 years, it can be assumed that establishing the connections has been completed. Thus the nerve tissue and particularly the brain tissue have completed their development relatively early compared with other tissues. From a functional point of view there is still enormous development to come.

The sensory and motor memories will be capable of storing ever-increasing amounts of information, and the regulation and controlling of all kinds of systems and functions will have an enormous increase in efficiency. Because of continued myelination (i.e. supplying with a lining of myelin) a faster conduction of impulses is achieved. Generally the sensory fibres are myelinated earlier than the motor fibres. Moreover, myelination proceeds in cranio-caudal and proximo-distal directions. This explains why the child has, successively more and more control of the movements of the head, trunk, arms and hands, legs and feet.

Finally the further ascent of the spinal cord in the vertebral canal must be mentioned (see Fig. 3.2.18). At the moment of birth this relative movement upward had proceeded till L_3/L_4; because of a slower and shorter period of growth of the spinal cord than the growth of the vertebrae etc, eventually the end-tip of the spinal cord will be at the level of L_1/L_2 in an adult. In the development of hormone production the most remarkable development is the increase in the quantities of sexual hormones at the onset of puberty.

(8) Sensory system

The quality of the sensory system of the neonate is already remarkably high. The capacity to smell is fully available and the baby recognizes the smell of the person feeding him and of the breast or bottle food. During the development to maturity the sense of smell does not improve very much further.

As far as taste is concerned the baby shows a preference for sweet and a slight dislike for the other tastes. Gradually the toddler learns how to distinguish between different flavours and some individual preferences become apparent.

The sense of touch in the skin (the sense of temperature, feel, pressure and pain) is completely present in the neonate and constitutes the most important means of contact between baby and external world. In the sensibility of the child the capacity to discriminate is greater than that of the adult. The explanation for this is that in a child the same number of sensors are placed under a far smaller surface of skin, thus more sensors per cm^2. The change in the capacity to discriminate during development can be illustrated (Fig. 3.4.6) by inflating a balloon (the skin) with dots (the sensors).

The neonate can distinguish between light and dark. An object moved to and fro at a distance of around 20 cm from the head of the neonate can be followed with the eyes. The fact that neonates direct their eyes to changes in light and dark is remarkable. Very often it is the hairline of the

Figure 3.4.6
Change of the capacity to discriminate in the course of development; the number of sensors per cm^2 decreases

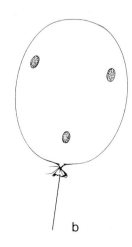

a   b

adult holding the baby which commands attention. In the course of the fourth month the capacity of accommodation of the eyes has become comparable to that of the adult, in other words the accommodation range is maximal. Within eight months the visual faculty is completely developed; the child sees depth, the differences in colour, and shapes.

From age 40, many people have to use reading glasses, because the lens cannot be made as convex as it used to be as a result of aging. The nearpoint is constantly approaching the far-point; the accommodation range diminishes.

The capacity to hear is already present in the uterus. There the child has already heard for weeks the sounds of the mother's heart and intestines. It is a well-known fact that the sounds of the heart beating can have a soothing effect on the baby. (It reminds him of the feeling of prenatal safety.) The sense of hearing of young people comprises a range of frequencies of about 20 Hz to around 20,000 Hz. As one grows older the flexibility of the basal membrane gradually diminishes. This becomes apparent mainly in a loss of capacity to hear high tones. The frequency range older people can hear has in many cases come down to a range of about 50 Hz as the lowest and around 8000 Hz as the highest boundary.

Balance is probably completely developed at an early stage; however, the movements necessary to maintain balance reach full capacity at a much slower rate.

(9) Motor system

Children are constantly growing. Trousers that are not yet worn out must be lengthened (never widened!), shoes recently bought are too small within a couple of months. For most people the development of a child is most clearly seen in the development of the motor system. On the one hand people are then concerned with changes in the measurements of the body and on the other hand with the control of the skeletal muscles: movement. The growth of the skeleton is the most important factor in the change in body measurements: height, width, depth and weight. The bones of the skeleton do not all grow at the same rate. The contribution of each

bone to the sum total of the increase in body height is therefore different during the different stages of development (Fig. 3.4.7). Young children have as yet a relatively large head and trunk compared to their limbs. As children have comparatively short legs, they usually need to trot or run in order to keep up with their parents during a walk. During puberty the arms and legs increase in (relative) length very quickly; giving the youngsters a feeling of strangeness. Hence: 'He doesn't know what to do with his arms and legs!'.

Many children measure their height from time to time: heels against the skirting board, a hand flat on the head, finger on the wall and then turn away, so that the height can be seen and then measured. Such measuring is, of course, very inaccurate and can only tell something about the height at that specific moment. In order to get a decisive picture of the growth in height an accurate measuring instrument is necessary and measuring must be repeated, for example every 6 months. The increase in height per time unit *(the growing speed)* for a cross-section of children is nowadays represented in graphic form (Fig. 3.4.8). This shows that the growth speed drops considerably from birth until about the third year. Then the growth speed remains practically the same (increase in height per annum: 5–6 cm), till an acceleration in growth, the *growth spurt* (at the age of c. 13). Whereas up to this moment the average difference in height between boys and girls of equal calendar ages has been negligible, at this moment a difference *does* arise. In girls the spurt of growth occurs between the ages of 10 and 14, and in boys it takes place between the 12th and the 16th birthday. Moreover in boys the spurt in growth lasts longer and occurs at a higher rate. Fully grown boys are therefore taller on average than fully grown girls.

In order to get a clear picture of growth, the weight and measurements are taken regularly (e.g. each month, and after the 1st birthday: every six months) and recorded as a graph and then compared to average values (Fig. 3.4.9).

In this comparison, the height is on the one hand related to the body weight in the weight-to-height diagram and on the other hand related to

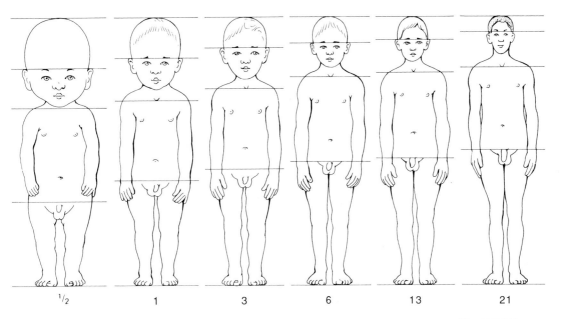

Figure 3.4.7
Proportional
growth in height.
The height of the
body has been
taken as the
standard

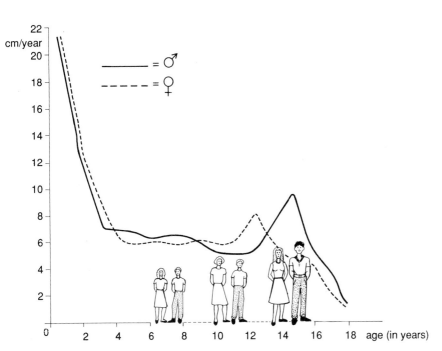

Figure 3.4.8
The average
growth rate of
today's children.
The growth spurt
begins, on
average, two
years earlier in
girls than in boys

the calendar age in the height-to-age diagram. The differences justify a distinction between sexes. Moreover there are inter-individual differences which must be considered. The variations in the different growth-curves are represented by means of percentage lines. The P50 line indicates the exact average growth. In other words 50% of children (or their measurements) can be found above this line, 50% below this line. Above the P90-line and below the P10-line only 10% of children are to be found.

Looking at the weight-to-height diagram it can be stated that those children with a weight recorded beneath the P10-line are very (maybe even *too*) thin, compared to other children of the same height.

It can also be stated that children with a weight

Figure 3.4.9
Diagrams of growth

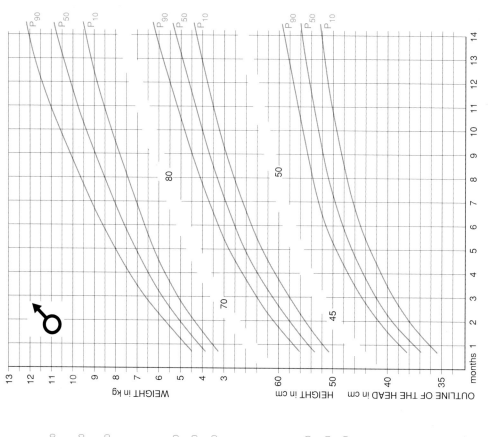

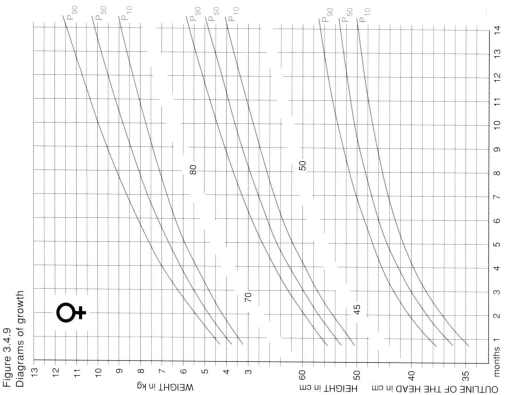

a. Weight, height and outline of the head (♀: 1–15 months)

b. Weight, height and outline of the head (♂: 1–15 months)

Figure 3.4.9 (cont.)

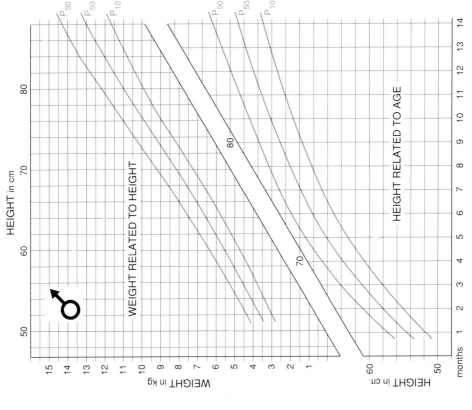

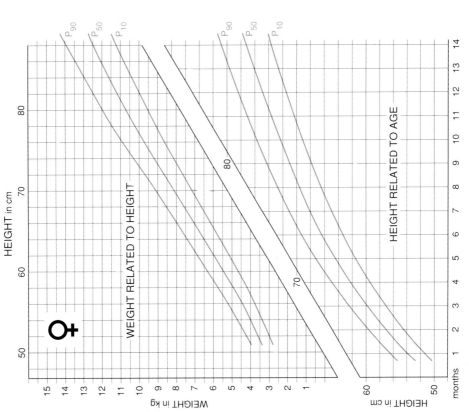

c. Weight related to height and height related to age (♀: 1–15 months)

d. Weight related to height and height related to age (♂: 1–15 months)

Figure 3.4.9 (cont.)

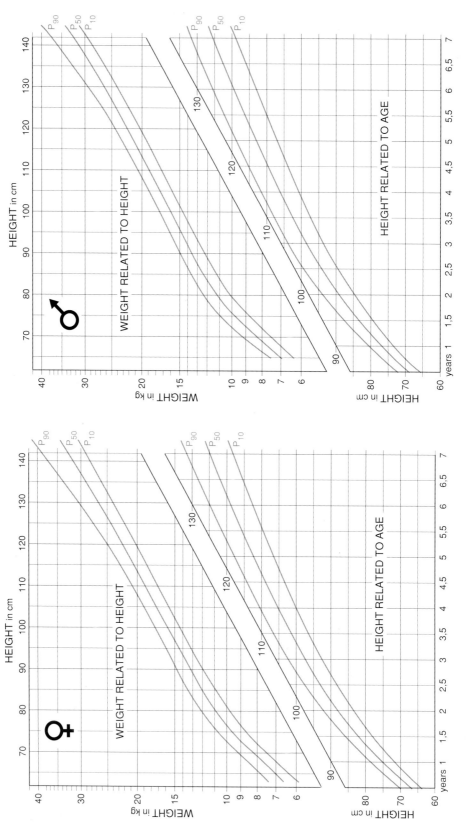

e. Weight related to height and height related to age ($♀$: $^1/_2$–7 years)

f. Weight related to height and height related to age ($♂$: $^1/_2$–7 years)

Figure 3.4.9 (cont.)

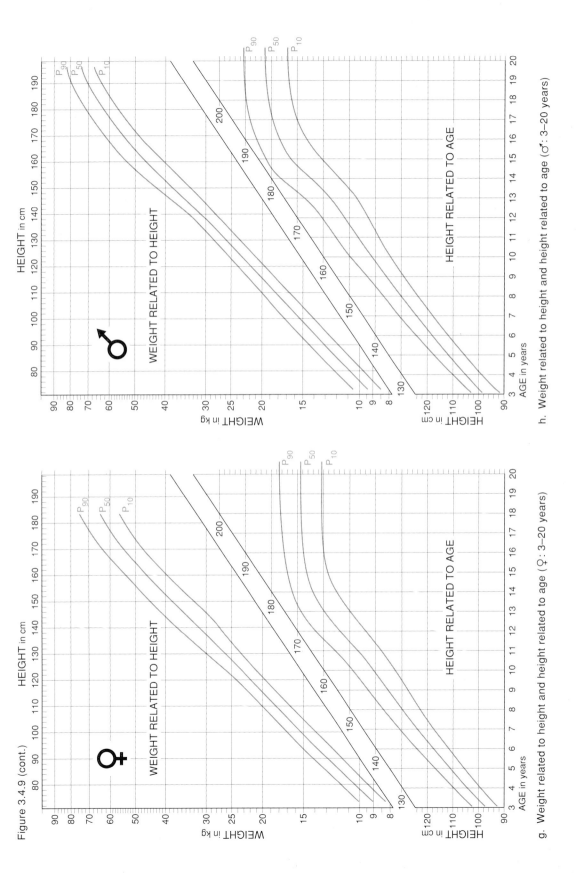

g. Weight related to height and height related to age (♀: 3–20 years)

h. Weight related to height and height related to age (♂: 3–20 years)

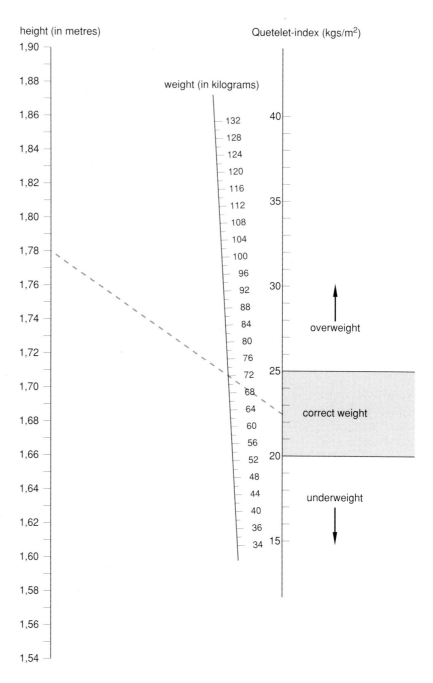

Figure 3.4.10
Quetelet-index (QI)
QI = kg/m².
The diagram makes it possible to establish the QI in a simple manner. Take the height and weight of a person (in the example respectively 1.78 m and 72 kg) on the vertical scales prepared for these figures. Draw a line between these two points and heighten this line to the index scale. This provides the QI-values.

above the P90-line are very (or even *too*) fat. Reading the 'height-to-age diagram' similar statements can be made.

A healthy child, who is regularly increasing in height and weight, will always have individual growth curves at about the same distance over or under the P50-lines: the child is following its individual percentage-lines.

After the ossification of the epiphysary discs the growth in height stops. The height of the body is reached and maintained during all of maturity. The body weight of the adult is by no means a fixed factor.

The changes in body weight are closely dependent on the ratio between the amount of calories taken in, and the amount of calories used up; the energy balance. If the intake is higher than the consumption, then the body weight increases and vice versa. What amounts to the ideal weight will always remain a topic of discussion. In the course of time numerous graphs have been devised to indicate whether an individual's height and weight are related to each other within acceptable borderlines.

In the Quetelet-index (Fig. 3.4.10), the weight is divided by the square of the height (QI = weight/(height)^2). A QI between 20 and 25 kg/m^2 indicates an acceptable weight.

Generally the weight one has at the end of growth in height – say by the 20th year of life – will be pretty near one's ideal weight.

Though the head in comparison with other parts of the body grows only very little in circumference (see Fig. 3.4.7), at the same time a remarkable change in the proportions of the head is seen (Fig. 3.4.11). The bones of the facial skull grow considerably more than those of the brain skull. The child – especially in the schoolchild phase – literally gets more face.

The small fontanelle (fontanelle posterior) has often already disappeared after two months; the

tough connective tissue has been substituted by bony tissue. The large fontanelle (the fontanelle anterior) is generally closed by the second birthday. The cranial sutures between the different flat plates of bone, however, remain in existence for a long time. The saggital suture (see Fig. 3.2.20b), disappears in one's early 30s on average, the coronal suture when one is about 40 years of age.

The development of one structure often influences the development of other structures (Fig. 3.4.12). The neonate shows hardly any frontal or dorsal curvature in the spinal column. This gradually develops. The first curve to arise is the lordosis in the neck area. The child lifts his head; the neck muscles develop. The next curve to come into being is the lordosis of the lumbar area as a result of gravity, and of the influence of the strong muscles of the back and of the iliopsoas muscle. The chest kyphosis ensures that there is a reasonable balance around the body median line.

The position of the pelvis (pelvic tilt) influences the depth of the curves in the spinal column.

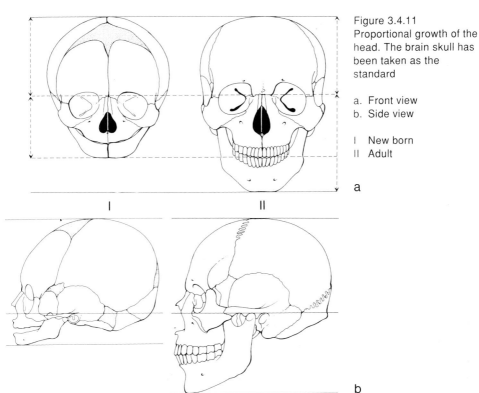

Figure 3.4.11
Proportional growth of the head. The brain skull has been taken as the standard

a. Front view
b. Side view

I   New born
II Adult

I

II

a

b

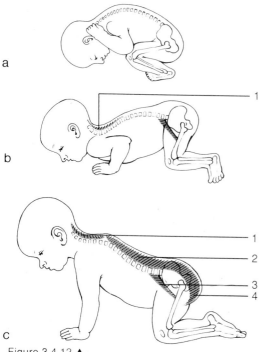

a

b

c

Figure 3.4.12 ▲
The development of the
curvatures in the spinal
column

a. Neonate
b. 1 month old
c. 3 months old

1. *muscles of the neck*
2. *muscles of the back*
3. *iliopsoas muscle*
4. *the 'buttock' muscles*

Figure 3.4.13 ▼
The 'toddler tummy'
(age 1.5–5 years). The
child stands upright with
an enhanced lumbar
lordosis and a
protruding abdomen

With toddlers and pre-school children a stronger loin lordosis is mostly seen, along with the characteristic protruding 'toddler's tummy' (Fig. 3.4.13). This gradually disappears during later development. The development in the posture of the legs in children is remarkable (Fig. 3.4.14). At first a baby is extremely bandy-legged, which gradually disappears until at the age of 18 months the legs are straight. Then knock-kneed legs develop: at the age of three the distance between the inner ankles can be as much as 8 cm (with the knees closed together). Between the ages of 3 and 6 the turned in legs grow straight again.

The motor function of the baby consists mainly of uncoordinated and undirected muscle movements and a large number of functional reflexes (among others: the sucking reflex, swallow reflex, sneeze reflex, cough reflex). The neonate cannot keep the head straight nor upright, for the muscles are still too weak for the relatively large head. In the course of the first year the baby successively learns to lift up the head and to keep it up, to turn over from abdomen to back and vice versa, to sit, to come from a lying position into a sitting position, to move forward 'on all fours', to crawl, to stand up and then to walk. These achievements are reached by every healthy child. They are a direct result of the development of the nervous and motor system.

During the toddler phase the child gradually acquires more and more control over the muscles; certain fine motor actions can already be performed. At the end of the pre-school phase the finer movements have to a large extent been developed. Muscular strength gradually increases up to the beginning of the growing spurt. Then, with boys particularly, the muscular strength increases markedly. On average the development of the motor system leads to a peak of strength and skill between the 20th and the 25th year of life. Past that age the mechanical qualities of the skeleton decline as do the joints, the muscles and tendons, each at their own pace (Fig. 3.4.15).

With advancing years comes a loss of elasticity in the bones. The percentage of collagen diminishes. At the same time the bone becomes

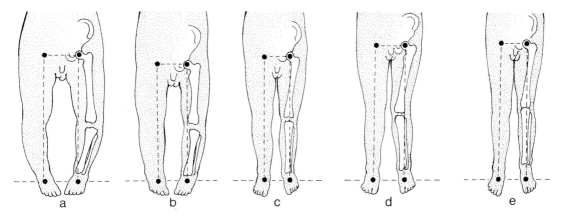

Figure 3.4.14
The normal development of
the posture of the legs

a. The neonate
b. A child of 6½ months
c. A child of 1½ years
d. A child of 2 years
e. A child of 5 years

more porous as a result of demineralization of
the substantia compacta.

There is a shift in the ratio of muscular tissue and
connective tissue towards the connective tissue
in the muscles, and it becomes tougher, more
stringy.

(10) Reproductive system

Generally the external genital organs of the
neonate look somewhat swollen. With newborn
boys an erection of the penis is no exception.

With newborn girls a kind of pseudo-menstru-
ation is sometimes seen, a discarding of blood
and white mucus. With both newborn boys and
girls the surroundings of the nipples may be
swollen and give off a colostrum-like liquid.
These extraordinary phenomena are a result of
maternal sex hormones still being present in the
baby's blood, as a remnant of the intrauterine
period. They disappear quite quickly.

With most newborn boys the testes have already
descended into the scrotum. If this is not the
case, it will take place *before* the first birthday.
The foreskin (prepuce) with the newborn, the
baby and the toddler is comparatively large and
has grown together with the glans.

Up to the beginning of puberty the development
of the reproductive organs is gradual. Then,
under the influence of a stronger production of

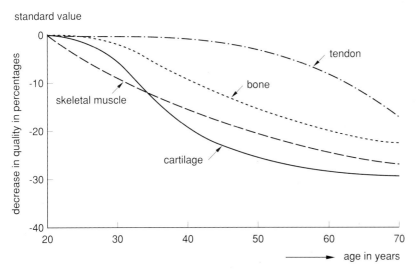

Figure 3.4.15
Rough representation of
the deterioration of the
mechanical qualities of
the parts of the motor
system. The mechanical
capacity at the age of 20
has been chosen as the
starting point (0)

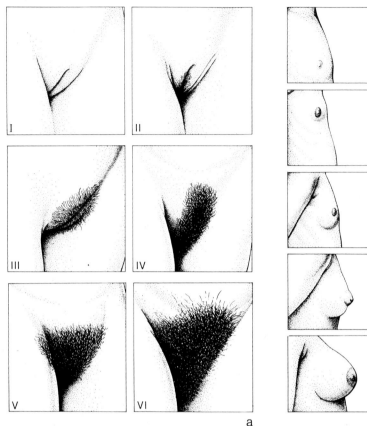

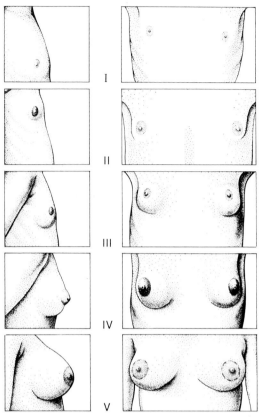

Figure 3.4.16
Sexual maturation

a. Development of pubic
   hair in girls
b. Development of breasts
   in girls
c. Development of pubic
   hair and genitals in boys

sex hormones, a growth spurt takes place in these organs. In a relatively short time the shape of boy and girl change to the shape of man and woman. The usual criteria for measuring sexual maturation are (Fig. 3.4.16):
– for girls: the growth of pubic hair, the development of the breasts and the calendar age at which the first menstruation, the menarche, occurs;
– for boys: the growth of the pubic hair and the development of the external genitals.

By the time many women reach 50 the lapses of time between menstruations gradually become longer. The woman comes to her menopause (climacterium). Because of changes in hormone production, namely the drop in oestrogen production, the breasts become slightly smaller and somewhat flaccid, the vagina loses some of its elasticity and there is less mucus produced in the vagina. Many women suffer from so-called flushes during the menopausal years. These are a kind of blushing, which may occur all over the body and may be accompanied by an intense perception of feeling hot. Usually the last menstruation (menopause) takes place at some time between the ages of 50 and 55. From then on the woman is no longer fertile. A new hormone balance is developing. The older man's production of sperm *does* diminish gradually, but as a rule the male remains fertile.

3. Dying and death

Every human being dies. Even the healthiest of persons ages and at the end of ageing death is waiting. From a biological point of view ageing and finally dying can be ascribed to the same

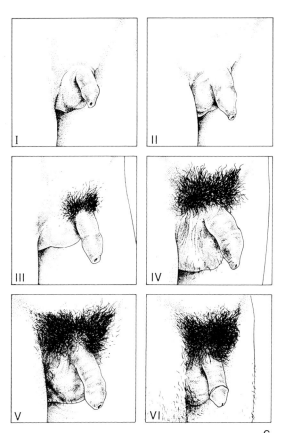

I

II

III

IV

V

VI

c

internal and external factors which determine the preceding stage of human development.

Dying can happen quickly, but can also be a very prolonged occurrence. With a dying person all kinds of phenomena can be observed which indicate the disintegration of the entire organism.

In most cases the sense of touch in the skin is the first sense that deteriorates; the perception of pain disappears. Eyesight lessens as a result of the cornea of the eyes becoming dull and because of a lack of tear fluid.

Hearing is the sense which remains intact longest.

The blood circulation of the dying person worsens. Those parts of the body that are farthest from the heart cool quickly: hands and feet, and also the nose. The blood pressure decreases and, immediately before death occurs, the heartbeat becomes rapid and irregular. Respiration occurs slowly, irregularly and is troublesome.

Mucus builds up in the respiratory tract, and can

no longer be expectorated. Breathing becomes rattled. The muscular tissue (both smooth and striated) slackens. Because of this the dying person can no longer maintain a good body posture. The head sinks to the chest, the mouth and the eyes are half open, the ability to swallow decreases and eventually is lost. The dying person also loses control over the anal and bladder sphincters (incontinence).

The dying person has a characteristic facial expression, which was described by Hippocrates (facies hippocratica): the features are sharper than normal, with flaccid motionless lips, flaccid sagging cheeks which move up and down with every motion of breath; a nose cold to the touch, glassy and deep-lying eyes and a pale to ashen skin.

When the dying person's breathing stops, the heart usually goes on beating for a short time. Death (of the heart) sets in because of lack of oxygen.

After the heart stops beating the blood sinks to the lowest parts of the body and solidifies there. As a result of this there arise local brown and purple discolourings of the skin: livores mortis. Two to three hours after life has come to an end, stiffening of the body (i.e. rigor mortis) is established. This is caused by calcium becoming free for the excitation of nerves and the presence of ATP in the muscles. The stiffening of the corpse disappears after about 30 hours; the store of ATP is used up.

About 24 hours after death, the body starts decomposing. The skin changes colour and the abdomen begins to distend. In the tissues enzymes are set free which instigate the self-disintegration of these tissues. Bacteria find their way into the body from the mouth cavity, the intestines, the respiratory and the urinary tract. They are no longer met by the defence systems at the disposal of a living human being, and a process of rotting ensues.

The body returns to the dust out of which life was created, perhaps many years ago.

5

Evolution

Introduction

The origin of mankind has been, and will continue to be, a subject of intense debate. It has been considered by philosophers, theologians, and anthropologists, as well as by the physical scientists. While it is not considered appropriate within the parameters of this book to pursue the theories expounded in philosophy, theology, or anthropology, it is apposite here to provide a very brief description of the events which are believed to have led to the development of life on earth in general, and to the specific evolution of mankind as an identifiable species.

The final chapter of this book on life sciences briefly examines the path which has led to the existence of man as we know him. This will offer us the opportunity to appreciate the enormous difference between the significance which we – as individuals – attach to our personal lives and the significance which our lives have as a tiny part of the phenomenon of evolution.

Learning outcomes

After studying this chapter you should have sufficient knowledge and understanding of:
- a number of general aspects of the evolution of the universe and of the solar system;
- the broad outline of the development of life, vegetable, animal and human;
- the dynamics of evolution.

1. In the beginning there was...

In the beginning there was a minute speck which exploded into the universe we now know by expanding trillions of times in a fraction of a second. Now, almost 15,000 million years later, ripples created by this explosion, the 'Big Bang', are being discovered at the edges of the universe by the Cosmic Background Explorer (COBE). These ripples are the cosmic remnants of the 'Big Bang' and are believed to be the largest and oldest structures in the universe. From this chaotic activity the immature universe expanded like a balloon (Fig. 3.5.1) from the point where the explosion had taken place. An ever-increasing space taken up by material came into existence. This material had been arranged as a spherical peel around the corporate primeval centre. Within this peel of material masses of elements, boiling gases, and dust happened to come together, and to acquire an ever-increasing density. About 10,000 million years ago these accumulations became the numerous galaxies, gigantic collections of material in rotation, often in the form of spiral-shaped tentacles (Fig. 3.5.2). About 5000

Figure 3.5.1
The Big Bang, the primary explosion

million years ago this material in our galaxy again started to become denser and denser; the stars came into existence. Our sun is one of the estimated 100 million stars in the Milky Way. The position of our sun is in a really remote, out-of-the-way place, far away from the centre. The sun can be compared to an enormous nuclear reactor.

The amount of energy which is being radiated by the sun is unimaginably large. It is assumed that the sun can continue to function for the next 10,000 million years. The gravitational force of the sun ensures that material is kept within orbits around the sun. About 4500 million years ago, this material gradually obtained a greater density because of its own gravitational forces, and started to develop into the planets which now circle round the sun in orbits (Fig. 3.5.3). In their turn some of these planets have accompanying bodies such as our moon.

Earth was at first a white-hot, liquid, eddying mass, due to the friction between the gathering, amalgamating masses of material from which the planet was being formed. After that the mantle of the earth began to cool down and solidify. The inner part of the earth, however, is still liquid and hot. This can be seen clearly in the eruption of a volcano.

The immature earth was subject to enormous forces of nature: earthquakes, eruptions of

Figure 3.5.2
A milky way system

volcanoes, incessant gale force winds, tremendous deluges of rain, terrible and never-ending flashes of lightning.
Slowly land masses emerged and seas, mountains, rivers, and an atmosphere of hydrogen, carbon dioxide, ammonia and methane evolved.

Under the influence of the electrical discharges of lightning more complex molecules could be composed out of, and in, this atmosphere.

Experiments have suggested that the formation of amino acids and certain fatty acids, among other substances, could have taken place under such circumstances. These building blocks for life eventually found themselves in the seas of that period. It was here that further chemical reactions took place, among which was the formation of sugar phosphates. The entire mixture of chemical elements and compounds in the primary ocean is often called the primary soupe.

2. Life

What is 'life'? In Module 1, Chapter 1 a number of characteristics of life have been stated: *self preservation, species preservation, flexibility,*

Figure 3.5.3
Our solar system

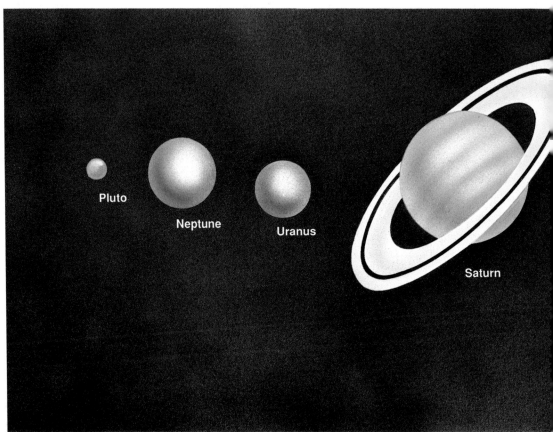

Pluto

Neptune

Uranus

Saturn

development. The period of time that the building blocks for life have had to travel before these criteria could be met, is millions of years long. Amino acids string together to become proteins; the building blocks for DNA-molecules (nuclein acids and bases) come into existence and fatty acids become arranged in doubled-up layers. These layers of fatty acids can be transformed into little bubbles. If, in such a bubble, amino acids, proteins and DNA are present a primitive *cell* exists. This probably happened for the first time around 3000 million years ago.

In the course of time since then, this first cell turned out to be capable of bringing about ever new types of protein. The DNA proved to be a very useful 'casting mould', on the basis of which the necessary amino acids could be shaped. In addition, the DNA was (and is) capable of self-replication. The copies were divided among the two daughter cells resulting from the self-division of the mother cell.

New proteins could be made, because within the casting mould, the DNA, there occurred by chance a small change. These tiny random changes, called *mutations*, were the result of errors made by the DNA during the copying phase, through the influence of radiation and through damage done to the DNA by other chemical substances within the cell. Mutations were, and are, extremely numerous. Most mutations are unfavourable or have no effect at all.

About one in a million of the mutations turned out to be useful. A living being with such a useful mutation (i.e. with a useful difference) has more chance of surviving than other species without this mutation. The 'improved' specimens of the primary cells were for that reason better *equipped for the struggle for life*. They could make better use of the possibilities offered by their environment and because of their success their number increased (more rapidly

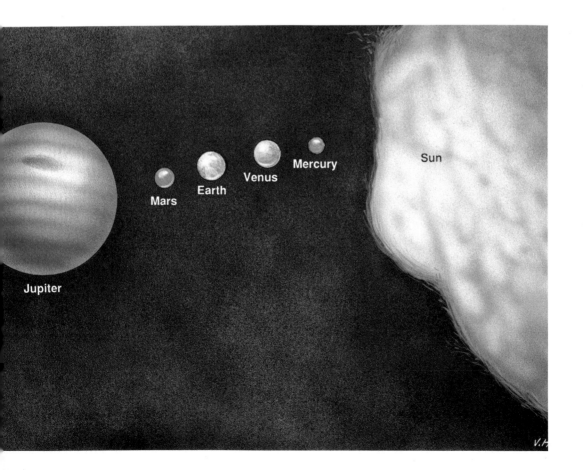

Jupiter   Mars   Earth   Venus   Mercury   Sun

V.H

than the numbers of the other types). Through cumulative selection there developed ever more varieties and increasingly sophisticated forms of life.

3. Vegetable and animal

The primary cells developed further. Chlorophyll, the substance giving plants their green colour, was incorporated into the cell wall. *Mutants* of this type could make sugar out of the abundantly available carbon dioxide; and water with the help of energy from the sugar is, of course, a source of energy par excellence.

This having been achieved, the first vegetable life on earth followed. This form of life was and is extremely successful. The primary seas became populated with all kinds of vegetable cells. The quality of being multicellular came into existence and this proved to be a very successful mutation. Fossils of such life-forms exist, which are over 2500 millions years old.

At a certain point in time there developed cells which had no chlorophyll at their disposal, but did have the capacity to 'eat' vegetable cells. At that moment the first form of animal life was achieved.

These animal cells – and later the multicellular animal creatures – provided for their energy needs by drawing on the stock of energy within the vegetable world.

4. From unicellular being to primate

Out of animal unicellulars the first primitive sea-creatures came into being. How long ago that took place is difficult to tell. The oldest fossils of sea-creatures date back to around 600 million years ago. In a relatively short time the plants and the animals had colonized the sea and consequently after that the land of the continents. About 500 million years ago molluscs and primitive fish had developed; 400 million years ago the first amphibians crawled ashore and around 350 million years ago the reptiles started their amazing rise. This species proved extremely successful on land that was by this time covered with a lush growth of vegetation.

Their hey day lasted about 250 millions years (till around 100 million years ago).

Gigantic forms of reptile life came to exist, such as the Tyrannosaurus Rex and the Diplodocus. For 125 million years these giant reptiles were lords and masters of the earth. The dinosaurs became extinct – practically overnight – probably due to serious cooling of the earth which were catastrophic for these cold-blooded animals. This drop in temperature is often ascribed to the consequences of the impact of a meteorite, which flung up enormous clouds of dust into the atmosphere and thus shielded off the sunlight for a long period of time.

About 100 million years ago the shrew-mouse-like ancestors of mammals came into being. Mammals are warm-blooded and have a fairly constant body temperature, they have hair on at least part of their body, they give birth to living offspring (viviparous), which they feed with milk from special cutaneous glands.

5. From primate to human being

Some 70 million years ago a class of mammals branched off and by their own choice went to live in the trees. They developed long hands and fingers, which were extremely well-suited to grasping and seizing, and good stereoscopic eyesight. This group of mammals, including man, is called *the primates*. A major feature in the development of the primates is the increase in brain volume.

Around 60 million years ago the first *half-apes* became active. They could leap from bough to bough. The half-apes, apes, man-apes and modern man probably have ancestors in common (Fig. 3.5.4). The development of the higher primates began in Africa around 40 million years ago. About 20 million years ago the ancestors of the *anthropoidea* developed. These anthropoid (human-like) primates gradually turned from a life in the trees to a life on the ground. A well-known example of this type of 'higher' apes is the Ramapithecus, which lived about 10 million years ago.

The Ramapithecus is considered to be the common ancestor of the man-apes and ape-men (*pithecanthropes*).

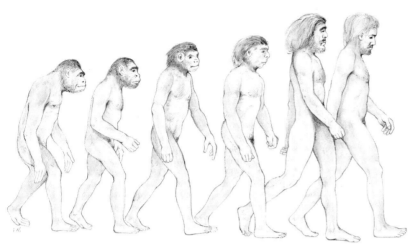

Figure 3.5.4
The forerunners of man

The Australopithecus is the first known anthropoid; he lived on the plains of Africa some 3 million years ago. The development of Australopithecus to the first human took place in the following 1.5 million years. Characteristic is the ever increasing brain volume, the erect gait and the wide usage of tools.

The first human beings dispersed out of Africa all over Europe and Asia.

Of *homo erectus*, who lived around 800,000 years ago, it is known that he used fire. Moreover the structure of his larynx had evolved to such an extent that a highly complicated form of speech communication had become possible.

Neanderthal man had his heyday from 100,000 to around 35,000 years ago. He made very laborious tools, hunted in groups and made drawings and paintings on the walls of the caves in which he lived.

Neanderthal men were the first human beings to bury their deceased. It is appropriate to call them *homo sapiens.*

The modern breeds of man (homo sapiens) have had prominence for over 30,000 years. Probably they quickly pushed aside the other types of homo and spread out over all the habitable continents. The beautiful paintings on the rock walls of the caves at Lascaux, in the Vézère valley in France, were painted about 20,000 years ago (Fig. 3.5.5).

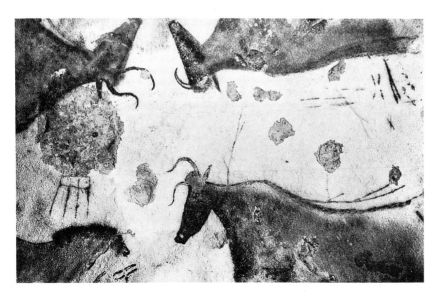

Figure 3.5.5
Painting in one of the caves at Lascaux

This expression of a refined culture has been followed by many, many other expressions of culture, with ever-increasing frequency. Mankind has built houses, cultivated land, constructed bridges across rivers, developed vehicles and vessels, tamed animals for his own uses, laid out roads, put machinery to work on numerous tasks, invented computers and has even visited another celestial body.

6. Where next?

Evolution has not come to an end. Man has changed and will change further. Man himself obtains more and more control over certain factors which steer evolution. Through bio-genetic manipulations new varieties of plants have come into existence, certain types of genes can be 'built' into bacteria, the creation of many identical multiplications by means of 'cloning' has proved to be possible, and the prospect of repairing defective or damaged genesis is in sight.

If man is willing to participate in the process of evolution – and it seems that way – then at the same time he makes himself responsible for the outcome of his actions. It is by no means certain that all this will lead to the benefit and the welfare of mankind.

Appendix 1
Latin and Greek affixes

Affix	Meaning	Example
a, ad-	to, towards	afferent
a, an-	absence of	anaemia
ab-	away from	abduction
-agon	making	glucagon
-algia	pain	neuralgia
ambi-	both	ambidextrous
ante-	before	antenatal
anti-	against	antihistamine
arth-	joint	arthritis
-asis	state of	haemostasis
aut-	self	autoregulation
bi-	twice, double	bifurcation
bio-	living	biochemistry
-bio	living	symbiosis
-blast	formative cell	cytoblast
brachi-	arm	brachial plexus
bronchi-	windpipe	bronchitis
capit-	head	capitellum
carcin-	cancer	carcinogen
cardio-	heart	cardiogram
cephal-	head	cephalic
cerebro-	brain	cerebrovascular
chroma-	colour	chromatin
con-	with	concentric
crani-	skull	cranial
cryo-	cold	cryotherapy
cut-	skin	cutaneous
cyst-	bladder	cystitis
cyto-	cell	cytolysis
de, des-	down	descending
de-	out of	detoxicate
derma-	skin	dermatology
di-	two	dichromatic
dia-	through	diabetes
dys-	painful	dysentery
e, ex-	out of	efferent
endo-	inside	endocrine
entero-	intestine	enterotomy

Affix	Meaning	Example
epi-	upon, above	epidermis
exo-	outside, beyond	exocrine
extra-	outside	extracellular
feto-	fetus	fetoscopy
fibro-	fibrous tissue	fibrocystic
galacto-	milk	galactose
gastro-	stomach	gastroenteritis
glossa-	tongue	glossitis
glyco-	sweet, sugar	glycogen
-graph	instrument for recording	electroencephalograph
gyno-	female	gynaecology
haema-	blood	haematology
hemi-	half	hemisphere
hepar-	liver	hepatitis
hetero-	another, different	heterolateral
histo-	tissue	histolysis
homo-	alike	homologous
hydro-	water	hydrophobia
hyper-	over, above	hyperactive
hypo-	below	hypotension
hyster-	uterus	hysterectomy
inter-	between, among	intercostal
intra-	inside	intracellular
-ism	condition	rheumatism
iso-	equal	isometrics
-itis	inflammation	fibrositis
kin-	movement	kinematics
kypho-	abnormal curvature	kyphosis
lacri-	tears	lacrimal
lacto-	milk	lactosuria
leuko-	white	leukaemia
lipo-	fat	lipolysis
-logy	study of	psychology
-lysis	breaking down	electrolysis
macro-	great	macrocyte
mal-	bad	malignant
media-	middle	median
mega-	large	megalocardia
meso-	middle	mesomorph
meta-	change	metaphase
micro-	small	microscope
mono-	one, single	monomolecular
myelo-	spinal cord	myelitis
myo-	muscle	myology
necro-	corpse	necrobiosis
neo-	new	neoplasm

Affix	Meaning	Example
neuro-	nerve	neurolgia
oculo-	eye	oculomotor
odon-	tooth	odontology
-oid	like, nearly	schizoid
-ology	knowledge of	pathology
ophthalm-	eye	ophthalmologist
ortho-	normal, correct	orthopaedics
os, osteo-	bone	osteomyelitis
-osis	development of	fibrosis
-osis	diseased condition	tuberculosis
paed-	child	paediatrician
para-	near, beside	parabiosis
patho-	disease	pathogen
per-	through, by	perforated
peri-	around	perineum
-phobia	fear	agrophobia
pneumo-	breath, air	pneumoconiosis
poly-	many	polygene
post-	after, behind	postmortem
pre, pro-	before	prenatal, prognosis
pseudo-	false	pseudoarthrosis
psycho-	spirit, mind	psychotic
quadri-	four	quadriceps
retro-	behind, backwards	retroflexion
rhino-	nose	rhinoscopy
-rrhagia	excessive flow	haemorrhagia
sclero-	hardening	sclerosis
-scope	instrument for visual or aural examination	stethoscope
semi-	half	semilunar valve
septic-	decaying	septicaemia
soma-	body	somatic
-stasis	motionless	homeostasis
sub-	under	subauricular
super-	above	superior
supra-	over	supraglottal
sym, syn-	together, with	sympathectomy, synapse
teno-	tendon	tenoplasty
terat-	congenital abnormality	teratology
tetra-	four	tetracycline
thalam-	thalamus	thalamotomy
therm-	heat	thermolysis
thorac	chest	thoracotomy
thym-	thymus	thymectomy
thyro-	thyroid	thyrocele
-tomy	incision, cutting	appendicectomy

Affix	Meaning	Example
tono-	pressure	tonography
toxic-	poison	toxicosis
trans-	across, through	transfusion
tri-	three	triceps
trich-	hair-like	trichosis
troph-	nourishment	trophoblast
-tropic	turning towards	hydrotrophic
ultra-	beyond	ultrasonic
uni-	one	unicellular
-uria	urine	polyuria
vago-	vagus nerve	vagotomy
vas-	vessel	vascular
veno-	vein	venography
ventro-	to the front	ventrosuspension
xantho-	yellow	xanthochromia
xeno-	different, foreign	xenograft
zygo-	pair, union	zygomatic arch
zymo-	fermentation	zymosis

Appendix 2
Abbreviations

a	artery
ACTH	adrenocorticotrophic hormone
ADH	antidiuretic hormone
ADP	adenosine diphosphate
ATP	adenosine triphosphate
C	cervical vertebrae
BMV	breath/minute volume
CCK-PZ	cholecystokinin-pancreozymin
CD	complete dilatation
COV	colloid-osmotic value
CNS	central nervous system
CT	computerized tomography
DNA	dioxyribonucleic acid
ECG	electrocardiogram
EEG	electroencephalogram
EMG	electromyogram
ER	endoplasmic reticulum
FSH	follicle stimulating hormone
HCG	human chorion gonadotrophin
HCS	human chorion somatomammotrophin
HLA	human leukocyte antigen (system)
HMV	heart/minute volume
ICSH	interstitial-cell-stimulating hormone
KPa	kilo Pascal
L	lumbar vertebrae
LH	luteinizing hormon
LTH	leuteotrophic hormone
LTM	long term memory
m	muscle
me	milieu extérieur
mi	milieu intérieur
mmHg	milimetre of mercury
MSH	melanocyte-stimulating hormone
n	nerve (spinal)
N	nerve (cranial)
Pa	Pascal units
PNS	peripheral nervous system
PTH	parathyroid hormone
QI	quetelet index
REF	renal erythropoietic factor

REM	rapid eye movement
RES	reticuloendothelial system
RH	rhesus (factor)
RNA	ribonucleic acid
S	sacral vertebrae
STM	short term memory
T	thoracic vertebrae
v	vein
VC	vital capacity

Index